Takeshi Takeda (Ed.)
Modern Carbonyl Olefination

Further Reading from Wiley-VCH

Grubbs, R. H. (Ed.)

Handbook of Metathesis, 3 Vols.

2003
3-527-30616-1

Otera, J.

Esterification

Methods, Reactions, and Applications

2003
3-527-30490-8

Nicolaou, K. C., Snyder, S. A.

Classics in Total Synthesis II

2003
3-527-30685-4 (Hardcover)
3-527-30684-6 (Softcover)

Marek, I. (Ed.)

Titanium and Zirconium in Organic Synthesis

2002
3-527-30428-2

Takeshi Takeda (Ed.)

Modern Carbonyl Olefination

WILEY-
VCH

WILEY-VCH Verlag GmbH & Co. KGaA

Editor:

Professor Dr. Takeshi Takeda
Department of Applied Chemistry
Tokyo University of Agriculture & Technology
2-24-16 Nakacho, Koganei, Tokyo 184-8588
Japan

■ This book was carefully produced. Nevertheless, editor, authors and publisher do not warrant the information contained therein to be free of errors. Readers are advised to keep in mind that statements, data, illustrations, procedural details or other items may inadvertently be inaccurate.

Library of Congress Card No.: applied for
A catalogue record for this book is available from the British Library.
Bibliographic information published by Die Deutsche Bibliothek
Die Deutsche Bibliothek lists this publication in the Deutsche Nationalbibliografie; detailed bibliographic data is available in the Internet at http://dnb.ddb.de

© 2004 WILEY-VCH Verlag GmbH & Co. KGaA, Weinheim
All rights reserved (including those of translation in other languages). No part of this book may be reproduced in any form – by photoprinting, microfilm, or any other means – nor transmitted or translated into machine language without written permission from the publishers. Registered names, trademarks, etc. used in this book, even when not specifically marked as such, are not to be considered unprotected by law.

Printed in the Federal Republic of Germany.
Printed on acid-free paper.

Typesetting Asco Typesetters, Hong Kong
Printing Strauss Offsetdruck GmbH, Mörlenbach
Bookbinding J. Schäffer GmbH & Co. KG, Grünstadt

ISBN 3-527-30634-X

Contents

Preface *xiii*

List of Authors *xv*

1 The Wittig Reaction 1
Michael Edmonds and Andrew Abell
1.1 Introduction 1
1.2 The "Classic" Wittig Reaction 2
1.2.1 Mechanism and Stereoselectivity 2
1.2.2 Nature of the Ylide and Carbonyl Compound 3
1.2.3 Reagents and Reaction Conditions 4
1.3 Horner–Wadsworth–Emmons Reaction 5
1.3.1 Mechanism and Stereochemistry 6
1.3.2 Reagents and Reaction Conditions 8
1.4 Horner–Wittig (HW) Reaction 9
1.4.1 Mechanism and Stereochemistry 9
1.4.2 Reagents and Reaction Conditions 14
1.5 Conclusion 15
References 16

2 The Peterson and Related Reactions 18
Naokazu Kano and Takayuki Kawashima
2.1 Introduction 18
2.2 Stereochemistry and the Reaction Mechanism of the Peterson Reaction 19
2.2.1 Stereochemistry and the Reaction Mechanism of the Peterson Reaction of β-Hydroxyalkylsilanes 19
2.2.1.1 Stepwise Mechanism 21
2.2.1.2 Reaction Mechanism via a 1,2-Oxasiletanide 22
2.2.2 Reaction Mechanism of the Addition Step of an α-Silyl Carbanion to a Carbonyl Compound 23
2.2.2.1 Approach Control of the Transition State 23
2.2.2.2 Concerted Mechanism 25
2.2.2.3 Chelation Control Mechanism 26

2.2.3	Theoretical Calculations on the Reaction Mechanism	29
2.2.4	Convergently Stereoselective Peterson Reactions	31
2.3	Generation of α-Silyl Carbanions and their Peterson Reactions	32
2.3.1	Generation of α-Silyl Carbanions from α-Silylalkyl Halides	32
2.3.1.1	Generation of α-Silyl Grignard Reagents from α-Silylalkyl Halides	32
2.3.1.2	Generation of α-Silyl Alkyllithium Reagents from α-Silylalkyl Halides	32
2.3.1.3	Synthesis of Terminal Alkenes by the Use of α-Silyl Carbanions Generated from α-Silylalkyl Halides	33
2.3.1.4	Reactions of α-Silyl Carbanions Generated from α-Silylalkyl Halides with Esters, Carboxylic Acids, and Acetals	35
2.3.1.5	The Reformatsky–Peterson Reactions of α-Silylalkyl Halides	37
2.3.2	Generation of α-Silyl Carbanions by Deprotonation of Alkylsilanes	37
2.3.2.1	Generation of α-Silyl Carbanions Bearing an Aryl or a Heteroaryl Group	37
2.3.2.2	Generation of α-Silyl Carbanions Bearing an Alkoxy Group	39
2.3.2.3	Generation of α-Silyl Carbanions Bearing a Nitrogen-Containing Group	41
2.3.2.4	Generation of α-Silyl Carbanions Bearing a Sulfur-Containing Group	41
2.3.2.5	Generation of α-Silyl Carbanions Bearing a Phosphorus-Containing Group	48
2.3.2.6	Generation of α-Silyl Carbanions Bearing a Halogen Group	50
2.3.2.7	Generation of α-Silyl Carbanions from α-Silyl Ketones	52
2.3.2.8	Generation of α-Silyl Carbanions Bearing an Ester Group	52
2.3.2.9	Generation of α-Silyl Carbanions Bearing a Lactone Group	53
2.3.2.10	Generation of α-Silyl Carbanions Bearing Thiocarboxylate or Dithiocarboxylate Ester Groups	53
2.3.2.11	Generation of α-Silyl Carbanions Bearing an Imine Group	53
2.3.2.12	Generation of α-Silyl Carbanions Bearing an Amide Group	54
2.3.2.13	Generation of α-Silyl Carbanions Bearing a Cyanide Group	55
2.3.2.14	Generation of α-Silyl Carbanions from Allylsilanes	56
2.3.2.15	Generation of α-Silyl Carbanions from Propargylsilanes	58
2.3.3	Generation of α-Silyl Carbanions by Substitution of a Heteroatom	58
2.3.3.1	Generation of α-Silyl Carbanions by Reduction of a Sulfanyl Group	58
2.3.3.2	Generation of α-Silyl Carbanions by Substitution of Selenium	59
2.3.3.3	Generation of α-Silyl Carbanions by Desilylation of Bis(trimethylsilyl)methane Derivatives	60
2.3.3.4	Generation of α-Silyl Carbanions by Tin–Lithium Transmetallation	60
2.3.4	Formation of β-Hydroxyalkylsilanes from Silyl Enol Ethers	60
2.3.5	Fluoride Ion Induced Peterson-Type Reactions	62
2.3.5.1	Generation of α-Silyl Carbanions by Fluoride Ion Induced Desilylation	62
2.3.5.2	Fluoride Ion Induced Peterson-Type Reactions of Bis(trimethylsilyl)methane Derivatives	64

2.3.5.3	Fluoride Ion Catalyzed Peterson-Type Reactions of Bis(trimethylsilyl)methylamine Derivatives *65*	
2.3.5.4	Fluoride Ion Catalyzed Peterson-Type Reactions with Elimination of Trimethylsilanol *68*	
2.3.6	Generation of α-Silyl Carbanions by Addition of Nucleophiles to Vinylsilanes *68*	
2.4	Synthetic Methods for β-Silyl Alkoxides and β-Hydroxyalkylsilanes *70*	
2.4.1	Reactions of α-Silyl Ketones, Esters, and Carboxylic Acids with Nucleophiles *70*	
2.4.2	Ring-Opening Reactions *72*	
2.4.2.1	Ring-Opening Reactions of Oxiranes *72*	
2.4.2.2	Ring-Opening Reactions of Cyclic Esters and Ethers *73*	
2.4.3	Hydroboration of 1-Silylallenes *74*	
2.4.4	Dihydroxylation of Vinylsilanes and Allylsilanes *76*	
2.5	Related Reactions *77*	
2.5.1	The Homo-Brook Rearrangement *77*	
2.5.2	Homo-Peterson Reaction *79*	
2.5.3	Vinylogous Peterson Olefination *80*	
2.5.4	Tandem Reactions and One-Pot Processes Involving the Peterson Reaction *81*	
2.5.5	The Germanium, Tin, and Lead Versions of the Peterson Reaction *85*	
2.5.5.1	The Germanium-Peterson Reaction *85*	
2.5.5.2	The Tin-Peterson Reaction *86*	
2.5.5.3	The Lead-Peterson Reaction *88*	
2.5.6	Synthesis of Carbon–Heteroatom Double-Bond Compounds by Peterson-Type Reactions *88*	
2.5.6.1	Synthesis of Imines *88*	
2.5.6.2	Synthesis of Phosphaalkenes *89*	
2.5.6.3	Synthesis of Silenes *90*	
2.5.6.4	Synthesis of Germenes *91*	
2.5.6.5	Synthesis of Sulfines *91*	
2.6	Conclusion *92*	
	References *93*	
3	**The Julia Reaction** *104*	
	Raphaël Dumeunier and István E. Markó	
3.1	Introduction *104*	
3.2	Historical Background *105*	
3.3	Coupling Between the Two Precursors of the Julia Reaction *106*	
3.3.1	Synthesis of Terminal Olefins *107*	
3.3.2	Preparation of 1,2-Disubstituted Olefins *109*	
3.3.3	Towards Trisubstituted Olefins *112*	
3.3.4	Towards Tetrasubstituted Olefins *114*	
3.3.5	Specific Considerations *115*	
3.3.5.1	Conjugated Olefins *115*	

3.3.5.2 Leaving Groups 115
3.3.5.3 Competitive Metallation on the Aromatic Ring of the Sulfone 118
3.4 Reductive Elimination 120
3.4.1 Sulfones Bearing Vicinal Hydroxyl Groups 122
3.4.2 Sulfones Bearing Vicinal Leaving Groups 127
3.4.3 Reverse Reductions 130
3.4.4 Reductions of Vicinal Oxygenated Sulfoxides 133
3.4.5 Reduction of Vinyl Sulfones 136
3.5 Second Generation Julia Reactions 136
3.6 Miscellaneous Julia Reactions 141
3.6.1 *gem*-Halogeno-Metal Electrophiles 141
3.6.2 Use of Sulfoximines 143
3.7 Conclusions 145
References 146

4 Carbonyl Olefination Utilizing Metal Carbene Complexes 151
Takeshi Takeda and Akira Tsubouchi
4.1 Introduction 151
4.2 Carbonyl Olefination with Titanocene-Methylidene and Related Reagents 152
4.2.1 Preparation of Titanocene-Methylidene 152
4.2.1.1 The Tebbe Reagent 152
4.2.1.2 β-Substituted Titanacyclobutanes as Precursors of Titanocene-Methylidene 159
4.2.1.3 Zinc and Magnesium Analogues of the Tebbe Reagent 161
4.2.2 Higher Homologues of Titanocene-Methylidene 161
4.3 Carbonyl Olefination with Dialkyltitanocenes 166
4.3.1 Methylenation with Dimethyltitanocene 166
4.3.2 Alkylidenation of Carbonyl Compounds with Dialkyltitanocenes and Related Complexes 172
4.3.3 Allenation of Carbonyl Compounds with Alkenyltitanocene Derivatives 176
4.3.4 Carbonyl Olefination Utilizing an Alkyl Halide–Titanocene(II) System 176
4.4 Carbonyl Olefination Utilizing a Thioacetal-Titanocene(II) System 178
4.4.1 Formation of Titanocene-Alkylidenes from Thioacetals and Titanocene(II) 178
4.4.2 Alkylidenation of Aldehydes, Ketones, and Carboxylic Acid Derivatives 179
4.4.3 α-Heteroatom-Substituted Carbene Complexes 182
4.4.4 Intramolecular Carbonyl Olefination 182
4.4.5 Related Olefinations Utilizing *gem*-Dihalides 184
4.5 Carbonyl Olefination Using Zirconium, Tantalum, Niobium, Molybdenum, and Tungsten Carbene Complexes 185
4.5.1 Zirconium Carbene Complexes 185

4.5.2	Tantalum and Niobium Carbene Complexes 188
4.5.3	Molybdenum Carbene Complexes 189
4.5.4	Tungsten Carbene Complexes 192
4.6	Conclusion 194
	References 194

5 Olefination of Carbonyl Compounds by Zinc and Chromium Reagents 200
Seijiro Matsubara and Koichiro Oshima

5.1	Introduction 200
5.2	Zinc Reagents 201
5.2.1	Methylenation Reactions 202
5.2.1.1	By $Zn-CH_2X_2$ 202
5.2.1.2	By $Zn-CH_2X_2-TiCl_n$ 203
5.2.2	Alkylidenation Reactions 208
5.2.2.1	From *gem*-Dihaloalkanes 208
5.2.2.2	Via Carbometallation 211
5.2.3	Alkenylsilane, -germane, and -borane Synthesis 212
5.3	Chromium Compounds 214
5.3.1	Alkylidenation 214
5.3.2	Preparation of Alkenylboranes, -silanes, and -stannanes with *E*-Configuration 215
5.3.3	Preparation of (*E*)-Haloalkenes 215
5.4	Applications in Natural Product Synthesis 217
5.4.1	$Zn-CH_2X_2-TiCl_4$ 218
5.4.2	CHX_3-CrCl_2 218
5.5	Conclusion 221
	References 221

6 The McMurry Coupling and Related Reactions 223
Michel Ephritikhine and Claude Villiers

6.1	Introduction 223
6.2	Scope of the McMurry Reaction 224
6.2.1	Intermolecular Coupling Reactions 224
6.2.1.1	Intermolecular Coupling of Aldehydes and Ketones 224
6.2.1.2	Intermolecular Coupling of Unsaturated Aldehydes and Ketones 226
6.2.1.3	Intermolecular Coupling of Aldehydes and Ketones with Functional Heteroatom Groups 228
6.2.1.4	Intermolecular Coupling of Organometallic Ketones and Aldehydes 235
6.2.1.5	The McMurry Reaction in Polymer Synthesis 237
6.2.1.6	Intermolecular Cross-Coupling Reactions 237
6.2.2	Intramolecular Coupling Reactions of Aldehydes and Ketones 240
6.2.2.1	Synthesis of Non-Natural Products 240
6.2.2.2	Synthesis of Natural Products 246
6.2.3	Tandem Coupling Reactions 249

6.2.4 Keto Ester Couplings *254*
6.2.4.1 Intermolecular Keto Ester Couplings *254*
6.2.4.2 Intramolecular Keto Ester Cyclizations; Synthesis of Cyclanones *255*
6.2.4.3 Intramolecular Cyclizations of Acyloxycarbonyl Compounds; Synthesis of Furans *255*
6.2.5 Intramolecular Couplings of Acylamidocarbonyl Compounds; Synthesis of Pyrroles and Indoles *256*
6.2.6 Reductive Coupling of Benzylidene Acetals *258*
6.3 Procedures and Reagents Used in the McMurry Reactions *259*
6.3.1 Procedures *259*
6.3.2 Reagents *260*
6.3.2.1 The $TiCl_4$- and $TiCl_3$-Reducing Agent Systems *260*
6.3.2.2 Effect of Additives on the $TiCl_4$- and $TiCl_3$-Reducing Agent Systems *261*
6.3.2.3 Other Systems for the McMurry Alkene Synthesis: Organotitanium Complexes, Titanium Oxides, Titanium Metal *266*
6.4 Mechanisms of the McMurry Reaction *266*
6.4.1 Nature of the Active Species *267*
6.4.2 Characterization of the Pinacolate Intermediates *268*
6.4.3 Evidence of Carbenoid Intermediates *272*
6.4.4 Mechanistic Analogies Between the McMurry, Wittig, and Clemmensen Reactions *273*
6.4.5 The Different Pathways of the McMurry Reaction *274*
6.5 Conclusion *275*
References *277*

7 Asymmetric Carbonyl Olefination *286*
Kiyoshi Tanaka, Takumi Furuta, and Kaoru Fuji
7.1 Introduction and Historical Aspects *286*
7.2 Strategies for Asymmetric Carbonyl Olefination *289*
7.3 Optically Active Phosphorus or Arsenic Reagents Used in Asymmetric Carbonyl Olefination *290*
7.4 Discrimination of Enantiotopic or Diastereotopic Carbonyl Groups *299*
7.4.1 Intermolecular Desymmetrization of Symmetrical Dicarbonyl Compounds *299*
7.4.2 Intramolecular Discrimination Reactions *303*
7.5 Discrimination of Enantiotopic or Diastereotopic Carbonyl π-Faces *306*
7.5.1 Reactions with Prochiral Carbonyl Compounds *306*
7.5.2 Reactions with Chiral Non-Racemic Carbonyl Compounds *311*
7.5.3 Reactions with Prochiral Ketenes to give Dissymmetric Allenes *313*
7.6 Kinetic Resolution *316*
7.6.1 Resolution of Racemic Carbonyl Compounds *316*
7.6.2 Resolution of Racemic Phosphorus Reagents *321*
7.6.3 Parallel Kinetic Resolution *321*
7.7 Dynamic Resolution *323*

7.8	Further Application of Asymmetric Wittig-type Reactions in Enantioselective Synthesis *325*	
7.8.1	Use of Asymmetric Wittig-Type Reactions in the Total Synthesis of Natural Products *325*	
7.8.2	Sequential HWE and Pd-Catalyzed Allylic Substitutions *327*	
7.8.3	Tandem Michael–HWE Reaction *329*	
7.9	Asymmetric Carbonyl Olefinations Without Usage of Optically Active Phosphorus Reagents *329*	
7.10	Asymmetric Carbonyl Olefination by Non-Wittig-Type Routes *331*	
7.11	Concluding Remarks and Future Perspectives *336*	
	References *338*	

Index *343*

Preface

This book summarizes the recent progress in the major methodologies of carbonyl olefination, which is one of the most fundamental transformations in organic synthesis. Carbonyl olefination has been extensively studied since Professor Georg Wittig discovered in 1953 the reaction of phosphonium ylides with carbonyl compounds, which has become known as the Wittig reaction. Since then, a variety of reagents have been developed for this transformation by a number of chemists. These reagents enable us to transform various carbonyl functions into carbon-carbon double bonds with different chemo- and stereoselectivities and are utilized in a variety of organic syntheses.

The mechanisms of these reactions bear marked similarities, in spite of the differences in their reactivities and selectivities. Thus, in certain cases, a four-membered intermediate similar to the 1,2-oxaphosphetane intermediate in the Wittig reaction appears in the Peterson reaction as a pentacoordinate 1,2-oxasiletanide. Reactions of transition metal carbene complexes with carbonyl compounds also proceed through the formation of a four-membered oxametallacycle, which was recently found to be an intermediate of some McMurry reactions. Carbonyl olefination utilizing dimetallic species of zinc or chromium is somewhat similar to the Julia reaction in that they both involve the process of β-elimination.

In this book, an effort has been made to provide comprehensive yet concise commentaries on the mechanisms of each reaction, as well as on their synthetic applications. These provide an accurate prescription for their use and should be useful for the development of a broader perspective on carbonyl olefination. The final chapter is concerned with asymmetric carbonyl olefination, which is one of the frontiers of organic synthesis. As this subject exemplifies, the established methodologies are not necessarily perfect and there still remain many problems to be solved in the field of carbonyl olefination. It is hoped that this book will be of wide use to all chemists engaged in organic synthesis, both in industrial laboratories and in academic institutions.

I would like to thank the authors of the individual chapters for their excellent contributions. This volume was only possible with the cooperation of the authors, who are experts in each field. Finally, I express my sincere gratitude to my wife, Yukiko, whose continuous encouragement was essential to the editing of this book.

Takeshi Takeda
Tokyo, 2003

List of Authors

Andrew D. Abell
Department of Chemistry
University of Canterbury
Private Bag 4800
Christchurch
New Zealand

Raphaël Dumeunier
Université catholique de Louvain
Département de Chimie
Unite de Chimie Organique et Medicinale
Bâtiment Lavoisier
Place Louis Pasteur 1
B-1348 Louvain-la-Neuve
Belgium

Michael Edmonds
School of Applied Science
Christchurch Polytechnic Institute of Technology
City Campus, Madras St
P.O. Box 540
Christchurch 8015
New Zealand

Michel Ephritikhine
Service de Chimie Moléculaire
Bât. 125
DSM, DRECAM, CNRS URA 331
CEA Saclay
91191 Gif-sur-Yvette
France

Kaoru Fuji
Kyoto Pharmaceutical University
Misasagi, Yamashina
Kyoto 607-8414
Japan

Takumi Furuta
School of Pharmaceutical Sciences
University of Shizuoka
Shizuoka 422-8526
Japan

Naokazu Kano
Department of Chemistry
Graduate School of Science
The University of Tokyo
7-3-1 Hongo, Bunkyo-ku
Tokyo 113-0033
Japan

Takayuki Kawashima
Department of Chemistry
Graduate School of Science
The University of Tokyo
7-3-1 Hongo, Bunkyo-ku
Tokyo 113-0033
Japan

Istvan E. Markó
Université catholique de Louvain
Département de Chimie
Unite de Chimie Organique et Medicinale
Bâtiment Lavoisier
Place Louis Pasteur 1
B-1348 Louvain-la-Neuve
Belgium

Seijiro Matsubara
Department of Material Chemistry
Graduate School of Engineering
Kyoto University
Yoshida, Sakyo-ku
Kyoto 606-8501
Japan

Koichiro Oshima
Department of Material Chemistry
Graduate School of Engineering
Kyoto University
Yoshida, Sakyo-ku

Kyoto 606-8501
Japan

Takeshi Takeda
Department of Applied Chemistry
Tokyo University of Agriculture & Technology
2-24-16 Nakacho, Koganei
Tokyo 184-8588
Japan

Kiyoshi Tanaka
School of Pharmaceutical Sciences
University of Shizuoka
Shizuoka 422-8526
Japan

Akira Tsubouchi
Department of Applied Chemistry
Tokyo University of Agriculture & Technology
2-24-16 Nakacho, Koganei
Tokyo 184-8588
Japan

Claude Villiers
Service de Chimie Moléculaire
Bât. 125
DSM, DRECAM, CNRS URA 331
CEA Saclay
91191 Gif-sur-Yvette
France

1
The Wittig Reaction

Michael Edmonds and Andrew Abell

1.1
Introduction

The reaction of a phosphorus ylide with an aldehyde or ketone, as first described in 1953 by Wittig and Geissler [1] (see Scheme 1.1), is probably the most widely recognized method for carbonyl olefination.

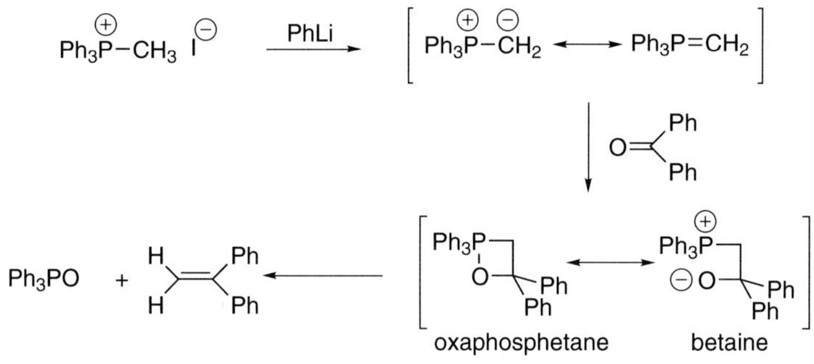

Scheme 1.1. The Wittig reaction.

This so-called Wittig reaction has a number of advantages over other olefination methods; in particular, it occurs with total positional selectivity (that is, an alkene always directly replaces a carbonyl group). By comparison, a number of other carbonyl olefination reactions often occur with double-bond rearrangement. In addition, the factors that influence *E*- and *Z*-stereoselectivity are well understood and can be readily controlled through careful selection of the phosphorus reagent and reaction conditions. A wide variety of phosphorus reagents are known to participate in Wittig reactions and the exact nature of these species is commonly used to divide the Wittig reaction into three main groups, namely the "classic" Wittig reaction of phosphonium ylides, the Horner–Wadsworth–Emmons reaction of phosphonate anions, and the Horner–Wittig reaction of phosphine oxide anions. Each of these reaction types has its own distinct advantages and limitations, and

Modern Carbonyl Olefination. Edited by Takeshi Takeda
Copyright © 2004 WILEY-VCH Verlag GmbH & Co. KGaA, Weinheim
ISBN: 3-527-30634-X

these must be taken into account when selecting the appropriate method for a desired synthesis.

1.2
The "Classic" Wittig Reaction [1–4]

The original work of Wittig and Geissler [1], as depicted in Scheme 1.1, provides a good example of a classic Wittig reaction in which a phosphonium ylide reacts with an aldehyde or ketone to afford the corresponding alkene and phosphine oxide. This reaction is very general and provides a convenient method for the preparation of a range of alkenes with good stereocontrol. The starting phosphonium ylides are themselves readily generated by the addition of a suitable base to the corresponding phosphonium salt (refer to Section 1.2.3).

1.2.1
Mechanism and Stereoselectivity

The mechanism of the Wittig reaction has long been considered to involve two intermediate species, a diionic betaine and an oxaphosphetane, as shown in Scheme 1.1. However, there has been much debate as to which of these two species plays the most important mechanistic role and also as to how each influences the stereochemical outcome under different reaction conditions. For many years, it was generally accepted that the betaine is the more important intermediate [5, 6]; however, recent low temperature ^{31}P NMR studies suggest that this may not be the case [7, 8]. This supposition is further supported by recent calculations that reveal that oxaphosphetanes are of lower energy than the corresponding betaines [9]. As such, the currently accepted mechanism for the Wittig reaction is as shown in Scheme 1.2 [4]. For a more detailed account of the evolution of the Wittig mechanism, the reader is referred to the excellent reviews by Vedejs and co-workers [4, 10].

The stereoselectivity of the Wittig reaction is directly linked to this mechanism. In particular, the reaction of a carbonyl compound with an ylide produces both the

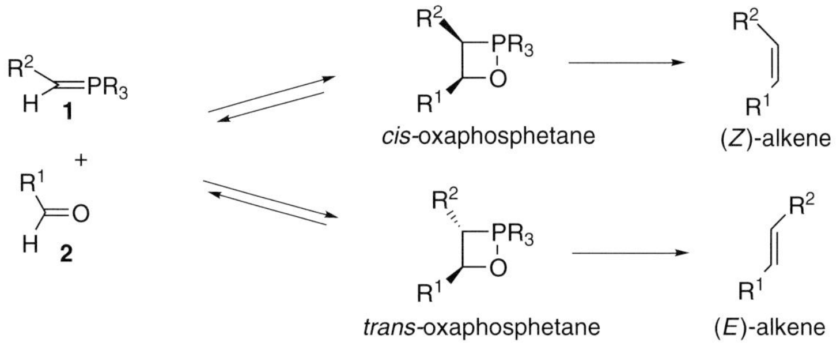

Scheme 1.2. The mechanism of the Wittig reaction.

cis and trans oxaphosphetanes (Scheme 1.2), which undergo stereospecific syn-elimination to give the corresponding E- and Z-alkenes, respectively. The Z-alkene tends to predominate under kinetic conditions, indicating an intrinsic preference for the cis-oxaphosphetane, a preference that as yet is not fully understood. However, under thermodynamic conditions, equilibration of the two oxaphosphetanes with reactants **1** and **2** allows predominant formation of the more stable trans-oxaphosphetane and hence the E-alkene. This is supported by reaction rate studies, which show that cis-oxaphosphetanes undergo retro-Wittig decomposition much faster than their trans counterparts [11–13]. It should be noted that the extent of the retro-Wittig decomposition is dependent on both the nature of the substituent on the oxaphosphetane and the reaction conditions used. A number of factors determine whether a Wittig reaction is under kinetic or thermodynamic control, one of the most important and readily controllable being the type of ylide used.

1.2.2
Nature of the Ylide and Carbonyl Compound

Stabilized ylides are those that possess an R substituent (Figure 1.1) that is anion-stabilizing/electron-withdrawing (e.g. CO_2CH_3, CN). Such ylides tend to be less reactive than other ylides and usually only react with aldehydes to give the E-alkene. This E-selectivity has been attributed to the fact that stabilized ylides react with aldehydes under thermodynamic control. Consequently, the less crowded and hence favored trans-oxaphosphetane gives rise to the observed E-alkene. Other factors that influence the E/Z ratio of alkenes in reactions of a stabilized ylide with an aldehyde are listed in Table 1.1.

In contrast, non-stabilized ylides possess an R substituent (Figure 1.1) that is anion-destabilizing/electron-releasing (e.g. alkyl groups). Reactions of these ylides with an aldehyde or ketone are generally under kinetic control and as a consequence give the Z-alkene. A number of factors influence the E/Z ratio of alkenes in reactions of non-stabilized ylides, and these are listed in Table 1.1.

The addition of a second equivalent of a strong base, usually an alkyllithium, to reactions of non-stabilized ylides facilitates the formation of the E-alkene (Scheme 1.3). The resulting β-oxido ylide is then quenched with acid under kinetic control to afford the E-alkene. This sequence is known as the Schlosser modification [14–16]. Lithium bases must be employed here since the presence of lithium ions is required to convert the oxaphosphetane **3** into the more acidic betaine.

The nature of the reactant carbonyl group in the substrate also influences the stereoselectivity of the Wittig reactions; for example, primary aliphatic aldehydes

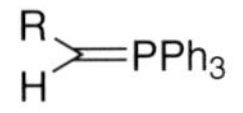

$$\underset{H}{\overset{R}{>}}{=}PPh_3$$

Triphenylphosphorane ylides are readily accessible by the reaction of triphenylphosphine with a suitable halide (See Section 1.2.3)

Fig. 1.1. R-Substituted ylides.

1 The Wittig Reaction

Tab. 1.1. Factors that influence the E/Z ratio in the "classic" Wittig reaction.

Stabilized Ylides	
Factors that favor the E-alkene	*Factors that disfavor the E-alkene*
Aprotic conditions	Li and Mg salts in DMF as solvent
A catalyic amount of benzoic acid[62]	Use of α-oxygenated aldehydes or methanol as solvent

Non-stabilized Ylides	
Factors that favor the Z-alkene	*Factors that disfavor the Z-alkene*
Bulky and/or aliphatic aldehydes	Small phosphorus ligands, cyclohexyl ligands
Bulky phosphorus ligands, e.g. phenyl ligands	Cyclic phosphorus ligands
Lithium-free conditions/lower temperatures	Aromatic or α,β-unsaturated aldehydes
Hindered ylides	Use of Schlosser modification
Low temperature	

favor Z-alkenes, while aromatic or α,β-unsaturated aldehydes tend to reduce Z-selectivity, especially in polar aprotic solvents. In addition, aryl alkyl ketones tend to give a higher Z-selectivity as compared to unsymmetrical dialkyl ketones.

1.2.3
Reagents and Reaction Conditions

Phosphonium ylides are typically prepared by the reaction of a phosphonium salt with a base. Non-stabilized ylides require a strong base (such as BuLi) under inert conditions, while stabilized ylides require a weaker base (for example, alkali metal hydroxides in aqueous solution). For more detail on the variety of bases and reaction conditions that can be used in the Wittig reaction, see references [2] and [5].

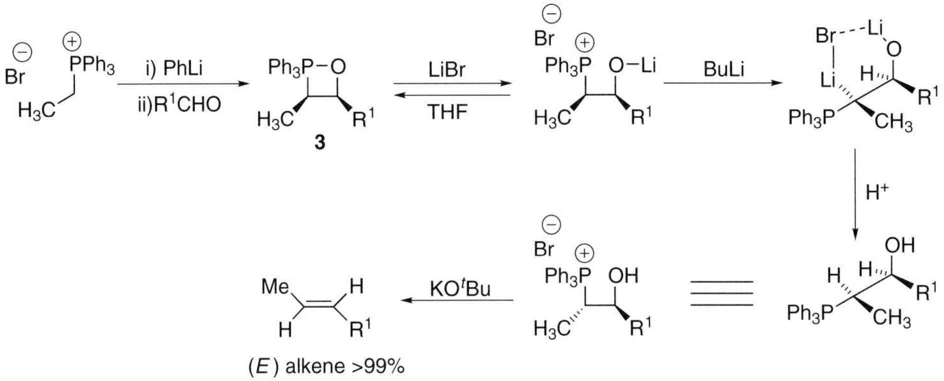

Scheme 1.3. The Schlosser modification.

The starting phosphonium salts are themselves readily obtained by reaction of triaryl- or trialkylphosphines with an alkyl halide (Scheme 1.4). In general, primary alkyl iodides and benzyl bromides are converted to the corresponding phosphonium salts by heating with triphenylphosphine at 50 °C in THF or CHCl$_3$, while primary alkyl bromides, chlorides, and branched halides usually require more vigorous conditions (for example, heating to 150 °C). These reactions can be carried out neat, or in acetic acid, ethyl acetate, acetonitrile or DMF solution.

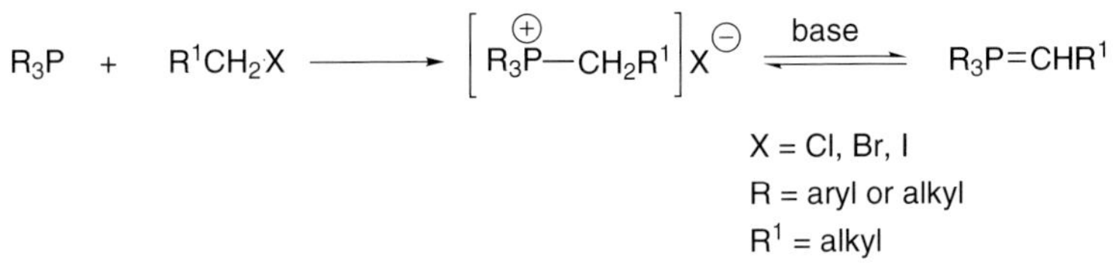

$$R_3P + R^1CH_2X \longrightarrow [R_3\overset{+}{P}-CH_2R^1]X^- \xrightleftharpoons{\text{base}} R_3P=CHR^1$$

X = Cl, Br, I
R = aryl or alkyl
R^1 = alkyl

Scheme 1.4. Preparation of phosphonium salts.

Phosphonium salts can also be prepared by alkylating existing phosphonium salts. This method has been used to incorporate interesting groups, such as silyl and stannyl functionalities [17, 18] (Scheme 1.5), into phosphonium salts. Such salts, upon reaction with base and aldehyde, produce synthetically useful allylsilanes and -stannanes.

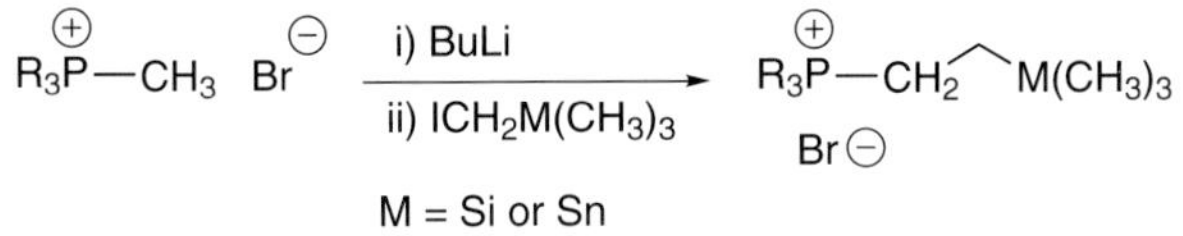

$$R_3\overset{+}{P}-CH_3\ Br^- \xrightarrow[\text{ii) ICH}_2\text{M(CH}_3)_3]{\text{i) BuLi}} R_3\overset{+}{P}-CH_2\frown M(CH_3)_3\ \ Br^-$$

M = Si or Sn

Scheme 1.5. Alkylation of phosphonium salts.

1.3
Horner–Wadsworth–Emmons Reaction [2, 3, 19]

The reaction of a phosphonate-stabilized carbanion with a carbonyl compound is referred to as the Horner–Wadsworth–Emmons reaction (HWE); see Scheme 1.6. The phosphonates used here are significantly more reactive than classical Wittig stabilized ylides and as such react with ketones and aldehydes. Another advantage of the HWE reaction over the classic Wittig reaction is that the phosphorus by-products are water-soluble and hence readily separated from the desired product. Phosphonates that lack a stabilizing α-substituent at R^2 (for example COO$^-$, COOR, CN, SO$_2$R, vinyl, aryl P(O)(OR)$_2$, OR or NR$_2$) generally result in a low yield of the product alkene. The starting phosphonates are readily prepared by means of the Arbuzov reaction of trialkyl phosphites with an organic halide (Section 1.3.2).

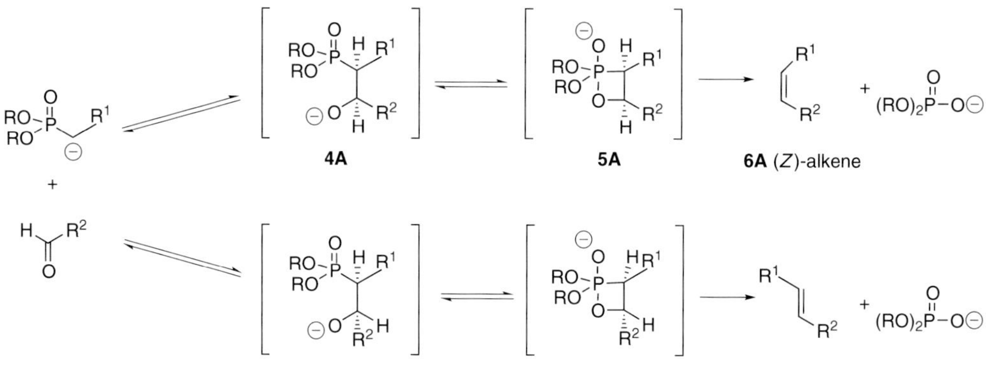

Scheme 1.6. The Horner–Wadsworth–Emmons reaction.

1.3.1
Mechanism and Stereochemistry

The commonly accepted mechanism for the Horner–Wadsworth–Emmons reaction is as depicted in Scheme 1.6. Here, reaction of the phosphonate stabilized carbanion with an aldehyde forms the oxyanion intermediates **4** under reversible conditions. Rapid decomposition of **4**, via the four-centered intermediates **5**, then affords alkenes **6**.

The stereochemical outcome of the Horner–Wadsworth–Emmons reaction is primarily dependent on the nature of the phosphonate used. In general, bulky substituents at both the phosphorus and the carbon adjacent to the carbanion favor formation of the *E*-alkene. This selectivity has been rationalized in terms of a lowering of steric strain in intermediate **5B** as compared to intermediate **5A**. *Z*-Selectivity in HWE reactions can, however, be achieved using the Still–Gennari modification [20]. Here, the use of a (2,2,2-trifluoroethyl) phosphonate enhances the rate of elimination of the originally formed adduct **5A** (Scheme 1.6) relative to equilibration of the intermediates **4** and **5**. An example of the Still–Gennari modification is illustrated in Scheme 1.7.

The so-called Ando method [21, 22] also provides access to a *Z*-alkene, as illustrated in Scheme 1.8 [23], where >99% *Z*-selectivity was obtained using

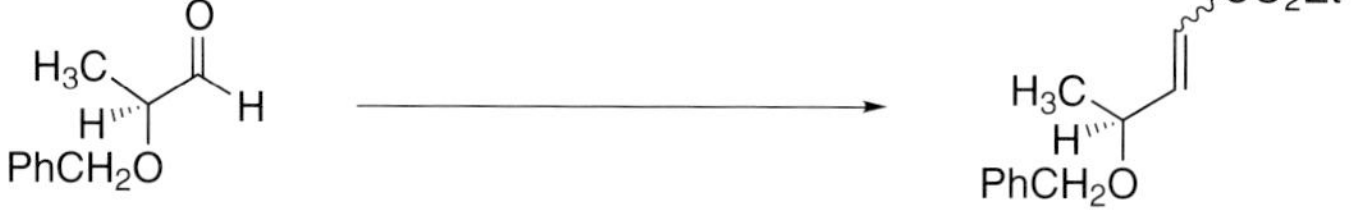

Reaction conditions
a) $(EtO)_2POCH_2CO_2Et$, NaH, THF   83%, *E*:*Z* = 12:1
b) $(CF_3CH_2O)_2POCH_2CO_2CH_3$, KH, THF   84% *E*:*Z* = 1:11

Scheme 1.7. The Still–Gennari modification.

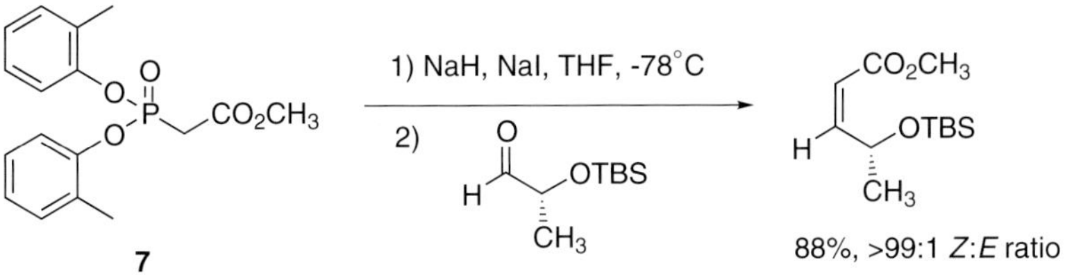

Scheme 1.8. The Ando method.

Ando's bis(o-methylphenyl) phosphonoacetate **7** in the presence of excess Na$^+$ ions (Scheme 1.8).

Some common factors that influence the stereochemical outcome of the HWE reaction are summarized in Table 1.2. In general, the addition of a phosphonate to a ketone occurs with moderate E-selectivity [24]. The reaction of an α-substituted phosphonate with an aldehyde also usually favors the E-alkene (Scheme 1.9), although some exceptions have been noted [25, 26]. The E-selectivity is further enhanced with the use of large phosphoryl and carbanion substituents [27, 28]. However, α-substituted phosphonates that bear a large alkyl chain give only modest E-selectivity. α-Fluoro phosphonates are reported to provide impressive E- or Z-stereoselectivity (Scheme 1.10), which has been attributed to electronic effects [29].

Substituents on the carbonyl compound can also influence the stereo-outcome of the HWE reaction. For example, carbonyl compounds that possess oxygenated groups at the α- or β-position generally favor the E-alkene [30, 31], although some exceptions have been observed [32].

The synthesis of tetrasubstituted alkenes using the HWE reaction generally proceeds with moderate selectivity [33, 34].

Tab. 1.2. Factors that influence the ratio E/Z alkenes in the Horner–Wadsworth–Emmons reaction.

Factors that favor the E-alkene	Factors that favor the Z-alkene
Bulky R groups on the phosphonate e.g. (RO)$_2$P(O)–CH(–)–R	Use of bis(2,2,2-trifluoroethyl) phosphonates (Still–Gennari modification)
Bulky R' groups adjacent to the carbanion e.g. (RO)$_2$P(O)–CH(–)–R'	Use of cyclic phosphonates such as **8**
Use of α-fluoro phosphonates	Use of (diarylphosphono)acetates (Ando method)/ excess Na$^+$ ions

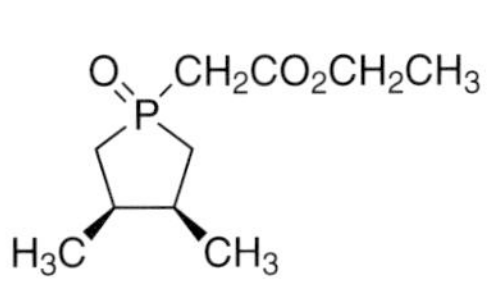

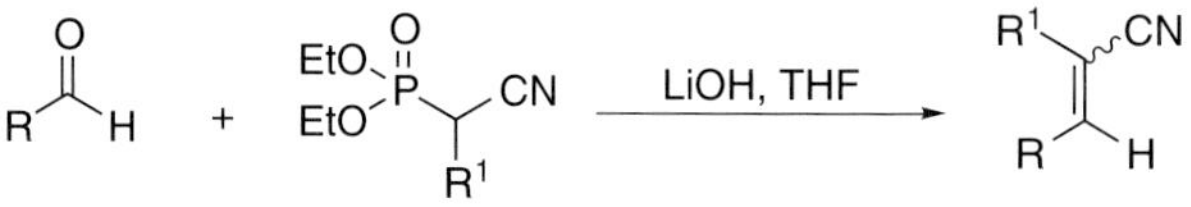

1 The Wittig Reaction

$$R-CHO + (EtO)_2P(O)-CH(R^1)-CN \xrightarrow{\text{LiOH, THF}} R^1\text{(CN)}C=CHR$$

R	R^1	E/Z, yield
Ph	H	90:10, 70%
Ph	CH$_3$	77:23, 85%
nC_7H_{15}	H	68:32, 74%
nC_7H_{15}	CH$_3$	58:42, 72%

Scheme 1.9. The effect of α-substituted phosphonate on E/Z ratio in the Horner–Wadsworth–Emmons reaction.

$$(EtO)_2P(O)-CF(CO_2R) \xrightarrow[\text{ii) cyclohexanecarbaldehyde}]{\text{i) base}} \text{cyclohexyl-CH=C(F)(CO}_2R)$$

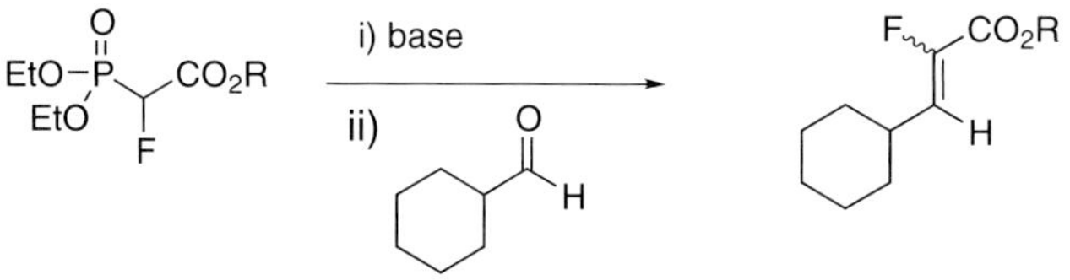

R	Base	E/Z, yield
Et	nBuLi	94:6, 83%
H	iPrMgBr	>1:99, 84%

Scheme 1.10. High stereoselectivity using α-fluorinated phosphonates in the Horner–Wadsworth–Emmons reaction.

1.3.2
Reagents and Reaction Conditions

The starting phosphonates are readily obtained by the Michaelis–Arbuzov reaction of trialkyl phosphites with an organic halide, and as would be expected, alkyl bromides are more reactive than the corresponding chlorides (Scheme 1.11).

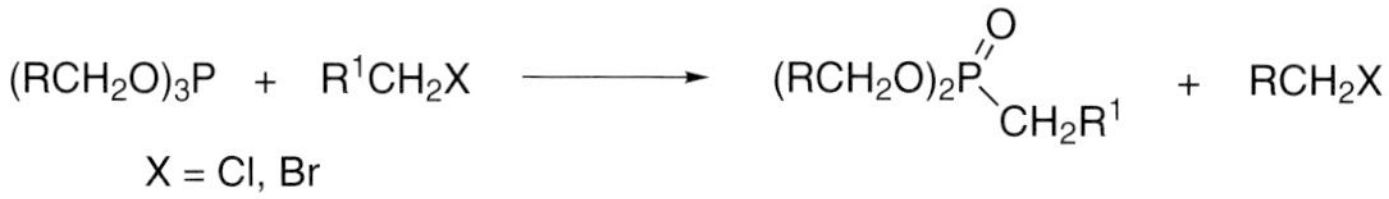

$$(RCH_2O)_3P + R^1CH_2X \longrightarrow (RCH_2O)_2P(O)CH_2R^1 + RCH_2X$$

X = Cl, Br

Scheme 1.11. The Michaelis–Arbuzov reaction.

An existing phosphonate can be further elaborated by alkylation or acylation of the phosphonate carbanion [35, 36] (Scheme 1.12). This provides a useful method for the preparation of β-keto phosphonates that are not generally available by means of the Michaelis–Arbuzov method. A range of β-keto phosphonates is also

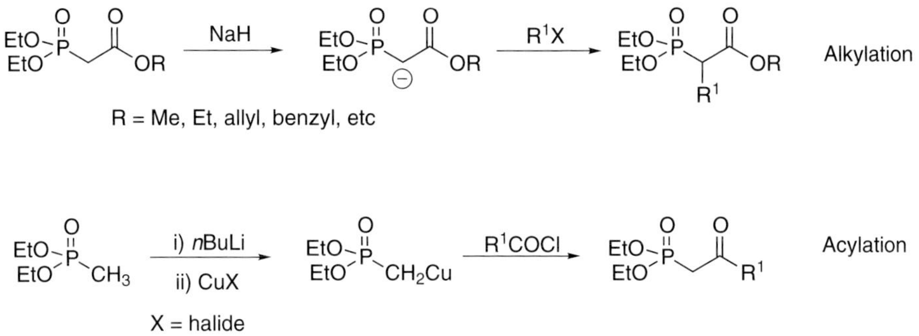

1.4 Horner–Wittig (HW) Reaction

Scheme 1.12. Modification of phosphonates by alkylation and acylation.

readily accessible through the acylation of TMS esters [37] or acid chlorides of dialkyl phosphonoacetic acids [38] (Scheme 1.13).

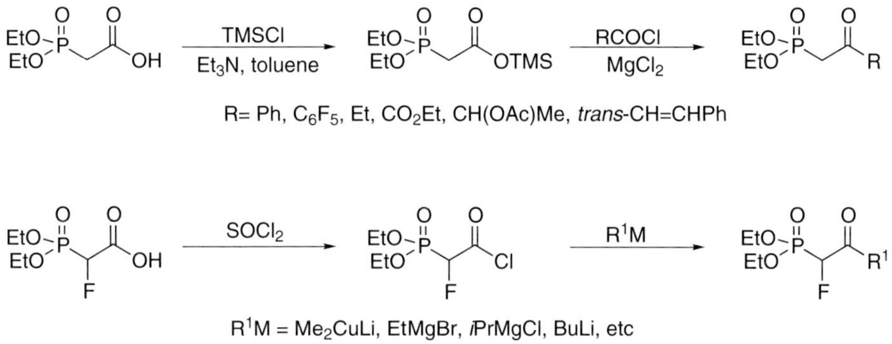

R= Ph, C_6F_5, Et, CO_2Et, CH(OAc)Me, trans-CH=CHPh

R^1M = Me_2CuLi, EtMgBr, iPrMgCl, BuLi, etc

Scheme 1.13. Preparation of β-ketophosphonates using TMS esters and acid chlorides.

As stated above, phosphonates react with both aldehydes and ketones, although ketones generally require more vigorous conditions. Long chain aliphatic aldehydes tend to be unreactive, while readily enolizable ketones usually give poor yields of the alkene.

A wide variety of bases and reaction conditions have been successfully applied to the HWE reaction, including phase-transfer catalysis.

1.4
Horner–Wittig (HW) Reaction [2, 3, 39]

Horner and co-workers were the first to describe the preparation of an alkene by treatment of a phosphine oxide with base followed by the addition of an aldehyde (Scheme 1.14) [40–42]. While initial experiments using bases such as potassium *tert*-butoxide produced the alkenes directly, it was quickly realized that the use of lithium bases allowed the intermediate β-hydroxy phosphine oxide diastereomers to be isolated and separated [40]. Each diastereomer can then be separately treated

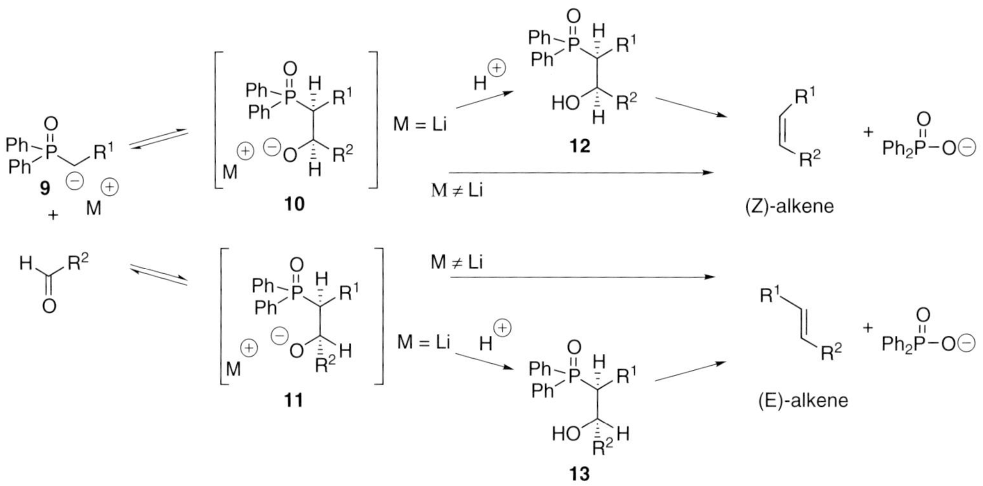

Scheme 1.14. The Horner–Wittig reaction.

with base to give the corresponding alkenes with high geometrical purity (Scheme 1.14).

Like the HWE reaction, the HW reaction gives rise to a phosphinate by-product that is water-soluble and hence readily removed from the desired alkene.

1.4.1
Mechanism and Stereochemistry

As mentioned above, the use of lithium bases in the HW reaction allows the reaction to be divided into two discrete steps [39]: 1) the **HW addition** of a lithiated phosphine oxide to an aldehyde (or ketone) to produce a β-hydroxy phosphine oxide, and 2) the **HW elimination** of a phosphinic acid to afford an alkene (Scheme 1.14). Careful manipulation of each step then allows control of the overall sequence. While the overall mechanism of the Horner–Wittig reaction is similar to that of the HWE reaction (Scheme 1.6), some additional discussion is required to understand its stereochemical outcome. The HW reaction can be carried out without isolation of the intermediate β-hydroxy phosphine oxides in cases where a non-lithium base is used and R^1 is able to stabilize the negative charge of the phosphorus α-carbanion 9. Under these conditions, reaction of an aldehyde with the phosphine oxide to give intermediates 10 and 11 is reversible. The E-alkene is then formed preferentially since elimination of intermediate 11 occurs much faster than that of 10.

However, if a lithium base is used and the reaction is carried out at low temperature, the intermediate β-hydroxy phosphine oxides 12 and 13 can be isolated (Scheme 1.14). Under these conditions, the *erythro* intermediate 12 predominates and this then undergoes *syn*-elimination to give the Z-alkene.

In cases where R^1 is unable to exert a stabilizing effect and a lithium base is used, the intermediates 10 and 11 do not undergo elimination and instead β-

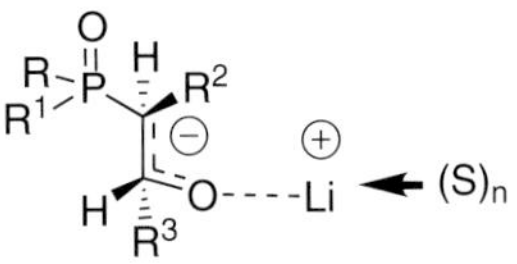

Fig. 1.2. Solvent-stabilized transition state model for the Horner–Wittig reaction.

hydroxy phosphine oxides **12** and **13** are readily isolated. These diastereomeric intermediates can then be separated by column chromatography or crystallization. When R^1 is non-stabilizing, HW addition reactions generally display limited stereocontrol, although under certain conditions (bulky R groups on the phosphine oxide and aldehyde; polar solvents) high Z-selectivity can be effected. It has been proposed that under these conditions the favored transition state has the solvent-stabilized oxido group positioned *anti* to the bulky phosphinyl group (Figure 1.2) [3].

Clayden and Warren have proposed the transition state model depicted in Figure 1.3 to explain the stereoselectivity observed for the HW reaction of a range of phosphine oxides, as summarized in Table 1.3 [39].

According to this model, both R^1 and R^2 occupy the less sterically demanding *exo* positions to give preferential formation of the Z-product. However, as the bulk of the R^1 group increases, the effect of steric interactions between R^1 and R^2, and R^1 and the *exo* Ph–P ring, is to make the R^1 group occupy the pseudo-equatorial *endo* position, thereby lowering the Z-selectivity (see Table 1.3). For moderately sized R^2 groups, the *exo* position (Figure 1.3) is favored since the *endo* position suffers from 1,3-diaxial interactions with the *endo* Ph–P ring. However, if R^1 is very small (e.g. Me), the lesser 1,3-diaxial interactions will reduce the Z-selectivity. Furthermore, if R^2 is large, its steric interaction with R^1 also reduces the Z-selectivity. The model depicted in Figure 1.3 suggests that the use of ketones or di-α-substituted phosphine oxides in the HW reaction would produce very little stereoselection. This is usually true unless there is a substantial difference in size between the two substituents on the ketone or di-α-substituted phosphine oxide. One such example is illustrated in Scheme 1.15 [43].

The presence of R^1 groups capable of coordinating to lithium (e.g. OMe, NR_3) also lowers the Z-selectivity, presumably by altering the structure of the transition state complex. Indeed, in at least one case (last entry of Table 1.3), significant E-selectivity is observed [44].

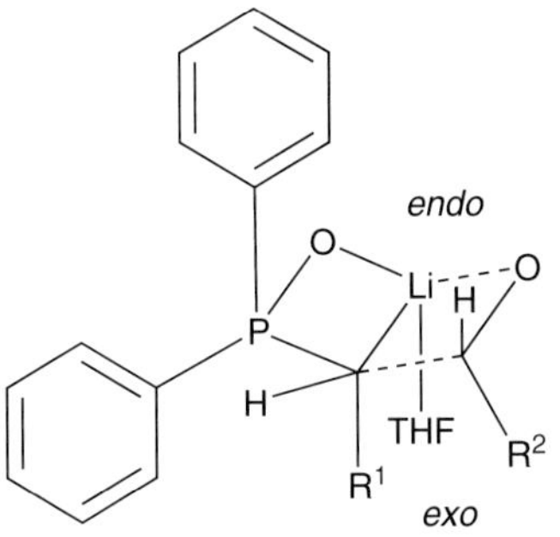

Fig. 1.3. Transition state model for the Horner–Wittig reaction.

Tab. 1.3. The effect of phosphonate and aldehyde substitution on Z/E stereoselectivity.

R^1	R^2	Yield (%)	Z/E ratio
Me	Ph	88	88:12
Bu	Ph	84	84:16
^{i}Bu	Ph	81	80:20
Ph	Ph	88	83:17
Me	cyclohexyl	86	79:21
Me	Me	93	75:25
Ph	Me	97	72:28
^{i}Pr	Pr	84	57:43
cyclohexyl	Pr	100	53:47
OMe	C_6H_{13}	79	50:50
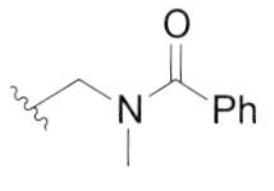 ⁓N(Ph)C(O)–	Me	68	55:45
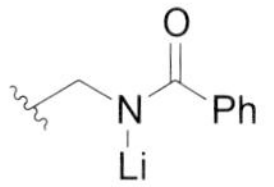 ⁓N(Li)C(O)Ph	Ph	80	5:95

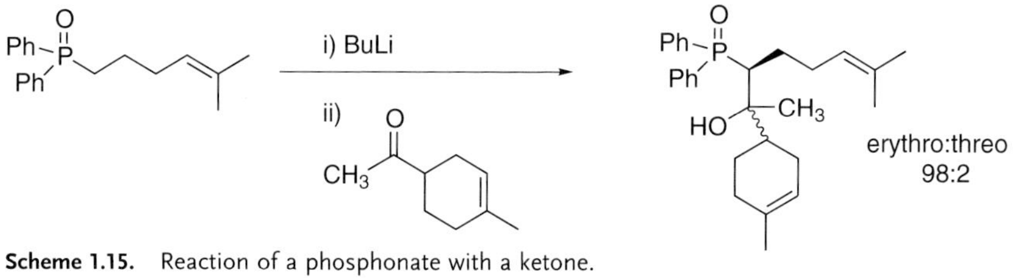

Scheme 1.15. Reaction of a phosphonate with a ketone.

erythro:threo 98:2

Since high E-selectivity is usually not possible using phosphine oxides that bear a non-stabilizing R group, the *threo* intermediate (and hence the E-alkene) must be obtained indirectly. Such an approach has been developed by Warren et al. (Scheme 1.16) [45, 46]. Oxidation of the 1,2-phosphinoyl alcohol **12** or acylation of

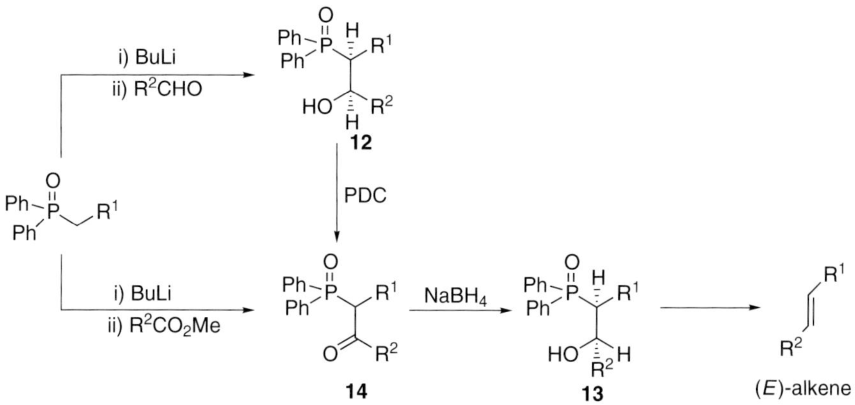

Scheme 1.16. Warren's method for indirect access to E-alkenes.

Tab. 1.4. Stereoselective reduction of β-keto phosphine oxide using different reducing conditions.

R¹	R²	NaBH₄ threo/erythro ratio (yield %)	NaBH₄/CeCl₃ threo/erythro ratio (yield %)
Me	Ph	89:11 (–)	70:30 (–)
^{i}Pr	Ph	83:17 (75)	5:95 (96)
^{i}Pr	Me	65:35 (71)	23:77 (83)
Cyclohexyl	Ph	76:24 (71)	5:95 (100)
(CH₂)₂OH	Me	67:33 (–)	>95:5

(–) = yield unavailable.

the phosphine oxide carbanion affords a β-keto phosphine oxide **14**, which is then reduced with NaBH₄ to afford predominantly the *threo* β-hydroxy phosphine oxide **13**. Treatment with base then affords *E*-alkenes in typically high yields.

While NaBH₄ reduction affords the *threo* β-hydroxy phosphine oxide **13**, other reducing conditions can provide a different stereo-outcome. Luche reduction (NaBH₄ in the presence of CeCl₃) [47–49] enhances the formation of the *erythro* β-hydroxy phosphine oxide (see Table 1.4) [39, 50, 51]. This shift in selectivity is particularly significant in cases where the R¹ group is branched at the β-position. However, in cases where a free OH occupies the γ-position, the *threo* β-hydroxy phosphine oxide again dominates. An explanation of this stereoselectivity is provided in Figure 1.4.

Ce^{III} chelation gives rise to a transition state in which the presence of a large, branched R¹ group forces the BH₄⁻ to approach from the opposite side of the carbonyl group (Figure 1.4A), resulting in *erythro* selectivity. The presence of a γ-positioned OH group in the R¹ group re-orientates the transition state to allow easier access to the carbonyl group from the same side (Figure 1.4B), ultimately resulting in *threo* selectivity.

The HW elimination of the β-hydroxy phosphine oxides also influences the stereo-outcome of the HW reaction. Treatment of *threo* β-hydroxy phosphine oxides **13** with base reliably leads to formation of the *E*-alkene; however, decomposition of

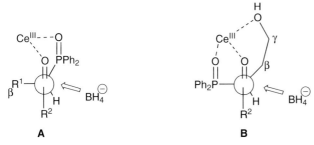

Fig. 1.4. Transition state justification for (A): *erythro* selectivity under normal Luche conditions, and (B): *threo* selectivity under Luche conditions with a free OH in the γ-position.

Tab. 1.5. Stereocontrol of the Horner–Wittig reaction.

Conditions that favor the E-alkene	Conditions that favor the Z-alkene
One-step HW reaction (R group of Ph_2POR is stabilizing; moderate temperature, non-Li base)	Two-step HW reaction (isolation and separation of predominantly *erythro* β-hydroxy phosphine oxide, then decomposition with base). Best if R group of Ph_2POR is not anion-destabilizing.
Oxidation of intermediate β-hydroxy phosphine oxide, followed by reduction with $NaBH_4$, then decomposition with base	Oxidation of intermediate β-hydroxy phosphine oxide, followed by reduction with $NaBH_4$/$CeCl_3$, then decomposition with base (Luche reduction). Best if R group of Ph_2POR is large and branched.

the *erythro* β-hydroxy phosphine oxides **12** is less stereoselective and lower yielding if the R^1 group in the α position is anion-stabilizing (refer to Scheme 1.14). This is attributed to reversible dissociation of the β-oxido phosphine oxide intermediates to the initial phosphine oxide anion and aldehyde. A summary of generalized conditions appropriate for obtaining either E- or Z-alkenes is presented in Table 1.5.

Other factors that have been used to enhance Z-selectivity include the use of low temperatures and a THF/TMEDA solvent system [46]. Work by Smith et al. on the synthesis of milbemycin has also shown that the choice of base counterion can have a significant effect, with sodium producing a 1:7 *E/Z* ratio compared to a 2:3 *E/Z* ratio when potassium was used as counterion [52].

While diphenylphosphine oxide is used in most HW reactions, attempts have been made to utilize other phosphine oxides in order to enhance Z-selectivity. For example, *ortho*-substituted diarylphosphine oxides such as $(anisyl)_2P(O)CHPr^-Li^+$ have been used to produce complete Z-selectivity [53]. Dibenzylphosphole oxides **15** have also been utilized by several researchers in an attempt to enhance stereoselectivity [51, 54, 55].

15

1.4.2
Reagents and Reaction Conditions

Phosphine oxides can be prepared by a number of methods (Scheme 1.17). One approach involves the reaction of organometallic reagents with halogenophosphines followed by oxidation (a). This method has proven to be popular due to the

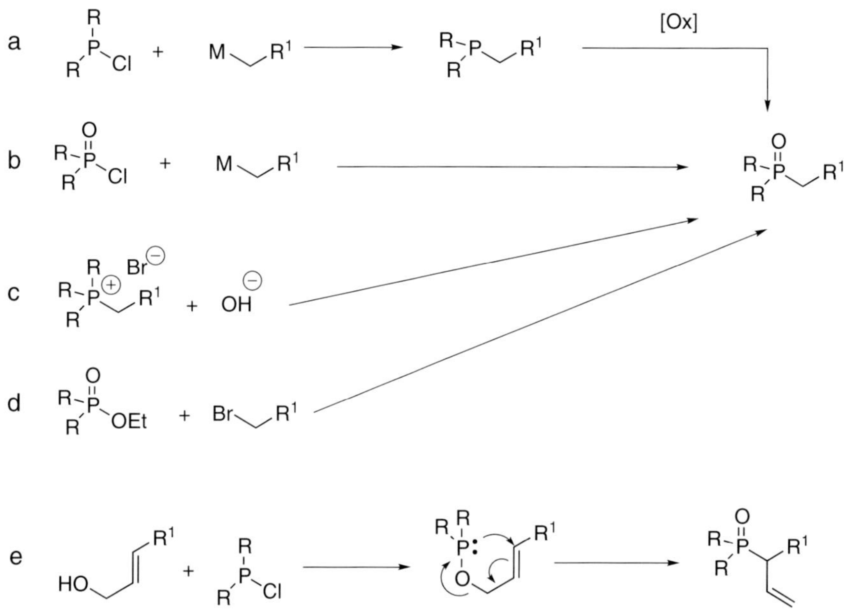

Scheme 1.17. Preparation of phosphine oxides.

ready availability of halogenophosphines such as Ph$_2$PCl [19]. Phosphine oxides have also been prepared by direct reaction of organometallics with phosphonyl halides [19] (b), hydrolysis of phosphonium salts [46] (c), and the Arbuzov reaction [56, 57] (d). In addition, allylic phosphine oxides are also available by means of the Arbuzov rearrangement [58] (e). More complicated phosphine oxides can also be prepared by deprotonation and acylation or alkylation of existing phosphine oxides [35, 36, 59–61].

1.5
Conclusion

Since its conception in 1953, the Wittig reaction (and its variants) has developed into a predictable and reliable method for the synthesis of a wide range of alkenes, often with high E- or Z-stereoselectivity. With much of the mechanism for this reaction now elucidated and a large literature base, the Wittig reaction can be applied with confidence to many alkene syntheses. In this chapter, we have described only a small proportion of the literature regarding the Wittig reaction; however, even with this small sample it is possible to see how the judicious selection of the type of phosphorus reagent, carbonyl compound, and reaction conditions (see, for example, Tables 1.1, 1.2, and 1.5) can be used to produce a desired compound with high E- or Z-selectivity.

References

1 G. Wittig, G. Geissler, *Liebigs Ann.* **1953**, *44*–57.
2 A. Maercker, *Org. React.* **1965**, *14*, 270–490.
3 B. E. Maryanoff, A. B. Reitz, *Chem. Rev.* **1989**, *89*, 863–927.
4 E. Vedejs, M. J. Peterson, *Top. Stereochem.* **1994**, *21*, 1–157.
5 I. Gosney, A. G. Rowley, in *Organophosphorus Reagents in Organic Synthesis* (Ed.: J. I. G. Cadogan), Academic Press, London, **1979**, pp. 17–153.
6 M. Schlosser, *Top. Stereochem.* **1970**, *5*, 1–30.
7 E. Vedejs, K. A. Snoble, *J. Am. Chem. Soc.* **1973**, *95*, 5778–5780.
8 E. Vedejs, G. P. Meier, K. A. Snoble, *J. Am. Chem. Soc.* **1981**, *103*, 2823–2831.
9 F. Ramirez, C. P. Smith, J. F. Pilot, *J. Am. Chem. Soc.* **1968**, *90*, 6722–6726.
10 E. Vedejs, M. J. Peterson, *Advances in Carbanion Chemistry* **1996**, 2.
11 B. E. Maryanoff, A. B. Reitz, M. S. Mutter, R. R. Inners, H. R. Almond Jr., R. R. Whittle, R. A. Olofson, *J. Am. Chem. Soc.* **1986**, *108*, 7664–7678.
12 E. Vedejs, T. Fleck, *J. Am. Chem. Soc.* **1989**, *111*, 5861–5871.
13 E. Vedejs, C. F. Marth, R. Ruggeri, *J. Am. Chem. Soc.* **1988**, *110*, 3940–3948.
14 M. Schlosser, K. F. Christmann, *Liebigs Ann.* **1967**, *708*, 1.
15 M. Schlosser, K. F. Christmann, *Angew. Chem. Int. Ed. Engl.* **1966**, *5*, 126.
16 N. Khiar, K. Singh, M. Garcia, M. Martin-Lomas, *Tetrahedron* **1999**, *40*, 5779–5782.
17 N. J. Lawrence, in *Preparation of Alkenes. A Practical Approach* (Ed.: J. M. J. Williams), **1996**, pp. 19–58.
18 D. Seyferth, K. R. Wursthorn, R. E. Mammarella, *J. Org. Chem.* **1977**, *42*, 3104–3106.
19 B. J. Walker, in *Organophosphorus Reagents in Organic Synthesis* (Ed.: J. I. G. Cadogan), Academic Press, London, **1979**, pp. 155–205.
20 W. C. Still, C. Gennari, *Tetrahedron Lett.* **1983**, *24*, 4405–4408.
21 K. Ando, *Tetrahedron Lett.* **1995**, *36*, 4105–4108.
22 K. Ando, *J. Org. Chem.* **1997**, *62*, 1934–1939.
23 P. M. Pihko, T. M. Salo, *Tetrahedron Lett.* **2003**, *44*, 4361–4364.
24 A. van der Klei, R. L. P. de Jong, J. Lugtenburg, A. G. M. Tielens, *Eur. J. Org. Chem.* **2002**, 3015–3023.
25 K. Stritzke, S. Schulz, R. Nishida, *Eur. J. Org. Chem.* **2002**, 3883–3892.
26 G. Etemad-Moghadam, J. Seyden-Penne, *Tetrahedron* **1984**, *40*, 5153–5166.
27 N. Minami, S. S. Ko, Y. Kishi, *J. Am. Chem. Soc.* **1982**, *104*, 1109–1111.
28 H. Nagaoka, Y. Kishi, *Tetrahedron* **1981**, *37*, 3873–3888.
29 S. Sano, R. Teranishi, Y. Nagao, *Tetrahedron Lett.* **2002**, *43*, 9183–9186.
30 T. Katsuki, A. W. M. Lee, P. Ma, V. S. Martin, S. Masamune, K. B. Sharpless, D. Tuddenham, F. J. Walker, *J. Org. Chem.* **1982**, *47*, 1373–1378.
31 A. Bernardi, S. Cardani, C. Scolastico, R. Villa, *Tetrahedron* **1988**, *44*, 491–502.
32 B. M. Trost, S. M. Mignani, *Tetrahedron Lett.* **1986**, *27*, 4137–4140.
33 S. Sano, T. Takehisa, S. Ogawa, K. Yokoyama, Y. Nagao, *Chem. Pharm. Bull.* **2002**, *50*, 1300–1302.
34 H. J. Bestmann, P. Ermann, H. Ruppel, W. Sperling, *Liebigs Ann.* **1986**, 479–498.
35 R. D. Clark, L. G. Kozar, C. H. Heathcock, *Synthesis* **1975**, 635–636.
36 P. Savignac, F. Mathey, *Tetrahedron Lett.* **1976**, 2829–2832.
37 D. Y. Kim, M. S. Kong, K. Lee, *J. Chem. Soc., Perkin Trans. 1* **1997**, 1361–1363.
38 P. Coutrot, C. Grison, *Tetrahedron Lett.* **1988**, *29*, 2655–2658.
39 J. Clayden, S. Warren, *Angew. Chem. Int. Ed. Engl.* **1996**, *35*, 241–270.

40 L. Horner, H. Hoffmann, H. G. Wippel, G. Klahre, *Chem. Ber.* **1959**, *92*, 2499.
41 L. Horner, H. Hoffmann, W. Klink, H. Ertel, V. G. Toscano, *Chem. Ber.* **1962**, *95*, 581.
42 L. Horner, *Pure Appl. Chem.* **1964**, *9*, 225.
43 A. D. Buss, N. Greeves, R. Mason, S. Warren, *J. Chem. Soc., Perkin Trans. 1* **1987**, 2569.
44 D. Cavalla, W. B. Cruse, S. Warren, *J. Chem. Soc., Perkin Trans. 1* **1987**, 1883–1898.
45 A. D. Buss, R. Mason, S. Warren, *Tetrahedron Lett.* **1983**, *24*, 5293–5296.
46 A. D. Buss, S. Warren, *J. Chem. Soc., Perkin Trans. 1* **1985**, 2307–2325.
47 J.-L. Luche, R. Hahn, P. Crabbe, *J. Chem. Soc., Chem. Commun.* **1978**, 601.
48 J.-L. Luche, *J. Am. Chem. Soc.* **1978**, *100*, 2226.
49 A. L. Gemal, J.-L. Luche, *J. Am. Chem. Soc.* **1981**, *103*, 5454.
50 G. Hutton, T. Jolliff, H. Mitchell, S. Warren, *Tetrahedron Lett.*, **1995**, *35*, 7905.
51 J. Elliot, D. Hall, S. Warren, *Tetrahedron Lett.* **1989**, *30*, 601–604.
52 S. R. Schow, J. D. Bloom, A. S. Thompson, K. N. Winzenberg, A. B. Smith, III, *J. Am. Chem. Soc.* **1986**, *108*, 2662–2665.
53 T. Kauffmann, P. Schwartze, *Chem. Ber.* **1986**, *119*, 2150.
54 T. G. Roberts, G. H. Whitham, *J. Chem. Soc., Perkin Trans. 1* **1985**, 1953–1955.
55 J. Elliot, S. Warren, *Tetrahedron Lett.* **1986**, *27*, 645–648.
56 J. Kallmerten, M. D. Wittman, *Tetrahedron Lett.* **1986**, *27*, 2443.
57 B. H. Bakker, D. S. Tijin A-Lim, A. Van der Gen, *Tetrahedron Lett.* **1984**, *25*, 4259.
58 S. K. Armstrong, E. W. Collington, J. G. Knight, A. Naylor, S. Warren, *J. Chem. Soc., Perkin Trans. 1* **1993**, 1433.
59 P. A. Grieco, C. S. Pogonowski, *J. Am. Chem. Soc.* **1973**, *95*, 3071–3072.
60 E. D'Incan, J. Seyden-Penne, *Synthesis* **1975**, 516–517.
61 J. Blanchard, N. Collognon, P. Savignac, H. Normant, *Synthesis* **1975**, 655–657.
62 J. G. Buchanan, A. R. Edgar, M. J. Power, P. D. Theaker, *Carbohydr. Res.* **1974**, *38*, C22–C24.

2
The Peterson and Related Reactions

*Naokazu Kano and Takayuki Kawashima**

2.1
Introduction

Alkene formation reactions are important tools in organic synthesis, and a variety of methodologies have been devised for the conversion of carbonyl compounds into alkenes. The Peterson reaction, developed by D. J. Peterson in 1968, is one of the most important and widely used methods for the conversion of carbonyl compounds into alkenes [1]. The Peterson reaction is usually defined as the formation of an alkene from an α-silyl carbanion and a carbonyl compound (Scheme 2.1). The intermediately formed β-hydroxyalkylsilanes are frequently isolated when carbanions without an anion-stabilizing group are used, and the subsequent alkene formation process from β-hydroxyalkylsilanes is specifically referred to as 'Peterson elimination'. The term 'the Peterson reaction' is often used to mean 'the Peterson elimination' process as well. The term 'Peterson olefination' is often employed as an alternative to 'Peterson reaction'.

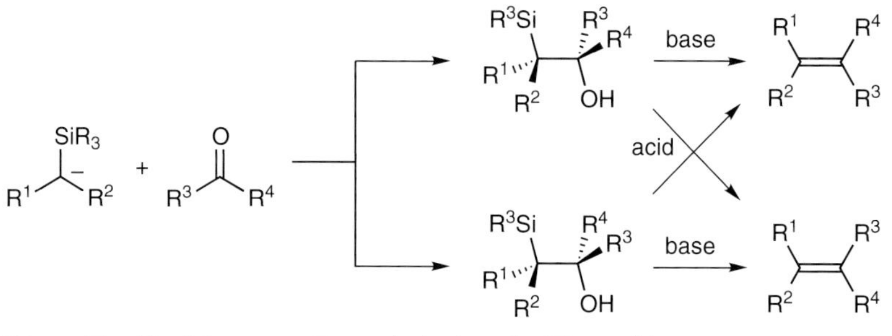

Scheme 2.1. The Peterson reaction under basic and acidic conditions.

Peterson reactions offer interesting and unique profiles, such as high stereoselectivity and sufficiently high reactivity to allow reactions with either ketones, aldehydes, or other carbonyl compounds bearing various functional groups. The

Modern Carbonyl Olefination. Edited by Takeshi Takeda
Copyright © 2004 WILEY-VCH Verlag GmbH & Co. KGaA, Weinheim
ISBN: 3-527-30634-X

stereoselectivity of the alkene products can be controlled by changing the reaction conditions, and various kinds of silicon reagents are available. Furthermore, the silanol eliminated as the by-product affords the disiloxane, which is volatile and thus readily removed as compared with the triphenylphosphine oxide generated in the Wittig reaction. Therefore, Peterson reactions have been utilized in the synthesis of many functionalized alkenes. There have already been several comprehensive reviews and accounts summarizing the Peterson reactions [2–10]. This chapter summarizes the present knowledge concerning these reactions and focuses on the issues of stereoselectivity, the reaction mechanisms, the precursors, applications, and comparison to related reactions. Most of the recent literature reports are covered.

2.2
Stereochemistry and the Reaction Mechanism of the Peterson Reaction

2.2.1
Stereochemistry and the Reaction Mechanism of the Peterson Reaction of β-Hydroxyalkylsilanes

The Peterson reaction of an α-silyl carbanion with an aldehyde, and further protonation of the β-silylalkoxide intermediate, affords the corresponding β-hydroxyalkylsilane when a metal cation is covalently bound to the oxygen. The β-hydroxyalkylsilane is usually obtained as a mixture of two possible diastereomers, which can be separated by column chromatography, and each of the diastereomers can then be subjected to the elimination process. The most important feature of the Peterson reaction is that either geometric isomer of an alkene can be prepared from a common single diastereomer of the β-hydroxyalkylsilane intermediate depending on the work-up conditions. When the *threo* isomer of 5-trimethylsilyl-4-octanol is treated with acid, the Z-isomer of 4-octene is obtained stereospecifically, whereas treatment with base affords the corresponding E-isomer (Scheme 2.2)

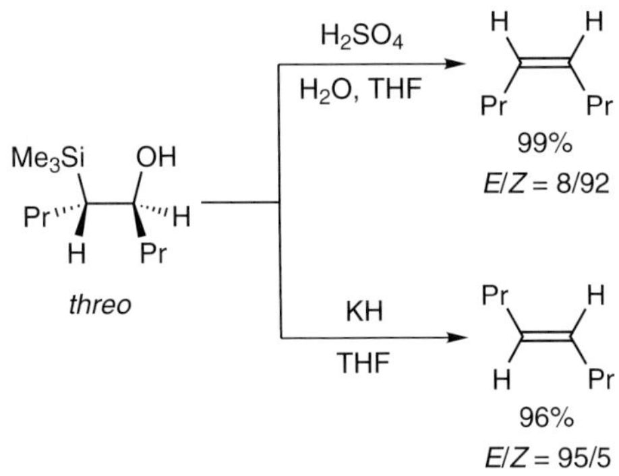

Scheme 2.2. Stereochemistry of the Peterson elimination of a β-hydroxyalkylsilane.

[11]. On the other hand, treatment of the *erythro* isomer of the β-hydroxyalkylsilane with acid or base gives the corresponding *E*- and *Z*-alkenes, respectively. Control experiments on the *threo* and *erythro* isomers of β-hydroxyalkylsilanes prepared either by ring opening of a 2-silyl oxirane or nucleophilic attack of an α-silyl ketone by an organometallic reagent, also show the same reactivities and stereospecificities. This feature is a useful advantage of the Peterson reaction over the Wittig reaction.

Considering the stereochemistry of the product, elimination of the trimethylsilanol moiety from the β-hydroxyalkylsilane under acidic conditions should proceed through protonation of the hydroxy group followed by simultaneous dehydration and desilylation in an *anti* manner (Scheme 2.3). The use of a Lewis acid likewise induces *anti*-elimination in a similar way [12].

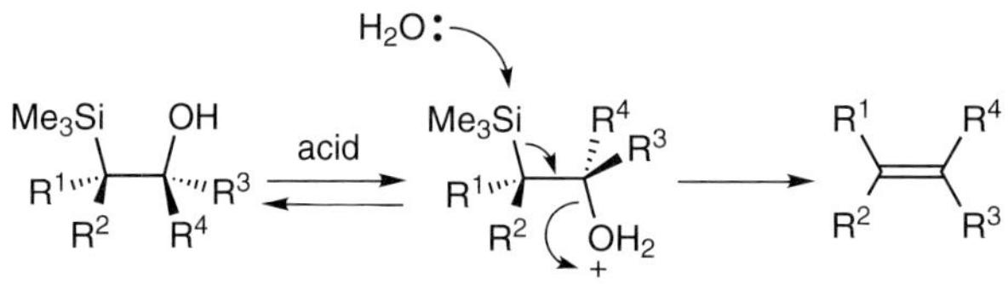

Scheme 2.3. Proposed mechanism of the Peterson elimination under acidic conditions.

On the other hand, elimination under basic conditions should proceed in a *syn* manner. Two possible pathways have been postulated for the elimination of a silyloxide moiety after deprotonation of a hydroxy group with an equimolar amount of base (Scheme 2.4). One is the stepwise 1,3-migration of a silyl group from carbon to oxygen, followed by elimination of a trimethylsilyloxide moiety. The other involves the formation of a pentacoordinate 1,2-oxasiletanide **2**, which is in equilibrium with the β-silylalkoxide anion **1**, and extrusion of the trimethylsilyloxide moiety therefrom. The reaction mechanism has not yet been entirely elucidated.

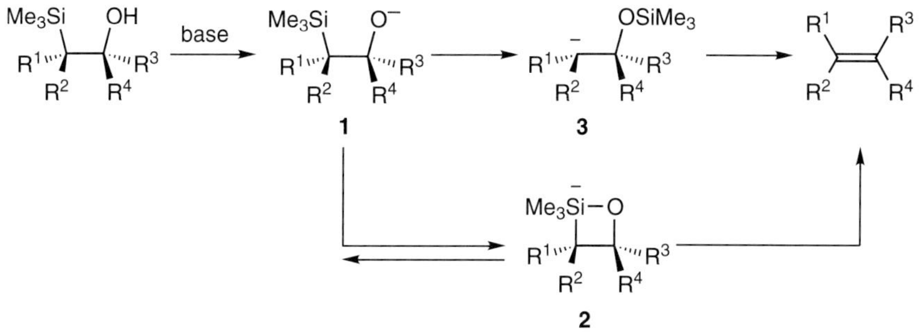

Scheme 2.4. Proposed mechanism of the Peterson elimination under basic conditions.

2.2.1.1 Stepwise Mechanism

In the stepwise mechanism, the stereospecificity of the elimination of the silyloxide from the β-hydroxyalkylsilane can be explained as follows. Both migration of the silyl group from carbon to oxygen and elimination of the trimethylsilyloxide moiety to form a double bond are so rapid that no rotation about the C–C bond is usually observed during these steps, giving the alkene with the same stereochemistry as that obtained by a concerted mechanism.

There are several reports in which it is postulated that the trimethylsilyloxide moiety is eliminated according to a stepwise mechanism. In the deprotonation of a β-hydroxyalkylsilane, the α-siloxyalkyl carbanion 3 generated by 1,3-silyl migration is in some cases protonated and further hydrolyzed to give the desilylated alcohol (Scheme 2.5) [13, 14] (see Section 2.5.1).

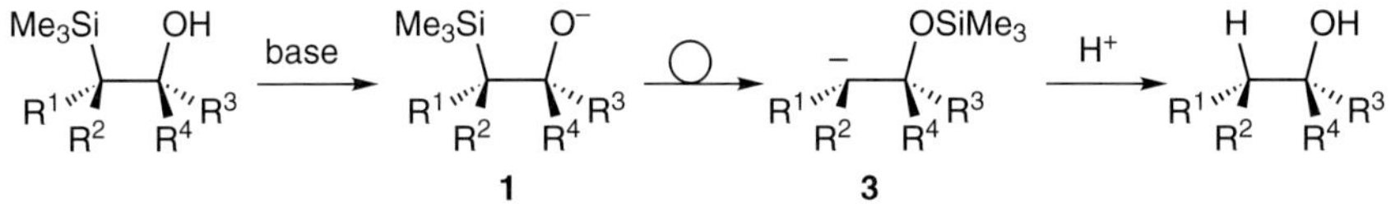

Scheme 2.5. Stepwise mechanism of the Peterson elimination under basic conditions.

In the reactions of α-silyl carbanions bearing an ester function at the α-position with carbonyl compounds, the ester enolates first attack to form the β-silylalkoxide anions 4, which give the corresponding siloxy ester enolates 5 by 1,3-silyl migration (Scheme 2.6). When the elimination of the trimethylsilyloxide anion is not so fast,

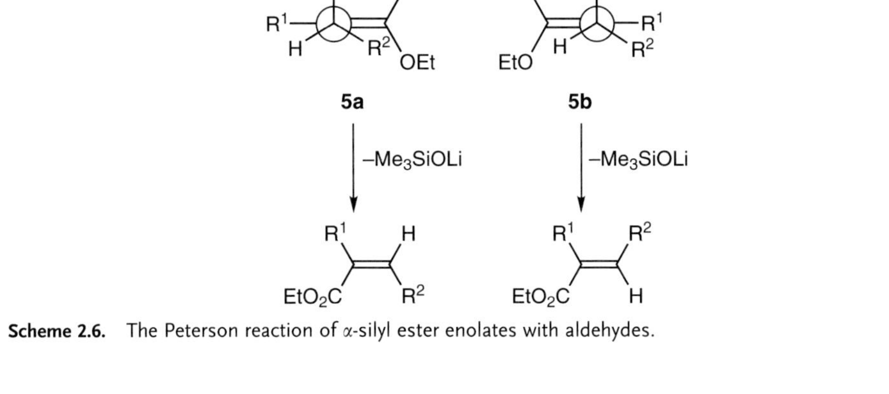

R^1 = SnBu$_3$; R^2 = alkyl, aryl

44–74%
E/Z = 37/63–69/31

Scheme 2.6. The Peterson reaction of α-silyl ester enolates with aldehydes.

the intermediately generated siloxy enolates tend to adopt the more stable conformation **5a** as opposed to **5b** [15–18]. Thus, the elimination should proceed in a stepwise rather than a concerted manner.

As a result, the alkene products do not have the stereochemistry that might have been expected from the starting materials in these cases. This principle allows the stereoselective synthesis of α,β-unsaturated esters without isolation of the two possible β-hydroxyalkylsilane intermediates formed upon acidic work-up (Scheme 2.7) [19].

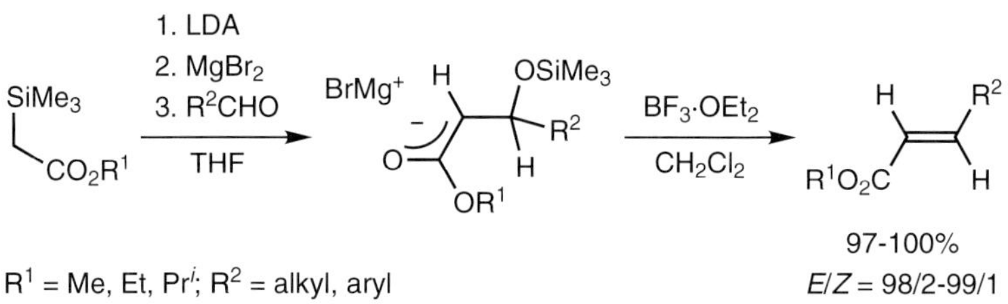

R^1 = Me, Et, Pr^i; R^2 = alkyl, aryl

Scheme 2.7. *E*-Stereoselective Peterson reaction to give α,β-unsaturated esters.

2.2.1.2 Reaction Mechanism via a 1,2-Oxasiletanide

As a pentacoordinate 1,2-oxaphosphetane is an intermediate in the Wittig reaction, a pentacoordinate 1,2-oxasiletanide has been postulated as an intermediate or a transition state of the Peterson reaction, considering the stereochemistry of the alkene products [14, 20–22]. In one report, it is suggested that a 1,2-oxasiletanide is formed in a concerted mechanism [23]. A few examples of stable 1,2-oxasiletanides **7** have been obtained by deprotonation of β-hydroxyalkylsilanes **6** bearing the Martin ligand with DBU, BuLi, or KH in the presence of a crown ether (Scheme 2.8) [24–28]. One of these has been crystallographically characterized, and it demonstrates a distorted trigonal bipyramidal structure around the pentacoordinate silicon with a strong Si–O interaction. The corresponding alkenes are formed by thermolysis of the 1,2-oxasiletanides **7**, and it has been demonstrated for the 1,2-oxasiletanide **7a** (R^1 = CH$_2$Bu^t, R^2 = H, R^3 = Ph) that the relative stereochemistry at the 3- and 4-positions is maintained during the alkene formation.

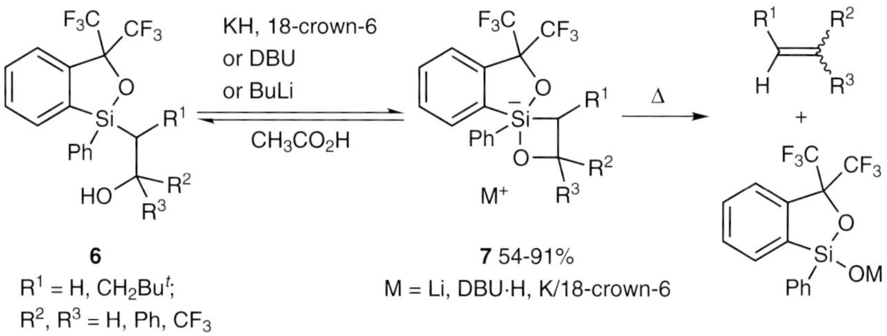

R^1 = H, CH$_2$Bu^t;
R^2, R^3 = H, Ph, CF$_3$

M = Li, DBU·H, K/18-crown-6

Scheme 2.8. Isolation and the olefin formation reaction of pentacoordinate 1,2-oxasiletanides.

2.2.2
Reaction Mechanism of the Addition Step of an α-Silyl Carbanion to a Carbonyl Compound

Peterson reactions of α-silyl carbanions and carbonyl compounds leading directly to the alkenes are generally not stereoselective since the β-hydroxyalkylsilane or β-silylalkoxide intermediates are usually formed as a mixture of *syn* and *anti* isomers, whereas Peterson elimination from a β-hydroxyalkylsilane proceeds in an exclusively stereospecific manner. Furthermore, the addition of an α-silyl carbanion to a carbonyl compound proceeds in an irreversible manner [29, 30]. Therefore, when the β-hydroxyalkylsilane intermediate can neither be isolated nor separated, the diastereomeric ratio of the alkene products of the Peterson reaction is determined in the addition step of the α-silyl carbanion and the carbonyl compound. Stereospecific preparation of the β-hydroxyalkylsilane is required for the utilization of the Peterson reaction in organic synthesis.

2.2.2.1 Approach Control of the Transition State

For the addition of an α-silyl carbanion to a carbonyl compound, some reaction models have been put forward that would indicate that the ratio of the resulting β-silylalkoxide anions is determined by the relative stabilities of the diastereoisomeric intermediates. This mechanism assumes that the carbanion attacks at 90° to the carbonyl framework [31]. For example, regarding the stereoselective formation of the alkene (Z)-**10** by the reaction of an α-silyl carbanion derived from methyl trimethylsilylacetate with 2-alkylcyclohexanone **8**, the stability of the intermediate β-silylalkoxide anion **9** is discussed (Scheme 2.9). The trimethylsilyl group occu-

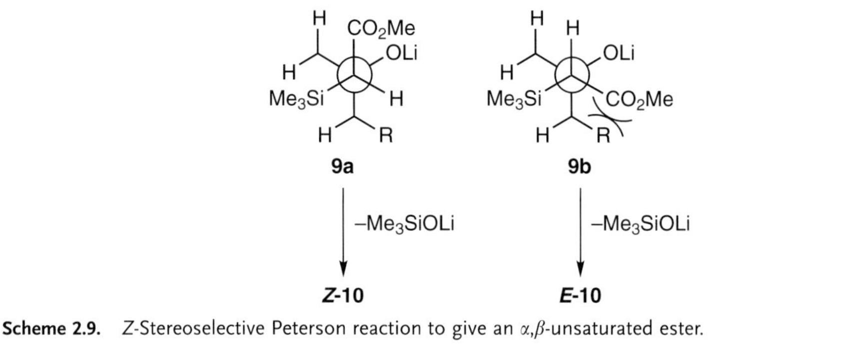

Scheme 2.9. Z-Stereoselective Peterson reaction to give an α,β-unsaturated ester.

pies the least crowded site and the methoxycarbonyl and side-chain moieties are staggered in the preferred geometry **9a**. Subsequent *syn*-elimination of trimethylsilyloxide then affords the alkene (Z)-**10** [32]. The major problem with this model is that it completely fails to account for the so-called 'erythro selectivity' of carbanion addition in the absence of a chelation control [33].

On the other hand, in the Peterson reaction between α-silyl benzyl anions **11** and benzaldehyde, the ratio of the *E*- and *Z*-isomers of the product, stilbene, depends on the size of the silyl group, and it is insensitive to the counterion, the solvent, added salts, variation of the carbanion-forming base, or the temperature (Scheme 2.10) [29, 34, 35].

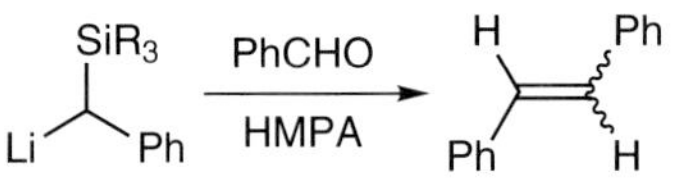

$R_3Si = Me_3Si$ E/Z = 57/43
Et_3Si 55/45
Me_2Bu^tSi 51/49
Ph_3Si 34/66

Scheme 2.10. The Peterson reaction of α-silylbenzyl anions with benzaldehyde.

Another approach control model has been presented to rationalize the relative transition state energies based on the above results, based on irreversible nucleophilic attack of the α-silyl carbanion at approximately 109° to the carbonyl carbon atom (Scheme 2.11). The smallest carbanion ligand, a hydrogen atom in this case, is disposed such that steric interactions are minimized. Attack occurs between the hydrogen and the substituent of the aldehyde so as to minimize steric repulsion.

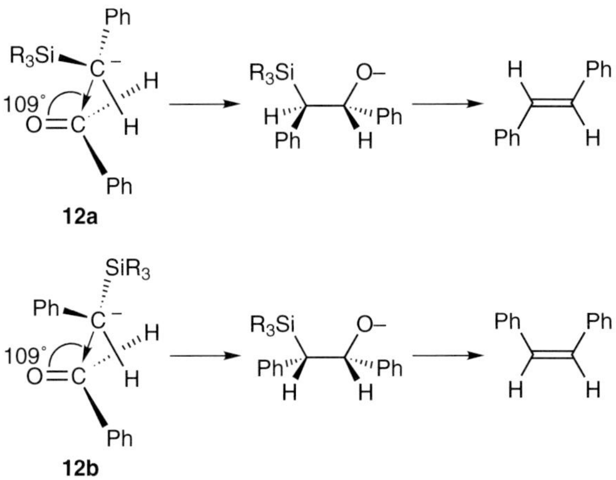

Scheme 2.11. Approach control model of the reaction of an α-silyl carbanion with benzaldehyde.

As regards the remaining groups, the largest group is positioned *anti* to the aldehyde substituent. The trimethylsilyl group is slightly smaller than a phenyl group in this model. As the size of the silyl group increases, transition state **12b** leading to (Z)-stilbene becomes more favorable than **12a** leading to (E)-stilbene.

2.2.2.2 Concerted Mechanism

In contrast to the stepwise addition mechanism, there is a report in which a concerted mechanism is proposed for the addition process, involving a 1,2-oxasiletanide intermediate. Peterson reaction of bis(trimethylsilyl)methyllithium with benzaldehyde gives a mixture of almost equimolar amounts of the E- and Z-isomers of styrylsilane, whereas Peterson elimination of the corresponding β-oxidosilane **14** generated by other methods affords the (E)-styrylsilane with very high isomeric purity (Schemes 2.12 and 2.13) [36]. The above results indicate that β-oxidosilane **14** is not formed as a major intermediate and that the Peterson reaction involves nearly simultaneous formation of C–C and Si–O bonds to give a 1,2-oxasiletanide intermediate **13** directly.

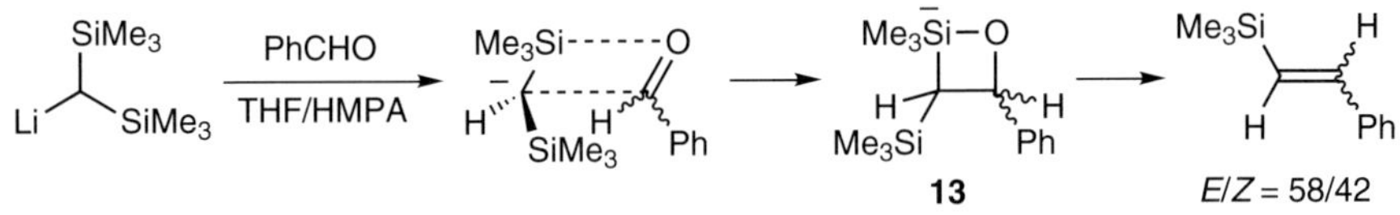

Scheme 2.12. Proposed concerted mechanism of the Peterson reaction.

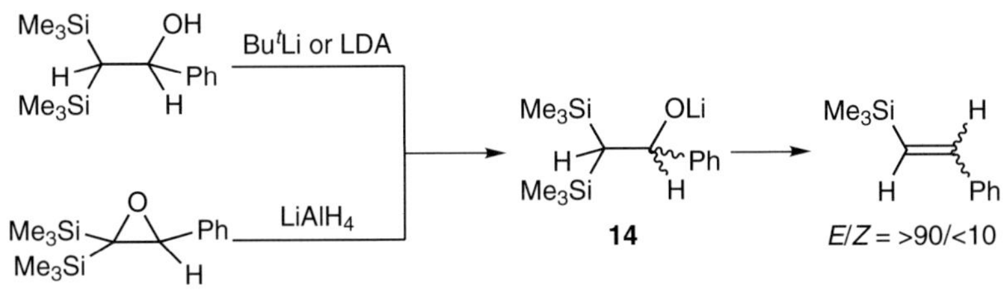

Scheme 2.13. The Peterson elimination of a β,β-bis(trimethylsilyl)alkoxide.

The importance of the Si–O interaction in the transition state has also been highlighted in another report [23]. In the reactions of α-triarylsilylbenzyl anions with benzaldehyde to give stilbenes, an increase in the Si–O interaction due to the presence of electron-withdrawing aryl groups on the silicon atom at the stage of the C–C bond formation leads to an increase in the proportion of (E)-stilbenes (Scheme 2.14). The silicon atom is rendered more electrophilic and this results in a stronger Si–O interaction and a smaller Si–C–C–O dihedral angle with an eclipsed conformation rather than a staggered conformation in the transition state **15**, which favors the less sterically hindered transition state **15a** leading to an increased proportion of the (E)-stilbene. The electronic character of the substituents should indeed influence the stereoselectivity of the Peterson reaction, which de-

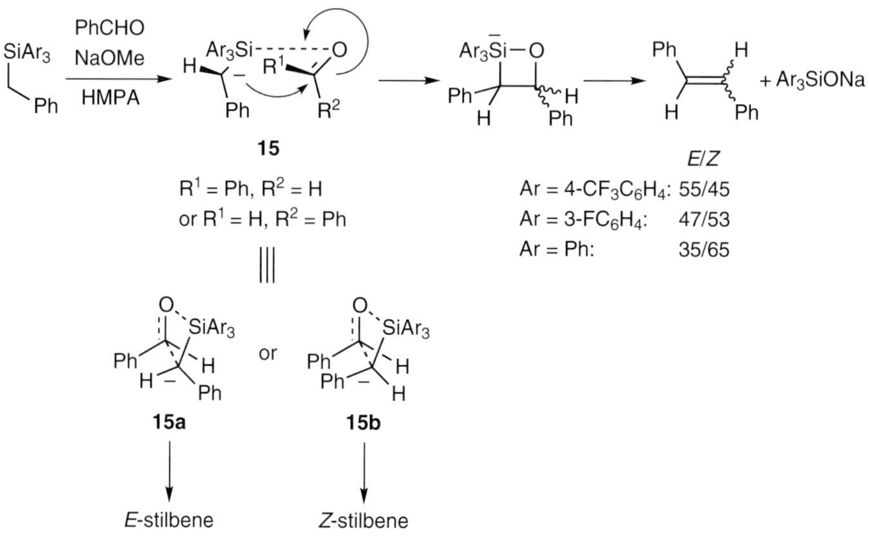

Scheme 2.14. The Peterson reaction of α-triarylsilylbenzyl anions with benzaldehyde.

pends on a delicate balance between the electronic and steric factors in the transition state.

2.2.2.3 Chelation Control Mechanism

Although acid- or base-mediated elimination from an isolated β-hydroxyalkylsilane is highly stereoselective, the direct addition processes of the Peterson reaction between an α-silyl carbanion and an aldehyde bearing a functional group are usually affected by several factors, including steric ones. The stereochemistry is especially influenced by the counter cation and the functional groups on both the carbanion and the carbonyl compounds [29]. When an α-silyl carbanion bears a functional group in the vicinity of the central carbon atom, the interaction between the counter cation and the functional group should be particularly important in the transition state because the stereochemistry of the products varies depending on the nature of the metal cation [37]. It is quite reasonable to assume that the reaction is under chelation control [38].

One example is given by the reactions of the dianions **16** derived from α'-(trimethylsilyl)enaminones with aldehydes to give α',β'-unsaturated enaminones **18** with regio- and stereocontrol of the new double bond (Scheme 2.15) [39]. Steric repulsion cannot be the sole decisive factor for the high stereoselectivity. There must be some contribution from the chelation of one lithium cation by two oxygen atoms of the aldehyde and enaminone moieties favoring the *erythro* configuration of aldehydes in the transition state *erythro*-**17** over the *threo* transition state. These interactions, which lock the conformation, should be involved in the stereodetermining transition state before the elimination of the siloxy group. Chelation control through Li–N interactions has also been reported [40].

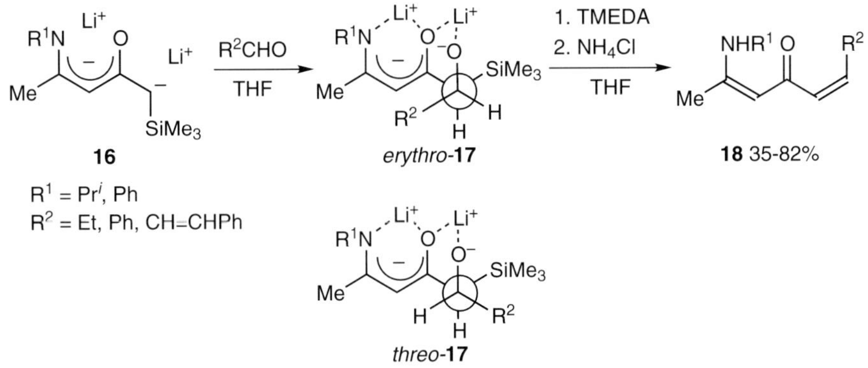

2.2 Stereochemistry and the Reaction Mechanism of the Peterson Reaction

Scheme 2.15. The Peterson reaction of α'-(trimethylsilyl)enaminones with aldehydes.

The importance of the cation–oxygen interaction has also been pointed out in highly Z-selective olefination reactions of the α-silyl-α-phosphoryl carbanion **19** (Scheme 2.16) [41, 42]. The chelating Li–O interactions are likely to determine the configuration of the carbanion **19** in the approach of the aldehyde. This transition state leads to the formation of the kinetically favored adducts **20**, which give the (Z)-alkenes **21** after syn-elimination of the oxophilic silylated moiety and hydrolysis.

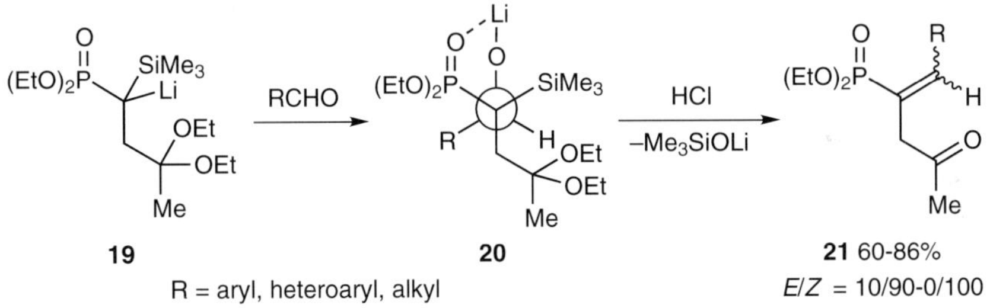

Scheme 2.16. Z-Selective Peterson reaction of an α-silyl-α-phosphoryl carbanion with aldehydes.

In some reports, it is claimed that the close interaction between the counter cation and the anionic carbon atom, as well as the oxygen atom of the carbonyl group, is maintained, with the implication of a closed four-centered transition state. The reactions of α-lithiated α-fluoro-α-trimethylsilylmethylphosphonate **22** with aromatic and heteroaromatic aldehydes preferably give the E-isomers of α-fluoroalkenylphosphonates **24**, whereas those with cyclic ketones bearing a substituent in the α-position give exclusively the Z-isomers (Schemes 2.17 and 2.18) [43]. A mechanistic pathway implying a closed transition state has been proposed to rationalize the results. The lithium cation is coordinated by an anionic carbon as well as the oxygen atoms of both the phosphoryl group and the aldehyde in the

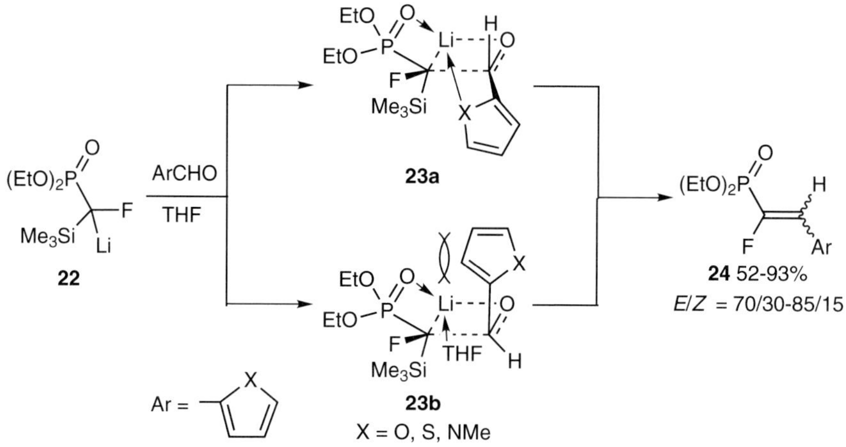

Scheme 2.17. The Peterson reaction of an α-lithiated α-fluoro-α-trimethylsilylmethylphosphonate with heteroaromatic aldehydes.

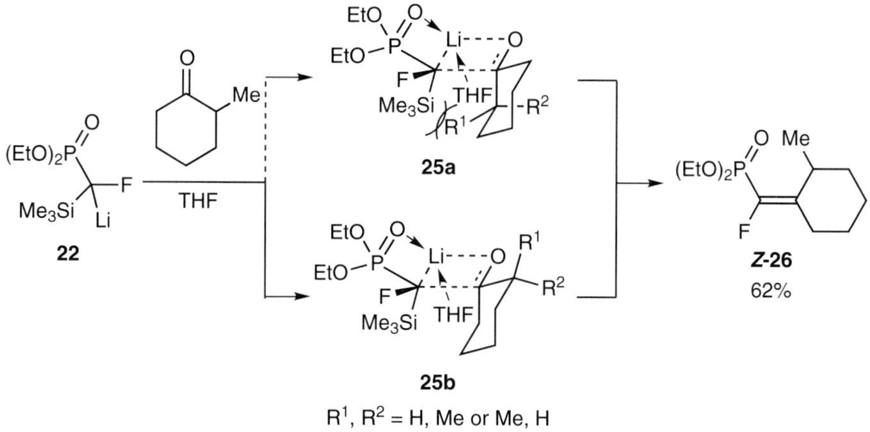

Scheme 2.18. The Peterson reaction of an α-lithiated α-fluoro-α-trimethylsilylmethylphosphonate with 2-methylcyclohexanone.

transition state, where the bulkier trimethylsilyl group takes up the outside position. This is known as a 'butterfly mechanism' owing to the spatial orientation in the butterfly-like transition state. With heteroaromatic aldehydes, transition state **23a** leading to the *E*-isomer of **24** is preferred since the heteroaromatic group is orientated in the outside position due to decreased steric repulsion with the phosphoryl group and an additional stabilization by intramolecular interaction between the heteroatom and lithium.

With α-substituted cyclohexanones, the transition state **25a** giving *E*-isomers suffers from severe steric repulsion, and hence *Z*-isomers such as (*Z*)-**26** are formed predominantly [43]. The butterfly mechanism is also invoked in the stereoselective Peterson reactions of silylated benzyl carbamates [44].

2.2.3
Theoretical Calculations on the Reaction Mechanism

Computational work has been limited to a few reports on the reaction pathway of the Peterson reaction. Theoretical calculations based on the CNDO method have been performed to elucidate the reaction pathway of ethylene formation from $H_3SiCH_2CH_2O^-$ 27 with elimination of a trihydrosilyloxide anion (Scheme 2.19) [45]. The calculations indicate that the silyl group of 27 migrates to give $H_3SiOCH_2CH_2^-$ 29 with concomitant cleavage of an Si–C bond. The 1,2-oxasiletanide 28 may represent an unstable intermediate during the course of the reaction, in contrast to the stability of the 1,2-oxaphosphetane intermediate in the Wittig reaction. It has been shown that the stepwise mechanism involving further elimination of trihydrosilyloxide anion is feasible for the Peterson reaction.

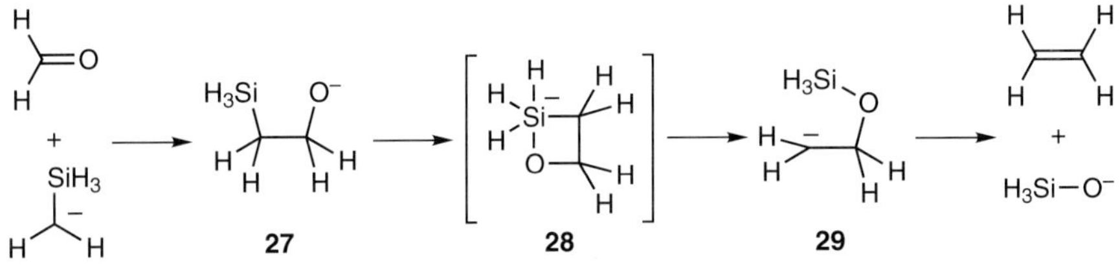

Scheme 2.19. The reaction pathway of a model compound based on CNDO calculations.

A theoretical study on the model Peterson reaction of acetone and a lithium enolate 30 derived from methyl α-(trimethylsilyl)acetate has recently been reported (Scheme 2.20) [46]. The geometries were optimized by means of density functional calculations with the B3LYP hybrid functional and the 6-31G* basis set, and the single-point energies were calculated with various sized basis sets (6-31G*, 6-311+G**, 6-311+G(2df,2p)) at the MP2 correlated level.

As observed experimentally, calculations show that the preferred configuration of lithium enolates 30 of a simple α-silyl ester is *trans*. The chelating effect of the lithium counter ion is found to be critical for the reaction, as judged by the analysis of reaction profiles. This reaction proceeds through four transition states (Figure 2.1). First, acetone coordinates to the lithium of the lithium enolate 30 to form a pre-complex (PC), and a new C–C bond is formed via a six-membered chair-like transition state, TS1. There is no Li–C coordination in the transition state, though a four-membered ring coordination of lithium has been proposed in the 'butterfly mechanism'. Rotation about the C–C bonds of the two rotamers 31a and 31b of the oxy anion intermediates takes place via a torsional transition state (TS2) and 31a leads to an intermediate 32 that is set up for direct elimination. The concerted breaking of the Si–C bond together with the formation of an Si–O bond in 31a causes the migration of the silyl group via a cyclic transition state (TS3) involving

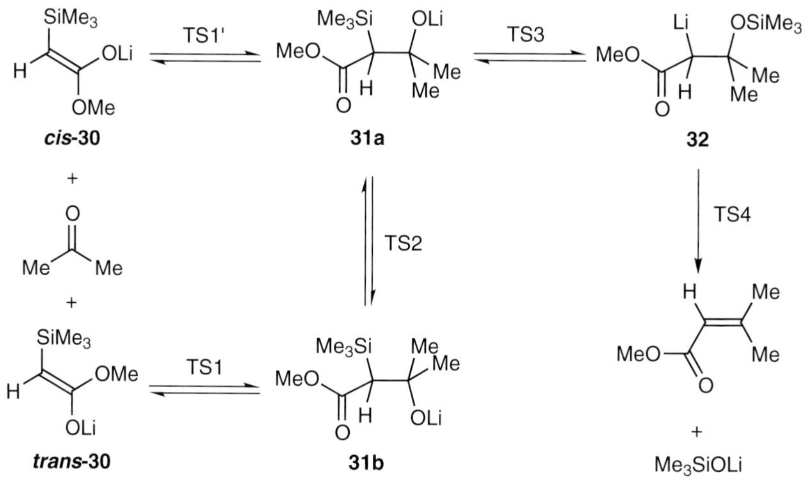

30 | *2 The Peterson and Related Reactions*

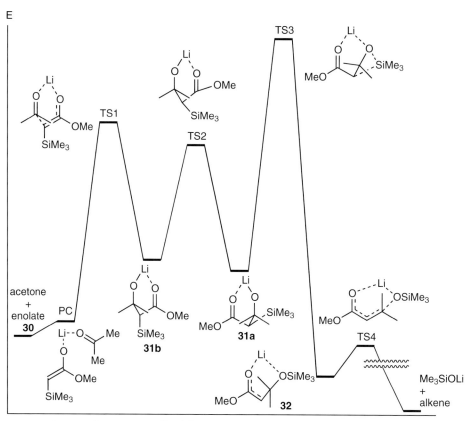

Scheme 2.20. The reaction pathway of a model compound based on theoretical calculations.

Fig. 2.1. The reaction profile of the Peterson reaction of acetone and an enolate.

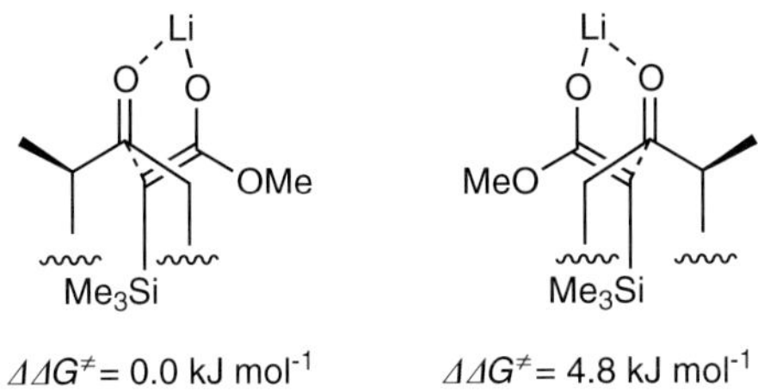

2.2 Stereochemistry and the Reaction Mechanism of the Peterson Reaction

$\Delta\Delta G^{\ddagger} = 0.0$ kJ mol^{-1} $\Delta\Delta G^{\ddagger} = 4.8$ kJ mol^{-1}

Fig. 2.2. Transition states of the reaction of 2-methylcyclohexanone and an enolate.

hypervalency at silicon. This process should be highly exothermic. Finally, lithium trimethylsilyloxide eliminates via a transition state, TS4, with a relatively low barrier, to give methyl 3-methyl-2-butenoate as the product without bond rotation and stereochemical scrambling. Considering the reaction profile, the Z-stereoselectivity, which is often observed in the Peterson reaction, would be determined by TS1.

This hypothesis is strongly supported by agreement between experimental and theoretical results. The barrier difference in the selectivity-determining step is estimated to be 4.2 kJ mol^{-1} from the Z:E product ratio of 93:7, which has been determined experimentally for the reaction with 2-methylcyclohexanone. On the other hand, the energy difference between the two transition states of TS1 is calculated to be 4.8 kJ mol^{-1}, which agrees very well with the above experimental value (Figure 2.2).

2.2.4
Convergently Stereoselective Peterson Reactions

There is another methodology for carrying out stereoselective Peterson reactions, which utilizes the difference in reactivity of the two diastereomers, *syn-* and *anti-β-*silylalkoxides, at low temperature. This is the convergent transformation from both diastereomers of *syn-* and *anti-β-*silylalkoxides to give one alkene (Scheme 2.21) [47]. When an α-silyl carbanion bears a bulky alkyl group on the anionic car-

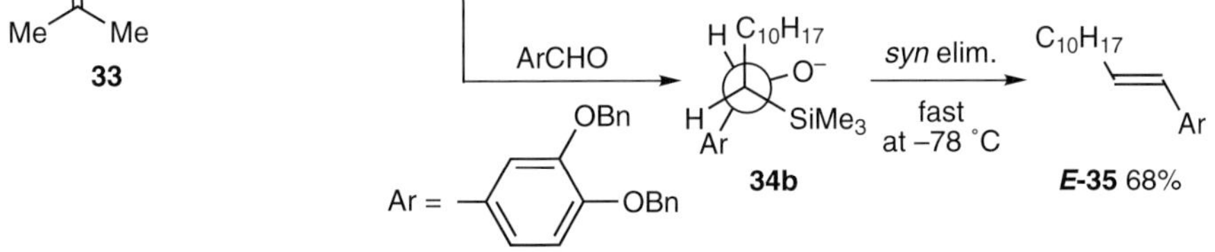

Scheme 2.21. Convergently stereoselective Peterson reaction.

bon atom, such as in **33**, the *syn-β*-silyl alkoxide **34b** eliminates stereospecifically at −78 °C to give the alkene (*E*)-**35**, whereas the *anti* isomer **34a** is unreactive at this temperature due to severe eclipsing interactions. The reaction mixture is then acidified with acetic acid and heated to induce stereospecific elimination of the *anti* isomer **34a** to give (*E*)-**35** via the complementary cationic pathway. Thus, the *E*-isomer of the alkene product **35** is obtained exclusively.

2.3
Generation of α-Silyl Carbanions and their Peterson Reactions

2.3.1
Generation of α-Silyl Carbanions from α-Silylalkyl Halides

2.3.1.1 Generation of α-Silyl Grignard Reagents from α-Silylalkyl Halides

α-Silyl Grignard reagents such as trimethylsilylmethylmagnesium chloride are prepared by treatment of α-silylalkyl chlorides with magnesium metal in diethyl ether (Scheme 2.22) [48–50]. Bromomethyltrimethylsilane and iodomethyltrimethylsilane are also converted to the respective Grignard reagents in a similar way [51, 52]. The resultant trimethylsilylmethylmagnesium halides are utilized in the synthesis of terminal alkenes (see Section 2.3.1.3).

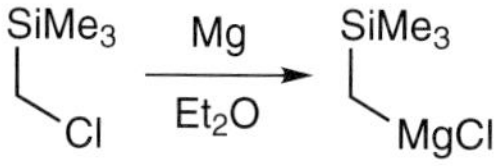

Scheme 2.22. Generation of an α-silyl Grignard reagent.

2.3.1.2 Generation of α-Silylalkyllithium Reagents from α-Silylalkyl Halides

α-Silylalkyllithium reagents are generated by lithium–halogen exchange of α-silylalkyl halides with alkyllithium reagents [53]. When 1-bromo-1-trimethylsilylcyclopropane **36** is allowed to react with butyllithium, the corresponding α-silyl cyclopropyllithium **37** is generated, which can be further converted to the alkenes **38** upon successive treatment with aldehydes and KH (Scheme 2.23) [54].

On the other hand, trimethylsilylmethyllithium is prepared by the reaction of chloromethyltrimethylsilane with lithium dispersion (Scheme 2.24) [55, 56]. The

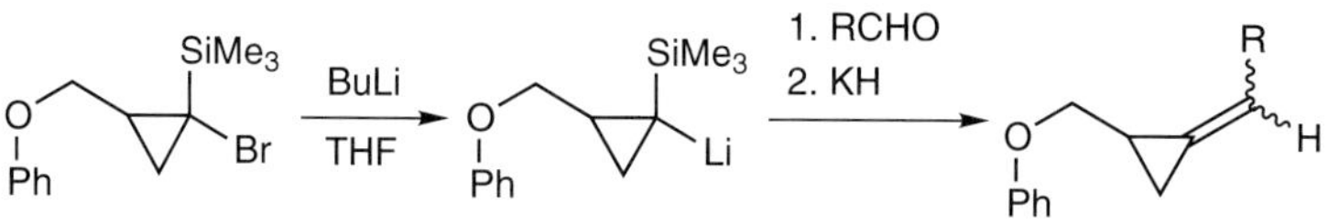

R = Ph, CH=CHMe

Scheme 2.23. Generation of an α-silylalkyllithium and its Peterson reaction.

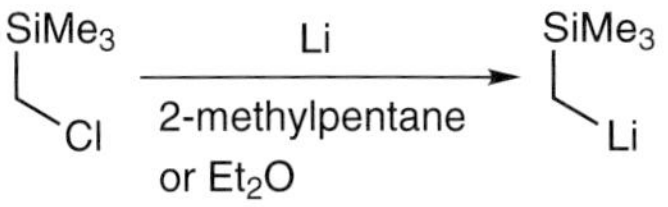

Scheme 2.24. Generation of trimethylsilylmethyllithium.

trimethylsilylmethyllithium thus generated is utilized in the synthesis of terminal alkenes (see Section 2.3.1.3).

2.3.1.3 Synthesis of Terminal Alkenes by the Use of α-Silyl Carbanions Generated from α-Silylalkyl Halides

Terminal alkenes are prepared by Peterson elimination of β-hydroxyalkylsilanes with no further substituents on the carbon atom bearing the silyl group. Thus, terminal alkenes are easily synthesized by the reaction of Me_3SiCH_2M reagents (M = MgCl, Li, $CeCl_2$, Cu, and so on) with carbonyl compounds, followed by Peterson elimination (Scheme 2.25). The most commonly used reagents for this purpose are trimethylsilylmethylmagnesium chloride and trimethylsilylmethyllithium, and Peterson reactions using these reagents have been employed in the total synthesis of some natural products [1]. Reaction of the former reagent with an appropriate carbonyl compound, followed by aqueous quench, affords the corresponding β-hydroxyalkylsilane **39**. These compounds are usually stable under the reaction conditions and can be isolated prior to their subsequent conversion. Thus, compounds **39** do not readily undergo the desired elimination process without the addition of further reagents. Both acids and bases are useful for this purpose since there is no need to consider the stereochemistry of the reaction. For example, treatment with NaH or KH at ambient or higher temperature is needed to complete the elimination. The elimination process can also be achieved by treatment with H_2SO_4 under thermal conditions.

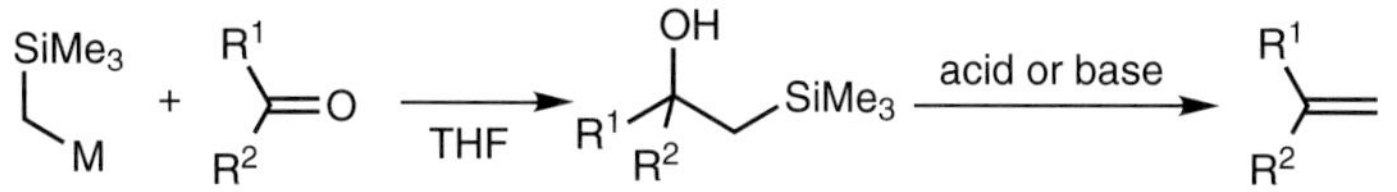

R^1, R^2 = alkyl, aryl
M = Li, MgCl, Li/$CeCl_3$, MgCl/$CeCl_3$
acid = H_2SO_4, HF, HF·pyridine
base = KH, NaH

Scheme 2.25. Synthesis of terminal alkenes by the Peterson reaction.

The lithium analogue, trimethylsilylmethyllithium, is also used for the synthesis of terminal alkenes [57, 58]. Its synthetic utility is sometimes limited by the high basicity and lack of chemoselectivity of lithium reagents.

Contamination with other double-bond isomers occurs in some cases due to isomerization under basic conditions or under severe thermal conditions such as in refluxing THF. Some modifications have been developed to allow the elimination process to be accomplished under mild conditions. For example, in situ treatment of β-hydroxyalkylsilane **40** with acetyl chloride or thionyl chloride gives the corresponding terminal alkene **41** without isomerization (Scheme 2.26) [50, 59, 60].

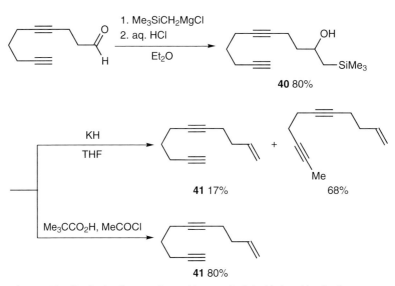

Scheme 2.26. Synthesis of an enediyne with a terminal double bond by the Peterson reaction.

Another convenient elimination protocol uses a solid polymeric perfluoroalkanesulfonic acid (Nafion-H®) catalyst [61]. The high-yielding reactions proceed at room temperature, and work-up involves only filtration of the catalyst. Non-enolizable, tertiary trimethylsilylmethyl-substituted alcohols can easily be used in this reaction.

Whereas trimethylsilylmethyllithium and trimethylsilylmethylmagnesium chloride have been employed in the total synthesis of natural products, better results are often obtained when these reagents are added to the carbonyl compounds in the presence of anhydrous CeCl$_3$ [62–65]. Trimethylsilylmethylcerium dichloride is presumed to take part in the reaction, and its low basicity and high nucleophilicity facilitate the reaction process. The cerium reagent is easily prepared by adding trimethylsilylmethyllithium to anhydrous CeCl$_3$ in THF at −78 °C. Addition of aldehydes and ketones to the reagent solution and further treatment with one equivalent of TMEDA per equivalent of CeCl$_3$ immediately prior to work-up affords the corresponding β-hydroxyalkylsilanes. This method is also applicable to enolizable aldehydes and ketones. The β-hydroxyalkylsilanes are converted to the terminal alkenes by treatment with KH or aqueous HF. The former reagent is found to cause double-bond isomerization in sensitive cases. Treatment with HF sometimes needs the presence of pyridine in order to achieve a good yield. In some cases, elution of the reaction solution through a column of silica gel affords the terminal alkene. Some natural products, such as (+)-taylorione **42**, have been synthesized by taking advantage of this methodology (Scheme 2.27) [66–68].

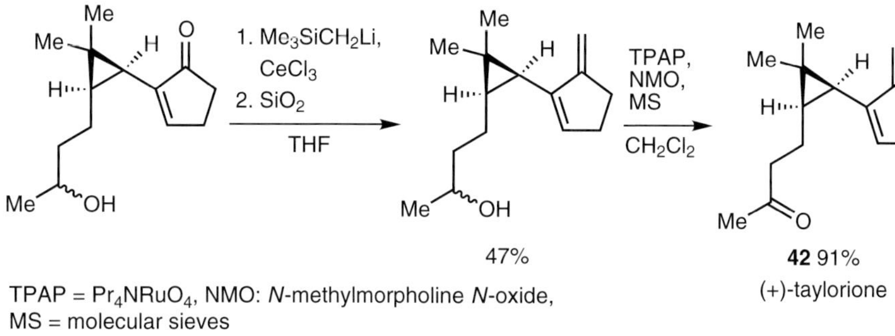

TPAP = Pr_4NRuO_4, NMO: *N*-methylmorpholine *N*-oxide,
MS = molecular sieves

Scheme 2.27. Synthesis of (+)-taylorione.

A combination of iodomethyltrimethylsilane and $Sm(OTf)_2$ also works as a methylenating reagent for carbonyl compounds (Scheme 2.28). When this reagent is applied to the reaction with ketones and aldehydes in THF/HMPA at room temperature, the carbonyl compounds are smoothly methylenated; this process is known as the samarium-Peterson reaction [69].

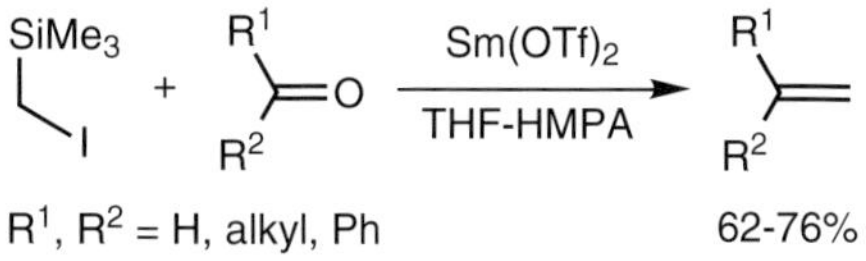

R^1, R^2 = H, alkyl, Ph   62-76%

Scheme 2.28. The samarium-Peterson reaction.

2.3.1.4 Reactions of α-Silyl Carbanions Generated from α-Silylalkyl Halides with Esters, Carboxylic Acids, and Acetals

The application of the methylenating reagents Me_3SiCH_2M as precursors of allylsilanes has been explored [70]. Their twofold addition to carboxylic acid derivatives, followed by deoxysilylation of the resulting bis(β-silylmethyl)alcohols **43**, provides ready access to the corresponding allylsilanes **44** (Scheme 2.29). Acid chlorides, esters, and lactones are also suitable substrates for this procedure [71–76]. A functionalized cerium reagent prepared from [(dimethylisopropoxysilyl)methyl]-magnesium chloride and $CeCl_3$ is also employed to perform the same reactions [77]. The cerium reagents generally give better results than Me_3SiCH_2Li or Me_3SiCH_2MgCl [78, 79].

Reactions of trimethylsilylmethyllithium with secondary and tertiary carboxylic acid methyl or ethyl esters give the corresponding α-trimethylsilyl ketones **45** in

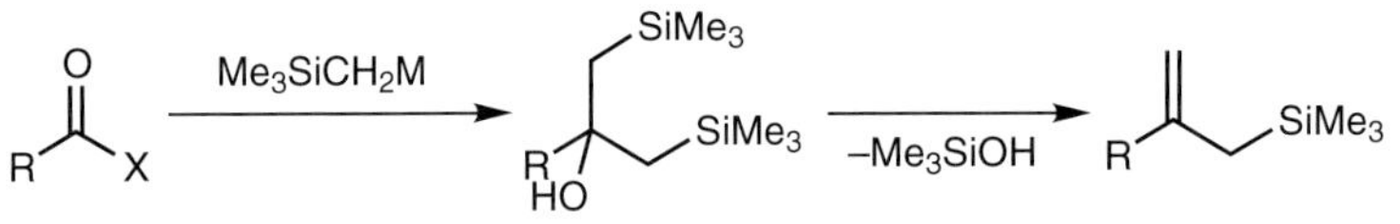

43 44 40-95%

R = alkyl, alkenyl, aryl, acetal; X = Cl, OR′; M = $CeCl_2$, Li, MgCl

Scheme 2.29. Synthesis of allylsilanes.

high yields, rather than the Peterson reaction products (Scheme 2.30) [80]. The α-silyl ketone intermediate resists further addition of the reagent, undergoing a kinetically preferred enolization. Primary esters similarly afford the α-trimethylsilyl ketones **45** in low yields. Methanolysis instead of aqueous work-up furnishes the corresponding methyl ketones **46** in high yields. Direct conversion of some secondary and tertiary esters to the corresponding methyl ketones is also achieved in the reactions of α-silyl carbanions [81–85].

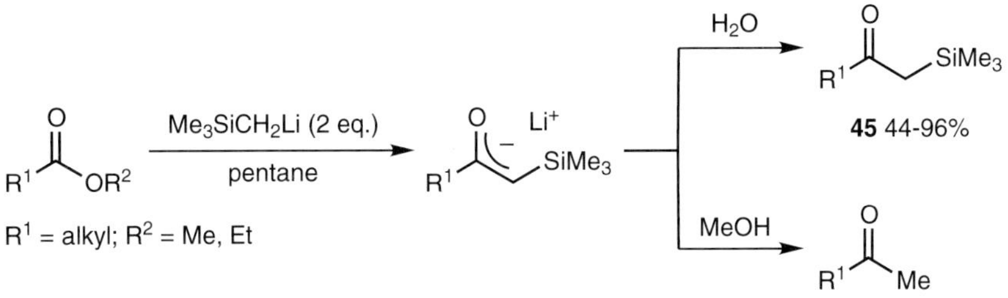

45 44-96%

R^1 = alkyl; R^2 = Me, Et

46 41-95%

Scheme 2.30. Synthesis of α-trimethylsilyl ketones and methyl ketones from esters.

In connection to the above reactions, methylenation of dimethyl acetals **48** instead of ketones can be conveniently performed by the reaction with Me_3SiCH_2-$Cu(PBu_3){\cdot}LiI$ **47**, which is prepared in situ from Me_3SiCH_2Li, CuI, and PBu_3 under the influence of $BF_3{\cdot}OEt_2$ (Scheme 2.31) [86]. Methylenation is successful with ketone acetals **48** even when they bear protected hydroxy functions, such as benzyl ether, benzoate, and *t*-butyldiphenylsilyl ether. The reactions proceed by way of Lewis acid promoted β-elimination of a β-alkoxysilane.

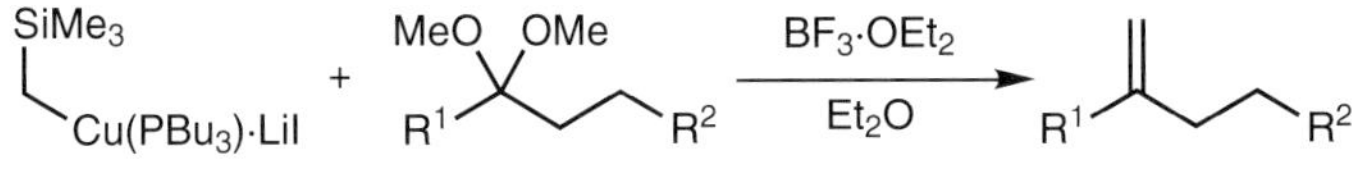

47 **48** 71-92%

R^1 = Me, Ph;
R^2 = Et, 4-$MeOC_6H_4$, $PhCH_2O$, $PhCO_2$, Bu^tPh_2SiO

Scheme 2.31. Methylenation of ketone acetals.

2.3.1.5 The Reformatsky–Peterson Reactions of α-Silylalkyl Halides

Reactions of α-silylalkyl halides with carbonyl compounds using activated zinc have been reported. Reactions of ethyl bromo(trimethylsilyl)acetate with aldehydes and ketones in the presence of zinc/silver-graphite rapidly afford the corresponding α,β-unsaturated esters (Scheme 2.32) [87]. In this case, the Reformatsky–Peterson reaction sequence proceeds under the mild conditions.

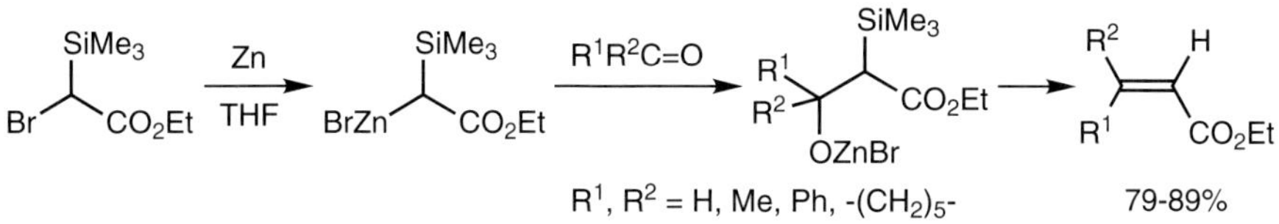

R^1, R^2 = H, Me, Ph, -(CH$_2$)$_5$-   79-89%

Scheme 2.32. Synthesis of α,β-unsaturated esters by the Reformatsky–Peterson reaction.

A stereoselective one-step synthesis of α,β-unsaturated nitriles through Reformatsky–Peterson reaction has also been reported [88]. A mixture of chloro(trimethylsilyl)acetonitrile, an aliphatic aldehyde, and activated zinc powder is refluxed in THF and quenched with aqueous ammonia to give the corresponding α,β-unsaturated nitrile with high Z-selectivity (Scheme 2.33).

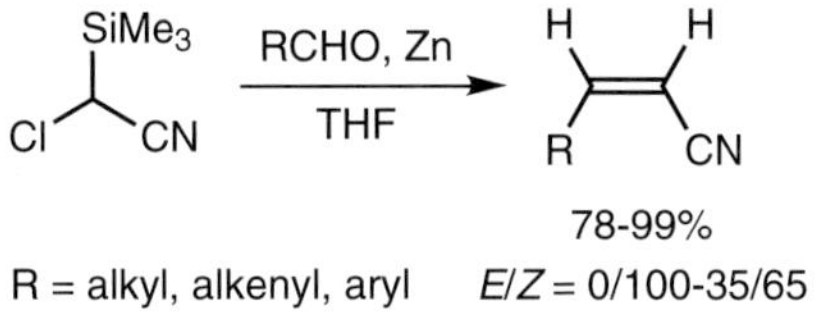

78-99%

R = alkyl, alkenyl, aryl   E/Z = 0/100-35/65

Scheme 2.33. Synthesis of α,β-unsaturated nitriles by the Reformatsky–Peterson reaction.

2.3.2 Generation of α-Silyl Carbanions by Deprotonation of Alkylsilanes

One of the simplest and most general methods for the preparation of α-silyl carbanions is by direct deprotonation at the α-position of alkylsilanes, Me$_3$SiCH$_2$R and Me$_3$SiCH(R)R′, with a suitable base. This process is feasible when the R or R′ groups can stabilize the resulting carbanion, for example when they are heteroatom substituents and/or electron-withdrawing groups.

2.3.2.1 Generation of α-Silyl Carbanions Bearing an Aryl or a Heteroaryl Group

It is usually difficult to deprotonate alkyl(trimethylsilyl)methanes, Me$_3$SiCH$_2$R, and dialkyl(trimethylsilyl)methanes, Me$_3$SiCH(R)R′, bearing only alkyl substituents (R, R′ = alkyl) at the α-positions [89]. Although a silyl group stabilizes the resultant carbanion, it does not accelerate the deprotonation.

On the other hand, benzyltrimethylsilane, although relatively unreactive towards BuLi alone, is readily deprotonated with BuLi in the presence of TMEDA in THF to provide the lithio derivative **49**, which is stabilized by both phenyl and trimethylsilyl groups (Scheme 2.34) [1, 48, 50, 90–92]. Additions of the α-silyl carbanion **49** to benzaldehyde and ketones result in the formation of the expected alkenes as mixtures of the *E*- and *Z*-isomers.

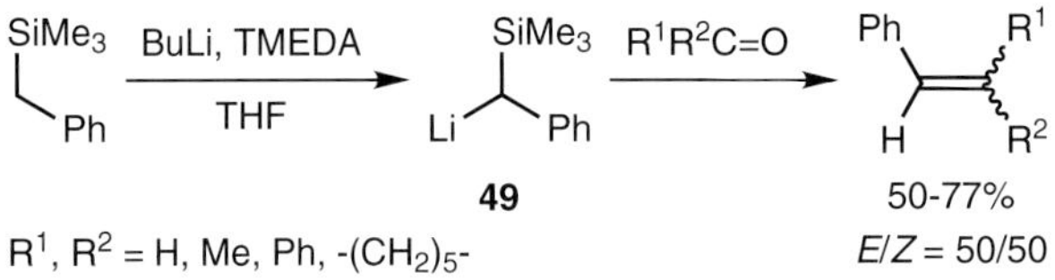

$R^1, R^2 = H, Me, Ph, -(CH_2)_5-$

Scheme 2.34. The Peterson reaction of α-trimethylsilylbenzyllithium with carbonyl compounds.

α-Silyl carbanions bearing a substituted aryl group at the α-position are utilized for organic synthesis. For example, lithiation of 2-(trimethylsilylmethyl)benzamide **50** with LDA, followed by addition of cyclobutanone, gives the corresponding cyclobutylidenemethylbenzamide **51** (Scheme 2.35) [93].

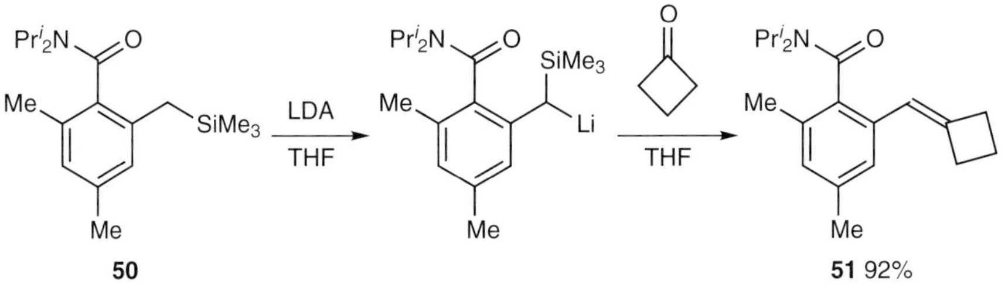

Scheme 2.35. The Peterson reaction of an α-silylbenzyl anion with cyclobutanone.

This procedure is also effective for the synthesis of α-silyl carbanions bearing a heteroaromatic group. Heteroaromatic compounds bearing a trimethylsilylmethyl group are easily lithiated with LDA in THF. Subsequent Peterson reactions of the resultant trimethylsilylmethyllithiums **52** bearing a heteroaromatic group with an amide give the corresponding enamines **53**, which are converted to 1,2,3,4-tetrahydro-β-carbolines **54** (Scheme 2.36) [94, 95].

Intramolecular versions of the Peterson reaction of α-silylbenzyl anions are utilized to construct heteroaromatic frameworks [96]. Lithiation of the *ortho*-substituted benzylsilanes **55** with LDA and subsequent intramolecular Peterson reaction of the derived carbanions **56** with the carbonyl moiety leads directly to intramolecular cyclization giving indole derivatives **57** (Scheme 2.37) [97].

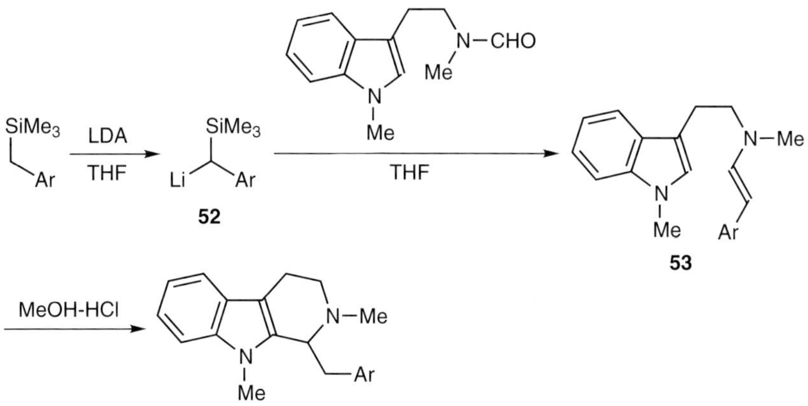

2.3 Generation of α-Silyl Carbanions and their Peterson Reactions

Ar = pyridin-4-yl, 3-ethylpyridin-4-yl, pyridin-2-yl, quinolin-4-yl, quinolin-2-yl, 3-methylquinolin-2-yl, 6,7-dimethoxyisoquinolin-1-yl

Scheme 2.36. Synthesis of 1,2,3,4-tetrahydro-β-carbolines by the Peterson reaction.

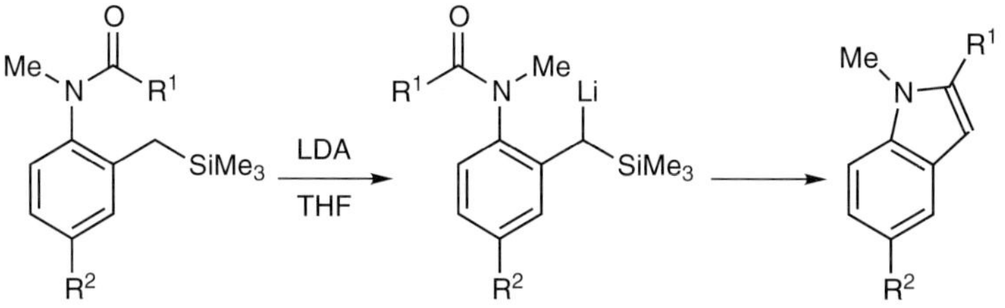

R^1 = Ph, PhCH=CH, p-ClC$_6$H$_4$; R^2 = OMe, Cl

Scheme 2.37. Synthesis of indole derivatives by the intramolecular Peterson reaction.

2.3.2.2 Generation of α-Silyl Carbanions Bearing an Alkoxy Group

Treatment of methoxymethyltrimethylsilane with Bu^{s}Li in THF gives methoxy-(trimethylsilyl)methyllithium, and its subsequent reactions with carbonyl compounds have been reported to afford the adducts, α-methoxy-β-hydroxyalkylsilanes **58** (Scheme 2.38). Although the initial adducts do not undergo elimination of a silyl group in situ, the corresponding enol ethers **59** are formed upon treating

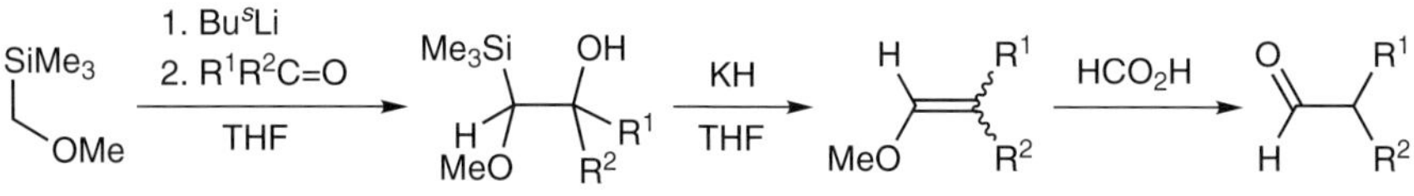

R^1, R^2 = H, alkyl

Scheme 2.38. Synthesis of aldehydes by the Peterson reaction.

2 The Peterson and Related Reactions

them with KH [98]. Treatment of the enol ether products **59** with aqueous formic acid gives the corresponding aldehydes in excellent yields.

α-Methoxybenzyltrimethylsilane can be similarly subjected to the Peterson reaction to afford the 1-alkenyl ethers bearing a phenyl group (Scheme 2.39) [99]. When the α-methoxybenzyltrimethylsilane is treated first with BuLi and TMEDA in THF, and then with aromatic aldehydes, heteroaromatic aldehydes, aliphatic aldehydes, α,β-unsaturated aldehydes, or ketones, the corresponding 1-alkenyl ethers **60** are directly produced as mixtures of the *E*- and *Z*-isomers in good yields. The 1-alkenyl ethers **60** thus obtained from aldehydes are readily hydrolyzed to give the benzoyl derivatives **61** under mild conditions, such as refluxing in methanol in the presence of a catalytic amount of hydrochloric acid.

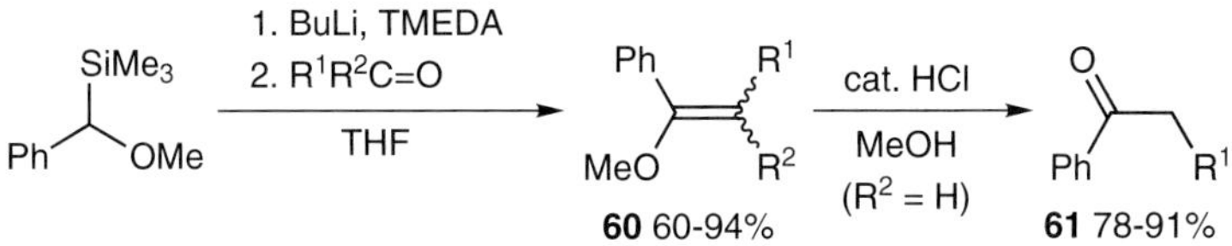

R^1 = H, alkyl, Ph;
R^2 = alkyl, alkenyl, aryl, heteroaryl

Scheme 2.39. Synthesis of benzoyl derivatives by the Peterson reaction.

Silyl(methoxy)benzotriazol-1-ylmethane **62** is lithiated with BuLi to give the corresponding anion, which undergoes Peterson reactions with carbonyl compounds (Scheme 2.40) [100, 101]. The products, 1-(1-methoxy-1-alkenyl)benzotriazoles **63**, are synthetically equivalent to an acylbenzotriazole synthon in which the carbonyl group is masked as an enol ether [102, 103]. Transformation of the alkenyl ethers into carboxylic acids is readily achieved by treatment with zinc bromide and hydrochloric acid in refluxing 1,4-dioxane [104].

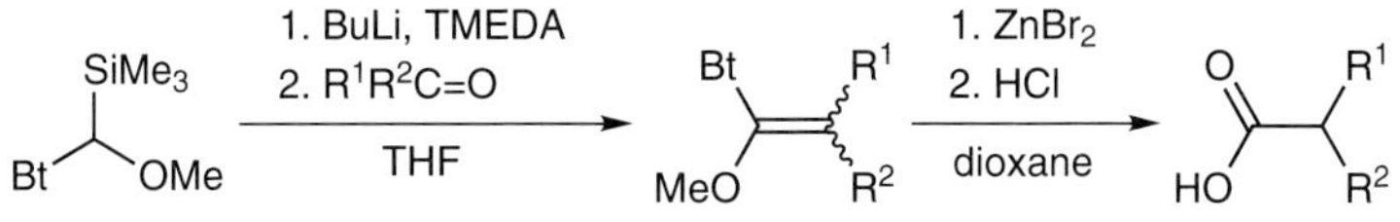

R^1, R^2 = H, alkyl, alkenyl, aryl
Bt = benzotriazol-1-yl

Scheme 2.40. Synthesis of carboxylic acids by the Peterson reaction.

The Peterson reactions of α-silylbenzyl carbamates provide a method for the preparation of aromatic alkenyl carbamates. Addition of carbonyl compounds to the carbanions derived from the α-silylbenzyl carbamates **64** with Bu^{t}Li leads directly to the corresponding alkenyl carbamates **65** with *Z*-selectivity (Scheme 2.41)

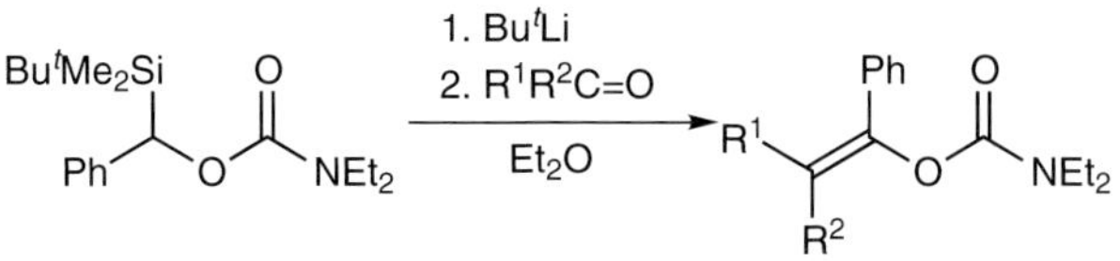

2.3 Generation of α-Silyl Carbanions and their Peterson Reactions | 41

Scheme 2.41. Synthesis of alkenyl carbamates by the Peterson reaction.

[44, 105]. Better selectivity is observed in diethyl ether than in more polar solvents such as THF, HMPA, or TMEDA.

2.3.2.3 Generation of α-Silyl Carbanions Bearing a Nitrogen-Containing Group

There have been a few reports on the simple route to enamidines offered by the Peterson reaction. Lithiation of α-silyl amidines such as 66 with Bu^{s}Li and subsequent Peterson reaction with carbonyl compounds gives the corresponding enamidines 67 (Scheme 2.42) [106, 107]. The enamidines 67 are converted to the amines by treatment with sodium borohydride in ethanol under slightly acidic conditions. These entire homologation processes can be performed without purification of the enamidine intermediates 67.

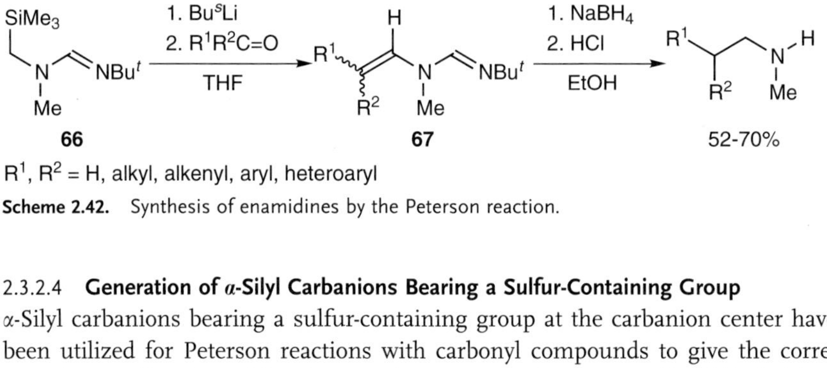

Scheme 2.42. Synthesis of enamidines by the Peterson reaction.

2.3.2.4 Generation of α-Silyl Carbanions Bearing a Sulfur-Containing Group

α-Silyl carbanions bearing a sulfur-containing group at the carbanion center have been utilized for Peterson reactions with carbonyl compounds to give the corresponding 1-alkenyl sulfides [108–113]. For example, the α-sulfanylalkyllithium 68 is prepared by treatment of phenyl trimethylsilylmethyl sulfide with Bu^{t}Li, and its subsequent reactions with carbonyl compounds produce the corresponding 1-alkenyl sulfides in moderate yields (Scheme 2.43) [114–116]. The 1-alkenyl sulfides are usually obtained as mixtures of E- and Z-isomers [117].

With cyclic α,β-unsaturated ketones such as 2-cyclohexen-1-one, the α-(phenylthio)alkyllithium 69 in THF adds in a 1,2-manner in the presence of TMEDA to give the corresponding dienyl sulfide 70, whereas the addition of HMPA to a THF solution of the carbanion 69 or the use of DME gives exclusively the 1,4-addition product 71 (Scheme 2.44) [118, 119].

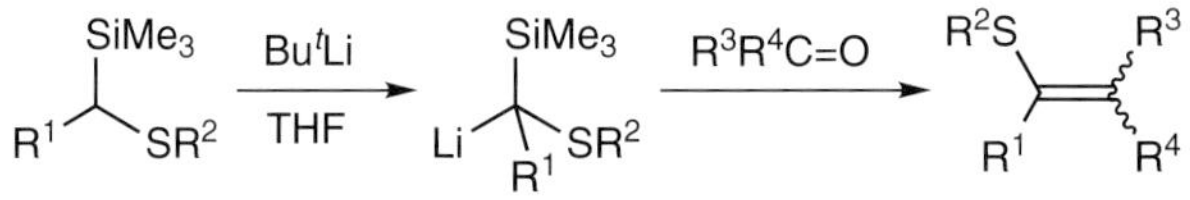

2 The Peterson and Related Reactions

Scheme 2.43. Synthesis of 1-alkenyl sulfides by the Peterson reaction.

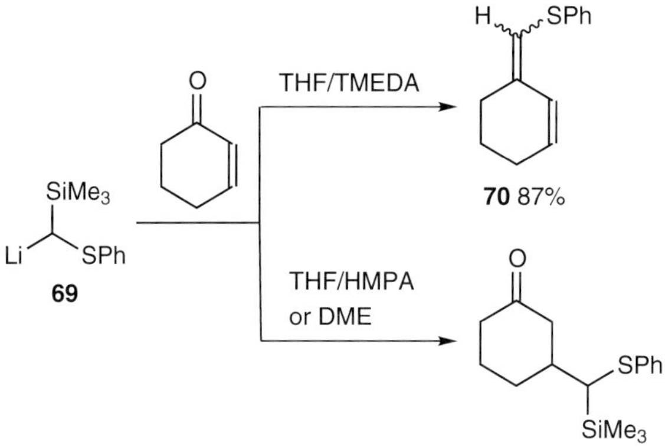

R^1 = H, alkyl, CN, SiMe$_3$; R^2 = Me, Ph, 2-thiazolyl;
R^3, R^4 = H, alkyl, alkenyl, aryl

Scheme 2.44. Reactions of an α-silyl carbanion with 2-cyclohexen-1-one.

The α-silyl carbanion **69** reacts with esters in the presence of TMEDA to give the adducts **72**, which undergo desilylation to afford α-(phenylthio)alkyl ketones **73** (Scheme 2.45). The use of acid chlorides or acid anhydrides gives complex mixtures [115].

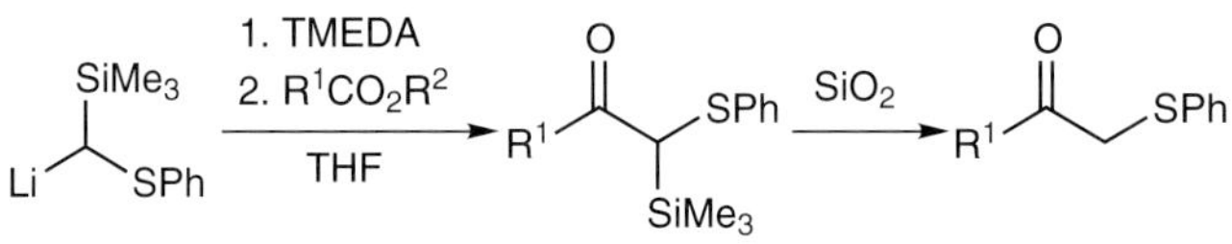

R^1 = alkyl, aryl; R^2 = alkyl

Scheme 2.45. Reactions of an α-silyl carbanion with esters.

Reactions of α-sulfanyltrimethylsilylmethyllithiums with several amides give vicinally N,S-disubstituted alkenes **74** (Scheme 2.46) [120, 121]. The E-isomers are predominant over the Z-isomers.

Aldehydes and ketones are efficiently converted to the ketene O,S-acetals **75** by Peterson reaction with methoxy(phenylthio)(trimethylsilyl)methyllithium, which is

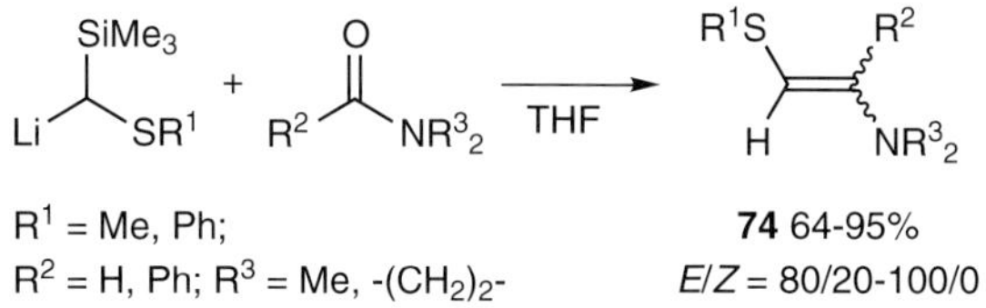

2.3 Generation of α-Silyl Carbanions and their Peterson Reactions

[Scheme 2.46 reaction]

R¹ = Me, Ph;
R² = H, Ph; R³ = Me, -(CH$_2$)$_2$-

74 64-95%
E/Z = 80/20-100/0

Scheme 2.46. Synthesis of vicinally N,S-disubstituted alkenes by the Peterson reaction.

generated by deprotonation of the corresponding silylated O,S-acetal with BuLi and utilized as the C-1 source (Scheme 2.47) [122]. The ketene O,S-acetal products **75**, which are obtained as mixtures in which the E-isomer predominates, are conveniently transformed into the corresponding carboxylic acid methyl esters, thiocarboxylic acid S-esters, and amides by acid-catalyzed methanolysis in the presence of hydrogen chloride and mercury(II) chloride, exposure to chlorotrimethylsilane and sodium iodide followed by filtration through alumina, and treatment with lithium methanethiolate followed by an amine, respectively [123]. The ketene O,S-acetals **75** thus synthesized are dihydroxylated with the Sharpless catalyst to afford the α-hydroxy methyl esters **76** with moderate to good enantioselectivities. However, the chemical yields of the α-hydroxy esters **76** from O,S-acetals **75** are low and unsatisfactory compared to those from the corresponding O,O-acetals [124].

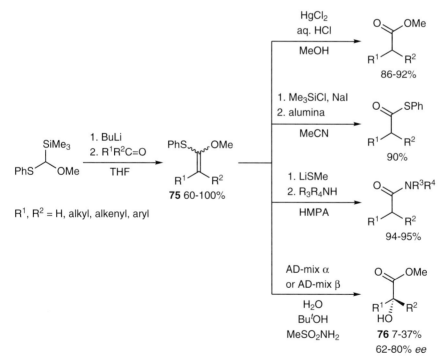

Scheme 2.47. Synthesis of ketene O,S-acetals by the Peterson reaction.

Peterson reactions of α-silyl carbanions **77** bearing two sulfanyl groups on the anionic carbon atom with ketones and aldehydes furnish the ketene dithioacetals **78** directly, in good to moderate yields (Scheme 2.48) [125].

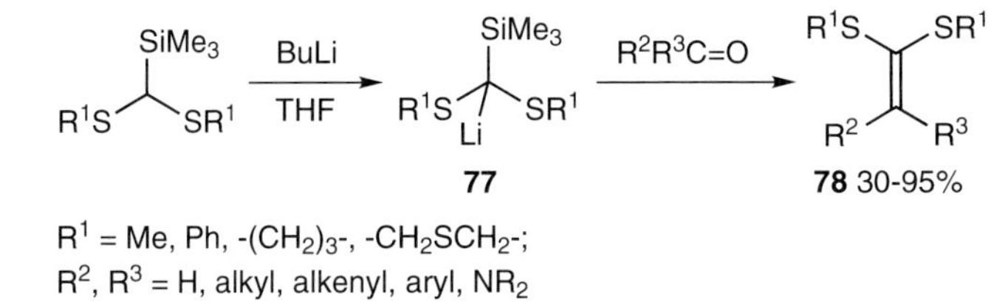

R^1 = Me, Ph, -(CH$_2$)$_3$-, -CH$_2$SCH$_2$-;
R^2, R^3 = H, alkyl, alkenyl, aryl, NR$_2$

Scheme 2.48. Synthesis of ketene dithioacetals by the Peterson reaction.

2-Lithio-2-trimethylsilyl-1,3-dithiane is the most widely utilized reagent for the conversion of ketones and aldehydes to the corresponding ketene dithioacetals (Scheme 2.49) [126–128]. It is used for the synthesis of functionalized 2-alkylidene-1,3-dithianes **79** [129–135]. The 2-alkylidene-1,3-dithianes **79** thus synthesized are useful synthetic intermediates, which are conveniently accessible by means of Peterson reactions, and they can be transformed into various compounds [136, 137]. For example, compounds **79** are converted to the corresponding carboxylic acids, aldehydes, and enones by hydrolysis, hydrogenation followed by hydrolysis, and deprotonation followed by alkylation and hydrolysis, respectively (Scheme 2.49) [138–140].

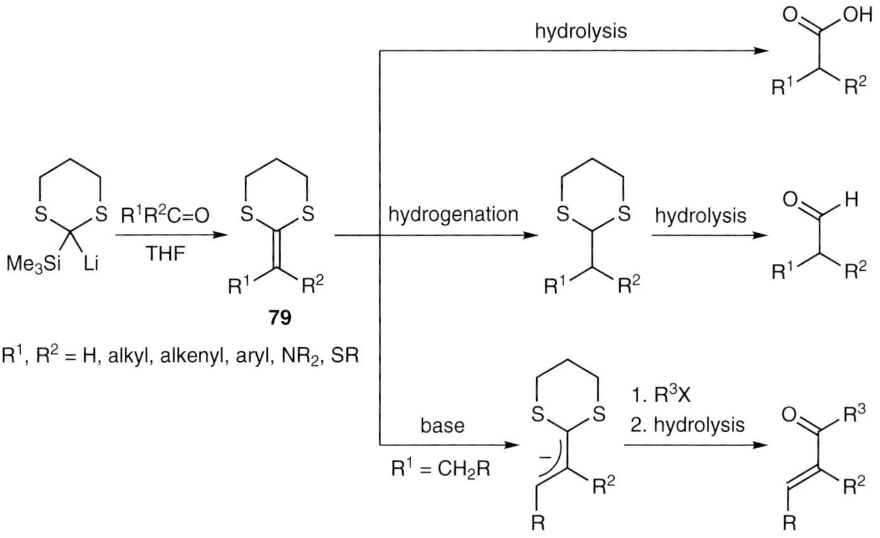

R^1, R^2 = H, alkyl, alkenyl, aryl, NR$_2$, SR

Scheme 2.49. Synthesis of 2-alkylidene-1,3-dithianes by the Peterson reaction.

2.3 Generation of α-Silyl Carbanions and their Peterson Reactions

On the other hand, the reactions of 2-lithio-2-trimethylsilyl-1,3-dithiane with esters and thioesters give not the ketene dithioacetals, but the 2-acyl-1,3-dithianes **81** (Scheme 2.50). The possibly formed silyl enol ethers **80** undergo subsequent hydrolysis to give **81** [141].

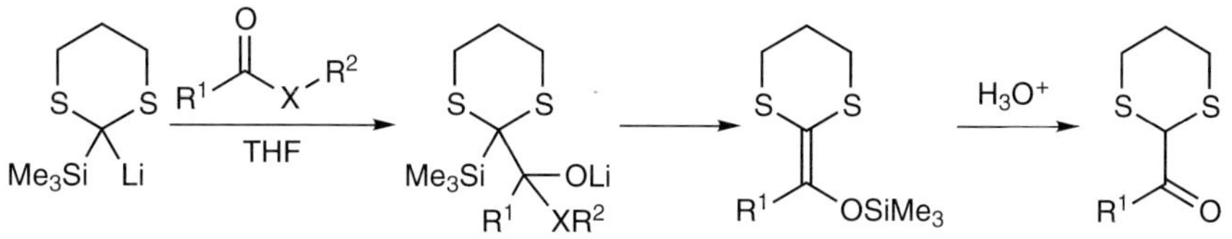

R^1 = Me, Ph; R^2 = Me, Et; X = O, S

Scheme 2.50. Reaction of 2-lithio-2-trimethylsilyl-1,3-dithiane with esters and thioesters.

Furthermore, ketene dithioacetals **83** prepared from unsaturated ketones **82** by means of Peterson reactions are converted to the bicyclic lactones **84** by treatment with cerium(IV) ammonium nitrate (CAN) in wet acetonitrile (Scheme 2.51) [142]. The oxidative cyclization occurs stereoselectively in moderate yields.

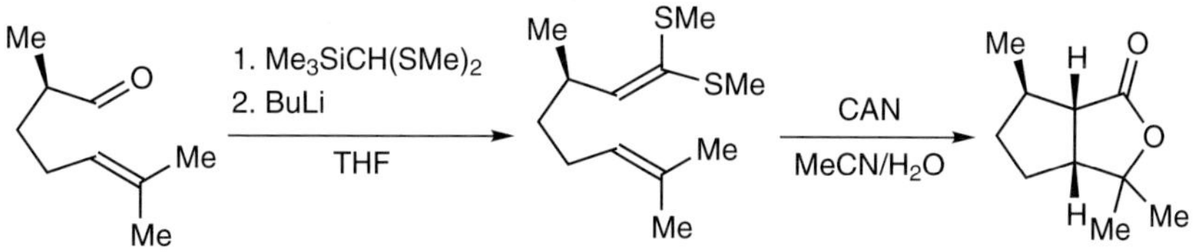

Scheme 2.51. Synthesis of a ketene dithioacetal by the Peterson reaction and its oxidative cyclization.

1-Alkenyl sulfoxides **86** are obtained by the condensation of α-silyl carbanion **85** with carbonyl compounds, although the alkene formation reactions are not stereoselective (Scheme 2.52) [143].

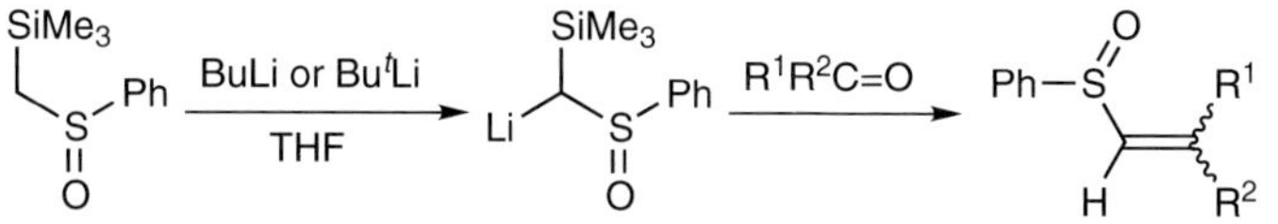

R^1, R^2 = H, Ph, alkyl, alkenyl

86 24-87%
E/Z = 34/66-66/34

Scheme 2.52. Synthesis of 1-alkenyl sulfoxides by the Peterson reaction.

2 The Peterson and Related Reactions

Treatment of trimethylsilylmethanesulfinamides **87** with MeLi or LDA, followed by addition of aromatic aldehydes or benzophenone, gives the corresponding 1-alkenyl sulfinamides **88** (Scheme 2.53) [144]. With aldehydes, the reactions provide exclusively the *E*-isomer of **88**.

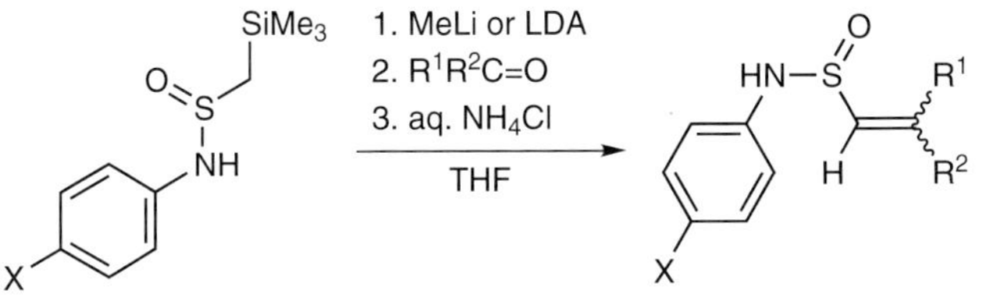

87
X = H, Me, Cl; R^1, R^2 = H, aryl

88 34–77%
E/Z = 90/10–100/0

Scheme 2.53. Synthesis of 1-alkenyl sulfinamides by the Peterson reaction.

α-Silyl carbanions bearing a sulfonyl group on the anionic carbon atom possibly give sulfonyl-substituted alkenes via the Peterson reaction and silylated alkenes via the Julia reaction [145, 146]. Since the Si–O interaction is so strong that only the Peterson reaction proceeds, the corresponding α,β-unsaturated sulfones are formed exclusively. The α-silyl carbanion **89** generated from phenyl trimethylsilylmethyl sulfone using BuLi readily reacts with both aldehydes and ketones to afford a stereo-isomeric mixture of the 1-alkenyl sulfones **90** on work-up (Scheme 2.54) [147–150]. Crucial to the success of the reaction is the use of DME as solvent in these cases. Other solvents such as THF are much less successful, because abstraction of an acidic α-proton from the carbonyl substrate may compete significantly with nucleophilic attack by the α-silyl carbanion in such solvents. Reactions of the α-silyl-α-sulfonyl carbanion **89** with amides giving 2-aminoalkenyl sulfones have also been reported [121].

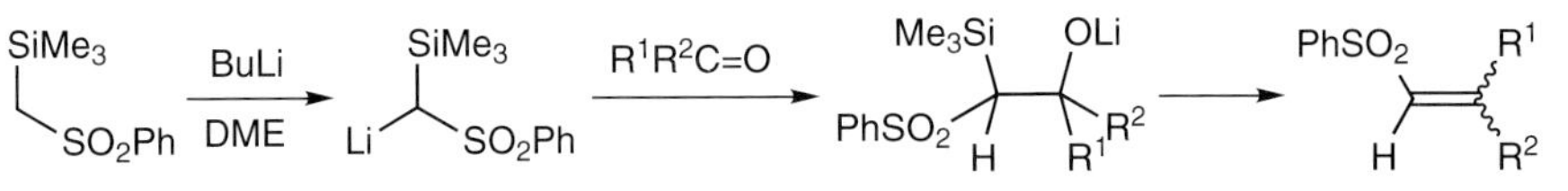

89
R^1, R^2 = H, alkyl, alkenyl, NR_2, etc.

90 50–92%
E/Z = 25/75–100/0

Scheme 2.54. Synthesis of 1-alkenyl sulfones by the Peterson reaction.

This method is also useful for the synthesis of alkenyl-substituted trifluoromethyl sulfone derivatives **91** (Scheme 2.55). Halogen–lithium exchange of iodomethyltrimethylsilane with Bu^{t}Li, followed by addition of 0.5 equivalents of triflic anhydride, provides the α-silyl carbanion bearing a trifluoromethanesulfonyl group. Reactions of the α-silyl carbanion with aldehydes give the corresponding

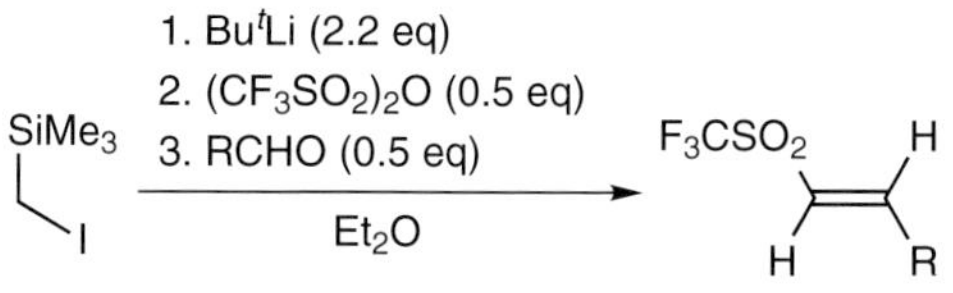

2.3 Generation of α-Silyl Carbanions and their Peterson Reactions

Scheme 2.55. Synthesis of 1-alkenyl trifluoromethyl sulfones by the Peterson reaction.

(E)-alkenes **91** stereoselectively [151]. The yields of **91** are in most cases better than those obtained by means of the Wittig reaction. A combination of the Peterson reaction of an in situ generated α-trimethylsilyl-α-sulfonyl-substituted carbanion with a carbonyl compound and sulfone elimination has also been reported [152, 153] (see Section 2.5.4).

α-Trimethylsilyl sulfonamides are easily converted into the corresponding carbanions **92** by deprotonation with LDA in THF or diethyl ether. The corresponding carbanions **92** are then reacted with non-enolizable aldehydes to yield the α,β-unsaturated sulfonamides **93** (Scheme 2.56) [154]. With R^1 = H, the Peterson reactions of **92** proceed with E-selectivity, whereas with R^1 = Ph the stereochemical outcome depends significantly on the substrates and the reaction conditions such as temperature and reaction time.

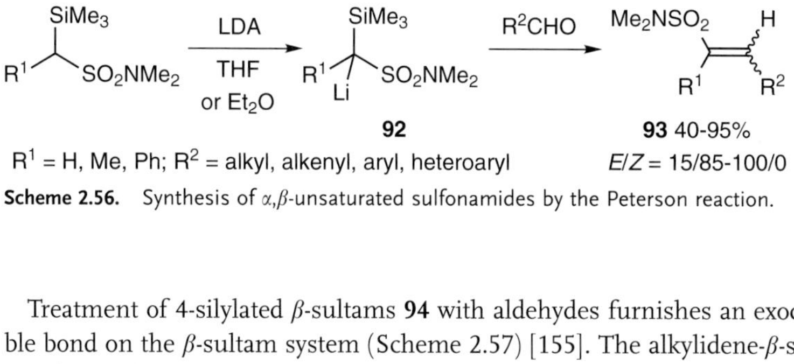

R^1 = H, Me, Ph; R^2 = alkyl, alkenyl, aryl, heteroaryl   **93** 40–95% E/Z = 15/85–100/0

Scheme 2.56. Synthesis of α,β-unsaturated sulfonamides by the Peterson reaction.

Treatment of 4-silylated β-sultams **94** with aldehydes furnishes an exocyclic double bond on the β-sultam system (Scheme 2.57) [155]. The alkylidene-β-sultams **95** are obtained as mixtures of geometric isomers.

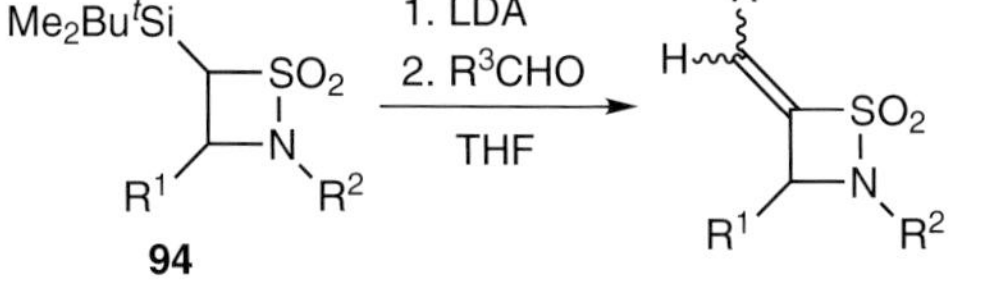

R^1 = H, Me; R^2 = Me, SiBu^{t}Me$_2$;   **95** 33–62%
R^3 = Me, aryl E/Z = 40/60–64/36

Scheme 2.57. Synthesis of alkylidene-β-sultams by the Peterson reaction.

2.3.2.5 Generation of α-Silyl Carbanions Bearing a Phosphorus-Containing Group

An α-silyl carbanion is generally more reactive than a phosphorus ylide, and so a relevant reagent has the advantage of being correspondingly more reactive. There are some reports of instances where the Wittig reaction fails but the Peterson methodology is successful [156–158]. For example, an α-silyl carbanion derived from ethyl trimethylsilylacetate undergoes the reaction with the cyclopentanone derivative **96** to give the corresponding alkene **97**, while the corresponding phosphorus ylide does not (Scheme 2.58) [159].

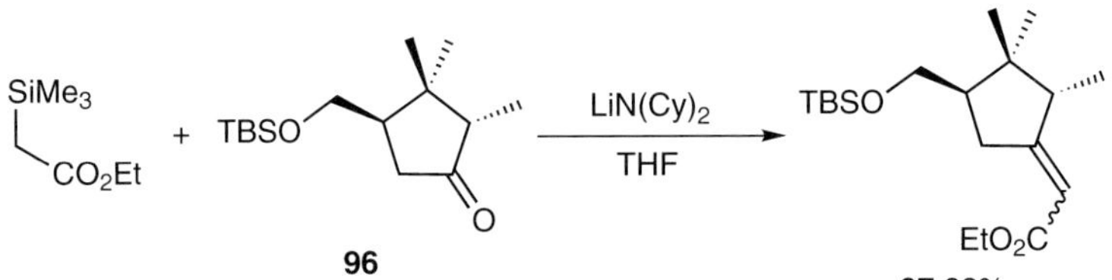

96

97 82%
E/*Z* = 7/93

Scheme 2.58. Synthesis of an alkylidenecyclopentane by the Peterson reaction.

In the Peterson reaction of an α-silyl carbanion bearing a phosphorus substituent on the anionic carbon atom, there are two possibilities for alkene formation, that is, the Peterson reaction to form phosphorus-substituted alkenes and the Wittig–Horner reaction and the Wadsworth–Emmons reaction to form silicon-substituted alkenes. Most of the reports on these competing reactions have been focused on the reactions of α-silyl phosphonates with carbonyl compounds. It is noteworthy that the alkenylphosphonates have been exclusively obtained in almost every case. That is, the Peterson reactions override the Wittig and Wadsworth–Emmons reactions.

In the cases of α-silylalkylphosphonates **98** bearing a hydro, fluoro, chloro or alkoxy group on the silyl-bearing α-carbon atom, the reactions of the corresponding carbanions with aldehydes give mixtures of the *E*- and *Z*-isomers of the alkenylphosphonates **99** (Scheme 2.59) [160–165]. Olefination reactions can also be achieved by using phosphonates, such as **100**, with ketones or, in some cases, esters as substrates (Scheme 2.60) [42, 166].

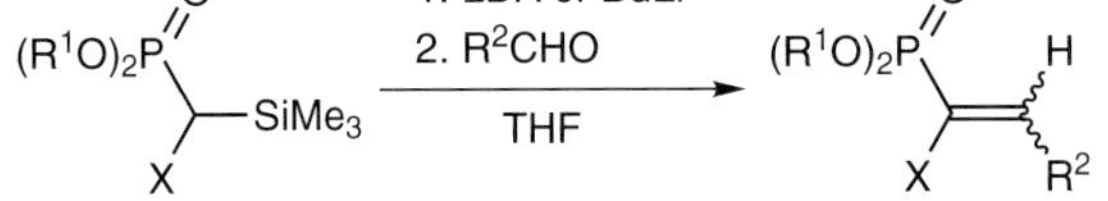

98

99 40-100%

R^1 = Me, Et; X = H, F, Cl, OR
R^2 = alkyl, alkenyl, aryl, heteroaryl

E/*Z* = 30/70-60/40

Scheme 2.59. Synthesis of 1-alkenylphosphonates by the Peterson reaction.

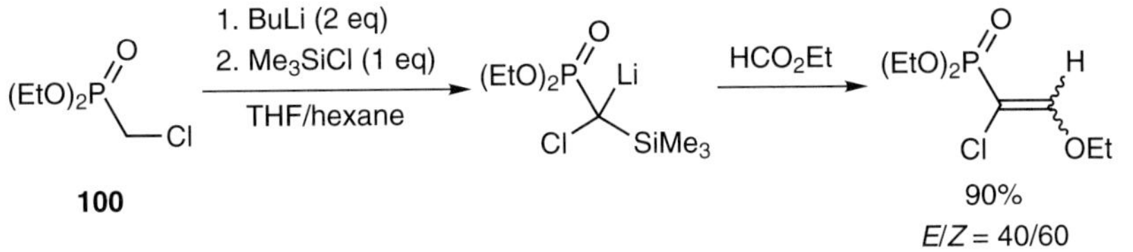

2.3 Generation of α-Silyl Carbanions and their Peterson Reactions

Scheme 2.60. The Peterson reaction of an α-silyl carbanion bearing a phosphoryl group with ethyl formate.

Reactions with aldehydes of carbanions bearing three different atoms on the anionic carbon atom prepared from α-silyl-α-sulfanylalkylphosphonate **101** give the corresponding (E)-1-alkenylphosphonates **102** with high selectivities (Scheme 2.61) [167]. Such high E-selectivity is also seen in the Peterson reactions of a few fluoromethylphosphonates such as **103** with trifluoroacetate esters (Scheme 2.62).

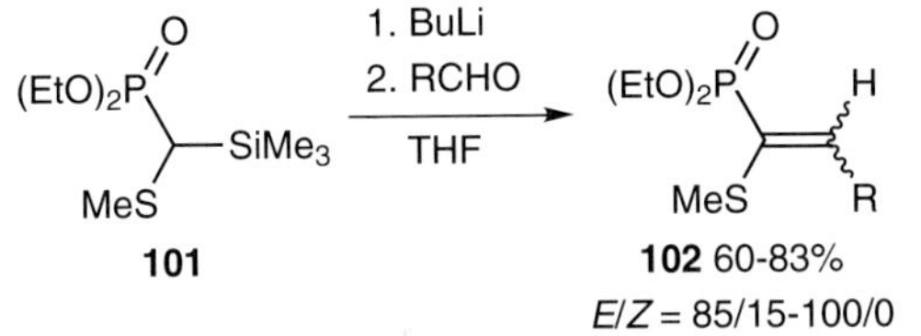

Scheme 2.61. The Peterson reaction of an α-silyl carbanion bearing both a phosphoryl and a sulfanyl group with aldehydes.

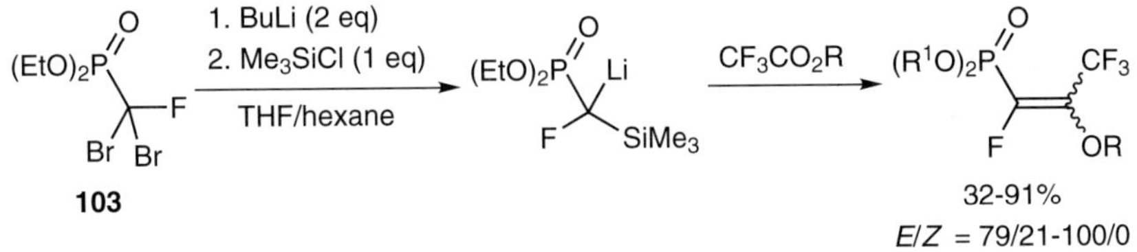

Scheme 2.62. The Peterson reaction of an α-silyl carbanion bearing a phosphoryl group and a fluorine atom with esters.

On the other hand, high Z-selectivity is seen in the olefination reactions of the carbanion **19** derived from 3,3-diethoxybutylphosphonate with aldehydes (Scheme 2.16) [41, 42]. Similarly, Z-selective Peterson reactions of the in situ generated α-phosphoryl-α-(trimethylsilyl)allyl anion **104** with aldehydes or alkyl formates to afford the 2-dienylphosphonates **105** have been reported (Scheme 2.63) [168, 169]. These methods allow access to (Z)-alkenylphosphonates, whereas Wittig–Horner reactions give the thermodynamic (E)-alkenes almost exclusively. These excellent Z-selectivities can be rationalized in terms of the chelation control mechanism (see Section 2.2.2.3).

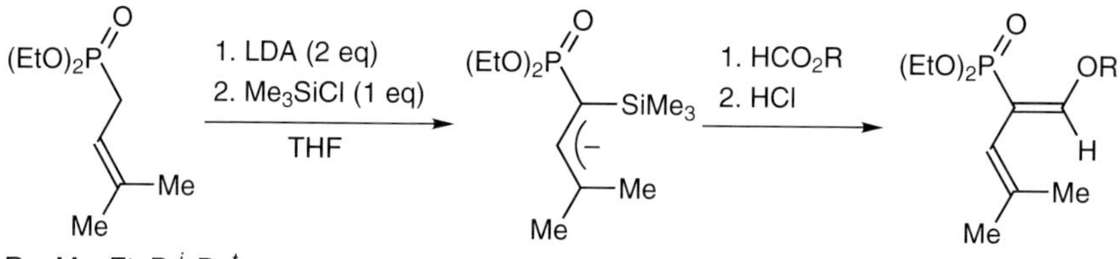

R = Me, Et, Pr^i, Bu^t **104** ***Z*-105** 72-94%

Scheme 2.63. Stereoselective synthesis of 2-dienylphosphonates by the Peterson reaction.

A few examples of the reactions of a (trimethylsilyl)methylenephosphorane with carbonyl compounds have been reported. Reactions of the silylated phosphorus ylide **106** with *α,β*-unsaturated aldehydes lead to the 1,3-dienylphosphonium salts **107** (Scheme 2.64) [170]. These reactions show that the silyl group is eliminated faster than the phosphorus group.

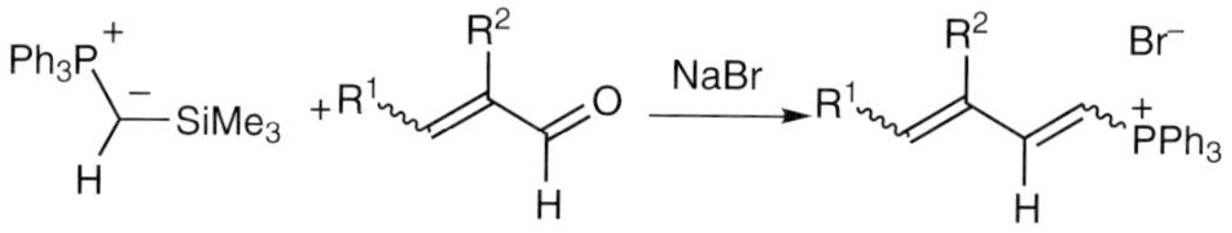

106 **107** 50-86%

R^1 = H, Me, Ph; R^2 = H, Me

Scheme 2.64. The Peterson reaction of an *α*-silyl carbanion bearing a phosphonio group with enals.

The carbanions of (trimethylsilylmethyl)iminophosphoranes **108** undergo Peterson reactions with aldehydes to give (1-alkenyl)iminophosphoranes **109** rather than aza-Wittig reactions (Scheme 2.65) [171]. The reactions usually afford a mixture of the *E*- and *Z*-isomers of **109**, with the *E*-products predominating. These carbanions undergo similar reactions with ketones.

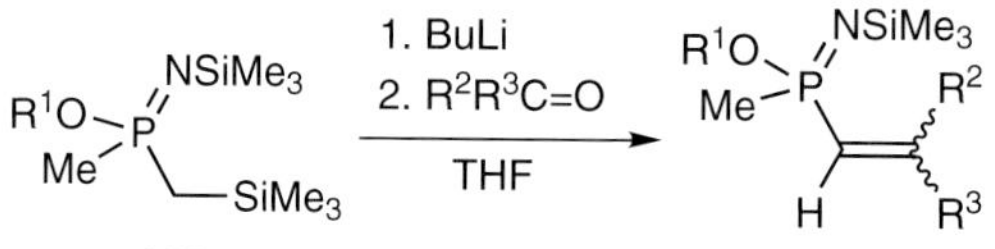

108 **109** 58-87%

R^1 = Ph, CH_2CF_3

R^2, R^3 = H, aryl, Bu^t, alkenyl

Scheme 2.65. The Peterson reaction of *α*-silyl carbanions bearing an iminophosphoryl group.

2.3.2.6 Generation of *α*-Silyl Carbanions Bearing a Halogen Atom

Reactions of an *α*-silyl carbanion bearing a halogen atom on the anionic carbon atom with carbonyl compounds afford the corresponding halohydrins, which can be isolated in some cases. Upon further treatment of the product with a base, two

2.3 Generation of α-Silyl Carbanions and their Peterson Reactions

possible types of reactions can ensue. One is the Peterson reaction to form the halogen-substituted alkene, and the other is the elimination of the halogen ion to form the oxirane [172]. The substituents and the reaction conditions, such as temperature, affect the type of product formed.

α,β-Epoxysilanes **110** are obtained by addition of a wide range of aldehydes and ketones to chloro(trimethylsilyl)methyllithium, which is prepared by treatment of chloromethyltrimethylsilane with Bu^{s}Li, followed by warming to room temperature (Scheme 2.66) [173, 174]. The chloride ion is eliminated in preference to attack of the alkoxide at silicon. Reaction of iodo(triphenylsilyl)methyllithium with benzaldehyde similarly provides the corresponding oxirane in a Z-selective manner [175].

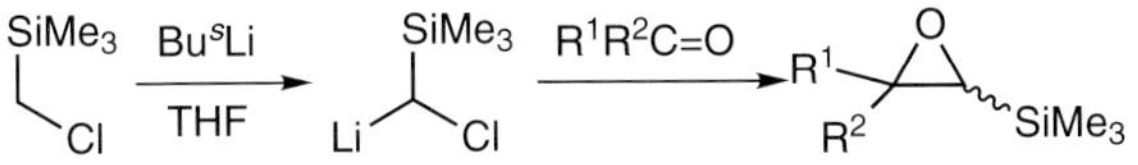

Scheme 2.66. Synthesis of α,β-epoxysilanes.

On the other hand, 1,1-dichloroalkenes are obtained by treatment of the 1,1-dichloro-2-hydroxyalkylsilanes **111** with boron trifluoride etherate or sulfuric acid (Scheme 2.67) [176].

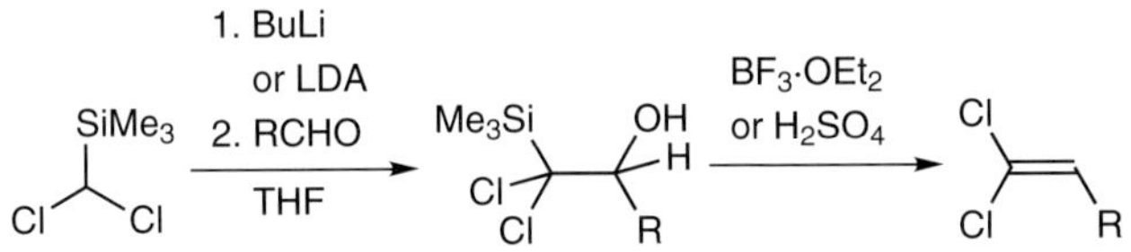

Scheme 2.67. Synthesis of 1,1-dichloroalkenes by the Peterson reaction.

The reactions of the carbanion derived from *tert*-butyl trimethylsilyl-α-chloroacetate **112** and LDA with carbonyl compounds yield the α-chloro-α,β-unsaturated esters **113** as mixtures of the E- and Z-isomers (Scheme 2.68). Quenching the reaction mixture with thionyl chloride enhances the yield of **113** [31, 177, 178].

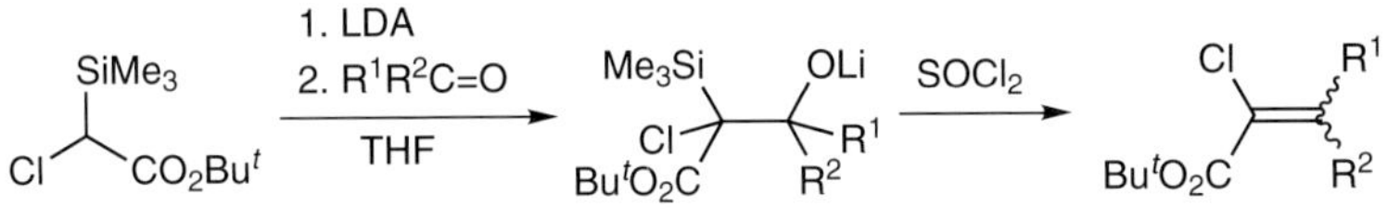

112
R^1, R^2 = H, alkyl

113 17-58%
E/Z = 51/49-82/18

Scheme 2.68. Synthesis of α-chloro-α,β-unsaturated esters by the Peterson reaction.

This methodology is applicable to the synthesis of α-fluoro- and α-bromo-α,β-unsaturated esters [179–183]. Sodium hexamethyldisilazide (NaHMDS) or LDA/KOBu^t can also be used as a base, and give better yields in some cases.

2.3.2.7 Generation of α-Silyl Carbanions from α-Silyl Ketones

Peterson reactions of α-silyl carbanions bearing a functional group at the α-position are especially promising methods for the construction of compounds bearing α,β-unsaturated functional groups. Treatment of α-silyl ketones such as **114** with a base affords the corresponding enolates **115**, which can be utilized for Peterson reactions. A hindered base with low nucleophilicity should be used to avoid both attack at the carbonyl carbon atom and the generation of another possible enolate anion of the ketone. Condensation of **115** with aldehydes and subsequent elimination of lithium trimethylsilyloxide from the cross-aldol condensation products gives the corresponding enones **116** in an E-selective manner (Scheme 2.69) [184–188].

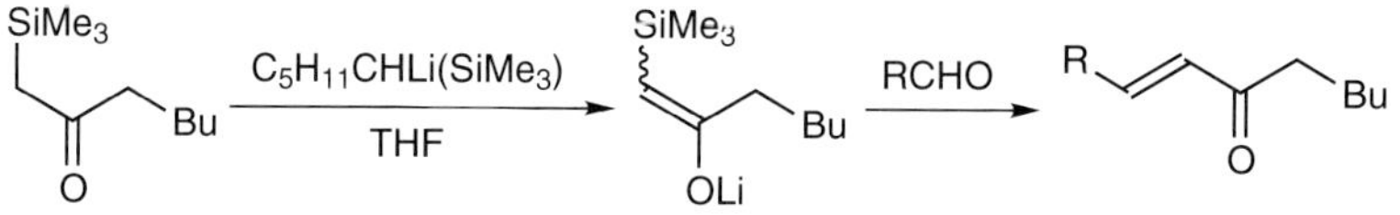

Scheme 2.69. Stereoselective synthesis of enones by the Peterson reaction of an enolate with aldehydes.

2.3.2.8 Generation of α-Silyl Carbanions Bearing an Ester Group

Treatment of α-silyl esters with a base readily affords the corresponding enolates, which can be utilized for Peterson reactions (Scheme 2.70) [189–196]. LDA is the most widely used base for the deprotonation of α-silyl esters. The carbonyl compounds used in the above reactions are aldehydes, saturated and unsaturated ketones, amides, lactones, and lactams. The products, α,β-unsaturated esters, are obtained as mixtures of the E- and Z-isomers in most cases. When another trimethylsilyl group is present on the anionic carbon atom, the reactions of the carbanion derived from the α,α-bis(trimethylsilyl) esters with ketones are unsuccessful, probably because of steric reasons, and result only in enolization [197].

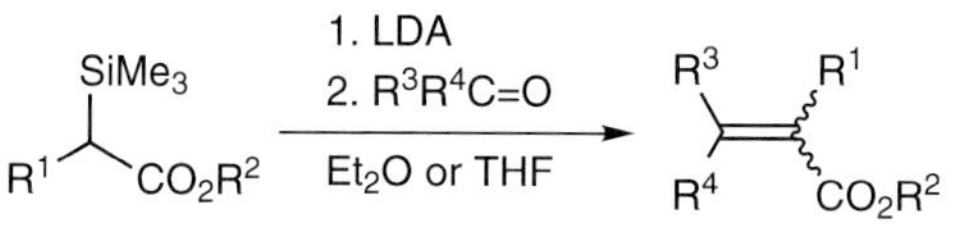

R^1 = H, Me, SiMe$_3$;
R^2, R^3, R^4 = H, alkyl, alkenyl, aryl, SiMe$_3$

60-90%
E/Z = 8/92-55/45

Scheme 2.70. The Peterson reaction of α-silyl carbanions bearing an ester group.

2.3.2.9 Generation of α-Silyl Carbanions Bearing a Lactone Group

The convenient synthetic methodology for the α-silyl esters has been extended to the synthesis of α-alkylidenelactones [198, 199]. Peterson reactions of the carbanion generated from α-silyl-γ-butyrolactones **117** with carbonyl compounds afford the corresponding α-alkylidene-γ-lactones **118** with high *E*-selectivity (Scheme 2.71). The yields are low when ketones are used as substrates.

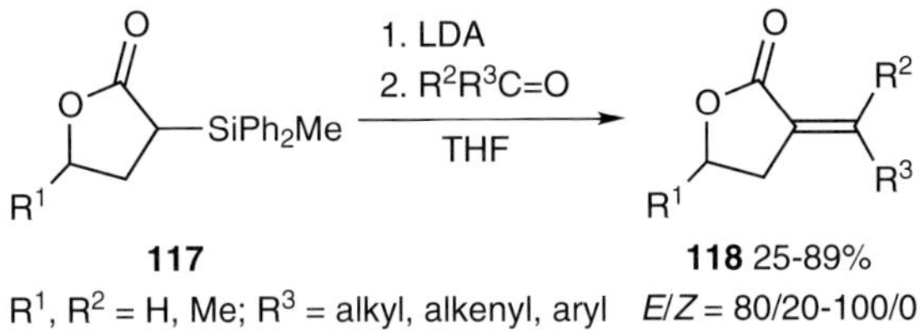
SiPh2Me
1. LDA
2. R2R3C=O
THF
R1
R2
R3
117
118 25-89%
R1, R2 = H, Me; R3 = alkyl, alkenyl, aryl   E/Z = 80/20-100/0

Scheme 2.71. The Peterson reaction of α-silyl carbanions bearing a lactone group.

2.3.2.10 Generation of α-Silyl Carbanions Bearing Thiocarboxylate or Dithiocarboxylate Ester Group

O-Alkyl α-(trimethylsilyl)thioacetates and *S*-alkyl α-(trimethylsilyl)dithioacetates **119** can be deprotonated and subjected to Peterson reactions with aldehydes to give the α,β-unsaturated thiocarboxylic acid *O*-esters and dithiocarboxylic acid *S*-esters **120**, respectively (Scheme 2.72). The alkenes are obtained *E*-selectively [200].

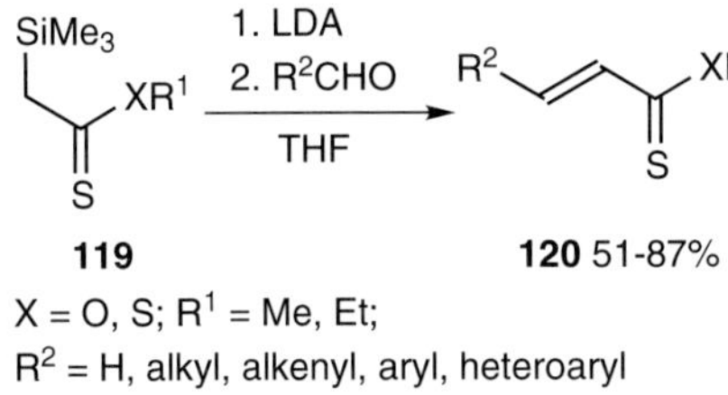
SiMe3
XR1
1. LDA
2. R2CHO
THF
R2
XR1
S
119
120 51-87%
X = O, S; R1 = Me, Et;
R2 = H, alkyl, alkenyl, aryl, heteroaryl

Scheme 2.72. The Peterson reaction of α-silyl carbanions bearing a thiocarboxylate or a dithiocarboxylate ester group.

2.3.2.11 Generation of α-Silyl Carbanions Bearing an Imino Group

Peterson reactions of α-silyl aldimines and further hydrolysis leads to the corresponding α,β-unsaturated aldehydes via the intermediary α,β-unsaturated imines [201–204]. α-Lithio α-silyl aldimines are generated by treatment of α-silyl aldimines **121** with LDA, and their subsequent reaction with aldehydes and ketones followed by treatment with oxalic acid or wet silica gel provides the corresponding α,β-unsaturated aldehydes **122** (Scheme 2.73). Since the intermediary unsaturated imines are easily hydrolyzed, vinylogation of the aldehydes to give the α,β-unsaturated aldehydes is readily achieved in good yields under mild conditions. Although the above reactions did not show good stereoselectivity, efficient conversion of a variety of aldehydes into the corresponding α,β-unsaturated aldehydes with high *E*-selectivity has also been reported [205–207].

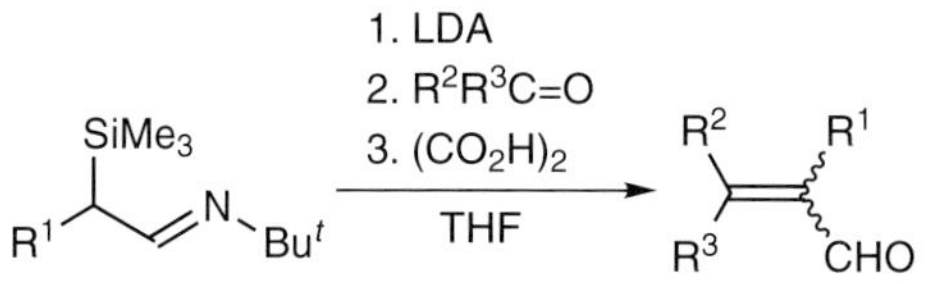

121 **122** 88-94%

R^1 = H, Me; E/Z = 50/50

R^2, R^3 = H, alkyl, alkenyl, aryl

Scheme 2.73. Synthesis of α,β-unsaturated aldehydes by the Peterson reaction of α-silyl carbanions bearing an imine group.

This methodology is also applicable to the conversion of α-silyl-α-ketimines **123** to the corresponding di- and trisubstituted α,β-unsaturated ketones (Scheme 2.74) [208, 209].

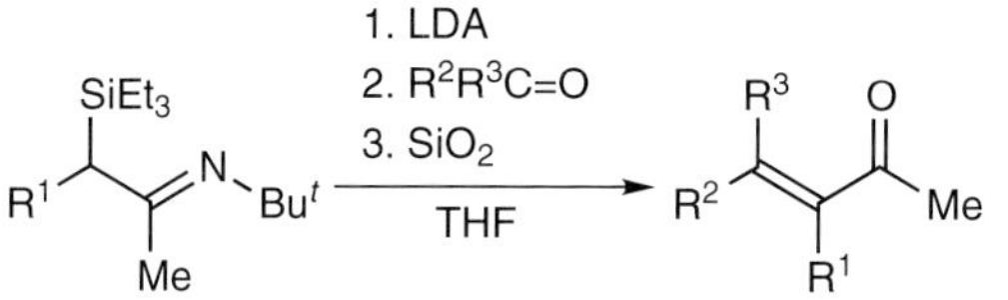

123 40-86%

R^1 = H, Me; E/Z = 17/83-68/32

R^2, R^3 = alkyl, alkenyl, aryl

Scheme 2.74. Synthesis of α,β-unsaturated ketones by the Peterson reaction of α-silyl carbanions bearing an imine group.

2.3.2.12 Generation of α-Silyl Carbanions Bearing an Amide Group

Peterson reactions of the lithium enolate **124** derived from *N*,*N*-dimethyl-(trimethylsilyl)acetamide with carbonyl compounds furnish the corresponding α,β-unsaturated amides (Scheme 2.75) [210]. Although the amides are obtained in good yields in the reactions with ketones and with non-enolizable aldehydes, the reactions with enolizable aldehydes give only negligible yields with predominant recovery of the starting amides. There appears to be little stereoselectivity in the

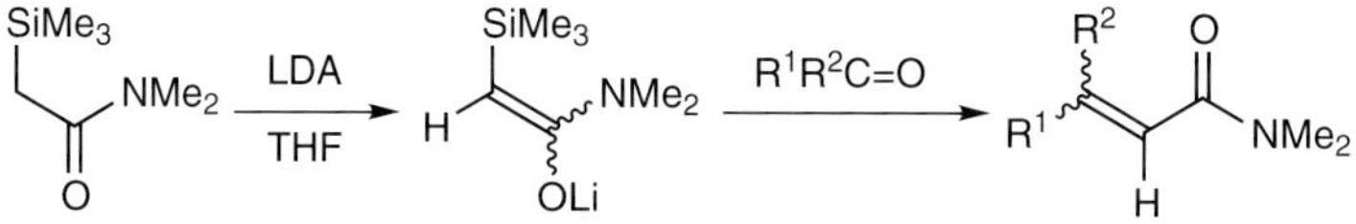

124 50-94%

R^1, R^2 = H, alkyl, alkenyl, aryl, NR_2, OR

Scheme 2.75. Synthesis of α,β-unsaturated amides by the Peterson reaction of an α-silyl carbanion bearing an amide group.

formation of the new double bond. Esters and amides can also be used as the substrates instead of aldehydes or ketones [211].

Highly Z-selective synthesis of α,β-unsaturated amides by Peterson reactions of N,N-dibenzyl-(triphenylsilyl)acetamide with aromatic aldehydes has also been reported (Scheme 2.76) [212, 213].

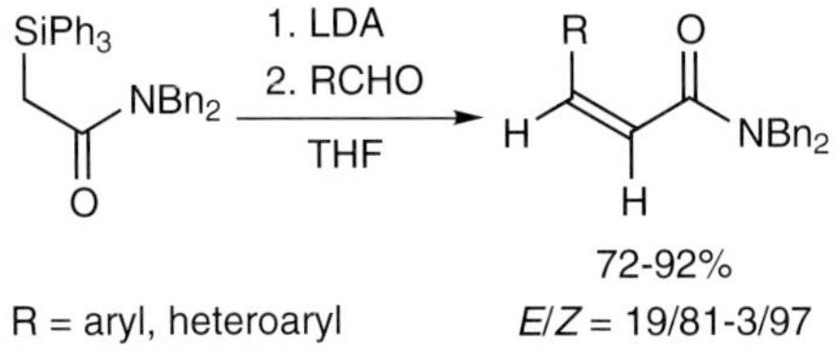

R = aryl, heteroaryl

72–92%
E/Z = 19/81–3/97

Scheme 2.76. Stereoselective synthesis of α,β-unsaturated amides by the Peterson reaction of an α-silyl carbanion bearing an amide group.

When silyl lactams such as the 3-silylazetidinones **125** are subjected to the Peterson reaction, the corresponding 3-alkylidene-β-lactams **126** are obtained as the major products, without any accompanying products of a ring-opening reaction (Scheme 2.77) [214–219].

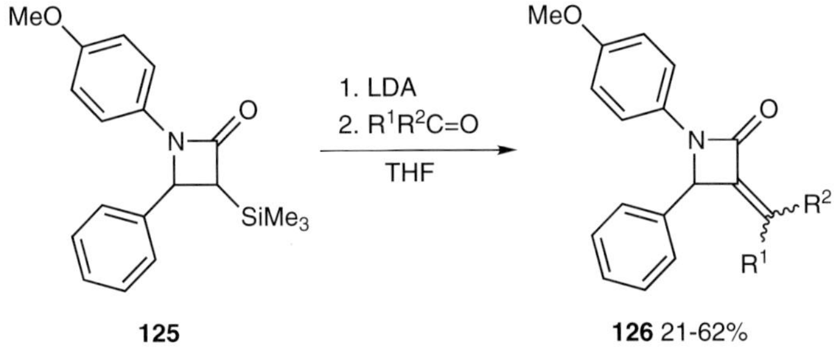

R^1, R^2 = H, alkyl, Ph, heteroaryl, CO_2R, OEt

Scheme 2.77. Synthesis of 3-alkylidene-β-lactams by the Peterson reaction.

2.3.2.13 Generation of α-Silyl Carbanions Bearing a Cyano Group

Peterson reactions of α-silyl nitriles provide facile access to α,β-unsaturated nitriles. The α-silyl carbanion derived from α-(trimethylsilyl)acetonitrile is treated with carbonyl compounds to give the corresponding α,β-unsaturated nitriles in good to moderate yields (Scheme 2.78) [220–222]. α-Alkyl-α-(trimethylsilyl)acetonitriles are also effective for this type of reaction [223, 224].

Although the stereoselectivity of the above reactions is usually not good, highly Z-selective reactions employing boron reagents have been reported, providing (Z)-α,β-unsaturated nitriles (Scheme 2.79) [225, 226].

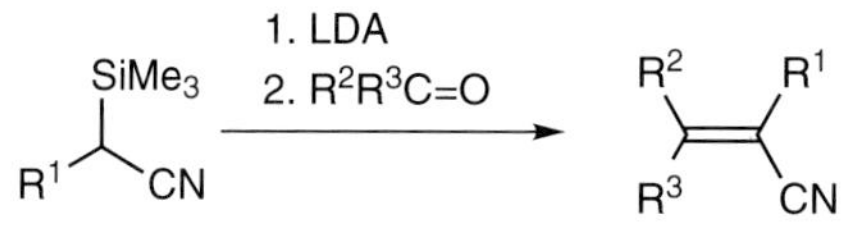

2 The Peterson and Related Reactions

Scheme 2.78. Synthesis of α,β-unsaturated nitriles by the Peterson reaction.

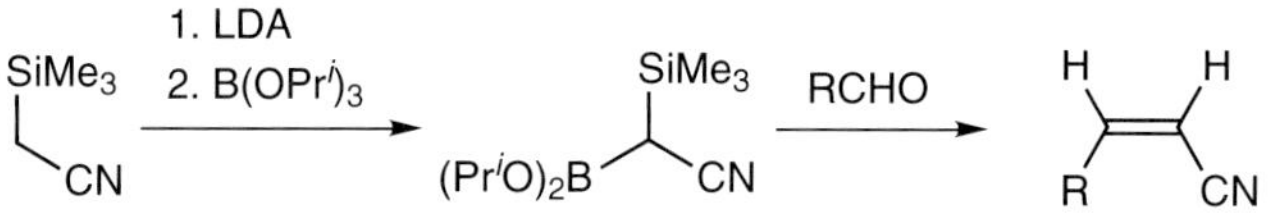

R^1 = H, alkyl;
R^2, R^3 = H, alkyl, alkenyl, aryl
59–94%
E/Z = 52/48–79/21

Scheme 2.79. Stereoselective synthesis of α,β-unsaturated nitriles by the Peterson reaction.

R = H, alkyl, alkenyl, Ph
68–90%
E/Z = 6/94–25/75

2.3.2.14 Generation of α-Silyl Carbanions from Allylsilanes

It has been found that the ambident α-(trimethylsilyl)allyl anion **127**, prepared by deprotonation of allyltrimethylsilane with Bu^{s}Li, reacts with many carbonyl compounds to give not the Peterson products, but the 4-hydroxy-1-alkenylsilanes **128** (Scheme 2.80). Because the ambident allyl anions can react at the γ-position, the γ-mode of reactivity prevents the Peterson reactions from occurring [227].

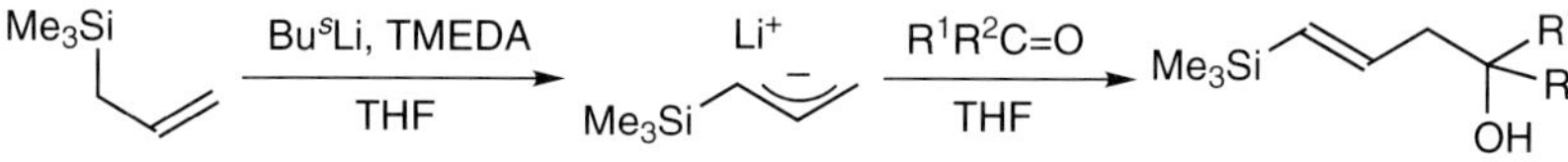

Scheme 2.80. Reaction of an α-silyl allyl anion with carbonyl compounds.

R^1, R^2 = H, alkyl, alkenyl, aryl

However, a change of the counter ion from lithium to titanium, a boron species, or magnesium would change both the regioselectivity and stereoselectivity. After the addition of Ti(OPr^i)$_4$ to α-silylallyl anions, the allyl anion reagents were found to react with several aldehydes at the α-position to give the corresponding β-hydroxy-α-vinylalkylsilanes **129** in a highly regio- and stereoselective manner (Scheme 2.81) [228]. Treatment of **129** with Bu^{t}OK and H$_2$SO$_4$ stereospecifically affords the corresponding (Z)- and (E)-dienes, respectively. Use of other additives such as (cyclopentyl)$_2$BCl and EtAlCl$_2$ instead of Ti(OPr^i)$_4$ gives similar results [229].

When carbanions bearing another trimethylsilyl group at the γ-position such as **130** are used, Peterson reaction gives the corresponding 1-trimethylsilyldienes **131** without metal exchange (Scheme 2.82) [230, 231].

Stereoselective synthesis of 1,4-disubstituted 1,3-dienes from aldehydes has also been reported [232, 233]. The stereoselective synthesis of β-hydroxy-α-(1-alkenyl)-

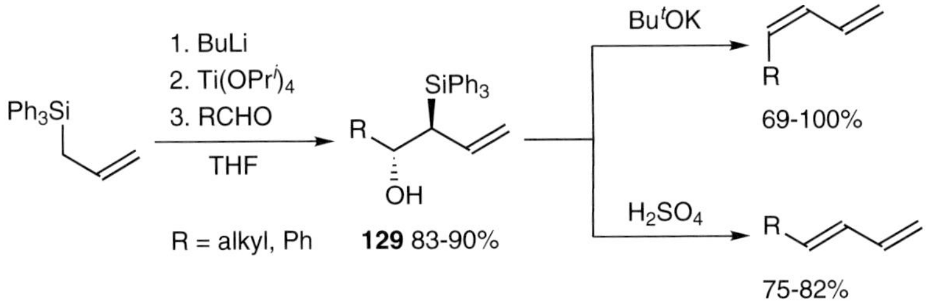

2.3 Generation of α-Silyl Carbanions and their Peterson Reactions

Scheme 2.81. Reaction of an α-silyl allyl anion with carbonyl compounds in the presence of titanium tetraisopropoxide.

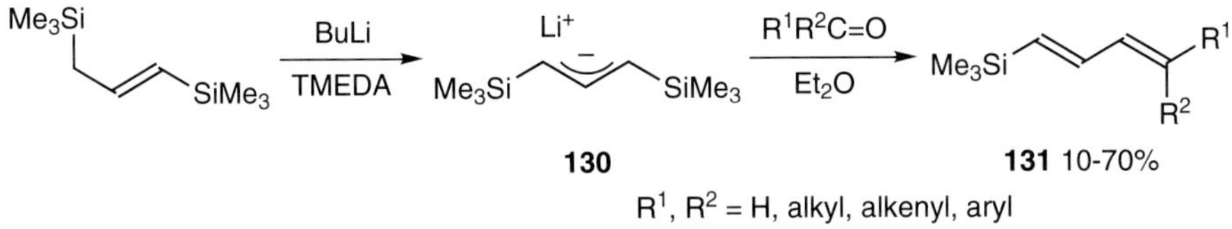

Scheme 2.82. The Peterson reaction of an α,γ-disilyl allyl anion with carbonyl compounds.

silanes **133** is achieved under the influence of the additives Ti(OPr^i)$_4$, MgBr$_2$, or B(OMe)$_3$, and further treatment with KH and H$_2$SO$_4$ gives the corresponding (1E,3Z)- and (1E,3E)-dienes, respectively, in good yields and with good stereoselectivity (Scheme 2.83). Compound **133** can be envisaged as being produced via the six-membered ring transition state **132**. Similarly, the stereoselective addition of an allylic chromium reagent to an aldehyde to give a β-hydroxy-α-(1-alkenyl)-silane, and subsequent Peterson elimination to give a (1E,3Z)-diene, has also been reported [234].

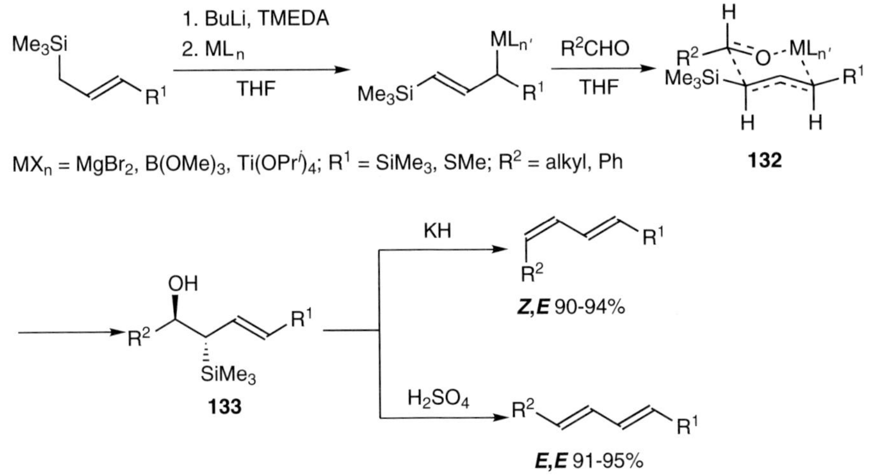

Scheme 2.83. The Peterson reaction of α-silyl allyl anions with carbonyl compounds in the presence of additives.

2.3.2.15 Generation of α-Silyl Carbanions from Propargylsilanes

Deprotonation of 1,3-bis(trimethylsilyl)propyne with Bu^tLi and subsequent addition of carbonyl compounds in the presence or absence of an additive such as $MgBr_2$ has been reported to furnish the enyne products with moderate to good Z-selectivity (Scheme 2.84) [235, 236]. It has been reasonably postulated that the reaction proceeds via a kinetically controlled pericyclic process as illustrated by intermediate **134**, where steric factors and the interaction between the counter ion and the oxygen atom play major roles. Other additives are also employed in Z-selective enyne formation reactions, such as $Ti(OPr^i)_4$, $B(OMe)_3$, and *B*-methoxy-9-borabicyclo[3.3.1]nonane (*B*-MeO-9-BBN) [225, 226, 237–240].

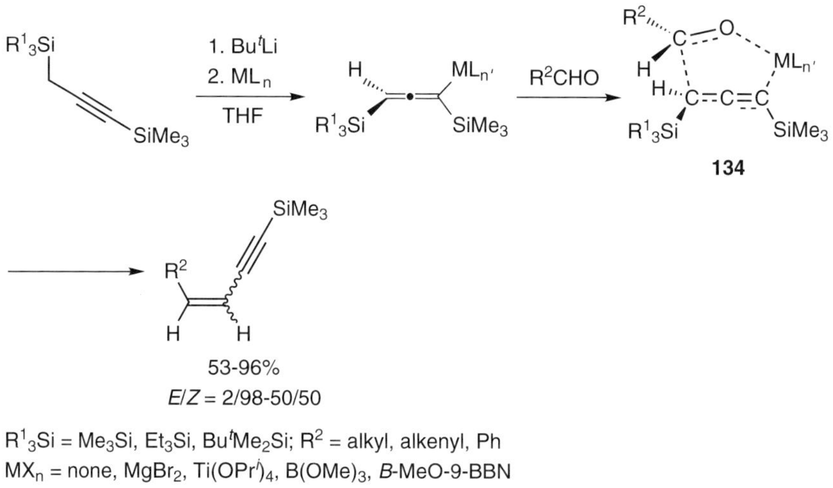
R¹₃Si
SiMe₃
1. BuᵗLi
2. MLₙ
THF
H
R¹₃Si
MLₙ'
SiMe₃
R²CHO
R²
C
O
H
MLₙ'
H
C=C=C
R¹₃Si
SiMe₃
134
SiMe₃
R²
H
H
53-96%
E/Z = 2/98-50/50
R¹₃Si = Me₃Si, Et₃Si, BuᵗMe₂Si; R² = alkyl, alkenyl, Ph
MXₙ = none, MgBr₂, Ti(OPrⁱ)₄, B(OMe)₃, B-MeO-9-BBN

Scheme 2.84. Synthesis of eneynes by the Peterson reaction.

2.3.3
Generation of α-Silyl Carbanions by Substitution of a Heteroatom

2.3.3.1 Generation of α-Silyl Carbanions by Reduction of a Sulfanyl Group

Reductive lithiation of α-silyl sulfides **135** by means of aromatic radical anions, such as lithium naphthalenide (LN), lithium 4,4′-di-*tert*-butylbiphenylide (LDBB) or lithium 1-(dimethylamino)naphthalenide (LDMAN), gives the corresponding α-silyl carbanions, which can be utilized in Peterson reactions (Scheme 2.85) [241–247]. LDMAN and LDBB usually offer higher reduction potentials than LN.

This protocol is especially useful for the synthesis of α-silylcyclopropyl anion **136**, which is difficult to generate by deprotonation of the corresponding silyl cyclopropane with a base such as Bu^sLi in the presence of TMEDA in THF (Scheme 2.86) [248]. Several alkylidene- and allenylidenecyclopropanes have been synthesized by the reactions of 1-silylcyclopropyllithiums generated in this way with aldehydes [249, 250].

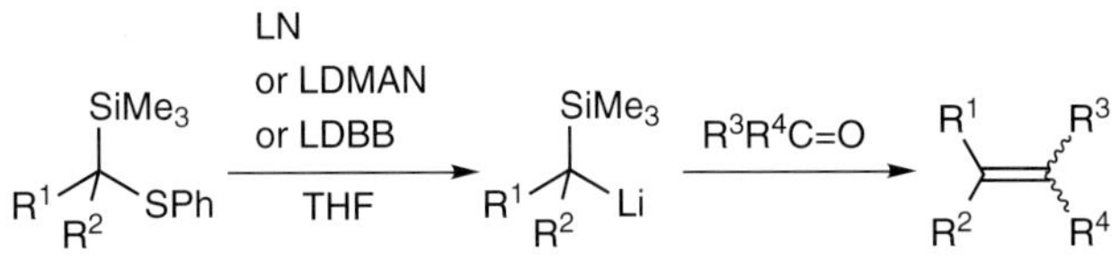

2.3 Generation of α-Silyl Carbanions and their Peterson Reactions

Scheme 2.85. The Peterson reaction of α-silyl carbanions generated by reduction of a sulfanyl group with aromatic radical anions.

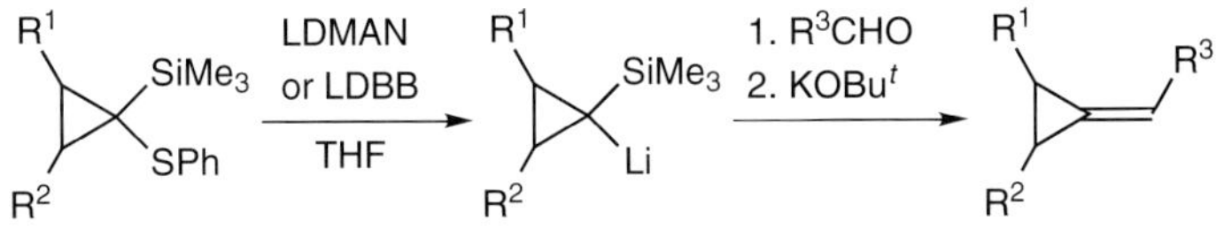

R^1, R^2 = H, alkyl, SiMe$_3$, SPh; R^3, R^4 = H, alkyl, aryl,

Scheme 2.86. Synthesis of alkylidenecyclopropanes by the Peterson reaction.

R^1, R^2 = H, alkyl, -(CH$_2$)$_4$- ; R^3 = H, alkyl, alkenyl, aryl

Reduction of α-silyl sulfides with tributylstannyllithium is also useful for the generation of α-silyl carbanions [251]. When α,α-bis(trimethylsilyl)alkyl phenyl sulfides **137** are treated sequentially with tributylstannyllithium and aldehydes, they give the corresponding alkenylsilanes (Scheme 2.87). This method has an advantage in that tributylstannyllithium is thermally stable and can be stored for several months.

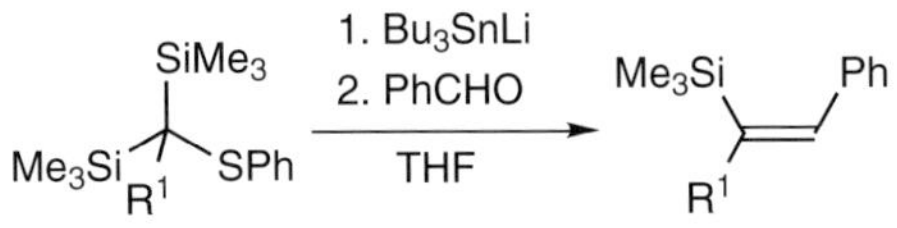

Scheme 2.87. The Peterson reaction of α-silyl carbanions generated by reduction of a sulfanyl group with tributylstannyllithium.

R^1 = Bu, Bn

2.3.3.2 Generation of α-Silyl Carbanions by Substitution of Selenium

α-Silyl carbanions can be generated by C–Se bond cleavage of α-silylalkyl selenides [252, 253]. Treatment of α-(trimethylsilyl)methyl selenides with BuLi at 0 °C leads to Se–Li transmetallation to give the corresponding α-silyl carbanions, which afford the corresponding alkenes upon sequential addition of carbonyl compounds and acid (Scheme 2.88).

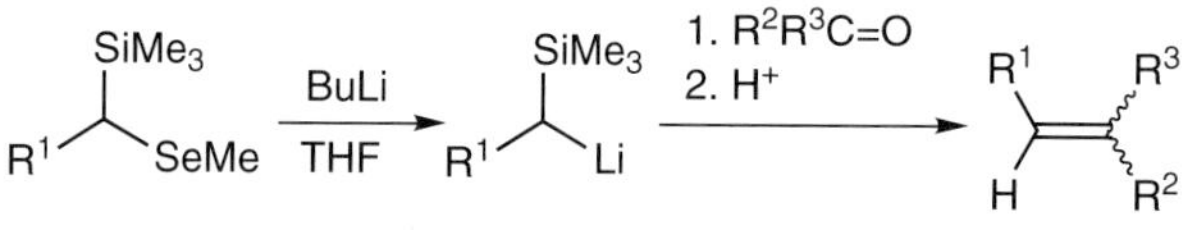

138 18-45%

R^1 = Me, Ph; R^2, R^3 = H, -$(CH_2)_5$-, Bn, alkyl

Scheme 2.88. The Peterson reaction of α-silyl carbanions generated by Se–Li transmetallation.

2.3.3.3 Generation of α-Silyl Carbanions by Desilylation of Bis(trimethylsilyl)methane Derivatives

Desilylation of *gem*-disilyl compounds produces α-silyl carbanions. Treatment of bis(trimethylsilyl)methanes such as **139** with potassium *tert*-butoxide in THF containing carbonyl compounds brings about the Peterson reaction (Scheme 2.89) [135, 254, 255]. Desilylation by means of aqueous KOH instead of $KOBu^t$ has also been reported [256].

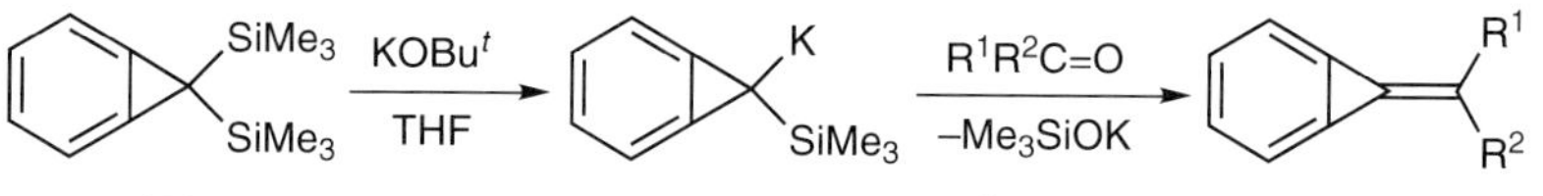

139

R^1 = Ar 10-33%

R^2 = H, Me, Ph

Scheme 2.89. The Peterson reaction of an α-silyl carbanion generated by desilylation of a bis(trimethylsilyl)methane derivative.

2.3.3.4 Generation of α-Silyl Carbanions by Tin–Lithium Transmetallation

Another approach to α-silyl carbanions involves Sn–Li exchange reaction of α-stannylsilanes using BuLi [257, 258]. Transmetallation of α-stannylalkylsilanes **140** with butyllithium affords the corresponding lithium reagents, which are then suitably predisposed for the Peterson reaction with aldehydes to give the corresponding alkenes (Scheme 2.90) [259].

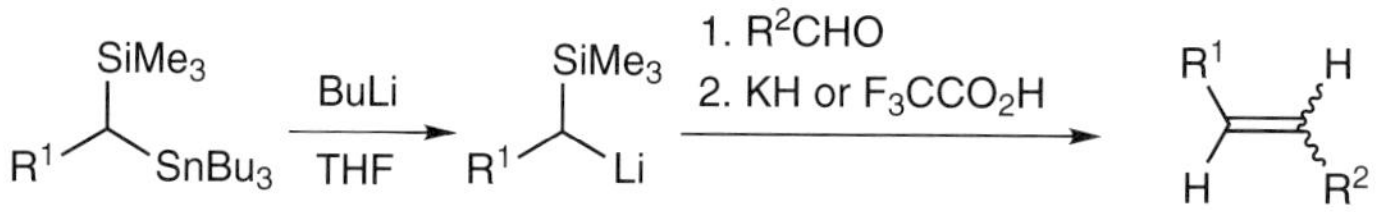

140 18-45%

R^1, R^2 = alkyl, aryl *E*/*Z* = 5/95-63/35

Scheme 2.90. The Peterson reaction of α-silyl carbanions generated by Sn–Li transmetallation.

2.3.4 Formation of β-Hydroxyalkylsilanes from Silyl Enol Ethers

Silyl enol ethers have been known to undergo aldol condensations with aldehydes in the presence of Lewis acids [260, 261]. When a silyl group occupies the β-

2.3 Generation of α-Silyl Carbanions and their Peterson Reactions

position of a silyl enol ether, the products, α-keto-β-hydroxyalkylsilanes, are suitable for Peterson elimination.

The O-methyl-C,O-bis(trimethylsilyl)ketene acetal **141** undergoes an aldol reaction with nonanal in the presence of an equimolar amount of a Lewis acid, and subsequent Peterson reaction gives the corresponding α,β-unsaturated carboxylic ester **143** (Scheme 2.91). The intermediately formed aldol adduct **142** can be isolated by quenching at low temperature. The silyl enol ether acts as an α-silyl carbanion equivalent in this case. The product **143** is obtained with high Z-stereoselectivity when TiCl$_4$ is used as the Lewis acid, while the use of AlCl$_3$ gives exclusively the E-isomer [262]. α,β-Unsaturated ketones are not suitable for this aldol-Peterson reaction sequence since the silyl enol ether undergoes Michael addition at the olefin moiety instead of at the carbonyl moiety of the enone [263].

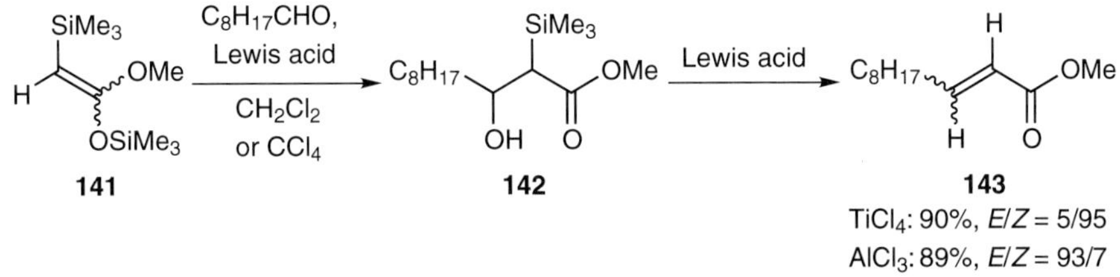

TiCl$_4$: 90%, E/Z = 5/95
AlCl$_3$: 89%, E/Z = 93/7

Scheme 2.91. Synthesis of an α,β-unsaturated ester by the Peterson reaction of a silyl enol ether.

Similarly to the above reactions, α,β-unsaturated aldehydes can be synthesized by aldol reaction of β-silyl enol ethers **144** with aromatic aldehydes in the presence of a catalytic amount of trimethylsilyl triflate (Scheme 2.92) [264]. The E-isomers are obtained exclusively.

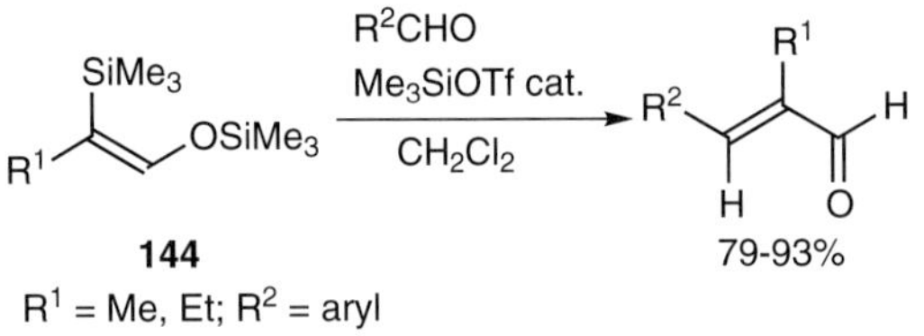

79-93%

R^1 = Me, Et; R^2 = aryl

Scheme 2.92. Synthesis of α,β-unsaturated aldehydes by the Peterson reaction of silyl enol ethers.

When C,O,O-tris(trimethylsilyl)ketene acetals such as **145** are subjected to the reaction with aldehydes in the presence of a catalytic amount of a Lewis acid followed by hydrolysis, the α,β-unsaturated carboxylic acids are formed in a highly E-selective manner (Scheme 2.93) [265, 266]. Fluoride ion catalysts such as NaF and CsF are also used instead of the Lewis acids.

These methodologies are also effective for the synthesis of α,β-unsaturated cyclic esters and amides [267, 268]. For example, α,β-unsaturated γ-lactone **149** is syn-

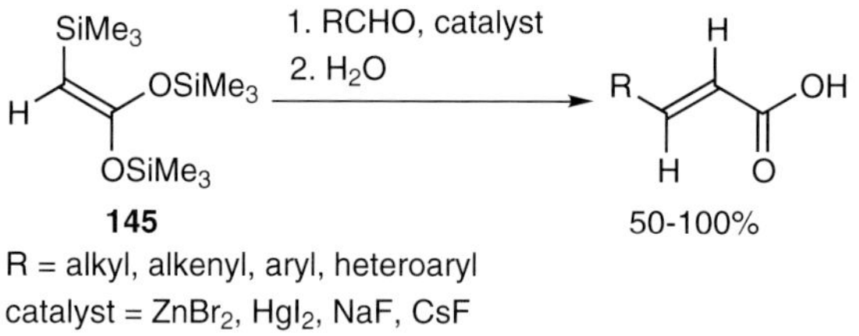

Scheme 2.93. Synthesis of α,β-unsaturated carboxylic acids by the Peterson reaction of silyl enol ethers.

thesized by the TiCl₄-mediated reaction of silyl enol ether **146** with acetaldehyde (Scheme 2.94) [15, 269, 270]. The aldol reaction is most likely to proceed via a chelated six-membered ring transition state **147**, which leads to the predominant formation of the silyl lactone **148**. The silyl (Z)-enolate **146** has such a rigid five-membered ring framework as to give rise to the stereoselective synthesis of the adduct **148**. Further reactions of **148** with BF₃·OEt₂ and LiN(SiMe₃)₂ give the corresponding alkenes **149** with high stereoselectivities.

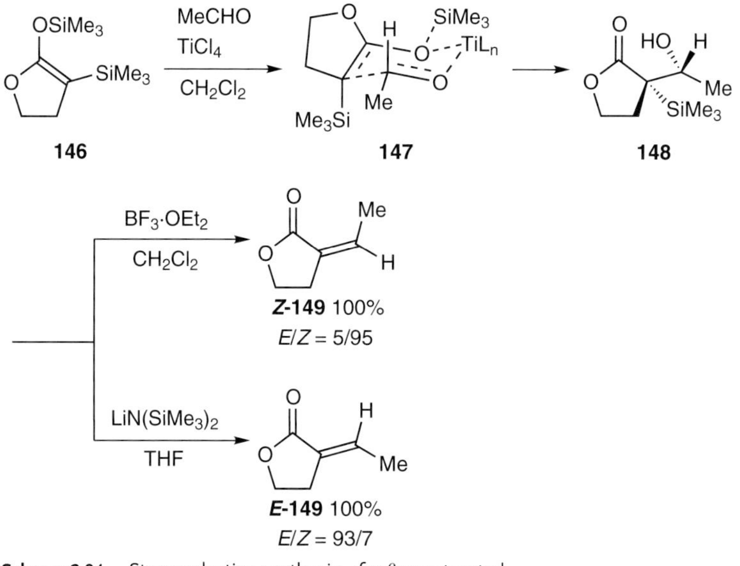

Scheme 2.94. Stereoselective synthesis of α,β-unsaturated lactones by the Peterson reaction of a silyl enol ether.

2.3.5
Fluoride Ion Induced Peterson-Type Reactions

2.3.5.1 Generation of α-Silyl Carbanions by Fluoride Ion Induced Desilylation

Fluoride ion induced desilylation of β-hydroxyalkylsilanes followed by elimination of the oxy group represents a modified methodology for carrying out Peterson re-

2.3 Generation of α-Silyl Carbanions and their Peterson Reactions

actions, although in some cases the hydroxy group is best transformed into another leaving group in advance [271]. For example, cyclopropene derivatives **151** are synthesized by treatment of *trans*-2-trimethylsilylcyclopropanols **150** with tetrabutylammonium fluoride (TBAF) after mesylation, whereas treatment with either KH or sulfuric acid resulted in ring opening of the cyclopropane ring (Scheme 2.95) [272].

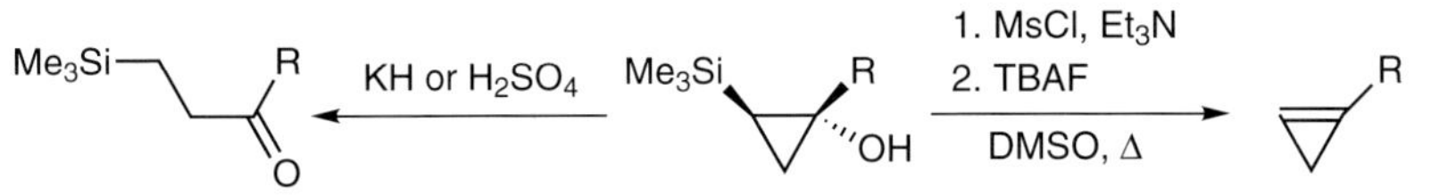

63–80%
R = C_9H_{19}, $(CH_2)_8CH=CH_2$

151 58–66%

Scheme 2.95. Synthesis of cyclopropenes by fluoride ion induced Peterson-type reaction.

Fluoride ion induced Peterson-type elimination of the silyl-γ-lactones **152** with TBAF in THF or DMSO leads to elimination of a silyl group and a carboxy group to afford the corresponding β,γ-unsaturated carboxylic acids **153** (Scheme 2.96) [273, 274]. Since this is an E2-type elimination, a mixture of two diastereomers **152a** and **152b** can be stereoconvergently converted to a single stereoisomer of **153** [275]. In these examples, a silyl group functions as a leaving group only for the generation of the carbanion and it does not migrate to the oxygen atom. The intermediary tetrabutylammonium 3-hexenoate **154** derivative can be trapped with methyl iodide (Scheme 2.97) [276].

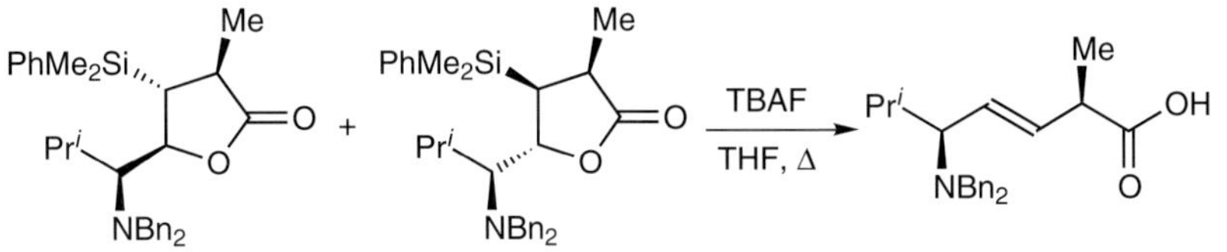

Scheme 2.96. Fluoride ion induced Peterson-type elimination of β-silyl-γ-lactones.

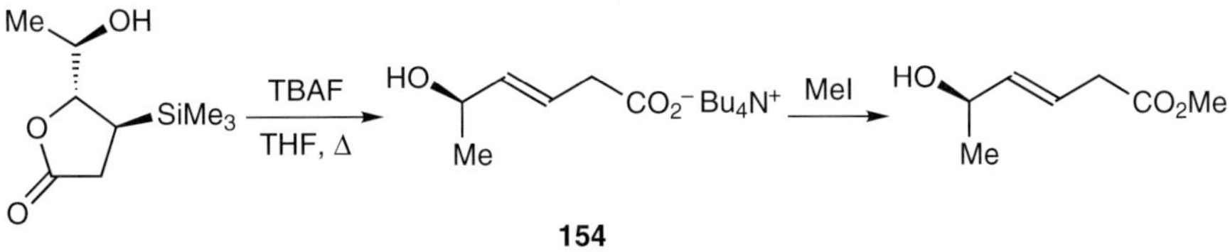

Scheme 2.97. Generation and trapping reaction of a tetrabutylammonium 3-hexenoate derivative.

Carbonyl compounds are converted into the corresponding allenes by reaction with α-lithioalkenylsilanes **155** and further transformation (Scheme 2.98). Since the direct Peterson reaction of alkenylsilanes **156** with a base to give allenes is prevented due to the difficulty of cleavage of the silicon–vinyl carbon bond in the vinyl system, the hydroxy group must first be converted into a better leaving group such as trifluoroacetate. Peterson elimination induced by fluoride salts such as tris(dimethylamino)sulfur (trimethylsilyl)difluoride (TASF) affords the corresponding allenes [277, 278]. Terminal allenes can also be synthesized by similar methods using α-lithiovinylsilanes [279, 280].

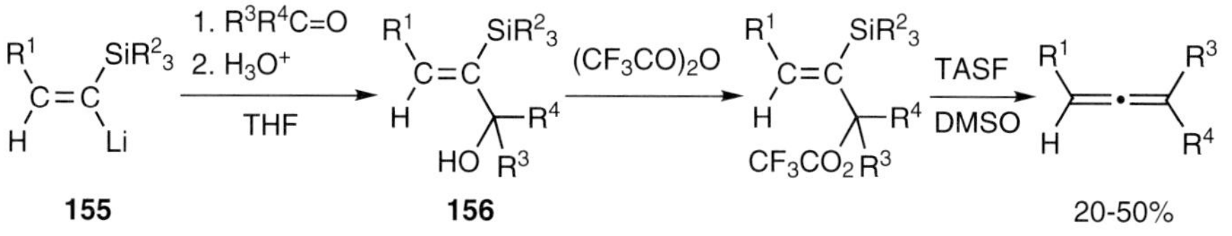

R^1 = Me, H; R^2 = Me, Ph; R^3, R^4 = alkyl, alkenyl, aryl

Scheme 2.98. Synthesis of allenes by fluoride ion induced Peterson-type reaction.

2.3.5.2 **Fluoride Ion Induced Peterson Reactions of Bis(trimethylsilyl)methane Derivatives**

There is another variation of fluoride ion induced Peterson-type reaction that utilizes two silyl groups. The reactions between bis(trimethylsilyl)methyl derivatives **157** and carbonyl compounds using fluoride ion as a catalyst give the corresponding alkenes (Scheme 2.99) [281–283]. This reaction is useful for many bis(trimethylsilyl)methane derivatives, bearing a variety of substituents, with many ketones and aldehydes. As regards the stereoselectivity of the reaction, *E*- and *Z*-isomers are mostly produced in equal amounts.

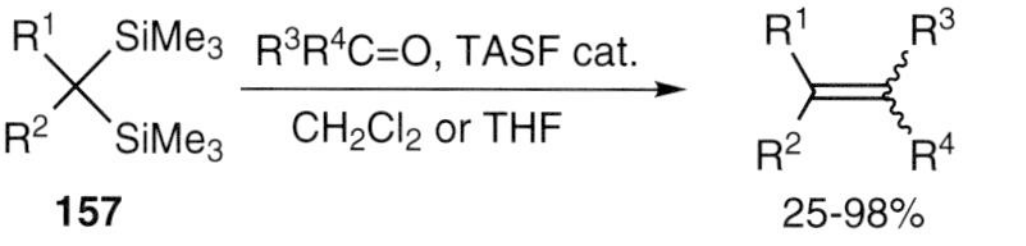

E/*Z* = 50/50–100/0

R^1 = Ph, CO_2Bu^t, CN, $P(O)(OEt)_2$, SPh, SO_2Ph, OMe, $SiMe_3$;
R^2 = H, SPh; R^3 = aryl, PhCH=CMe, Pr^i, Bu^t; R^4 = H, Ph, Me

Scheme 2.99. Fluoride ion induced Peterson-type reaction of bis(trimethylsilyl)methane derivatives.

The reaction is initiated by fluoride ion to generate fluorotrimethylsilane and the α-silyl carbanion, and then the latter reacts with the carbonyl compound to afford the corresponding β-silyl alkoxide (Scheme 2.100). Migration of the trimethylsilyl group to the oxygen atom and further elimination of trimethylsilyloxide affords

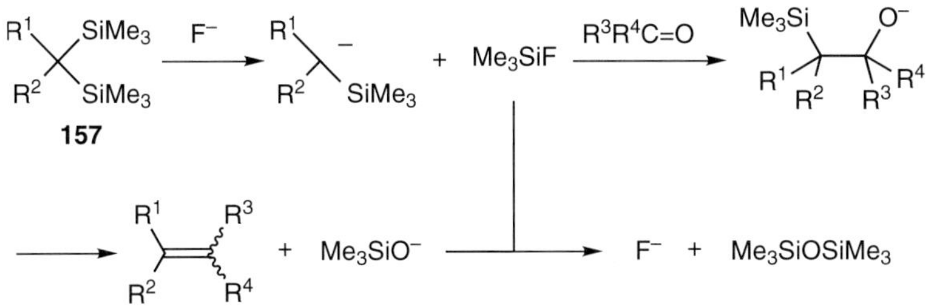

Scheme 2.100. Proposed mechanism for the fluoride ion induced Peterson-type reaction.

the corresponding alkene with regeneration of the catalyst. One silyl group is used to generate an anion on the silyl-bearing carbon atom and the other is used for the Peterson elimination. A significant feature of this reaction is that it proceeds under mild and almost neutral conditions via metal-free carbanion species and can be applied to compounds possessing base- or acid-sensitive functionalities, such as alkoxy, alkoxycarbonyl, and cyano groups. Only a catalytic amount of fluoride ion is effective in the above reactions, rather than a stoichiometric or excess amount of base or acid as used in the conventional Peterson reactions.

This variant of the Peterson reaction can be applied in forming the nitrophenyl-substituted methylenecyclopropa[*b*]naphthalene **158** by employing KF with a catalytic amount of TBAF as the fluoride ion transfer reagents in acetonitrile (Scheme 2.101) [284]. This method is applicable to nitro compounds that are sensitive to strong bases.

Scheme 2.101. Synthesis of an alkylidenecyclopropa[*b*]naphthalene by fluoride ion induced Peterson-type reaction.

2.3.5.3 Fluoride Ion Catalyzed Peterson Reactions of Bis(trimethylsilyl)methylamine Derivatives

The fluoride ion induced Peterson-type reaction is most widely and effectively utilized in the synthesis of enamines from *N*-bis(trimethylsilyl)methylamide derivatives such as **159** [285]. It has been especially exploited to prepare *N*-1-alkenyl phthalimides and *N*-1-alkenyl amides **160** (Scheme 2.102) [286–288].

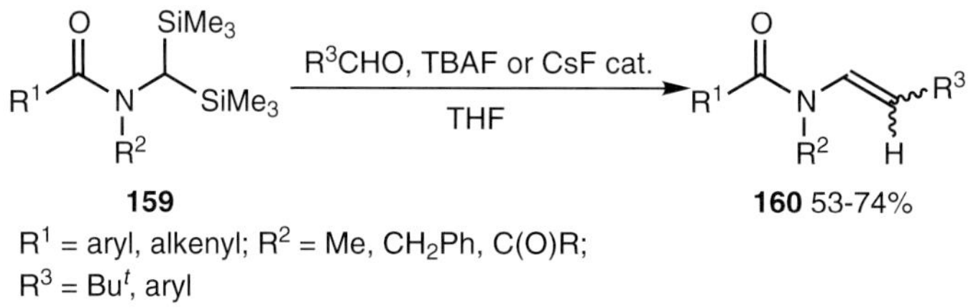

2 The Peterson and Related Reactions

Scheme 2.102. Fluoride ion induced Peterson reaction of N-bis(trimethylsilyl)methylamide derivatives.

159 → 160 53-74%

R^1 = aryl, alkenyl; R^2 = Me, CH_2Ph, C(O)R; R^3 = Bu^t, aryl

Reagents: R^3CHO, TBAF or CsF cat., THF

Fluoride ion induced Peterson reactions of the bis(trimethylsilyl)methyl derivatives **161**, in which the nitrogen atom is doubly bonded to a carbon atom, give the corresponding 2-aza-1,3-dienes **162** under mild conditions (Scheme 2.103) [289].

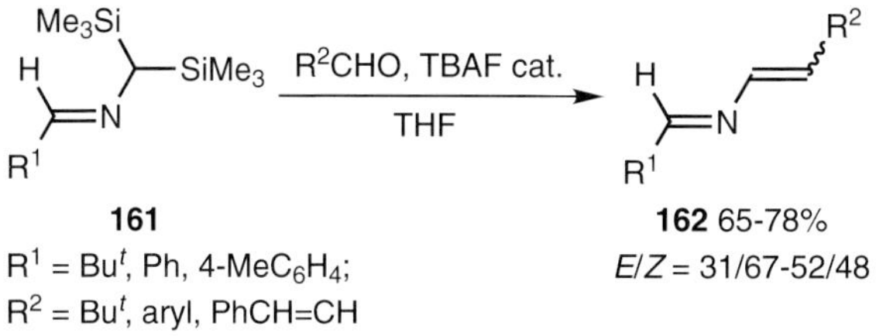

Scheme 2.103. Synthesis of 2-aza-1,3-dienes by fluoride ion induced Peterson reaction.

161 → 162 65-78%; E/Z = 31/67-52/48

R^1 = Bu^t, Ph, 4-MeC_6H_4; R^2 = Bu^t, aryl, PhCH=CH

Reagents: R^2CHO, TBAF cat., THF

Bis(trimethylsilyl)methyl isothiocyanate **163** also yields the corresponding N-styryl isothiocyanate **164**, together with 5-phenyl-4-trimethylsilyloxazolidine-2-thione **165** (Scheme 2.104) [290, 291].

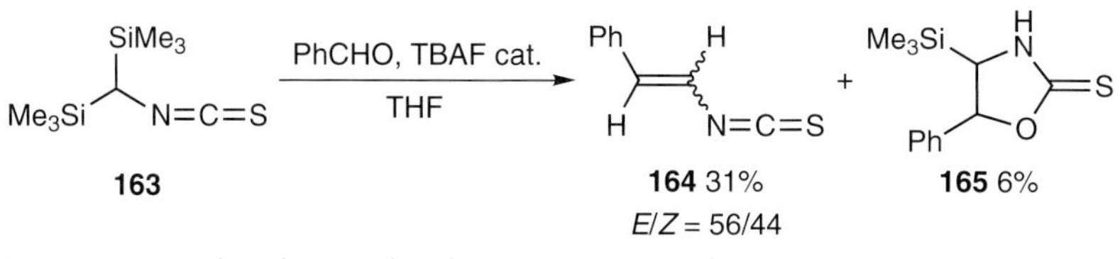

Scheme 2.104. Fluoride ion induced Peterson reaction of bis(trimethylsilyl)methyl isothiocyanate.

163 → 164 31% (E/Z = 56/44) + 165 6%

Reagents: PhCHO, TBAF cat., THF

This reaction is also used for substrates in which the nitrogen atom of the bis(trimethylsilyl)methylamine moiety forms part of a ring system. Fluoride ion catalyzed Peterson reactions of the bis(trimethylsilyl)methyl-substituted benzo-1,2,3-triazoles **166** with carbonyl compounds give the N-1-alkenyl derivatives **167** with low stereoselectivity (Scheme 2.105) [292, 293]. The reaction of a 1,2,4-triazole derivative leads to a similar result [294].

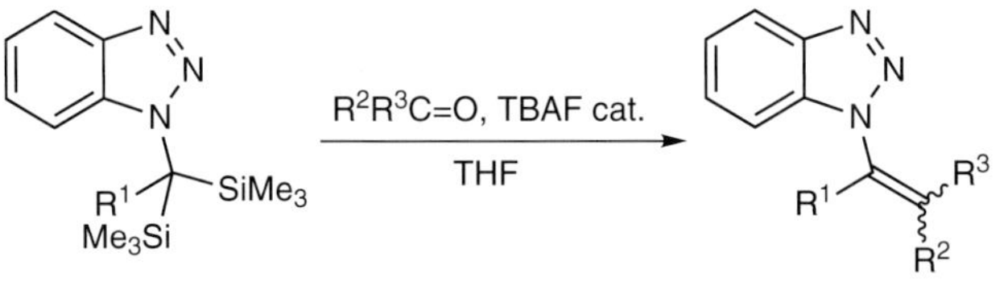

2.3 Generation of α-Silyl Carbanions and their Peterson Reactions

Scheme 2.105. Fluoride ion induced Peterson reaction of bis(trimethylsilyl)methyl-substituted benzo-1,2,3-triazoles.

166
R^1 = H, Me, CH$_2$Ph, SiMe$_3$;
R^2 = Me, PhCH=CH, aryl; R^3 = H, Me, Ph

167 23–98%

An intramolecular version of this reaction offers a facile synthesis of nitrogen-containing heterocycles such as the oxacephem skeleton **168** (Scheme 2.106) [295]. Attempted cyclization under Peterson reaction conditions with LDA or butyllithium proved unsuccessful.

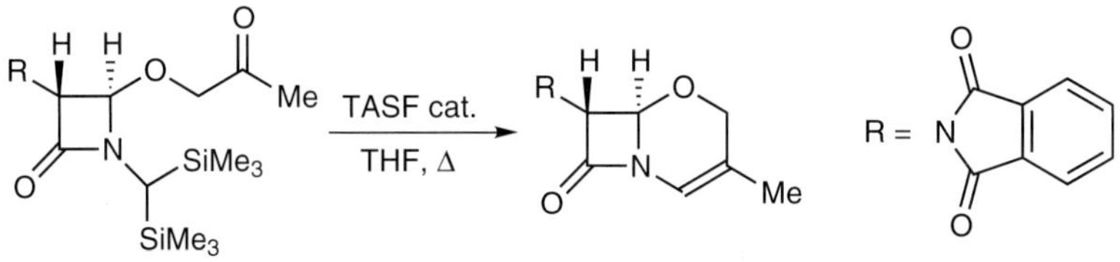

Scheme 2.106. Construction of the oxacephem skeleton by fluoride ion induced Peterson reaction.

168 50%

In other reports, stoichiometric amounts of TBAF or cesium fluoride have been used for the construction of 1(2H)-isoquinolones **169** (Scheme 2.107) [287, 296].

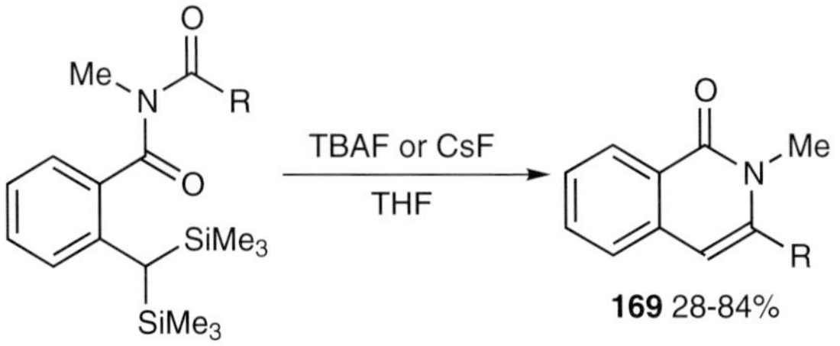

169 28–84%

R = H, aryl, heteroaryl, cyclohexyl, Pr^i

Scheme 2.107. Synthesis of 1(2H)-isoquinolones by fluoride ion induced Peterson reaction.

Application in the construction of an eight-membered ring has also been reported in the synthesis of the benzazocinone derivative **170** using cesium fluoride (Scheme 2.108) [288, 297].

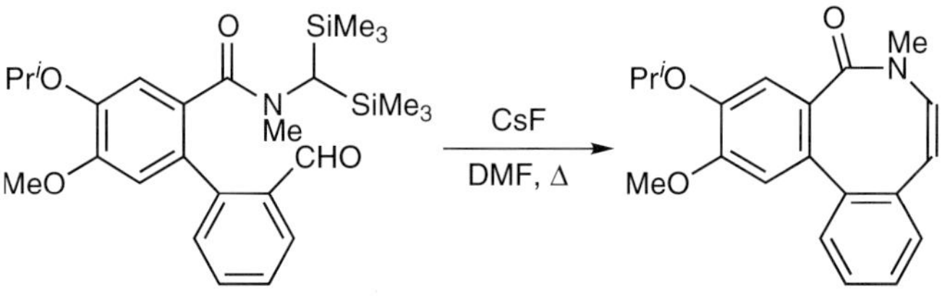

Scheme 2.108. Synthesis of a benzazocinone derivative by fluoride ion induced Peterson reaction.

2.3.5.4 Fluoride Ion Catalyzed Peterson-Type Reactions with Elimination of Trimethylsilanol

A new version of the Peterson reaction of trimethylsilylmethane derivatives bearing an electron-withdrawing group with several carbonyl compounds has been developed by taking advantage of fluoride ion catalysis. Addition of α-(trimethylsilyl)-acetates to non-enolizable aldehydes and further heating of the generated β-siloxyacetates with the same catalyst affords the corresponding α,β-unsaturated esters with high E-stereoselectivity (>98%) under formal elimination of trimethylsilanol (Scheme 2.109) [298–302]. This method is also applicable to the synthesis of α,β-unsaturated imines. A combination of anhydrous cesium fluoride and DMSO is important in this case. With enolizable aldehydes, the alkenes are obtained, accompanied by the self-condensation by-products. This methodology demonstrates that fluoride represents another choice of catalyst besides acid or base in the Peterson-type reactions, and offers high stereoselectivity.

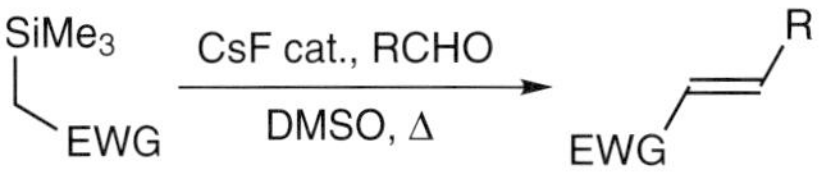

EWG = CO_2Et, CH=NBu^t, C(Me)=NBu^t; R = Ph, PhCH=CH, 2-furyl, C_7H_{15}

Scheme 2.109. Fluoride ion catalyzed Peterson-type reaction with elimination of trimethylsilanol.

An intramolecular version of this type of reaction of an α-silylacetate in DMF at 0 °C has been utilized for the synthesis of the lactone **171** (Scheme 2.110) [303].

2.3.6
Generation of α-Silyl Carbanions by Addition of Nucleophiles to Vinylsilanes

The addition of an alkyllithium to a vinylsilane preferentially leads to the α-silyl carbanion, which can react with a carbonyl compound to give the correspond-

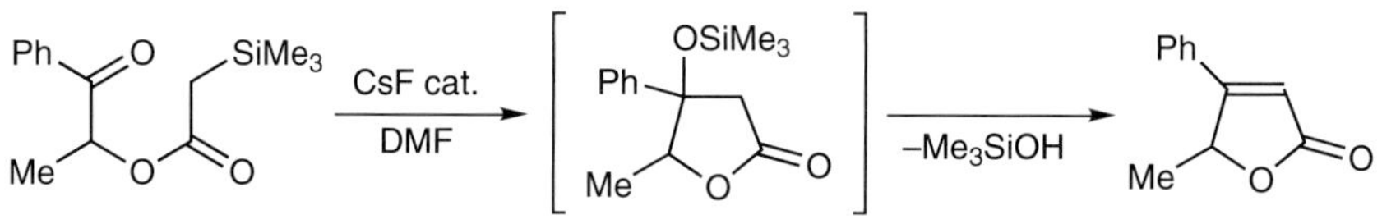

2.3 Generation of α-Silyl Carbanions and their Peterson Reactions

Scheme 2.110. An intramolecular version of fluoride ion catalyzed Peterson-type reaction with elimination of trimethylsilanol.

ing alkene [11, 304]. An α-silyl carbanion is formed regioselectively since the β-carbon is less sterically encumbered and the resulting carbanion is stabilized by an α-silyl group. For example, treatment of trimethylvinylsilane with ethyllithium affords the α-silylalkyllithium **172**, reaction of which with butanal gives the β-hydroxyalkylsilane **173** (Scheme 2.111). Peterson reaction of the alcohol yields 4-octene as a mixture of the *E*- and *Z*-isomers.

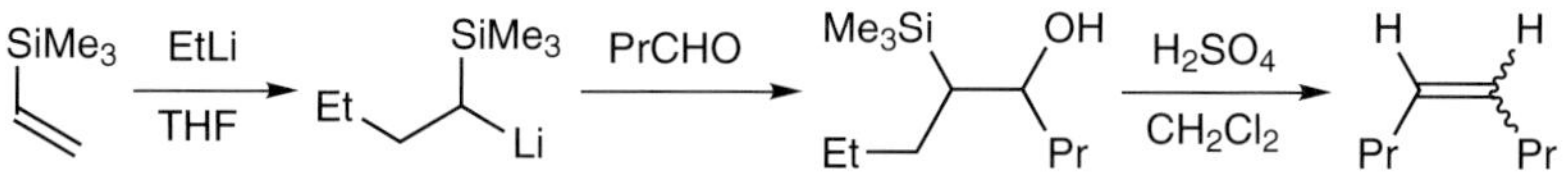

Scheme 2.111. Generation of an α-silyl carbanion by addition of ethyllithium to a vinylsilane.

This methodology has been applied to the synthesis of alkenyl sulfides. Addition of alkyllithiums to phenyl 1-trimethylsilylethenyl sulfide followed by addition of carbonyl compounds affords the corresponding functionalized alkenyl sulfides **174** (Scheme 2.112) [243, 305]. Alkene formation occurs even with enolizable ketones.

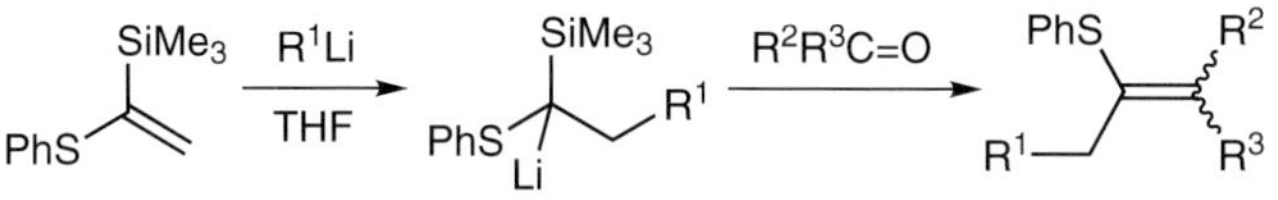

R^1 = Et, Pen, Bn; R^2, R^3 = H, alkyl, Ph

Scheme 2.112. Generation of α-silyl carbanions by addition of alkyllithiums to a 1-trimethylsilylethenyl sulfide.

Conjugate addition reactions of Grignard reagents or organolithium reagents to the 1-amidoethenylsilane **175** afford the silyl dianions **176**, which are found to readily undergo Peterson reaction with pentanal (Scheme 2.113) [306]. The intermediary β-hydroxyalkylsilanes are not isolated. Yields are found to be better with lithium salts than with magnesium salts. Mixtures of stereoisomeric enones **177** are obtained with the *E*-isomer predominating in most cases.

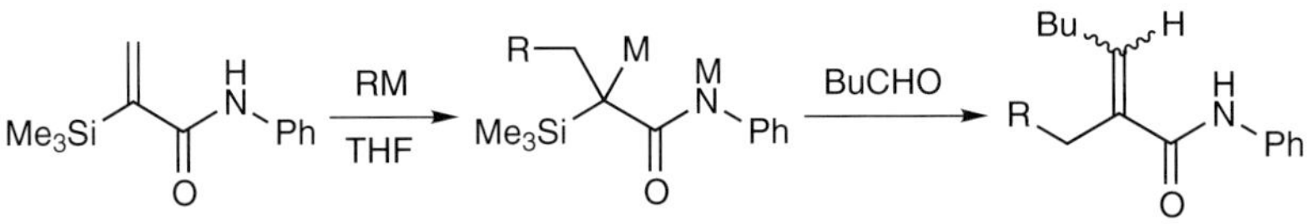

175 **176** **177** 57-77%

M = Li, MgBr E/Z = 50/50-75/25

Scheme 2.113. The Peterson reaction of silyl dianions generated by conjugate addition of organometallic reagents to a 1-amidoethenylsilane.

1-Silylethenyl ketones undergo smooth Michael addition with Grignard reagents or alkyllithiums to give the enolates **178**, which are then trapped with benzaldehyde to afford the *E*- and *Z*-isomers of enones **179** after the Peterson reaction (Scheme 2.114) [307]. (*E*)-Alkenes become the major products when the condensation with benzaldehyde is carried out under thermodynamic control at room temperature in diethyl ether, while *Z*-isomers are more favored as kinetically controlled products at −78 °C in THF. 1-Silyl acrylates show reactivities similar to those of 1-silylethenyl ketones [308].

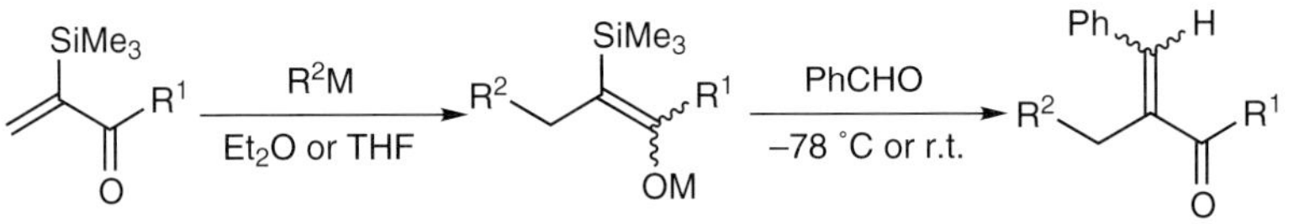

178 **179** 40-77%

R^1 = Me, OMe; R^2 = Me, Pr^i, Bu^t, Ph; M = MgBr, Li E/Z = 15/85-100/0

Scheme 2.114. The Peterson reaction of enolates generated by conjugate addition of alkyllithiums to 1-silylethenyl ketones.

2.4 Synthetic Methods for *β*-Silyl Alkoxides and *β*-Hydroxyalkylsilanes

2.4.1 Reactions of *α*-Silyl Ketones, Esters, and Carboxylic Acids with Nucleophiles

β-Hydroxyalkylsilanes can be prepared by nucleophilic attack of various nucleophiles on the carbonyl carbon of α-silyl aldehydes, ketones, or esters followed by aqueous work-up [199, 245, 309–314]. Further elimination of a silyloxide moiety gives the corresponding alkenes. The reactions of α-(*tert*-butyldiphenylsilyl)alkyl aldehydes **180** with organolithium reagents or Grignard reagents stereoselectively afford the *β*-hydroxyalkylsilanes **181** in addition reactions involving transition states conforming to Cram's or Felkin–Anh's rules (Scheme 2.115) [315, 316]. The addition takes place with a high degree of diastereoselectivity to form almost ex-

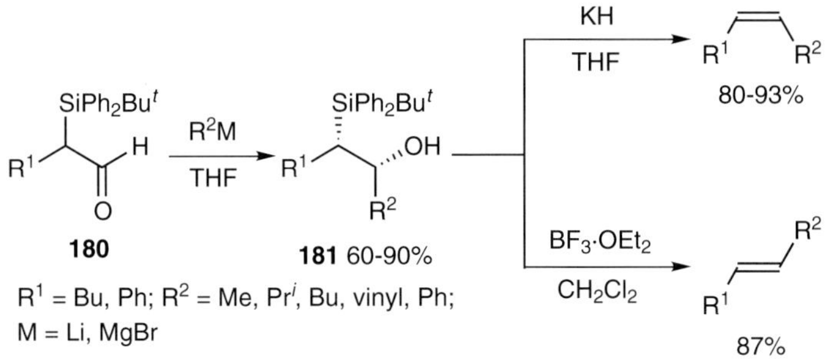

Scheme 2.115. The Peterson elimination of β-hydroxyalkylsilanes generated by addition of nucleophiles to α-silyl aldehydes.

clusively *erythro*-β-hydroxyalkylsilanes **181**. Facile β-elimination selectively affords the corresponding (Z)- or (E)-alkenes under standard basic (KH) or acidic (BF$_3$) conditions. These processes can also be carried out as one-pot reactions [317, 318].

Stereoselective reduction of α-silyl ketones such as **182** with metal hydride reagents such as DIBAH, LiAlH$_4$, NaBH$_4$, and L-Selectride has been reported to give the corresponding β-hydroxyalkylsilanes **183** (Scheme 2.116) [11, 304, 316, 319]. Peterson reaction of the 2-silyl-1,3-diol **183** with KH gives a mixture of the corresponding alkene **184** and its isomer **185**.

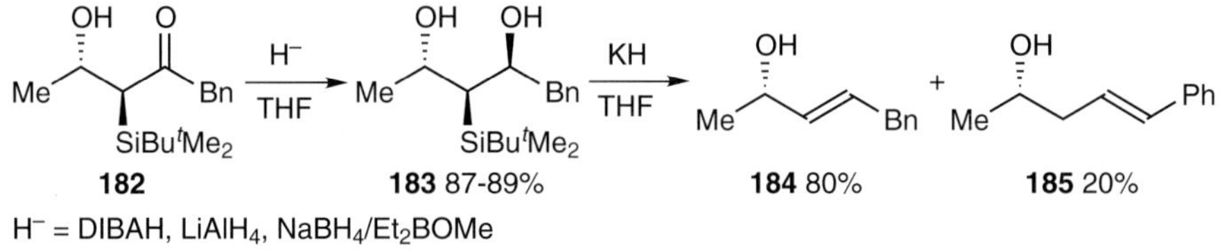

Scheme 2.116. The Peterson elimination of a β-hydroxyalkylsilane generated by hydride reduction of α-silyl ketones.

α-Methyldiphenylsilyl esters **186** are reacted with organomagnesium halides or a sequential combination of an organomagnesium halide and an organolithium reagent followed by elimination to give tri- and tetrasubstituted alkenes (Scheme 2.117) [320, 321]. The α-silyl esters are therefore synthetically equivalent to vinyl dications. The yield depends on the steric bulk of substituents on the ester **186** and on the organometallic reagents.

Terminal alkenes are prepared by the Peterson elimination of β-hydroxyalkylsilanes having no further substituents on the carbon atom bearing the hydroxy group. The β-hydroxyalkylsilane **188** is prepared from a β-silyl carboxylic acid **187** by reduction with LAH, and further mesylation initiates the Peterson elimination affording the terminal alkene **189** (Scheme 2.118) [322].

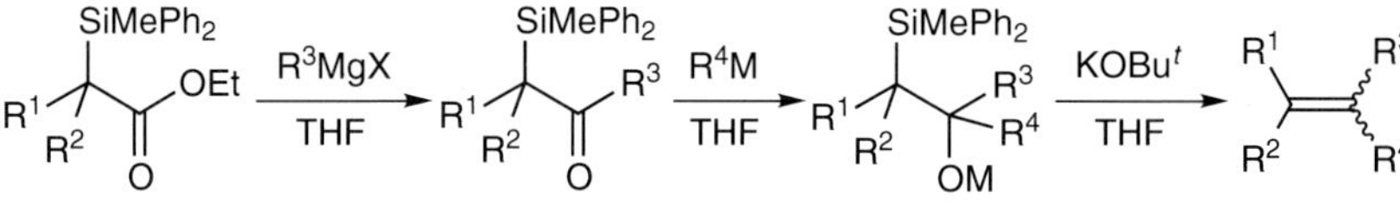

Scheme 2.117. The Peterson elimination of β-silyl alkoxides generated by attack of nucleophiles on α-silyl esters.

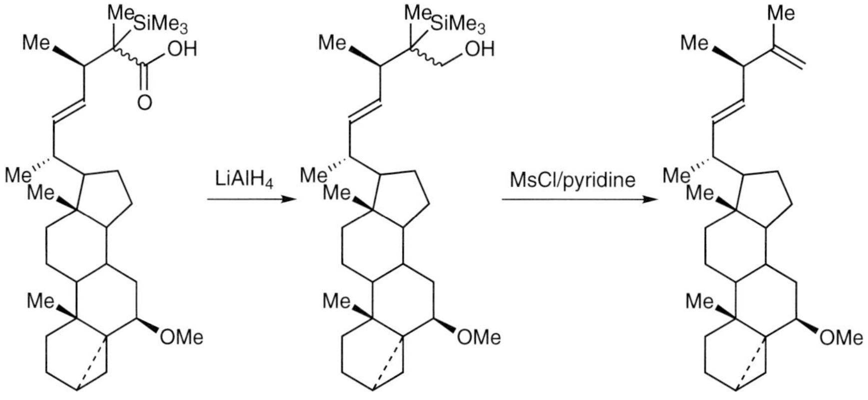

R^1 = H, Me; R^2 = Me, Et, C_8H_{17}, CH_2=CH(CH_2)$_7$
R^3, R^4 = alkyl, Ph; M = MgX, Li

Scheme 2.118. The Peterson elimination of a β-hydroxyalkylsilane generated by reduction of an α-silyl carboxylic acid.

2.4.2
Ring-Opening Reactions

2.4.2.1 Ring-Opening Reactions of Oxiranes

β-Silyl alkoxide anions can be prepared by the ring opening of two types of oxirane, namely 2-silyloxiranes and oxiranes without a silyl group. Ring opening of the former with carbon nucleophiles leads to β-silyl oxide anions suitable for the Peterson reaction, while that of the latter with silyl nucleophiles also provides the corresponding oxide anions (Scheme 2.119).

A variety of α,β-epoxy silanes react with organocopper or organolithium reagents in a regio- and stereospecific manner at the carbon adjacent to silicon to form diastereomerically pure β-hydroxyalkylsilanes [20, 323–327]. Further stereospecific β-elimination with a base or an acid affords the corresponding alkenes. Similarly, treatment of α,β-epoxy silanes with various nucleophiles such as organosodium reagents, Grignard reagents, organoaluminum reagents, organocerium reagents, and amines causes α-ring opening of the epoxide to give the corresponding oxide anions regioselectively and stereoselectively [36, 328–335]. The reasons for the

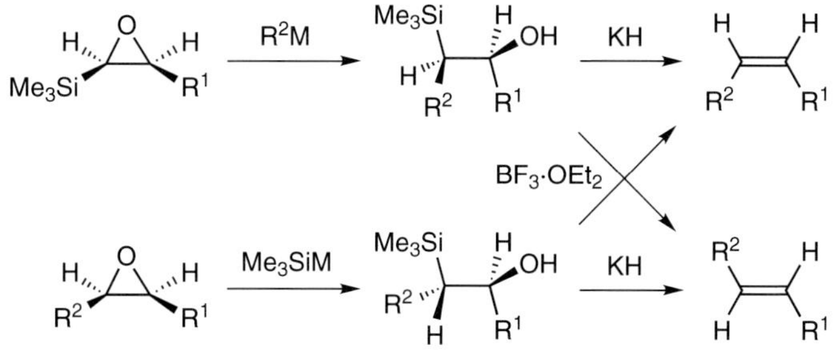

2.4 Synthetic Methods for β-Silyl Alkoxides and β-Hydroxyalkylsilanes

[Scheme 2.119 diagram]

R^1 = alkyl; R^2 = alkyl, alkenyl, aryl, alkynyl, PhS, Bu$_3$Sn, N$_3$, H
Metal reagents = R$_2$CuLi, R$_2$Cu(CN)Li$_2$, RLi, RNa, RMgBr, RCeCl$_2$, LiAlH$_4$, Et$_2$AlCN, R$_2$NH/alumina

Scheme 2.119. The Peterson elimination of β-hydroxyalkylsilanes generated by ring opening of oxiranes.

preference for α-opening of the α,β-epoxy silanes have not been immediately obvious [336].

Ring-opening of α,β-epoxysilane **190** to give the halohydrins **191** and subsequent Peterson reaction furnishing an alkene bearing a halogen atom such as **192** has also been reported (Scheme 2.120) [337–340].

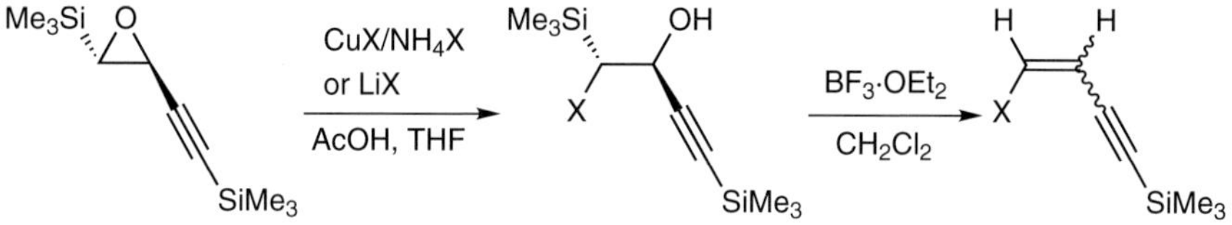

[Scheme 2.120 diagram]

190 → **191** 64-81%, X = Cl, Br, I → **192** 72-80%, E/Z = 1/99-11/89

Scheme 2.120. The Peterson elimination of β-hydroxyalkylsilanes generated by ring opening of α,β-epoxysilanes.

Treatment of an oxirane bearing no silyl group with a silyl nucleophile leads to the same result. Both trimethylsilylpotassium and dimethylphenylsilyllithium effect smooth conversion of oxiranes into alkenes with inversion of configuration, nucleophilic ring opening being followed by spontaneous *syn*-β-elimination (Scheme 2.121) [341, 342].

2.4.2.2 Ring-Opening Reactions of Cyclic Esters and Ethers

Similarly to the ring opening of oxiranes, the nucleophilic attack that accompanies the ring opening of lactones and cyclic carbonates bearing a silyl group induces Peterson reactions [343]. Fluoride ion mediated ring opening of the lactone **193** with TBAF gives the corresponding β,γ-unsaturated carboxylate **194**, which produces an unsaturated δ-lactone **195** on acidification (Scheme 2.122) [344]. On

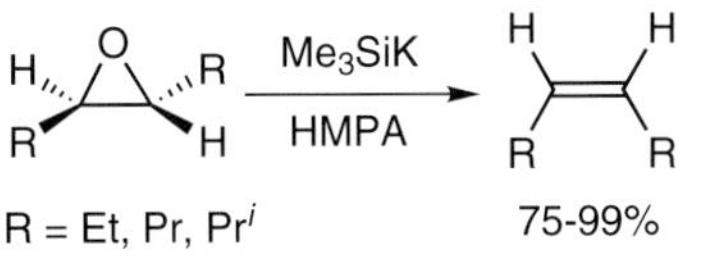

Scheme 2.121. The Peterson elimination of β-silyl alkoxides generated by ring opening of oxiranes with trimethylsilylpotassium.

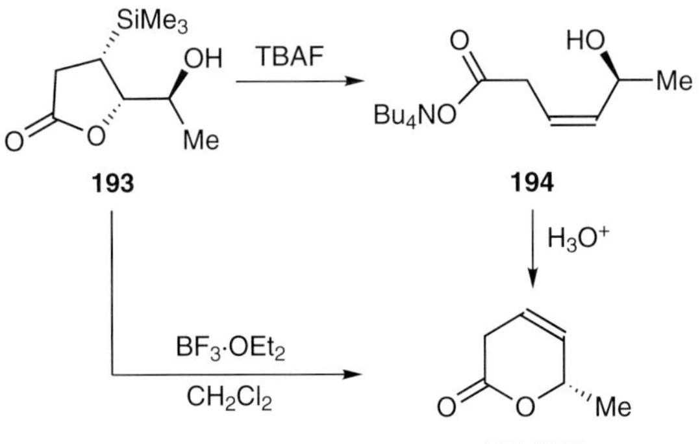

Scheme 2.122. The Peterson elimination accompanying ring opening of a lactone.

treatment with $BF_3 \cdot OEt_2$, the γ-lactone **193** gives **195** directly. Such ring-opening/elimination processes are also observed in the reaction of the carbonate **196** with TBAF (Scheme 2.123) [331].

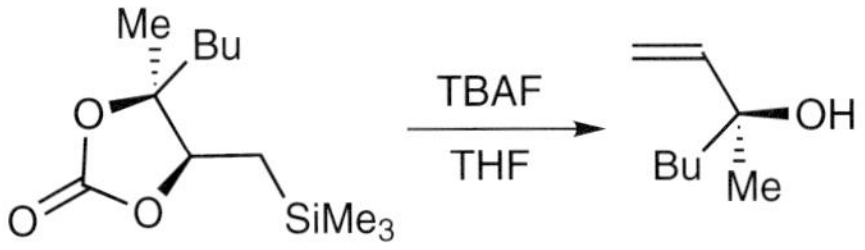

Scheme 2.123. The Peterson elimination accompanying ring opening of a cyclic carbonate.

Treatment of the cyclic siloxane **197** with BuLi results in transformation to the corresponding alkene **198** through nucleophilic cleavage of the Si–O bonds of the cyclic siloxane followed by Peterson-type *syn*-elimination (Scheme 2.124) [345].

2.4.3
Hydroboration of 1-Silylallenes

Hydroboration of allenes has long been known to give the corresponding allylboranes, which are useful for the preparation of allyl alcohols [346–348]. When 1-

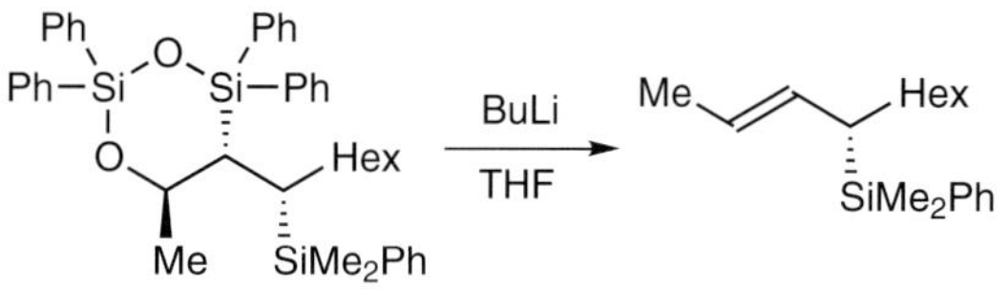

Scheme 2.124. The Peterson elimination accompanying ring opening of a cyclic siloxane.

trimethylsilylallenes are used as the starting materials, their hydroboration and the subsequent allylation of carbonyl compounds provides the α-(1-alkenyl)-β-boryloxy-alkylsilanes, which are good precursors for Peterson reactions.

Hydroboration of 1-trimethylsilylallenes with 9-borabicyclo[3.3.1]nonane (9-BBN) affords the corresponding γ-silyl allylboranes **199**, which have predominantly the E-geometry. Subsequent condensation of **199** with aldehydes proceeds to give the corresponding 1,3-butadienes with high diastereoselectivity after work-up (Scheme 2.125) [349–351]. This method has been applied to the conversion of 1-alkoxy- or 1-phenylthio-1-trimethylsilylallenes to the corresponding 1,3-dienes in both an E- and a Z-stereoselective manner, while the use of titanium reagents allows Z-selectivity or only low E-selectivity [352, 353].

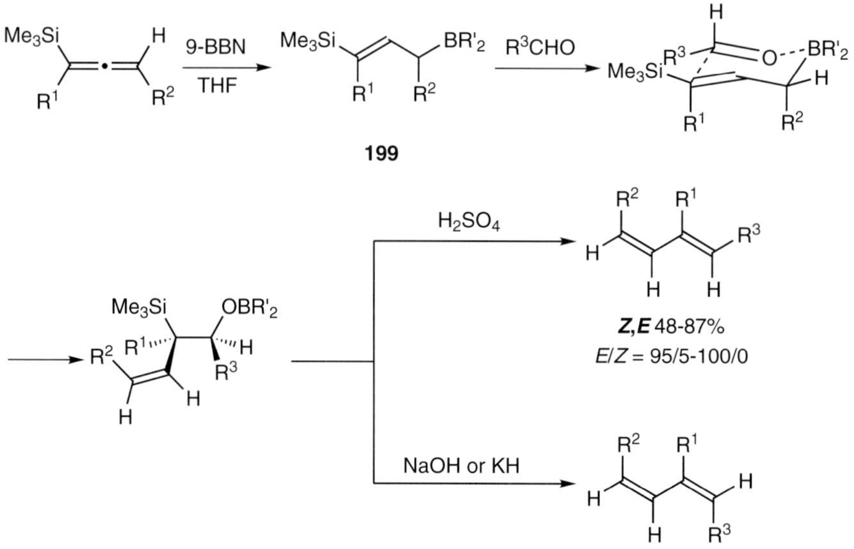

R^1 = alkyl, OR, SPh, OC(O)NPr$^i_2$; R^2 = H, alkyl; R^3 = alkyl, alkenyl, aryl, heteroaryl, dienyl
BR'$_2$ = borabicyclo[3.3.1]nonyl

Z,E 48-87%
E/Z = 95/5-100/0

Z,Z 52-92%
E/Z = 0/100-3/97

Scheme 2.125. Hydroboration of 1-silylallenes with 9-BBN followed by the Peterson reaction.

The geometry of the other double bond of the diene product, originally between the 2- and 3-positions of the 1-trimethylsilylallenes, can be controlled by selecting either 9-BBN or dicyclohexylborane as the hydroborating reagent (Scheme 2.126). Consequently, all four geometric isomers of several representative internal 1,3-dienes can be synthesized with high isomeric purity by utilizing different combinations of the hydroborating reagents and work-up conditions [354, 355].

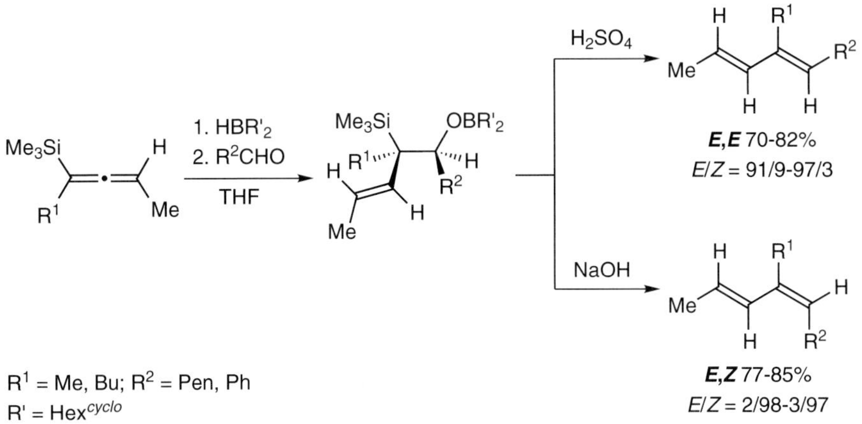

R^1 = Me, Bu; R^2 = Pen, Ph
R' = Hex^{cyclo}

Scheme 2.126. Hydroboration of 1-silylallenes with dicyclohexylborane followed by the Peterson reaction.

2.4.4
Dihydroxylation of Vinylsilanes and Allylsilanes

Transformation of an alkenylsilane to a 1,2-dihydroxyalkylsilane, followed by the Peterson reaction after conversion of its α-hydroxy group to methoxy group, provides a useful method for the stereoselective synthesis of an alkene. 1-Methoxy-*trans*-cyclooctene **201** is prepared by *syn*-elimination from β-hydroxyalkylsilane **200**, which is derived from *syn*-dihydroxylation of 1-trimethylsilyl-*cis*-cyclooctene, although the product **201** is thermally unstable and easily isomerizes to the *cis*-cyclooctene (Scheme 2.127, for NMO, see Scheme 2.27) [356].

Chiral secondary allylic alcohols **202** are synthesized from allylsilanes and vinylsilanes by way of successive asymmetric dihydroxylation (AD) and Peterson elimination; although the Peterson reaction process does not constitute asymmetric induction, it results in the loss of an asymmetric point (Scheme 2.128) [357]. The allylsilane undergoes efficient dihydroxylation using AD-mix-α, AD-mix-β, or the modified Sharpless AD reagents, although enantiomeric excess is dependent on the structure of the alkene [358, 359].

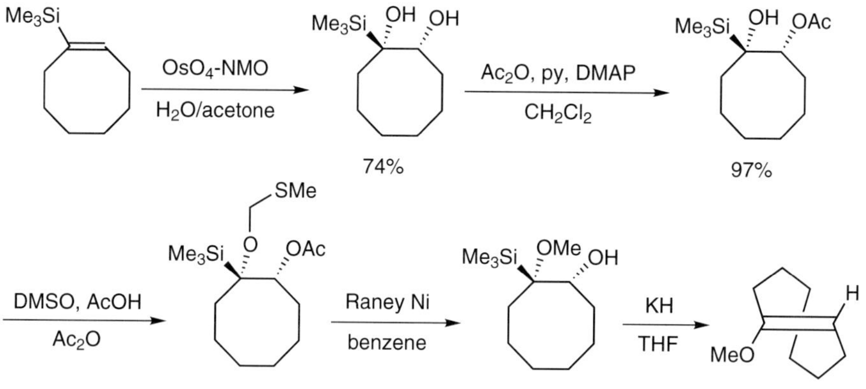

72% **200** 79% **201** 91%

Scheme 2.127. Synthesis of 1-methoxy-*trans*-cyclooctene by Peterson elimination.

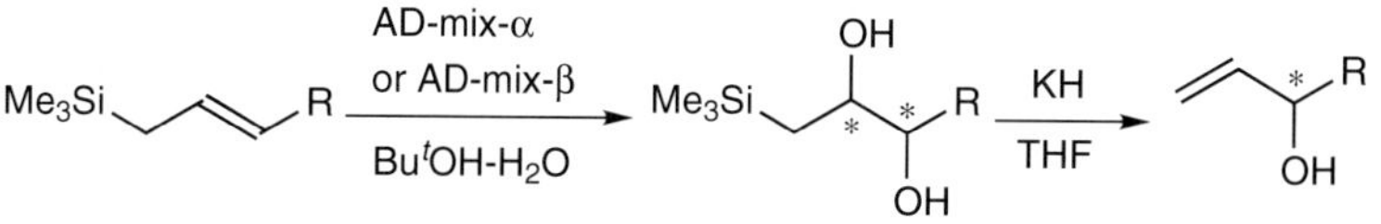

R = Pr^i, Pen, Hex^{cyclo}, Ph 46-98% **202** 77-85% 62-96% ee

Scheme 2.128. Synthesis of chiral allyl alcohols by Peterson elimination.

2.5
Related Reactions

2.5.1
The Homo-Brook Rearrangement

The Peterson reaction under basic conditions necessitates 1,3-migration of a silyl group from carbon to oxygen. Instead of the subsequent elimination of a siloxide anion to furnish an alkene, protonation of the carbanion gives the silyl ether. When the silyl ether is hydrolyzed, protodesilylation (or protiodesilylation) occurs and the silicon is replaced by hydrogen [13, 14] (see Section 2.2.1.1). This type of reaction has been called the homo-Brook rearrangement in analogy with the Brook rearrangement.

Treatment of α-methoxy-β-hydroxyalkylsilanes **203a** and **203b** with $KOBu^t$ in aqueous DMSO yields the methoxy alcohols **204a** and **204b**, respectively, with high isomeric purity (Scheme 2.129) [20]. The protodesilylation involved in the homo-Brook rearrangement takes place with completely stereospecific retention of configuration at carbon.

The homo-Brook rearrangement readily occurs when 1-(1-hydroxyalkyl)-1-alkenylsilanes **205** are treated with NaH in HMPA (Scheme 2.130). The re-

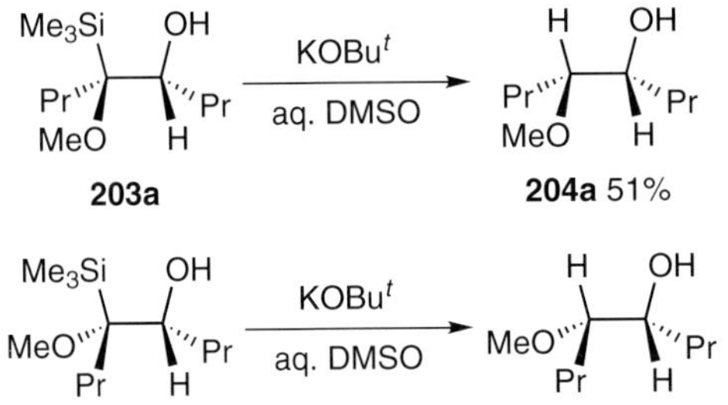

2 The Peterson and Related Reactions

Scheme 2.129. Protodesilylation of β-hydroxyalkylsilanes.

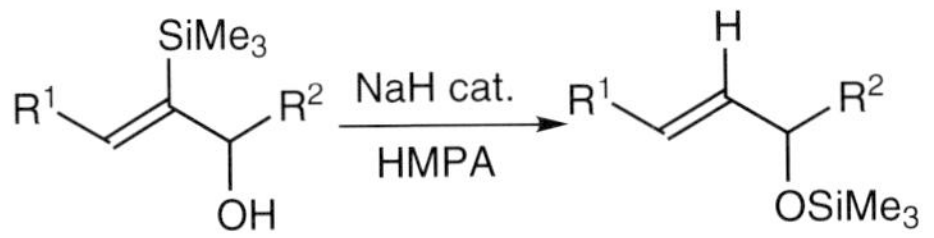

Scheme 2.130. The homo-Brook rearrangement of β-hydroxyalkylsilanes.

R^1 = H, Bu^t; R^2 = H, alkyl, alkenyl, aryl

arrangement involves migration of silicon from an sp² carbon to oxygen [360, 361].

The protodesilylation should take place via a postulated 1,2-oxasiletanide transition state, which leads to an alkene in the absence of any proton source. Protonation of the isolated pentacoordinate 1,2-oxasiletanides **7b** actually gives the desilylated alcohol **206** after hydrolysis in competition with the Peterson reaction (Scheme 2.131) [25, 28] (see Section 2.2.1.2). The decisive factors with regard to the

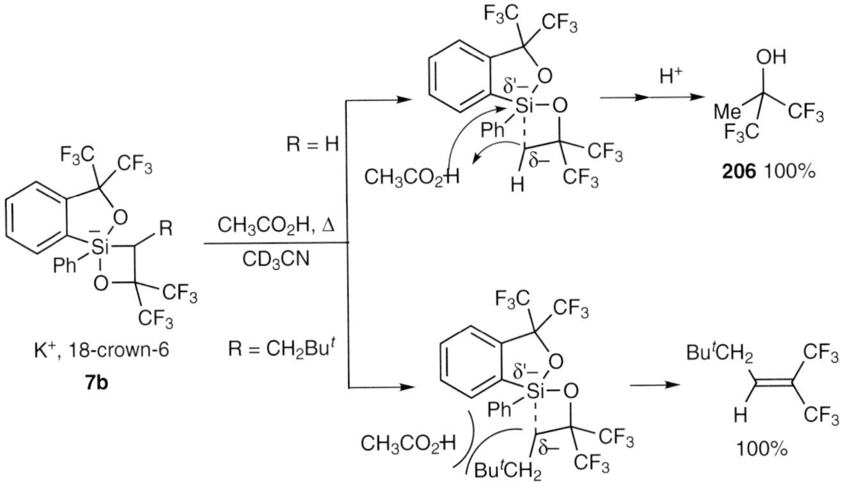

Scheme 2.131. Two reaction modes of 1,2-oxasiletanides.

two reaction modes of the four-membered ring silicate **7b**, the Peterson reaction and the homo-Brook rearrangement, can be identified as the acidity of the proton source and the steric hindrance of the substituent at the 3-position of the 1,2-oxasiletanide **7b**. These factors control the protonation at the α-carbon atom. Similar arguments have been presented in relation to a pentacoordinate spirobis [1,2-oxasiletanide] [26, 362].

Treatment of the resulting carbanion with various electrophiles instead of protonation affords the corresponding adducts [363–365]. Successive addition of the α-silyldibromomethyllithium **207** and HMPA to esters causes the homo-Brook rearrangement providing carbanions of the alkyl silyl mixed acetals **208** (Scheme 2.132). Quenching with various electrophiles effectively yields the corresponding alkyl silyl acetals **209**.

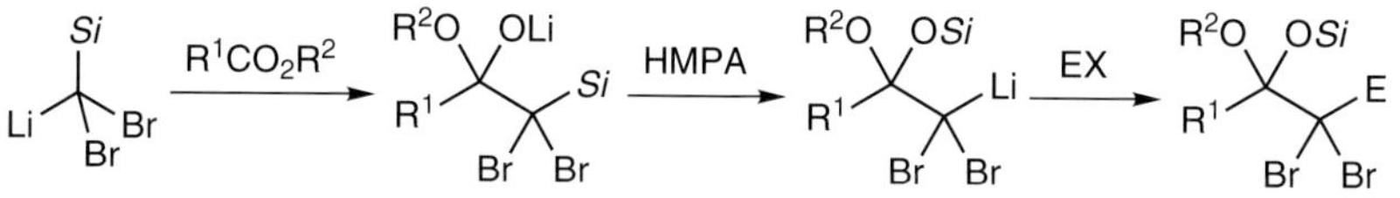

Si = Bu^{t}Me$_2$Si; R^1 = H, Ph; R^2 = Et, Pr^i
EX = MeOH, MeI, BnBr, CH$_2$=CHCH$_2$Br

Scheme 2.132. Homo-Brook rearrangement of β-silyl alkoxides.

2.5.2
Homo-Peterson Reaction

There have been a few reports on the homo-Peterson reaction, that is the reaction of an α-silyl carbanion with an oxirane to afford the corresponding cyclopropane. The reaction of tris(trimethylsilyl)methyllithium with styrene oxide forms the corresponding cyclopropane **210** (Scheme 2.133) [366]. Appropriate selection of the oxirane is important for this reaction. Both propene oxide and cyclohexene oxide protonate the carbanion, while stilbene oxide is inert towards it.

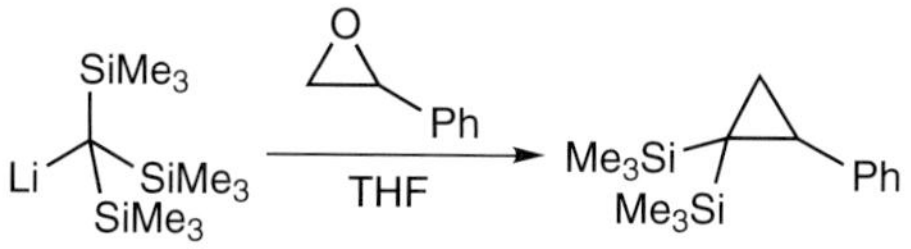

Scheme 2.133. Homo-Peterson reaction of tris(trimethylsilyl)methyllithium with styrene oxide.

Carbanions **211** with arylthio or alkylthio substituents at the attacking carbon are used for homo-Peterson reactions (Scheme 2.134) [367]. The intermediary alkoxide **212** can transfer a silyl group from the carbon to oxygen, and subsequent attack of the carbanion **213** gives the cyclopropanes **214**.

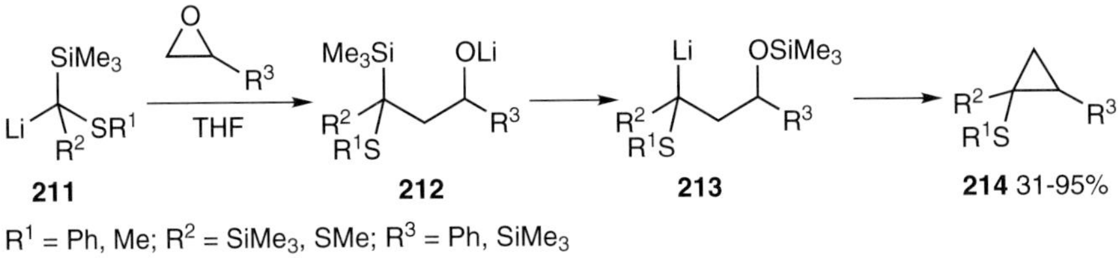

Scheme 2.134. Homo-Peterson reaction of α-silyl alkyllithiums with epoxides.

R^1 = Ph, Me; R^2 = SiMe$_3$, SMe; R^3 = Ph, SiMe$_3$

2.5.3
Vinylogous Peterson Olefination

The vinylogous Peterson reaction, a vinylogous counterpart of the Peterson reaction of 4-hydroxy-2-alkenylsilanes 215 with KH or SnCl$_4$ giving the corresponding dienes, has been reported (Scheme 2.135). When the terminal double bond is not subject to geometrical isomerism, the single isomer with E-selectivity in the formation of the internal double bond is obtained [368].

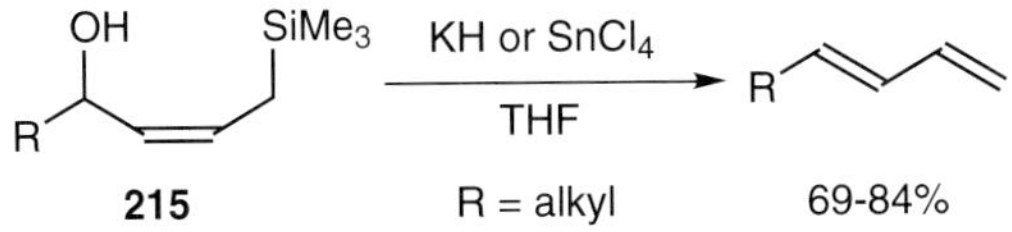

Scheme 2.135. Synthesis of terminal dienes by the vinylogous Peterson reaction.

Further study on the stereochemistry of a base-catalyzed vinylogous Peterson reaction showed that *syn*-elimination occurs and that the *cis* double bond is constructed selectively at the positions adjacent to the carbon atom that originally bear the hydroxy group (Scheme 2.136) [369, 370].

Vinylogous Peterson reactions of the E-isomers of 4-hydroxy-2-alkenylsilanes 216 with a catalytic amount of aqueous HCl give the corresponding 1,3-dienes with high E-selectivity (Scheme 2.137) [371].

Similarly, 1,4-elimination from 4-hydroxy-2-butynylsilanes 217 induced by fluoride ion gives the corresponding 1,2,3-butatrienes (Scheme 2.138) [372, 373]. The hydroxy group should be converted into a good leaving group such as acetate or mesylate before elimination.

The benzyl analogue of the vinylogous Peterson reaction is called the benzo-Peterson reaction [374]. Reaction of the o-(hydroxymethyl)benzylsilane 218 with an excess of maleic anhydride in refluxing toluene yields the Diels–Alder cycloadduct 219 with *endo* selectivity (Scheme 2.139) [375, 376]. It is assumed that the reaction proceeds with formation of the benzyl cation 220, conversion to the o-quinodimethane 221 by the action of water, and cycloaddition of the latter with maleic anhydride in a stereoselective fashion. The initial steps are regarded as an acid-catalyzed benzyl analogue of the vinylogous Peterson reaction. It is noteworthy that the elimination of the silyl group is carried out under weakly acidic conditions.

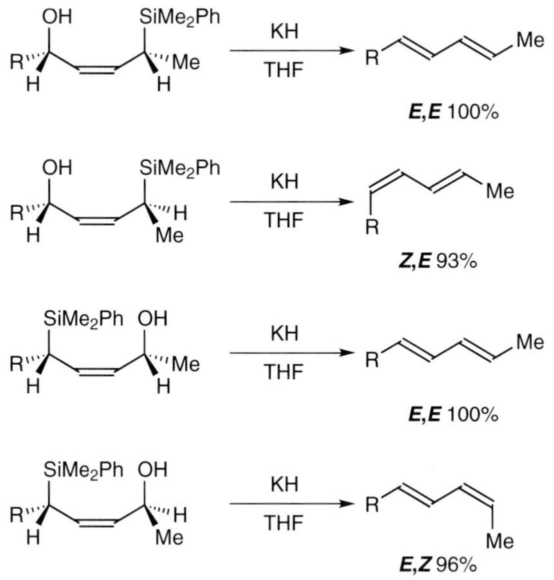

R = Hex^{cyclo}

Scheme 2.136. Stereochemistry of the vinylogous Peterson reaction.

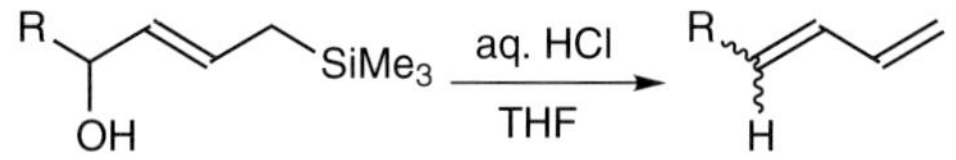

216 quant.

R = alkyl, aryl $E/Z = 91/9$

Scheme 2.137. The vinylogous Peterson reaction under acidic conditions.

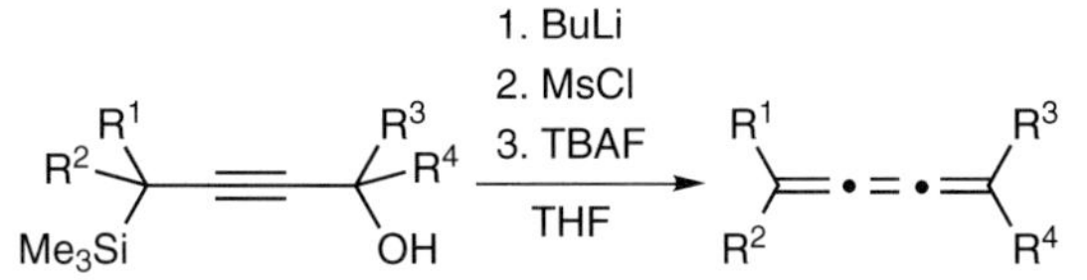

217 52-84%

R^1, R^2, R^3, R^4 = H, Me, Bu, Ph

Scheme 2.138. Synthesis of 1,2,3-butatrienes by fluoride ion induced 1,4-elimination.

2.5.4
Tandem Reactions and One-Pot Processes Involving the Peterson Reaction

There are two types of tandem reactions; one involves reaction to generate a β-silyl alkoxide followed by Peterson elimination to give the corresponding alkene, and the other involves an initial Peterson reaction followed by addition to the generated

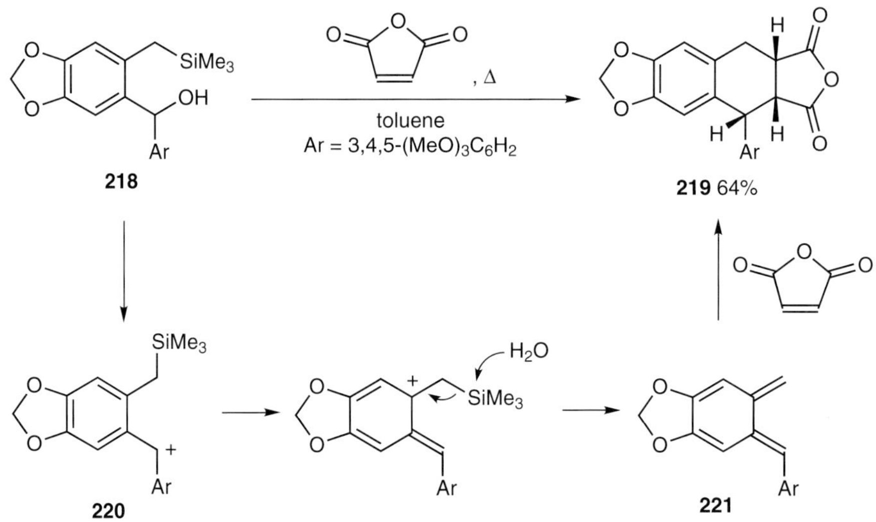

Scheme 2.139. The benzo-Peterson reaction forming an o-quinodimethane.

alkene. In both cases, once the precursor has been appropriately designed, each reaction successively triggers the next one.

A consecutive reaction generating a β-silyl alkoxide causes the tandem reaction involving Peterson elimination. Treatment of 7-oxabicyclo[2.2.1]heptenes **222** with PhMe$_2$SiLi or PhMe$_2$SiCu·LiCN gives the dienes **223** (Scheme 2.140). The proposed mechanism for conversion to the diene involves a silametallation of the highly strained alkene **222** from the *exo* face, followed by ring opening to the β-silyl alkoxide **224** and its Peterson elimination. The former addition–ring-opening processes facilitate the latter Peterson elimination [377].

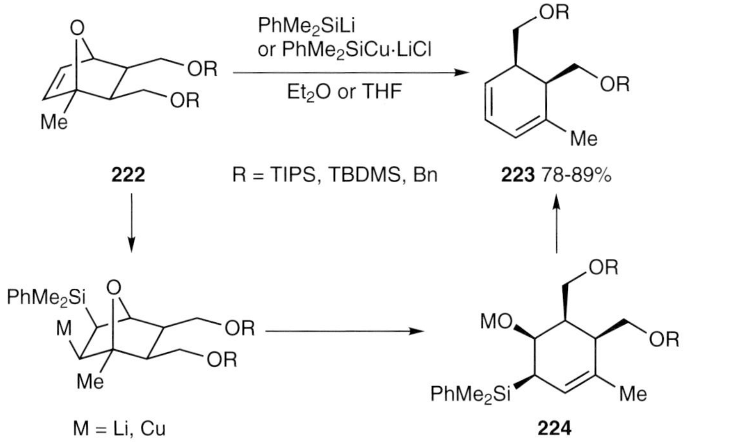

Scheme 2.140. Tandem reaction involving silametallation and the Peterson reaction.

The [2,3]-Wittig rearrangement of bis-allylic ethers such as **225** with lithium dicyclohexylamide or butyllithium followed by Peterson elimination results in the formation of an E/Z mixture of trienes (Scheme 2.141) [378, 379].

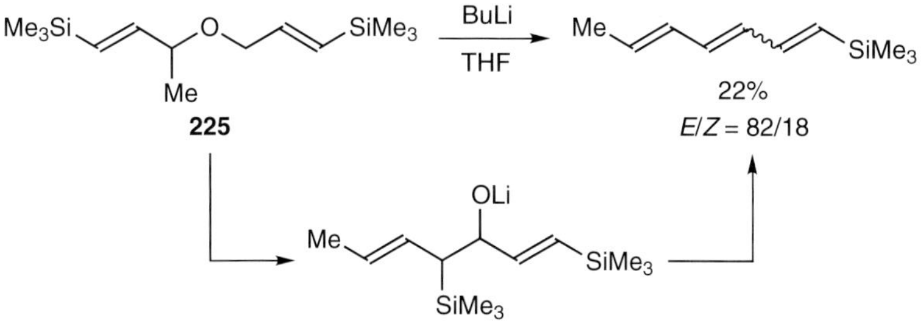

Scheme 2.141. Tandem reaction involving [2,3]-Wittig rearrangement and the Peterson reaction.

A tandem reaction comprising of an initial Peterson reaction and subsequent cycloaddition has been reported. Treatment of the disilylated β-hydroxyacetals **226** with furans **227** in the presence of TiCl$_4$ gives rise to [4+3]-cycloadducts **228** of the furans and the silylated allyl cations. The latter are obtained in moderate yields as single stereoisomers by Peterson reaction and elimination of the alkoxide, followed by desilylation (Scheme 2.142) [380].

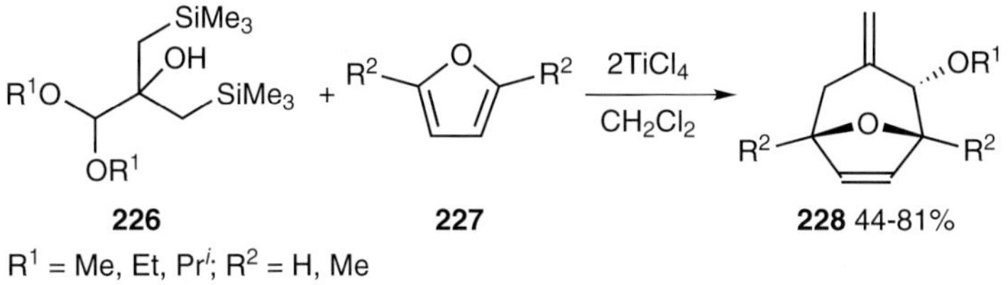

Scheme 2.142. Tandem reaction involving the Peterson reaction, [4+3]-cycloaddition, and desilylation.

Another example is a tandem Peterson–Michael reaction using α-silylalkylphosphine chalcogenides. The sequential reactions of α-silylalkylphosphine oxides and sulfides **229** with BuLi and the lithio derivative of 2-hydroxytetrahydropyran give the corresponding α-(2-tetrahydropyranyl)alkylphosphine chalcogenides **230** (Scheme 2.143). The first step is a Peterson reaction of the α-phosphonyl-α-silyl carbanion with acyclic aldehyde **231**, which is in equilibrium with the lithium derivative of 2-hydroxytetrahydropyran. The second step is an intramolecular Michael addition of oxide anion **232** to the ethenylphosphine chalcogenide moiety to afford the α-lithio derivative **233**, which is easily quenched to give the final products [381]. A one-pot reaction involving consecutive Michael addition, aldol reaction, and Peterson reaction processes has also been reported (Scheme 2.114) [382].

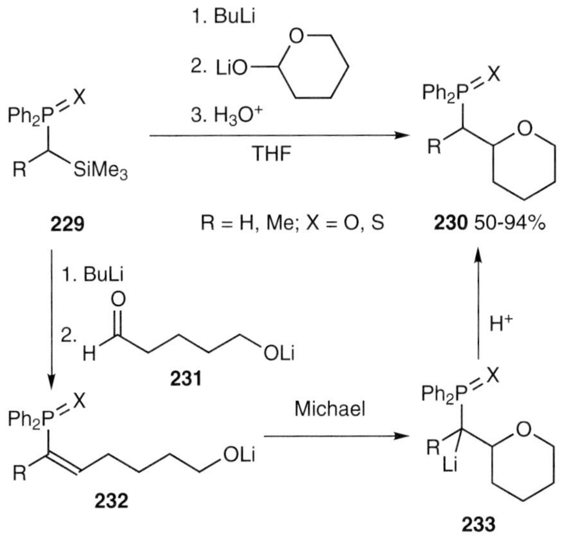

Scheme 2.143. Tandem reaction involving the Peterson reaction and the Michael reaction.

When the reaction conditions employed in all the steps are the same as those for the Peterson reaction, multi-step processes involving the Peterson reaction can be integrated in a one-pot process. Successive treatment of allyl sulfones **234** with BuLi, Me$_3$SiCl, BuLi, aldehydes, and lithium hexamethyldisilazide (LiHMDS) gives the enynes in good to moderate overall yields for the four-step reactions (silylation, carbon–carbon bond formation, Peterson elimination, and sulfone elimination) (Scheme 2.144). The Peterson and sulfone eliminations can be successfully combined since both processes are carried out under basic conditions [152, 153].

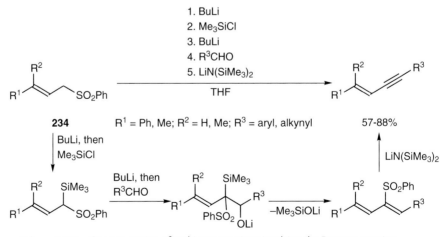

Scheme 2.144. One-pot version of multi-step reactions involving the Peterson reaction.

2.5.5
The Germanium, Tin, and Lead Versions of the Peterson Reaction

2.5.5.1 The Germanium-Peterson Reaction

Some examples of germanium versions of the Peterson reaction, such as syntheses of terminal alkenes, have been reported [383]. Peterson-type reactions of (trimethylgermyl)acetate esters **235** with some aldehydes give the corresponding (E)-alkenes stereoselectively, whereas the use of the silyl analogues gives a mixture of the geometrical isomers (Scheme 2.145) [384, 385].

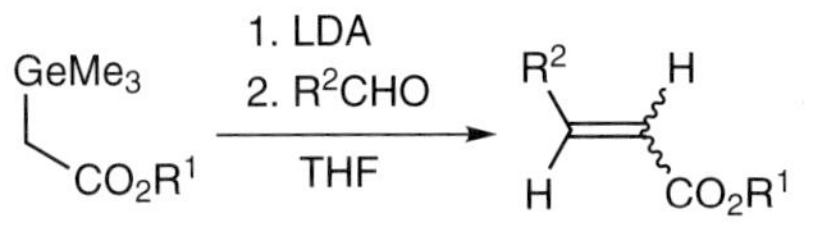

235
R^1 = Et, Bu^t;
R^2 = Ph, Octyl, Bu^t, PrCH=CH, etc.

41-92%
E/Z = 90/10-99/1

Scheme 2.145. The germanium-Peterson reaction.

The Peterson reaction of α-(trimethylgermyl)-α-(trimethylsilyl)acetate **236** with aldehydes selectively yields α-trimethylgermyl-α,β-unsaturated esters **237**, showing the lability of the silyl group over the germyl group in the Peterson reaction (Scheme 2.146). This is probably due to the high affinity of silicon for oxygen. Similar reactivity is observed in the cases of trimethylgermyl(trimethylsilyl)-acetonitrile and trimethylgermyl(trimethylsilyl)acetamide [386, 387].

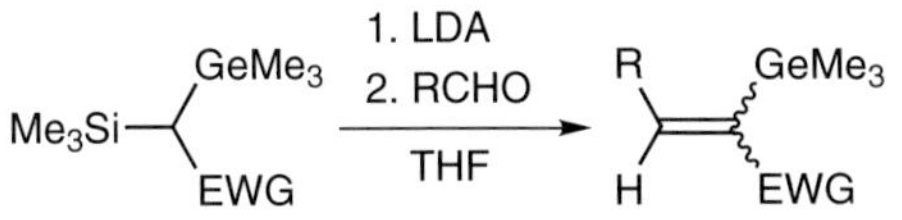

236
R = Ph, alkyl, heteroaryl;
EWG = CO_2Et, CN, $CONMe_2$

237 38-87%
E/Z = 55/45-99/1

Scheme 2.146. The Peterson reaction of α-germyl-α-silyl carbanions with aldehydes.

Although it is thought that the reaction proceeds by syn-elimination of the trimethylgermyloxide anion, there have been only a few reports of studies on the intermediate [385]. The β-germyl alkoxide intermediate **239** with a strong Ge–O interaction, which affords the double-bond migrated olefin **240** on thermolysis, can be isolated in the reaction of β-hydroxyalkylgermane **238** with KH in the presence of 18-crown-6 (Scheme 2.147) [28, 388]. A pentacoordinate 1,2-oxagermetanide intermediate can be isolated in the same reaction of a β-hydroxyalkylgermane bearing the Martin ligand [389].

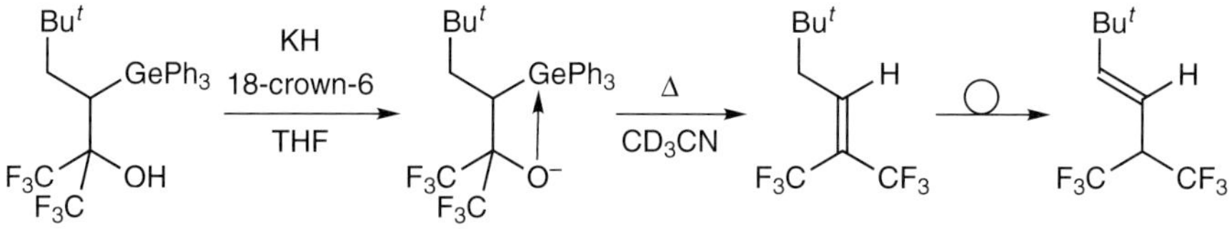

238 K+, 18-crown-6 239 quant. 240 quant.

Scheme 2.147. Isolation of a β-germyl alkoxide intermediate with a strong Ge–O interaction.

2.5.5.2 The Tin-Peterson Reaction

The synthesis of simple terminal alkenes has been reported by the reaction of (triphenylstannyl)methyllithium with carbonyl compounds [390]. (Phenylthio)-triphenylstannylmethyllithium and benzaldehyde can be selectively transformed into (E)-β-(phenylthio)styrene or, alternatively, its Z-isomer via β-hydroxyalkylstannane **241** by means of a tin-Peterson reaction (Scheme 2.148) [391]. LDA is usually used as the base in the tin-Peterson reaction. Butyllithium is not a suitable base for abstraction of the α-proton because lithium–tin exchange may occur [259].

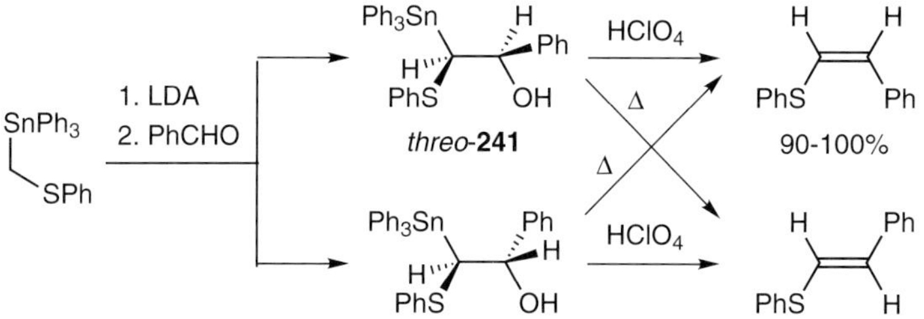

Scheme 2.148. The tin-Peterson reaction of an α-stannyl carbanion with benzaldehyde.

Stable pentacoordinate 1,2-oxastannetanides **243** have been isolated by deprotonation of β-hydroxyalkylstannanes **242** with KH in the presence of 18-crown-6, and were found to give the corresponding alkenes upon thermolysis, as in the case of the germanium compound (Scheme 2.149) [392].

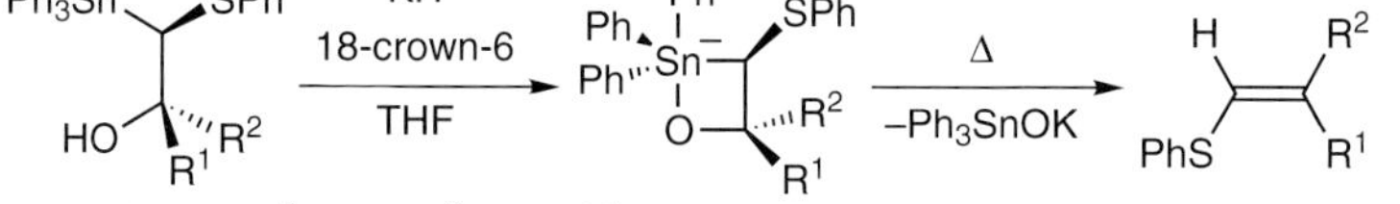

242 R^1 = Ph, R^2 = H, R^1 = R^2 = CF_3; K^+, 18-crown-6 **243** quant.; 82-95%

Scheme 2.149. Isolation of pentacoordinate 1,2-oxastannetanide intermediates in the tin-Peterson elimination.

The stannyl substituent SnR^1_3 sometimes has a substantial influence on the stereochemistry of alkene formation from α-lithiostannylalkylphosphonates **244** with aldehydes (Scheme 2.150) [393, 394]. The crowded tributylstannyl group favors the formation of the thermodynamic *E*-isomer with high stereoselectivity, whereas the less hindered triphenylstannyl group favors the kinetic *Z*-isomer. The less hindered triphenylstannyl group promotes faster decomposition of the *erythro* intermediate to give the *Z*-isomer, whereas the less reactive tributylstannyl group induces a conformational change prior to elimination and gives predominantly the *E*-isomer in the elimination step from the β-oxidophosphonate intermediate. Semiempirical (PM3) calculations on the adducts of another α-lithio(triphenylstannyl)-alkylphosphonate with benzaldehyde formed in the addition step accounted for the highly *Z*-selective course of the tin-Peterson reaction [395].

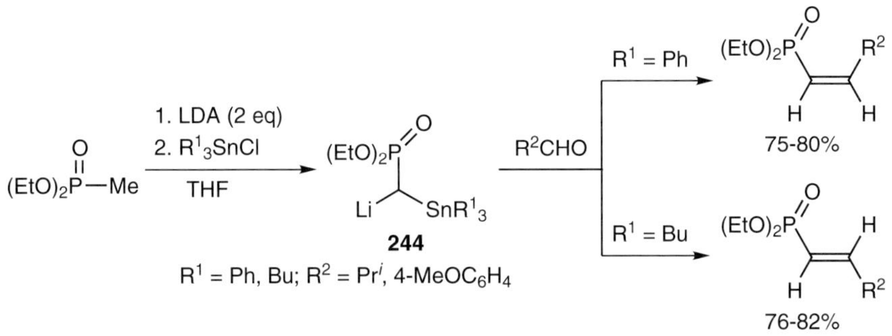

Scheme 2.150. The tin-Peterson reaction of α-stannyl carbanions bearing a phosphoryl group with aldehydes.

The Peterson reaction of α-tributylstannyl-α-(trimethylsilyl)acetate **245a** with benzaldehyde selectively yields the α-trimethylstannyl-α,β-unsaturated alkenoate **246**, while that of α-tributylstannyl-α-(trimethylgermyl)acetate **245b** yields the α-trimethylgermyl-α,β-unsaturated alkenoate **247** (Scheme 2.151) [18, 396, 397]. The

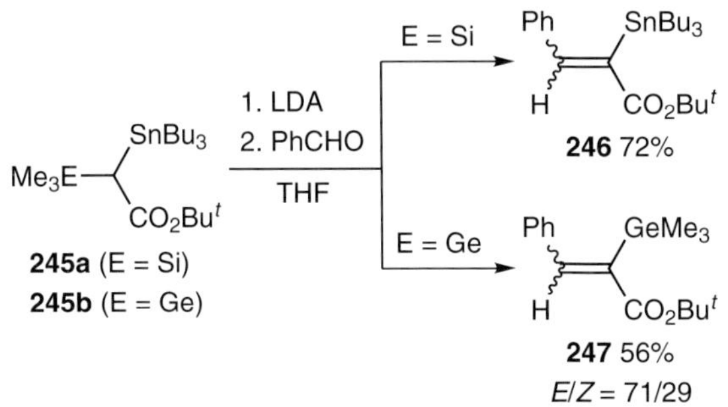

Scheme 2.151. The Peterson reaction and the tin-Peterson reaction of α-stannyl carbanions bearing another group 14 element moiety.

tendency for the group 14 element moiety to be eliminated in the Peterson-type reaction decreases in the following order: $Me_3Si > Bu_3Sn > Me_3Ge$ [398].

2.5.5.3 The Lead-Peterson Reaction

Lead-Peterson olefination is achieved similarly to the germanium and tin versions, though the lead version is not advantageous over the latter reactions in most cases. A much simpler and more gentle elimination method applicable to the lead compounds is to pass the solution of the bis(triphenylplumbyl)ethanol derivative **249**, which is synthesized from triplumbylmethane **248** and benzaldehyde, through a silica gel column (Scheme 2.152) [399].

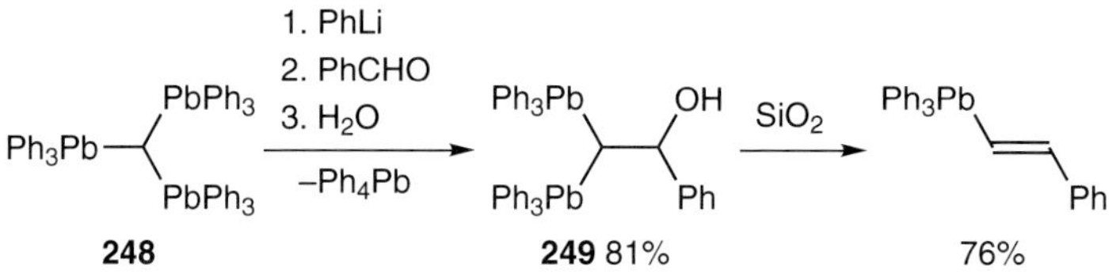

Scheme 2.152. The lead-Peterson reaction of an α,α-diplumbyl carbanion with benzaldehyde.

2.5.6
Synthesis of Carbon–Heteroatom Double-Bond Compounds by Peterson-Type Reactions

2.5.6.1 Synthesis of Imines

The modification of the Peterson reaction using an N-trimethylsilylamide anion instead of an α-silyl carbanion offers a promising route to the corresponding imines. Treatment of N-(p-tolyl)-N-trimethylsilylamide anion with carbonyl compounds yields the corresponding ketimines [400]. In particular, LiHMDS has been utilized for the preparation of N-trimethylsilylimines, which are useful as masked imine derivatives in the synthesis of β-lactam antibiotics [401–407]. Reactions of LiHMDS with non-enolizable aldehydes, enolizable aldehydes, ketones, a diketone, and α-keto esters give the respective imines (Scheme 2.153) [408–413]. Chlorotrimethylsilane is added to convert the generated lithium trimethylsilanolate into hexamethyldisiloxane.

R^1 = H, Bu^t, aryl;
R^2 = alkyl, alkenyl, alkynyl, aryl, C(O)R, CO_2R

Scheme 2.153. Synthesis of imines by the modified Peterson reaction using LiHMDS.

Enantiomerically pure *E*-sulfineimines **252** are formed by successive treatment of *R*- or *S*-menthyl *p*-toluenesulfinate **250** with LiHMDS and aldehydes (Scheme 2.154). This protocol involves Peterson-type reaction of the silylsulfinamide anion intermediate **251** with the aldehyde [414].

Scheme 2.154. Synthesis of (*E*)-sulfinimines by the modified Peterson reaction.

2.5.6.2 Synthesis of Phosphaalkenes

Phosphaalkenes, phosphorus–carbon double-bond compounds, which are usually unstable, can be synthesized by means of a Peterson-type reaction, the so-called "phospha-Peterson reaction" [415, 416]. One synthetic route to phosphaalkene **254** is by treatment of a lithium silylphosphide **253** bearing a bulky aryl substituent with an aldehyde, with elimination of a lithium silanolate (Scheme 2.155) [417, 418]. This reaction is found to be *E*-selective because of steric repulsion. Some ketones are also amenable to the phospha-Peterson reactions [419–422]. Phosphorus-containing cumulative double-bond compounds such as 1-phosphaallenes **255**, 1,3-azaphosphaallenes **256**, and 1,3-diphosphaallene **257** are also synthesized by way

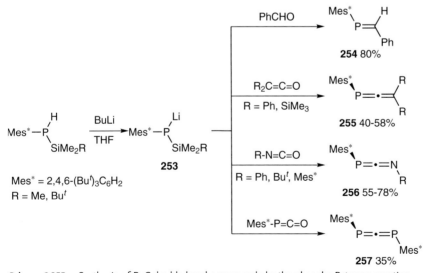

Scheme 2.155. Synthesis of P=C double-bond compounds by the phospha-Peterson reaction.

of Peterson-type reactions [423–425]. Use of isocyanates and a phosphaketene instead of a ketene affords 1,3-azaphosphaallenes **256** and a 1,3-diphosphaallene **257**, respectively [426–428].

1-Phosphaallenes **255a** are alternatively obtained by the usual Peterson reactions of 1-silyl-2-phosphaethenyllithium **258** with benzophenone and benzaldehyde, respectively (Scheme 2.156) [429].

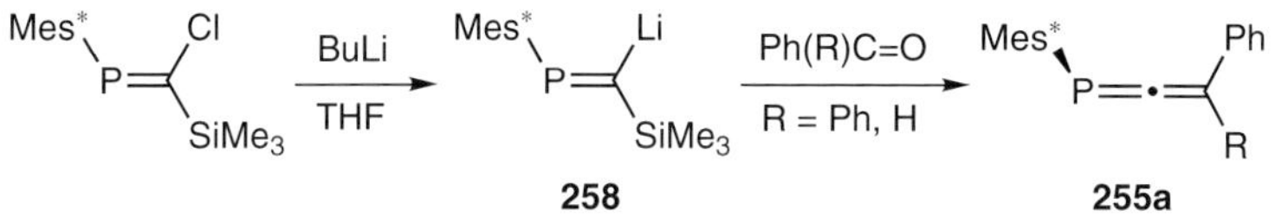

Scheme 2.156. Synthesis of 1-phosphaallenes by the Peterson reaction.

2.5.6.3 Synthesis of Silenes

The modifications of the Peterson olefination that involve elimination of lithium silanolate by reaction of tris(trialkylsilyl)silyllithium **259** with a carbonyl compound or by deprotonation of a (1-hydroxyalkyl)tris(trialkylsilyl)silane **260** with base are promising routes to silenes **262**, silicon–carbon double-bond compounds (Scheme 2.157) [430–436]. These types of reactions are sometimes called "sila-Peterson reactions"; the α-carbon atom in the 2-hydroxyalkylsilanes in the original Peterson reaction is replaced by a silicon atom. Silenes are formed through the same intermediate when acyl(trimethylsilyl)silanes **261** are treated with organolithium reagents [437, 438]. Although the silenes readily dimerize in head-to-tail or head-to-head manners in the absence of scavenger reagents, some stable silenes can be isolated by taking advantage of the kinetic stabilization offered by bulky silyl groups [439, 440].

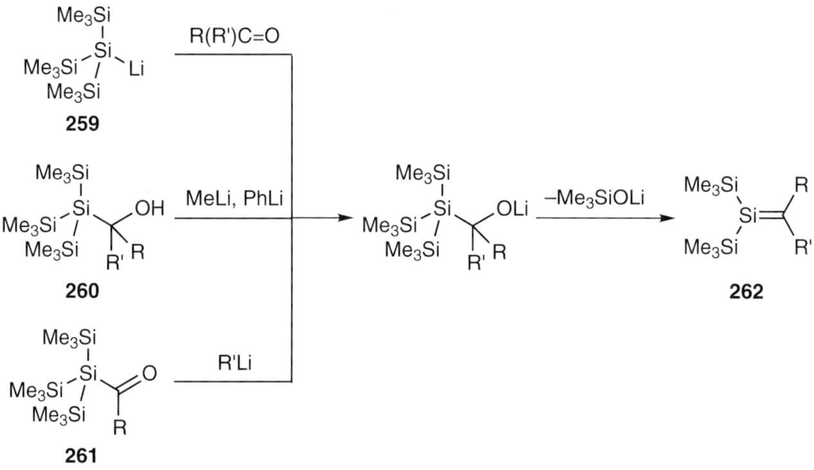

Scheme 2.157. Synthesis of silenes by the sila-Peterson reaction.

An acid-induced amino-sila-Peterson olefination of geminal (dimethylamino)-bis(hypersilyl)methanes **263** with concentrated H_2SO_4 gives a transient 1,1,2-trisilylsilene **264** (Scheme 2.158) [441]. The silene **264** is trapped in the presence of excess H_2SO_4, and further hydrolytic work-up affords the silanol **265**.

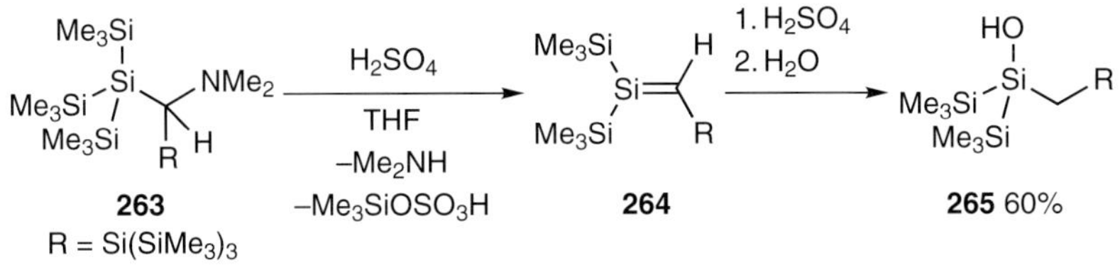

Scheme 2.158. Synthesis of silenes by the amino-sila-Peterson reaction.

2.5.6.4 Synthesis of Germenes

A germene, that is a germanium–carbon double-bond compound, can be generated by a Peterson-type reaction akin to the sila-Peterson reaction. The reaction of tris(trimethylsilyl)germyllithium·3THF with 2-adamantanone yields the germene **266**, which, in the absence of trapping reagents, gives the corresponding head-to-head dimer **267** (Scheme 2.159) [442].

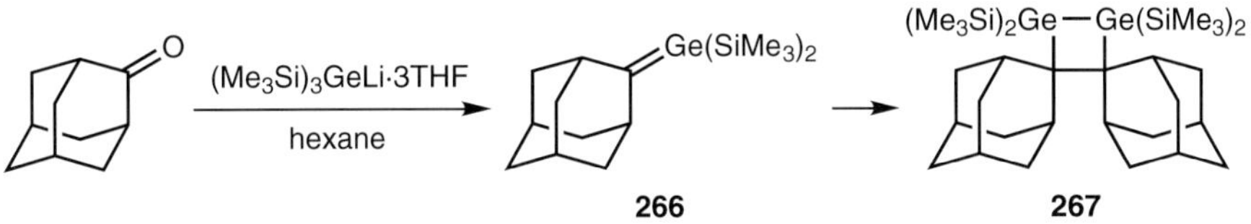

Scheme 2.159. Synthesis of a germene by a Peterson-type reaction.

2.5.6.5 Synthesis of Sulfines

Peterson-type reaction of α-silyl carbanions with sulfur dioxide gives sulfines **268**. In this case, a C=S double bond is formed with elimination of lithium silanolate (Scheme 2.160) [443, 444]. This method can be applied to the synthesis of phosphorylated sulfines and α,β-unsaturated sulfines [445, 446].

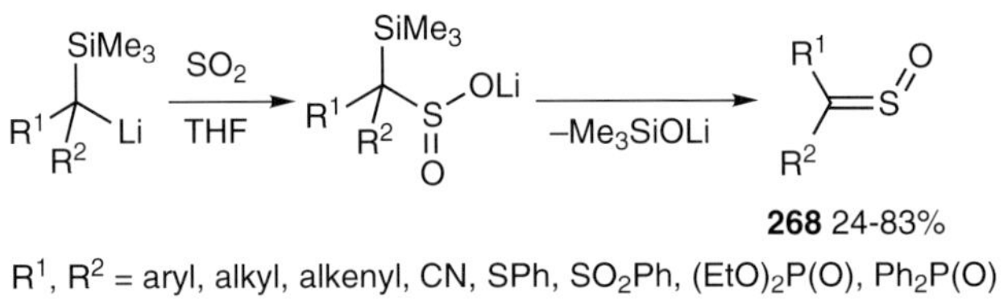

R^1, R^2 = aryl, alkyl, alkenyl, CN, SPh, SO_2Ph, $(EtO)_2P(O)$, $Ph_2P(O)$

Scheme 2.160. Synthesis of sulfines by a Peterson-type reaction.

Further Peterson-type reaction of α-silyl carbanions with the sulfines **268** affords the intermediary thiocarbonyl S-ylides **269**, which lead to the alkenes (Scheme 2.161) [447].

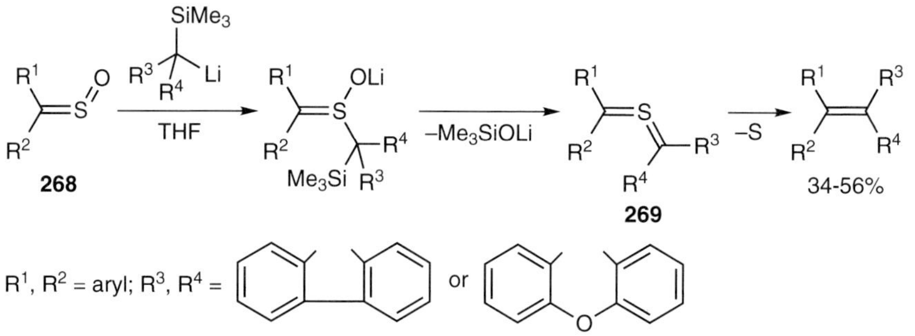

Scheme 2.161. Peterson-type reaction of α-silyl carbanions with sulfines giving the corresponding alkenes.

2.6
Conclusion

In this chapter, we have reviewed the Peterson reaction of various types of α-silyl carbanions with a variety of carbonyl compounds, as well as related reactions, and the relevant reaction mechanisms. The Peterson reaction, the silicon variant of the Wittig reaction, gives a diastereomeric mixture of the intermediately formed β-hydroxyalkylsilanes, and these can be separated and isolated, and in some cases further subjected to treatment with acid or base to stereospecifically afford either the E- or Z-isomer of the corresponding alkene. The higher reactivity of an α-silyl carbanion towards carbonyl compounds and the stereospecificity of the elimination from β-hydroxyalkylsilanes are both superior to those of the phosphorus ylide in the Wittig reaction. Direct alkene formation from α-silyl carbanions and carbonyl compounds also plays a prominent role in organic synthesis since several stereoselective versions have been developed. Various synthetic methods for obtaining the α-silyl carbanions and β-silyl oxide anions are now available, and these have expanded the utility of the Peterson reaction. Thus, the Peterson reaction has been widely employed as one of the most valuable tools in organic synthesis permitting alkene formation, thereby providing a variety of functionalized molecules. Several natural products and biologically active molecules have also been synthesized by means of the Peterson methodology. However, mechanistic studies have not reached a consistent conclusion that adequately accounts for the whole range of reaction systems because the results obtained vary depending on both the reaction conditions and the substituents on the substrates; nevertheless, pentacoordinate 1,2-oxasiletanides have been synthesized and isolated in a particular case. To further develop the Peterson reaction, which is amenable to various substrates and can proceed with both high *E*- and *Z*-stereoselectivities and high yields, full elucidation of the reaction mechanism by means of various experimental studies as well as theoretical calculations is absolutely crucial.

References

1 D. J. Peterson, *J. Org. Chem.* **1968**, *33*, 780–784.
2 D. J. Peterson, *Organomet. Chem. Rev. A* **1972**, *7*, 295–358.
3 T. H. Chan, *Acc. Chem. Res.* **1977**, *10*, 442–448.
4 W. P. Weber, *Silicon Reagents for Organic Synthesis*, Springer-Verlag, New York, **1983**, pp. 58–73.
5 E. W. Colvin, *Silicon in Organic Synthesis*, Butterworths, London, **1981**, pp. 141–152.
6 D. J. Ager, *Synthesis* **1984**, 384–398.
7 D. J. Ager, *Org. React.* **1990**, *38*, 1–223.
8 S. E. Kelly, *Comprehensive Organic Synthesis*, Vol. 1 (Eds.: B. M. Trost, I. Fleming), Pergamon Press, Oxford, **1991**, pp. 731–737.
9 A. G. M. Barrett, J. M. Hill, E. M. Wallace, J. A. Flygare, *Synlett* **1991**, 764–770.
10 L. F. van Staden, D. Gravestock, D. J. Ager, *Chem. Soc. Rev.* **2002**, *31*, 195–200.
11 P. F. Hudrlik, D. Peterson, *J. Am. Chem. Soc.* **1975**, *97*, 1464–1468.
12 S. Akai, Y. Tsuzuki, S. Matsuda, S. Kitagaki, Y. Kita, *J. Chem. Soc., Perkin Trans. 1* **1992**, 2813–2820.
13 P. F. Hudrlik, P. E. Holmes, A. M. Hudrlik, *Tetrahedron Lett.* **1988**, *29*, 6395–6398.
14 K. Yamamoto, T. Kimura, Y. Tomo, *Tetrahedron Lett.* **1985**, *26*, 4505–4508.
15 K. Yamamoto, Y. Tomo, S. Suzuki, *Tetrahedron Lett.* **1980**, *21*, 2861–2864.
16 G. L. Larson, F. Quiroz, J. Suárez, *Synth. Commun.* **1983**, *13*, 833–844.
17 G. L. Larson, C. F. de Kaifer, R. Seda, L. E. Torres, J. R. Ramirez, *J. Org. Chem.* **1984**, *49*, 3385–3386.
18 A. Zapata, C. Fortoul R., C. Acuña A., *J. Organomet. Chem.* **1993**, *448*, 69–74.
19 M. Larchevêque, A. Debal, *J. Chem. Soc., Chem. Commun.* **1981**, 877–878.
20 P. F. Hudrlik, A. M. Hudrlik, A. K. Kulkarni, *J. Am. Chem. Soc.* **1982**, *104*, 6809–6811.
21 A. Pelter, D. Buss, E. Colclough, B. Singaram, *Tetrahedron* **1993**, *49*, 7077–7103.
22 K. Yamamoto, Y. Tomo, *Tetrahedron Lett.* **1983**, *24*, 1997–2000.
23 A. R. Bassindale, R. J. Ellis, P. G. Taylor, *J. Chem. Research (S)* **1996**, 34–35.
24 T. Kawashima, N. Iwama, R. Okazaki, *J. Am. Chem. Soc.* **1992**, *114*, 7598–7599.
25 K. Naganuma, T. Kawashima, R. Okazaki, *Chem. Lett.* **1999**, 1139–1140.
26 T. Kawashima, K. Naganuma, R. Okazaki, *Organometallics* **1998**, *17*, 367–372.
27 T. Kawashima, R. Okazaki, *Synlett* **1996**, 600–608.
28 T. Kawashima, *J. Organomet. Chem.* **2000**, *611*, 256–263.
29 A. R. Bassindale, R. J. Ellis, P. G. Taylor, *Tetrahedron Lett.* **1984**, *25*, 2705–2708.
30 G. L. Larson, J. A. Prieto, A. Hernández, *Tetrahedron Lett.* **1981**, *22*, 1575–1578.
31 M. J. Crimmin, P. J. O'Hanlon, N. H. Rogers, *J. Chem. Soc., Perkin Trans. 1* **1985**, 541–547.
32 S. P. Green, D. A. Whiting, *J. Chem. Soc., Chem. Commun.* **1992**, 1753–1755.
33 W. A. Kleselinck, C. T. Buse, C. H. Heathcock, *J. Am. Chem. Soc.* **1977**, *99*, 247–248.
34 A. R. Bassindale, R. J. Ellis, J. C.-Y. Lau, P. G. Taylor, *J. Chem. Soc., Chem. Commun.* **1986**, 98–100.

35 A. R. Bassindale, R. J. Ellis, J. C.-Y. Lau, P. G. Taylor, *J. Chem. Soc., Perkin Trans. 2* **1986**, 593–597.
36 P. F. Hudrlik, E. L. O. Agwaramgbo, A. M. Hudrlik, *J. Org. Chem.* **1989**, 54, 5613–5618.
37 R. K. Boekman, Jr., R. L. Chinn, *Tetrahedron Lett.* **1985**, 26, 5005–5008.
38 D. Bell, E. A. Crowe, N. J. Dixon, G. R. Geen, I. S. Mann, M. R. Shipton, *Tetrahedron* **1994**, 50, 6643–6652.
39 R. Dalpozzo, A. De Nino, E. Iantorno, G. Bartoli, L. Sambri, *Tetrahedron* **1997**, 53, 2585–2590.
40 K. Itami, T. Nokami, J. Yoshida, *Tetrahedron* **2001**, 57, 5045–5054.
41 A. Mohamed-Hachi, E. About-Jaudet, J.-C. Combret, N. Collignon, *Synthesis* **1997**, 653–656.
42 Y. Zanella, S. Berté-Verrando, R. Dizière, P. Savignac, *J. Chem. Soc., Perkin Trans. 1* **1995**, 2835–2838.
43 R. Waschbüsch, J. Carran, P. Savignac, *Tetrahedron* **1996**, 52, 14199–14216.
44 L. F. van Staden, B. Bartels-Rahm, J. S. Field, N. D. Emslie, *Tetrahedron* **1998**, 54, 3255–3278.
45 C. Trindle, J.-T. Hwang, F. A. Carey, *J. Org. Chem.* **1973**, 38, 2664–2669.
46 M. B. Gillies, J. E. Tønder, D. Tanner, P.-O. Norrby, *J. Org. Chem.* **2002**, 67, 7378–7388.
47 J. B. Perales, N. F. Makino, D. L. Van Vranken, *J. Org. Chem.* **2002**, 67, 6711–6717.
48 A. G. Brook, J. M. Duff, D. G. Anderson, *Can. J. Chem.* **1970**, 48, 561–569.
49 L. H. Sommer, G. M. Goldberg, J. Gold, F. C. Whitmore, *J. Am. Chem. Soc.* **1947**, 69, 980.
50 T. H. Chan, E. Chang, *J. Org. Chem.* **1974**, 39, 3264–3268.
51 F. C. Whitmore, L. H. Sommer, *J. Am. Chem. Soc.* **1946**, 68, 481–484.
52 C. R. Hauser, C. R. Hance, *J. Am. Chem. Soc.* **1952**, 74, 5091–5096.
53 B. L. Chenard, C. Slapak, D. K. Anderson, J. S. Swenton, *J. Chem. Soc., Chem. Commun.* **1981**, 179–180.
54 T. Hiyama, A. Kanakura, Y. Morizawa, H. Nozaki, *Tetrahedron Lett.* **1982**, 23, 1279–1280.
55 W. Connolly, G. Urry, *Inorg. Chem.* **1963**, 2, 645–646.
56 S. Murai, I. Ryu, J. Iriguchi, N. Sonoda, *J. Am. Chem. Soc.* **1984**, 106, 2440–2442.
57 W. Huang, T. T. Tidwell, *Synthesis* **2000**, 457–470.
58 B. L. Hayes, T. A. Adams, K. A. Pickin, C. S. Day, M. E. Welker, *Organometallics* **2000**, 19, 2730–2740.
59 E. Dunach, R. L. Halterman, K. P. C. Vollhardt, *J. Am. Chem. Soc.* **1985**, 107, 1664–1671.
60 B. Faust, F. Mitzel, *J. Chem. Soc., Perkin Trans. 1* **2000**, 3746–3751.
61 G. A. Olah, V. P. Reddy, G. K. S. Prakash, *Synthesis* **1991**, 29–30.
62 A. D. Lebsack, L. E. Overman, R. J. Valentekovitch, *J. Am. Chem. Soc.* **2001**, 123, 4851–4852.
63 K. Takao, M. Hara, T. Tsujita, K. Yoshida, K. Tadano, *Tetrahedron Lett.* **2001**, 42, 4665–4668.
64 K. Takao, T. Tsujita, M. Hara, K. Tadano, *J. Org. Chem.* **2002**, 67, 6690–6698.
65 C. Johnson, B. D. Tait, *J. Org. Chem.* **1987**, 52, 281–283.
66 C. Johnstone, W. J. Kerr, U. Lange, *J. Chem. Soc., Chem. Commun.* **1995**, 457–458.
67 C. Johnstone, W. J. Kerr, U. Lange, *Tetrahedron* **1996**, 52, 7391–7420.
68 P. Chalard, R. Remuson, Y. Gelas-Mialhe, J.-C. Gramain, I. Canet, *Tetrahedron Lett.* **1999**, 40, 1661–1664.
69 S. Fukuzawa, K. Mutoh, T. Tsuchimoto, T. Hiyama, *J. Org. Chem.* **1996**, 61, 5400–5405.
70 T. K. Sarkar, *Synthesis* **1990**, 1101–1111.
71 B. A. Narayanan, W. H. Bunnelle, *Tetrahedron Lett.* **1987**, 28, 6261–6264.
72 T. V. Lee, J. A. Channon, C. Cregg, J. R. Porter, F. S. Roden, H. T.-L. Yeoh, *Tetrahedron* **1989**, 45, 5877–5886.
73 T. V. Lee, J. R. Porter, F. S. Roden, *Tetrahedron Lett.* **1988**, 29, 5009–5012.
74 M. Ochiai, E. Fujita, M. Arimoto, H. Yamaguchi, *J. Chem. Soc., Chem. Commun.* **1982**, 1108–1109.
75 T. Yamazaki, N. Ishikawa, *Chem. Lett.* **1984**, 521–524.

76 V. G. S. Box, D. P. Brown, *Heterocycles* **1991**, *32*, 1273–1277.
77 T. J. Mickelson, J. L. Koviach, C. J. Foryth, *J. Org. Chem.* **1996**, *61*, 9617–9620.
78 M. B. Anderson, P. L. Fuchs, *Synth. Commun.* **1987**, *17*, 621–635.
79 I. Fleming, A. Pearce, *J. Chem. Soc., Perkin Trans. 1* **1981**, 251–255.
80 M. Demuth, *Helv. Chim. Acta* **1978**, *61*, 3136–3138.
81 J.-M. Durgnat, A. Warm, P. Vogel, *Synth. Commun.* **1992**, *22*, 1883–1893.
82 A.-C. Guevel, D. J. Hart, *Synlett* **1994**, 169–170.
83 A.-C. Guevel, D. J. Hart, *J. Org. Chem.* **1996**, *61*, 473–479.
84 M. A. Evans, J. P. Morken, *J. Am. Chem. Soc.* **2002**, *124*, 9020–9021.
85 J. Mulzer, A. Mantoulidis, E. Öhler, *J. Org. Chem.* **2000**, *65*, 7456–7467.
86 T. Suzuki, T. Oriyama, *Synlett* **2000**, 859–861.
87 A. Fürstner, *J. Organomet. Chem.* **1987**, *336*, C33–C36.
88 C. Palomo, J. M. Aizpurua, N. Aurrekoetxea, *Tetrahedron Lett.* **1990**, *31*, 2209–2210.
89 D. J. Peterson, *J. Organomet. Chem.* **1967**, *9*, 373–374.
90 T. Konakahara, Y. Takagi, *Heterocycles* **1980**, *14*, 393–396.
91 R. I. Papasergio, B. W. Skelton, P. Twiss, A. H. White, C. L. Raston, *J. Chem. Soc., Dalton Trans.* **1990**, 1161–1172.
92 T. H. Chan, E. Chang, E. Vinokur, *Tetrahedron Lett.* **1970**, 1137–1140.
93 J. Clayden, P. Johnson, J. H. Pink, *J. Chem. Soc., Perkin Trans. 1* **2001**, 371–375.
94 R. Vohra, D. B. MacLean, *Tetrahedron Lett.* **1993**, *34*, 7673–7676.
95 R. Vohra, D. B. MacLean, *Can. J. Chem.* **1994**, *72*, 1660–1667.
96 A. Couture, H. Cornet, P. Grandclaudon, *J. Organomet. Chem.* **1992**, *440*, 7–13.
97 G. Bartoli, M. Bosco, R. Dalpozzo, P. E. Todesco, *J. Chem. Soc., Perkin Trans. 1* **1988**, 807–808.
98 P. Magnus, G. Roy, *J. Chem. Soc., Chem. Commun.* **1979**, 822–823.
99 S. Kanemasa, J. Tanaka, H. Nagahama, O. Tsuge, *Bull. Chem. Soc. Jpn.* **1985**, *58*, 3385–3386.
100 A. R. Katritzky, X. Zhao, I. V. Shcherbakova, *J. Chem. Soc., Perkin Trans. 1* **1991**, 3295.
101 D. P. M. Pleynet, J. K. Dutton, A. P. Johnson, *Tetrahedron* **1999**, *55*, 11903–11926.
102 A. P. Johnson, J. K. Dutton, D. P. M. Pleynet, *Heterocycles* **1994**, *37*, 1913–1932.
103 J. K. Dutton, R. W. Steel, A. S. Tasker, V. Popsavin, A. P. Johnson, *J. Chem. Soc., Chem. Commun.* **1994**, 765–766.
104 A. R. Katritzky, D. Toader, L. Xie, *Synthesis* **1996**, 1425–1427.
105 L. F. van Staden, B. Bartels-Rahm, N. D. Emslie, *Tetrahedron Lett.* **1997**, *38*, 1851–1852.
106 A. I. Meyers, G. E. Jagdmann, Jr., *J. Am. Chem. Soc.* **1982**, *104*, 877–879.
107 A. I. Meyers, P. D. Edwards, T. R. Bailey, G. E. Jagdmann, Jr., *J. Org. Chem.* **1985**, *50*, 1019–1026.
108 T. Yamazaki, N. Ishikawa, *Bull. Chim. Soc. Fr.* **1986**, 937–943.
109 Y. Terao, M. Aono, I. Takahashi, K. Achiwa, *Chem. Lett.* **1986**, 2089–2092.
110 A. R. Katritzky, W. Kuzmierkiewicz, J. M. Aurrecoechea, *J. Org. Chem.* **1987**, *52*, 844–849.
111 D. I. Han, D. Y. Oh, *Synth. Commun.* **1988**, *18*, 2111–2116.
112 M. Isobe, Y. Ichikawa, Y. Funabashi, S. Mio, T. Goto, *Tetrahedron* **1986**, *42*, 2863–2872.
113 D. Roche, S. Danoun, M. Madesclaire, *Synth. Commun.* **1994**, *24*, 3213–3218.
114 F. A. Carey, A. S. Court, *J. Org. Chem.* **1972**, *37*, 939–943.
115 D. J. Ager, *Tetrahedron Lett.* **1981**, *22*, 2803–2806.
116 M. Pellet, F. Huet, *Tetrahedron* **1988**, *44*, 4463–4468.
117 J. A. Prieto, G. L. Larson, P. Gonzalez, *Synth. Commun.* **1989**, *19*, 2773–2778.
118 C. A. Brown, A. Yamaichi, *J. Chem. Soc., Chem. Commun.* **1979**, 100–101.
119 D. J. Ager, M. B. East, *J. Org. Chem.* **1986**, *51*, 3983–3992.

120 T. Agawa, M. Ishikawa, M. Komatsu, Y. Ohshiro, *Chem. Lett.* **1980**, 335–336.

121 T. Agawa, M. Ishikawa, M. Komatsu, Y. Ohshiro, *Bull. Chem. Soc. Jpn.* **1982**, *55*, 1205–1208.

122 A. de Groot, B. J. M. Jansen, *Synth. Commun.* **1983**, *13*, 985–990.

123 S. Hackett, T. Livinghouse, *J. Org. Chem.* **1986**, *51*, 879–885.

124 H. Monenschein, G. Dräger, A. Jung, A. Kirschning, *Chem. Eur. J.* **1999**, *5*, 2270–2280.

125 J. Butter, R. H. Meijer, R. M. Kellogg, *Tetrahedron Lett.* **1998**, *39*, 6357–6360.

126 P. F. Jones, M. F. Lappert, *J. Chem. Soc., Chem. Commun.* **1972**, 526.

127 F. A. Carey, A. S. Court, *J. Org. Chem.* **1972**, *37*, 1926–1929.

128 D. Seebach, M. Kolb, B.-T. Gröbel, *Chem. Ber.* **1973**, *106*, 2277–2290.

129 R. Gompper, W. Reiser, *Tetrahedron Lett.* **1976**, 1263–1264.

130 R. S. Brinkmeyer, *Tetrahedron* **1979**, 207–210.

131 C. Reinhard, B.-V. Herget, M. Schulz, W. Massa, S. Peschel, *Tetrahedron Lett.* **1989**, *30*, 3521–3524.

132 G. Foulard, T. Brigaud, C. Portella, *J. Org. Chem.* **1997**, *62*, 9107–9113.

133 P. C. B. Page, M. J. McKenzie, D. R. Buckle, *Tetrahedron* **1998**, *54*, 14581–14596.

134 J. Mlynarski, A. Banaszek, *Tetrahedron* **1999**, *55*, 2785–2794.

135 G. Fritzsche, R. Gleiter, H. Irngartinger, T. Oeser, *Eur. J. Org. Chem.* **1999**, 73–81.

136 B.-T. Gröbel, D. Seebach, *Synthesis* **1977**, 357–402.

137 M. Kolb, *Synthesis* **1990**, 171–190.

138 D. Seebach, B.-Th. Gröbel, A. K. Beck, M. Braun, K.-H. Geiss, *Angew. Chem. Int. Ed. Engl.* **1972**, *11*, 443–444.

139 R. L. Funk, P. M. Novak, M. M. Abelman, *Tetrahedron Lett.* **1988**, *29*, 1493–1496.

140 D. Seebach, M. Kolb, B.-T. Gröbel, *Tetrahedron Lett.* **1974**, 3171–3174.

141 P. C. B. Page, M. B. van Niel, D. Westwood, *J. Chem. Soc., Perkin Trans. 1* **1988**, 269–275.

142 B. B. Snider, B. Shi, C. A. Quickley, *Tetrahedron* **2000**, *56*, 10127–10132.

143 F. A. Carey, O. Hernandez, *J. Org. Chem.* **1973**, *38*, 2670–2675.

144 A. R. Katritzky, P. Lue, N. Malhotra, *Gazz. Chim. Ital.* **1991**, *121*, 471–474.

145 M. Julia, J.-M. Paris, *Tetrahedron Lett.* **1973**, 4833–4836.

146 P. Kocienski, *Comprehensive Organic Synthesis*, Vol. 6 (Eds.: B. M. Trost, I. Fleming), Pergamon Press, Oxford, **1991**, pp. 987–1000.

147 D. Craig, S. V. Ley, N. S. Simpkins, *J. Chem. Soc., Perkin Trans. 1* **1985**, 1949–1952.

148 S. V. Ley, N. S. Simpkins, *J. Chem. Soc., Chem. Commun.* **1983**, 1281–1282.

149 D. J. Ager, *J. Chem. Soc., Chem. Commun.* **1984**, 486–488.

150 K. Schank, F. Schroeder, *Liebigs Ann. Chem.* **1977**, 1676–1682.

151 A. Mahadevan, P. L. Fuchs, *Tetrahedron Lett.* **1994**, *35*, 6025–6028.

152 A. Orita, N. Yoshioka, J. Otera, *Chem. Lett.* **1997**, 1023–1024.

153 A. Orita, N. Yoshioka, P. Struwe, A. Braier, A. Beckmann, J. Otera, *Chem. Eur. J.* **1999**, *5*, 1355–1363.

154 M. Mladenova, F. Gaudemar-Bardone, *Phosphorus, Sulfur, and Silicon* **1991**, *62*, 257–267.

155 M. Müller, H.-H. Otto, *Liebigs Ann. Chem.* **1992**, 687–692.

156 R. K. Boeckman, Jr., S. M. Silver, *Tetrahedron Lett.* **1973**, 3497–3500.

157 R. K. Boeckman, Jr., S. M. Silver, *J. Org. Chem.* **1975**, *40*, 1755–1759.

158 K. Shimoji, H. Taguchi, K. Oshima, H. Yamamoto, H. Nozaki, *J. Am. Chem. Soc.* **1974**, *96*, 1620–1621.

159 J.-M. Galano, G. Audran, H. Monti, *Tetrahedron Lett.* **2001**, *42*, 6125–6128.

160 P. Savignac, M.-P. Teulade, *J. Organomet. Chem.* **1987**, *323*, 135–144.

161 A. Kenny, J. Nieschalk, D. O'Hagan, *J. Fluorine Chem.* **1996**, *80*, 59–62.

162 G. M. Blackburn, M. J. Parratt, *J. Chem. Soc., Perkin Trans. 1* **1986**, 1425–1430.

163 L. Schmitt, N. Cavusoglu, B. Spiess, G. Schlewer, *Tetrahedron Lett.* **1998**, *39*, 4009–4012.

164 R. Diziere, P. Savignac, *Tetrahedron Lett.* **1996**, *37*, 1783–1786.
165 J. Binder, E. Zbiral, *Tetrahedron Lett.* **1986**, *27*, 5829–5832.
166 M.-P. Teulade, P. Savignac, *J. Organomet. Chem.* **1988**, *338*, 295–303.
167 M. Mikolajczyk, P. Balczewski, *Synthesis* **1989**, 101–106.
168 H. Al-Badri, E. About-Jaudet, N. Collignon, *Tetrahedron Lett.* **1996**, *37*, 2951–2954.
169 H. Al-Badri, E. About-Jaudet, N. Collignon, *J. Chem. Soc., Perkin Trans. 1* **1996**, 931–938.
170 F. Plénat, *Tetrahedron Lett.* **1981**, *22*, 4705–4708.
171 G. M. Schelde, R. H. Neilson, *Organometallics* **1989**, *8*, 1987–1991.
172 J. Villieras, R. Tarhouni, B. Kirschleger, M. Rambaud, *Bull. Chim. Soc. Fr.* **1985**, 825–830.
173 F. Cooke, G. Roy, P. Magnus, *Organometallics* **1982**, *1*, 893–897.
174 C. Burford, F. Cooke, G. Roy, P. Magnus, *Tetrahedron* **1983**, *39*, 867–876.
175 T. Kauffmann, R. König, M. Wensing, *Tetrahedron Lett.* **1984**, *25*, 637–640.
176 A. Hosomi, M. Inaba, H. Sakurai, *Tetrahedron Lett.* **1983**, *24*, 4727–4728.
177 T. H. Chan, M. Moreland, *Tetrahedron Lett.* **1978**, 515–518.
178 N. A. Braun, I. Klein, D. Spitzner, B. Vogler, S. Braun, H. Borrmann, A. Simon, *Liebigs Ann.* **1995**, 2165–2169.
179 J. T. Welch, J. Lin, *Tetrahedron* **1996**, *52*, 291–304.
180 J. T. Welch, R. W. Herbert, *J. Org. Chem.* **1990**, *55*, 4782–4784.
181 J. Lin, J. T. Welch, *Tetrahedron Lett.* **1998**, *39*, 9613–9616.
182 A. Zapata, F. G. Ferrer, *Synth. Commun.* **1986**, *16*, 1611–1615.
183 T. Hudlicky, L. Radesca-Kwart, L.-G. Li, T. Bryant, *Tetrahedron Lett.* **1988**, *27*, 3283–3286.
184 I. Matsuda, H. Okada, S. Sato, Y. Izumi, *Tetrahedron Lett.* **1984**, *25*, 3879–3882.
185 J. A. Miller, G. Zweifel, *J. Am. Chem. Soc.* **1981**, *103*, 6217–6219.
186 T. Inoue, T. Sato, I. Kuwajima, *J. Org. Chem.* **1984**, *49*, 4671–4674.
187 O. Tsuge, S. Kanemasa, T. Suzuki, K. Matsuda, *Bull. Chem. Soc. Jpn.* **1986**, *59*, 2851–2860.
188 O. Tsuge, S. Kanemasa, T. Otsuka, T. Suzuki, *Bull. Chem. Soc. Jpn.* **1988**, *61*, 2897–2908.
189 H. Taguchi, K. Shimoji, H. Yamamoto, H. Nozaki, *Bull. Chem. Soc. Jpn.* **1974**, *47*, 2529–2531.
190 S. L. Hartzell, D. F. Sullivan, M. W. Rathke, *Tetrahedron Lett.* **1974**, 1403–1406.
191 P. Albaugh-Robertson, J. A. Katzenellenbogen, *J. Org. Chem.* **1983**, *48*, 5288–5302.
192 L. Strekowski, M. Visnick, M. A. Battiste, *Tetrahedron Lett.* **1984**, *25*, 5603–5606.
193 G. Sauvé, P. Deslongchamps, *Synth. Commun.* **1985**, *15*, 201–212.
194 G. L. Larson, J. A. Soderquist, M. R. Claudio, *Synth. Commun.* **1990**, *20*, 1095–1104.
195 G. B. Gill, G. D. James, K. V. Oates, G. Pattenden, *J. Chem. Soc., Perkin Trans. 1* **1993**, 2567–2579.
196 M. Majewski, N. M. Irvine, J. MacKinnon, *Tetrahedron: Asymmetry* **1995**, *6*, 1837–1840.
197 S. L. Hartzell, M. W. Rathke, *Tetrahedron Lett.* **1976**, 2737–2740.
198 G. L. Larson, R. M. B. de Perez, *J. Org. Chem.* **1985**, *50*, 5257–5260.
199 G. L. Larson, *Pure & Appl. Chem.* **1990**, *62*, 2021–2026.
200 K. Hartke, O. Kunze, *Liebigs Ann. Chem.* **1989**, 321–330.
201 E. J. Corey, D. Enders, M. G. Bock, *Tetrahedron Lett.* **1976**, 7–10.
202 R. H. Schlessinger, M. A. Poss, S. Richardson, P. Lin, *Tetrahedron Lett.* **1985**, *26*, 2391–2394.
203 M. G. Silvestri, P. J. Bednarski, E. Kho, *J. Org. Chem.* **1985**, *50*, 2798–2799.
204 D. P. Provencal, J. W. Leahy, *J. Org. Chem.* **1994**, *59*, 5496–5498.
205 R. Desmond, S. G. Mills, R. P. Volante, I. Shinkai, *Tetrahedron Lett.* **1988**, *29*, 3895–3898.
206 N. Y. Grigorieva, O. N. Yudina,

A. M. Moiseenkov, *Synthesis* **1989**, 591–595.
207 R. Baudouy, P. Prince, *Tetrahedron* **1989**, *45*, 2067–2074.
208 A. A. Croteau, J. Termini, *Tetrahedron Lett.* **1983**, *24*, 2481–2484.
209 D. Mead, A. E. Asato, M. Denny, R. S. H. Liu, *Tetrahedron Lett.* **1987**, *28*, 259–262.
210 R. P. Woodbury, M. W. Rathke, *J. Org. Chem.* **1978**, *43*, 1947–1949.
211 R. P. Woodbury, M. W. Rathke, *Tetrahedron Lett.* **1978**, 709–712.
212 S. Kojima, H. Inai, T. Hidaka, K. Ohkata, *Chem. Commun.* **2000**, 1795–1796.
213 S. Kojima, H. Inai, T. Hidaka, T. Fukuzaki, K. Ohkata, *J. Org. Chem.* **2002**, *67*, 4093–4099.
214 S. Gürtler, H.-H. Otto, *Arch. Pharm.* **1989**, *322*, 3–10.
215 S. Gürtler, H.-H. Otto, *Arch. Pharm.* **1989**, *322*, 105–109.
216 H.-J. Bergmann, R. Mayrhofer, H.-H. Otto, *Arch. Pharm.* **1986**, *319*, 203–216.
217 S. Gürtler, S. Ruf, M. Johner, H.-H. Otto, *Pharmazie* **1996**, *51*, 811–815.
218 E. L. Williams, *Synth. Commun.* **1992**, *22*, 1017–1021.
219 K. Suda, K. Hotoda, F. Iemura, T. Takanami, *J. Chem. Soc., Perkin Trans. 1* **1993**, 1553–1555.
220 I. Matsuda, S. Murata, Y. Ishii, *J. Chem. Soc., Perkin Trans. 1* **1979**, 26–30.
221 J. Matsubara, K. Nakao, Y. Hamada, T. Shioiri, *Tetrahedron Lett.* **1992**, *33*, 4187–4190.
222 B. Mauze, L. Miginiac, *Synth. Commun.* **1990**, *20*, 2251–2258.
223 I. Ojima, M. Kumagai, *Tetrahedron Lett.* **1974**, 4005–4008.
224 B. Mauze, L. Migniac, *Synth. Commun.* **1992**, *22*, 2229–2235.
225 R. Haruta, M. Ishiguro, K. Furuta, A. Mori, N. Ikeda, H. Yamamoto, *Chem. Lett.* **1982**, 1093–1096.
226 K. Furuta, M. Ishiguro, R. Haruta, N. Ikeda, H. Yamamoto, *Bull. Chem. Soc. Jpn.* **1984**, *57*, 2768–2776.
227 E. Ehlinger, P. Magnus, *J. Am. Chem. Soc.* **1980**, *102*, 5004–5011.

228 Y. Ikeda, H. Yamamoto, *Bull. Chem. Soc. Jpn.* **1986**, *59*, 657–658.
229 Y. Yamamoto, Y. Saito, K. Maruyama, *J. Chem. Soc., Chem. Commun.* **1982**, 1326–1328.
230 M. J. Carter, I. Fleming, A. Percival, *J. Chem. Soc., Perkin Trans. 1* **1981**, 2415–2434.
231 R. Corriu, N. Escudie, C. Guerin, *J. Organomet. Chem.* **1984**, *264*, 207–216.
232 T.-H. Chan, J.-S. Li, *J. Chem. Soc., Chem. Commun.* **1982**, 969–970.
233 Y. Ikeda, J. Ukai, N. Ikeda, H. Yamamoto, *Tetrahedron Lett.* **1987**, *43*, 731–741.
234 I. Paterson, A. Schlapbach, *Synlett* **1995**, 498–500.
235 Y. Yamakado, M. Ishiguro, N. Ikeda, H. Yamamoto, *J. Am. Chem. Soc.* **1981**, *103*, 5568–5570.
236 E. J. Corey, C. Rücker, *Tetrahedron Lett.* **1982**, *23*, 719–722.
237 K. K. Wang, Z. Wang, Y. G. Gu, *Tetrahedron Lett.* **1993**, *34*, 8391–8394.
238 K. K. Wang, B. Liu, J. L. Petersen, *J. Am. Chem. Soc.* **1996**, *118*, 6860–6867.
239 K. K. Wang, C. Shi, J. L. Petersen, *J. Org. Chem.* **1998**, *63*, 4413–4419.
240 W. Friebolin, W. Eberbach, *Helv. Chim. Acta* **2001**, *84*, 3822–3836.
241 D. J. Ager, *Tetrahedron Lett.* **1981**, *22*, 2923–2926.
242 D. J. Ager, *J. Org. Chem.* **1984**, *49*, 168–170.
243 D. J. Ager, *J. Chem. Soc., Perkin Trans. 1* **1986**, 183–194.
244 T. Cohen, J. P. Sherbine, J. R. Matz, R. R. Hutchins, B. M. McHenry, P. R. Willey, *J. Am. Chem. Soc.* **1984**, *106*, 3245–3252.
245 P. A. Brown, R. V. Bonnert, P. R. Jenkins, M. R. Selim, *Tetrahedron Lett.* **1987**, *28*, 693–696.
246 D. J. Ager, *Tetrahedron Lett.* **1984**, *24*, 419–422.
247 W. G. Dauben, B. A. Kowalczyk, *Tetrahedron Lett.* **1990**, *31*, 635–638.
248 L. A. Paquette, K. A. Horn, G. J. Wells, *Tetrahedron Lett.* **1982**, *23*, 259–262.
249 T. Cohen, S.-H. Jung, M. L. Romberger, D. W. McCullough, *Tetrahedron Lett.* **1988**, *29*, 25–26.
250 C. A. Shook, M. L. Romberger, S.-H.

Jung, M. Xiao, J. P. Sherbine, B. Zhang, F.-T. Lin, T. Cohen, *J. Am. Chem. Soc.* **1993**, *115*, 10754–10773.
251 T. Takeda, K. Ando, A. Mamada, T. Fujiwara, *Chem. Lett.* **1985**, 1149–1152.
252 W. Dumont, A. Krief, *Angew. Chem. Int. Ed. Engl.* **1976**, *15*, 161.
253 S. Halazy, W. Dumont, A. Krief, *Tetrahedron Lett.* **1981**, *22*, 4737–4740.
254 B. Halton, C. J. Randall, G. J. Gainsford, P. J. Stang, *J. Am. Chem. Soc.* **1986**, *108*, 5949–5956.
255 B. Halton, M. J. Cooney, T. W. Davey, G. S. Forman, Q. Lu, R. Boese, D. Blaser, A. H. Maulitz, *J. Chem. Soc., Perkin Trans. 1* **1995**, 2819–2827.
256 B. Halton, C. S. Jones, *J. Chem. Soc., Perkin Trans. 2* **1998**, 2505–2508.
257 D. E. Seitz, A. Zapata, *Tetrahedron Lett.* **1980**, *21*, 3451–3454.
258 D. E. Seitz, A. Zapata, *Synthesis* **1981**, 557–558.
259 A. G. M. Barrett, J. M. Hill, *Tetrahedron Lett.* **1991**, *32*, 3285–3288.
260 T. Mukaiyama, *Angew. Chem. Int. Ed. Engl.* **1977**, *16*, 817–826.
261 T. Mukaiyama, *Org. React.* **1982**, *28*, 203–331.
262 I. Matsuda, Y. Izumi, *Tetrahedron Lett.* **1981**, *22*, 1805–1808.
263 I. Matsuda, *J. Organomet. Chem.* **1987**, *321*, 307–316.
264 L. Duhamel, J. Gralak, A. Bouyanzer, *Tetrahedron Lett.* **1993**, *34*, 7745–7748.
265 M. Bellassoued, N. Lensen, M. Bakasse, S. Mouelhi, *J. Org. Chem.* **1998**, *63*, 8785–8789.
266 N. Lensen, S. Mouelhi, M. Bellassoued, *Synth. Commun.* **2001**, *31*, 1007–1011.
267 G. Simchen, G. Siegel, *Synthesis* **1989**, 945–946.
268 G. Simchen, G. Siegel, *Liebigs Ann. Chem.* **1992**, 607–613.
269 K. Yamamoto, Y. Tomo, *Chem. Lett.* **1983**, 531–534.
270 Y. Tomo, K. Yamamoto, *Tetrahedron Lett.* **1985**, *26*, 1061–1064.
271 Y. Takeda, T. Matsumoto, F. Sato, *J. Org. Chem.* **1986**, *51*, 4728–4731.
272 R. Mizojiri, H. Urabe, F. Sato, *J. Org. Chem.* **2000**, *65*, 6217–6222.
273 M. J. Daly, R. A. Ward, D. F. Thompson, G. Procter, *Tetrahedron Lett.* **1995**, *36*, 7545–7548.
274 K. Kim, S. Okamoto, Y. Takayama, F. Sato, *Tetrahedron Lett.* **2002**, *43*, 4237–4239.
275 F. Rehders, D. Hoppe, *Synthesis* **1992**, 865–870.
276 G. Procter, A. T. Russell, P. J. Murphy, T. S. Tan, A. N. Mather, *Tetrahedron* **1988**, *44*, 3953–3973.
277 E. Torres, G. L. Larson, G. J. McGarvey, *Tetrahedron Lett.* **1988**, *29*, 1355–1358.
278 G. L. Larson, E. Torres, C. B. Morales, G. J. McGarvey, *Organometallics* **1986**, *5*, 2274–2283.
279 T. H. Chan, W. Mychajlowskij, B. S. Ong, D. N. Harpp, *J. Org. Chem.* **1978**, *43*, 1526–1532.
280 T. H. Chan, W. Mychajlowskij, *Tetrahedron Lett.* **1974**, 171–174.
281 C. Palomo, J. M. Aizpurua, J. M. García, I. Ganboa, F. P. Cossio, B. Lecea, C. López, *J. Org. Chem.* **1990**, *55*, 2498–2503.
282 H. Wetter, *Helv. Chim. Acta* **1978**, *61*, 3072–3075.
283 W. Meichle, H.-H. Otto, *Arch. Pharm.* **1989**, *322*, 263–270.
284 B. Halton, Q. Lu, P. J. Stang, *Aust. J. Chem.* **1990**, *43*, 1277–1282.
285 J.-P. Picard, *Can. J. Chem.* **2000**, *78*, 1363–1379.
286 V. Snieckus, *Pure Appl. Chem.* **1990**, *62*, 671–680.
287 C. Palomo, J. M. Aizpurua, M. Legido, J. P. Picard, J. Dunogues, T. Constantieux, *Tetrahedron Lett.* **1992**, *33*, 3903–3906.
288 J.-C. Cuevas, P. Patil, V. Snieckus, *Tetrahedron Lett.* **1989**, *30*, 5841–5844.
289 J. Lasarte, C. Palomo, J. P. Picard, J. Dunogues, J. M. Aizpurua, *J. Chem. Soc., Chem. Commun.* **1989**, 72–74.
290 T. Hirao, A. Yamada, Y. Ohshiro, T. Agawa, *Angew. Chem. Int. Ed. Engl.* **1981**, *20*, 126–127.
291 T. Hirao, A. Yamada, K.-i. Hayashi, Y. Ohshiro, T. Agawa, *Bull. Chem. Soc. Jpn.* **1982**, *55*, 1163–1167.

292 A. R. Katritzky, R. J. Offerman, P. Cabildo, M. Soleiman, *Recl. Trav. Chim. Pays-Bas* **1988**, *107*, 641–645.
293 D. P. M. Pleynet, J. K. Dutton, M. Thornton-Pett, A. P. Johnson, *Tetrahedron Lett.* **1995**, *36*, 6321–6324.
294 S. Shimizu, M. Ogata, *J. Org. Chem.* **1987**, *52*, 2314–2315.
295 C. Palomo, J. M. Aizpurua, J. M. García, J. P. Picard, J. Dunogues, *Tetrahedron Lett.* **1990**, *31*, 1921–1924.
296 A. Couture, H. Cornet, E. Deniau, P. Grandclaudon, S. Lebrun, *J. Chem. Soc., Perkin Trans. 1* **1997**, 469–476.
297 P. A. Patil, V. Snieckus, *Tetrahedron Lett.* **1998**, *39*, 1325–1326.
298 M. Bellassoued, N. Ozanne, *J. Org. Chem.* **1995**, *60*, 6582–6584.
299 M. Bellassoued, E. Reboul, M. Salemkour, *Synth. Commun.* **1995**, *25*, 3097–3108.
300 M. Bellassoued, E. Reboul, M. Salemkour, *Tetrahedron* **1996**, *52*, 4607–4624.
301 M. Bellassoued, M. Salemkour, E. Reboul, *Synth. Commun.* **1997**, *27*, 3103–3117.
302 M. Bellassoued, M. Salemkour, *Bull. Chim. Soc. Fr.* **1997**, *14*, 115–124.
303 Y. Kita, J. Sekihachi, Y. Hayashi, Y.-Z. Da, M. Yamamoto, S. Akai, *J. Org. Chem.* **1990**, *55*, 1108–1112.
304 P. F. Hudrlik, D. J. Peterson, *Tetrahedron Lett.* **1974**, 1133–1136.
305 S. Kanemasa, H. Kobayashi, J. Tanaka, O. Tsuge, *Bull. Chem. Soc. Jpn.* **1988**, *61*, 3957–3964.
306 M. P. Cooke, Jr., C. M. Pollock, *J. Org. Chem.* **1993**, *58*, 7474–7481.
307 J. Tanaka, H. Kobayashi, S. Kanemasa, O. Tsuge, *Bull. Chem. Soc. Jpn.* **1989**, *62*, 1193–1197.
308 J. Tanaka, S. Kanemasa, Y. Ninomiya, O. Tsuge, *Bull. Chem. Soc. Jpn.* **1990**, *63*, 466–475.
309 P. F. Hudrlik, D. Peterson, *Tetrahedron Lett.* **1972**, 1785–1787.
310 K. Utimoto, M. Obayashi, H. Nozaki, *J. Org. Chem.* **1976**, *41*, 2940–2941.
311 M. Obayashi, K. Utimoto, H. Nozaki, *Bull. Chem. Soc. Jpn.* **1979**, *52*, 1760–1764.
312 P. F. Hudrlik, A. K. Kulkarni, *J. Am. Chem. Soc.* **1981**, *103*, 6251–6253.
313 P. A. Brown, R. V. Bonnert, P. R. Jenkins, N. J. Lawrence, M. R. Selim, *J. Chem. Soc., Perkin Trans. 1* **1991**, 1893–1894.
314 G. Bartoli, G. Palmieri, M. Petrini, *Tetrahedron* **1990**, *46*, 1379–1384.
315 A. Barbero, Y. Blanco, C. García, F. J. Pulido, *Synthesis* **2000**, 1223–1228.
316 J. Fässler, A. Linden, S. Bienz, *Tetrahedron* **1999**, *55*, 1717–1730.
317 A. G. Barrett, J. A. Flygare, *J. Org. Chem.* **1991**, *56*, 638–642.
318 A. G. M. Barrett, J. M. Hill, E. M. Wallace, *J. Org. Chem.* **1992**, *57*, 386–389.
319 D. Enders, S. Nakai, *Chem. Ber.* **1991**, *124*, 219–226.
320 D. Hernández, G. L. Larson, *J. Org. Chem.* **1984**, *49*, 4285–4287.
321 G. L. Larson, D. Hernández, *Tetrahedron Lett.* **1982**, 1035–1038.
322 V. A. Khripach, V. N. Zhabinskii, O. V. Konstantinova, N. B. Khripach, *Tetrahedron Lett.* **2000**, *41*, 5765–5767.
323 P. F. Hudrlik, D. Peterson, R. J. Rona, *J. Org. Chem.* **1975**, *40*, 2263–2264.
324 J. A. Soderquist, B. Santiago, *Tetrahedron Lett.* **1989**, *30*, 5693–5696.
325 D. C. Chauret, J. M. Chong, Q. Ye, *Tetrahedron: Asymmetry* **1999**, *10*, 3601–3614.
326 S.-S. P. Chou, H.-L. Kuo, C.-J. Wang, C.-Y. Tsai, C.-M. Sun, *J. Org. Chem.* **1989**, *54*, 868–872.
327 K. Matsumoto, Y. Takeyama, K. Oshima, K. Utimoto, *Tetrahedron Lett.* **1991**, *32*, 4545–4548.
328 S. Schabbert, E. Schaumann, *Eur. J. Org. Chem.* **1998**, 1873–1878.
329 Y. Zhang, J. A. Miller, E. Negishi, *J. Org. Chem.* **1989**, *54*, 2043–2044.
330 P. F. Hudrlik, A. M. Hudrlik, R. N. Misra, D. Peterson, G. P. Withers, A. K. Kulkarni, *J. Org. Chem.* **1980**, *45*, 4444–4448.
331 S. Okamoto, T. Yoshino, H. Tsujiyama, F. Sato, *Tetrahedron Lett.* **1991**, *32*, 5793–5796.
332 A. Fürstner, C. Brehm, Y. Cancho-

Grande, *Org. Lett.* **2001**, *3*, 3955–3957.
333 Y. Ukaji, A. Yoshida, T. Fujisawa, *Chem. Lett.* **1990**, 157–160.
334 T. Kauffmann, R. Koenig, R. Kriegesmann, M. Wensing, *Tetrahedron Lett.* **1984**, *25*, 641–644.
335 P. Cuadrado, A. M. González-Nogal, *Tetrahedron Lett.* **2000**, *41*, 1111–1114.
336 P. F. Hudrlik, D. M. R. S. Bhamidipati, A. M. Hudrlik, *J. Org. Chem.* **1996**, *61*, 8655–8658.
337 F. Babudri, V. Fiandanese, G. Marchese, A. Punzi, *Tetrahedron* **2001**, *57*, 549–554.
338 P. F. Hudrlik, A. M. Hudrlik, R. J. Rona, R. N. Misra, G. P. Withers, *J. Am. Chem. Soc.* **1977**, *99*, 1993–1996.
339 S. Okamoto, T. Shimazaki, Y. Kobayashi, F. Sato, *Tetrahedron Lett.* **1987**, *28*, 2033–2036.
340 M. Shimizu, H. Yoshioka, *Tetrahedron Lett.* **1989**, *30*, 967–970.
341 P. B. Dervan, M. A. Shippey, *J. Am. Chem. Soc.* **1976**, *98*, 1265–1267.
342 M. T. Reetz, M. Plachky, *Synthesis* **1976**, 199–200.
343 S. R. Wilson, M. J. Di Grandi, *J. Org. Chem.* **1991**, *56*, 4766–4772.
344 A. T. Russell, G. Procter, *Tetrahedron Lett.* **1987**, *28*, 2041–2044.
345 M. Suginome, T. Iwanami, A. Matsumoto, Y. Ito, *Tetrahedron: Asymmetry* **1997**, *8*, 859–862.
346 H. C. Brown, R. Liotta, G. W. Kramer, *J. Am. Chem. Soc.* **1979**, *101*, 2966–2970.
347 H. C. Brown, P. V. Ramachandran, *Pure Appl. Chem.* **1991**, *63*, 307–316.
348 W. R. Roush, in *Comprehensive Organic Synthesis*, Vol. 2 (Eds.: B. M. Trost, I. Fleming), Pergamon Press, Oxford, **1991**, pp. 1–53.
349 C. Liu, K. K. Wang, *J. Org. Chem.* **1986**, *51*, 4736–4737.
350 K. K. Wang, C. Liu, Y. G. Gu, F. N. Burnett, P. D. Sattsangi, *J. Org. Chem.* **1991**, *56*, 1914–1922.
351 P. D. Sattsangi, K. K. Wang, *Tetrahedron Lett.* **1992**, *33*, 5025–5028.
352 W. H. Pearson, K.-C. Lin, Y.-F. Poon, *J. Org. Chem.* **1989**, *54*, 5814–5819.
353 W. H. Pearson, J. M. Schkeryantz, *Synthesis* **1991**, 342–346.
354 K. K. Wang, Y. G. Gu, C. Liu, *J. Am. Chem. Soc.* **1990**, *112*, 4424–4431.
355 Y. W. Andemichael, K. K. Wang, *J. Org. Chem.* **1992**, *57*, 796–798.
356 M. J. Prior, G. H. Whitham, *J. Chem. Soc., Perkin Trans. 1* **1986**, 683–688.
357 S. Okamoto, K. Tani, F. Sato, K. B. Sharpless, D. Zargarian, *Tetrahedron Lett.* **1993**, *34*, 2509–2512.
358 A. R. Bassindale, P. G. Taylor, Y. Xu, *J. Chem. Soc., Perkin Trans. 1* **1994**, 1061–1067.
359 J. A. Soderquist, A. M. Rane, C. J. Lopez, *Tetrahedron Lett.* **1993**, *34*, 1893–1896.
360 F. Sato, Y. Tanaka, M. Sato, *J. Chem. Soc., Chem. Commun.* **1983**, 165–166.
361 S. R. Wilson, G. M. Geogiadis, *J. Org. Chem.* **1983**, *48*, 4143–4144.
362 T. Kawashima, *Bull. Chem. Soc. Jpn.* **2003**, *76*, 471–483.
363 H. Shinokubo, K. Miura, K. Oshima, K. Utimoto, *Tetrahedron Lett.* **1993**, *34*, 1951–1954.
364 H. Shinokubo, K. Oshima, K. Utimoto, *Chem. Lett.* **1995**, 461–462.
365 H. Shinokubo, K. Miura, K. Oshima, K. Utimoto, *Tetrahedron* **1996**, *52*, 503–514.
366 I. Fleming, C. D. Floyd, *J. Chem. Soc., Perkin Trans. 1* **1981**, 969–976.
367 E. Shaumann, C. Friese, *Tetrahedron Lett.* **1989**, *30*, 7033–7036.
368 A. G. Angoh, D. L. J. Clive, *J. Chem. Soc., Chem. Commun.* **1984**, 534–536.
369 I. Fleming, I. T. Morgan, A. K. Sarkar, *J. Chem. Soc., Chem. Commun.* **1990**, 1575–1577.
370 I. Fleming, I. T. Morgan, A. K. Sarkar, *J. Chem. Soc., Perkin Trans. 1* **1998**, 2749–2763.
371 R. Angell, P. J. Parsons, A. Naylor, E. Tyrrell, *Synlett* **1992**, 599–600.
372 H.-F. Chow, X.-P. Cao, M.-k. Leung, *J. Chem. Soc., Chem. Commun.* **1994**, 2121–2122.
373 K. K. Wang, B. Liu, Y.-d. Lu, *J. Org. Chem.* **1995**, *60*, 1885–1887.
374 S. Takano, S. Otaki, K. Ogasawara, *Heterocycles* **1985**, *23*, 2811–2814.
375 S. Takano, S. Otaki, K. Ogasawara,

J. Chem. Soc., Chem. Commun. **1985**, 485–487.
376 S. Takano, N. Sato, S. Otaki, K. Ogasawara, *Heterocycles* **1987**, *25*, 69–73.
377 M. Lautens, S. Ma, R. K. Belter, P. Chiu, A. Leschziner, *J. Org. Chem.* **1992**, *57*, 4065–4066.
378 N. Kishi, T. Maeda, K. Mikami, T. Nakai, *Tetrahedron* **1992**, *48*, 4087–4090.
379 K. Mikami, T. Maeda, T. Nakai, *Tetrahedron Lett.* **1986**, *27*, 4189–4190.
380 M. Harmata, D. E. Jones, *Tetrahedron Lett.* **1997**, *38*, 3861–3862.
381 T. Kawashima, M. Nakamura, N. Inamoto, *Phosphorus, Sulfur, and Silicon* **1992**, *69*, 293–297.
382 O. Tsuge, S. Kanemasa, Y. Ninomiya, *Chem. Lett.* **1984**, 1993–1996.
383 T. Kauffmann, *Angew. Chem. Int. Ed. Engl.* **1982**, *21*, 410–429.
384 S. Inoue, Y. Sato, *Organometallics* **1988**, *7*, 739–743.
385 S. Inoue, Y. Sato, *J. Org. Chem.* **1991**, *56*, 347–352.
386 S. Inoue, Y. Sato, *Organometallics* **1986**, *5*, 1197–1201.
387 S. Urayama, S. Inoue, Y. Sato, *J. Organomet. Chem.* **1988**, *354*, 155–160.
388 T. Kawashima, N. Iwama, N. Tokitoh, R. Okazaki, *J. Org. Chem.* **1994**, *59*, 491–493.
389 T. Kawashima, Y. Nishiwaki, R. Okazaki, *J. Organomet. Chem.* **1995**, *499*, 143–146.
390 T. Kauffmann, *Tetrahedron Lett.* **1978**, 4399–4402.
391 T. Kauffmann, R. Kriegesmann, A. Hamsen, *Chem. Ber.* **1982**, *115*, 1818–1824.
392 T. Kawashima, N. Iwama, N. Tokitoh, R. Okazaki, *J. Am. Chem. Soc.* **1993**, *115*, 2507–2508.
393 N. Minouni, E. About-Jaudet, N. Collignon, P. Savignac, *Synth. Commun.* **1991**, *21*, 2341–2348.
394 N. Minouni, H. A. Badri, E. About-Jaudet, N. Collignon, *Synth. Commun.* **1995**, *25*, 1921–1932.
395 M. C. Fernández, M. Ruiz, V. Ojea, J. M. Quintela, *Tetrahedron Lett.* **2002**, *43*, 5909–5912.
396 A. Zapata, C. Fortoul R., C. Acuña A., *Synth. Commun.* **1985**, *15*, 179–184.
397 N. Kanemoto, Y. Sato, S. Inoue, *J. Organomet. Chem.* **1988**, *348*, 25–31.
398 D. J. Ager, G. E. Cooke, M. B. East, S. J. Mole, A. Rampersaud, V. J. Webb, *Organometallics* **1986**, *5*, 1906–1908.
399 A. Rensing, K.-J. Echsler, T. Kauffmann, *Tetrahedron Lett.* **1980**, *21*, 2807–2810.
400 W. Verboom, M. R. J. Hamzink, D. N. Reinhoudt, R. Visser, *Tetrahedron Lett.* **1984**, *25*, 4309–4312.
401 D. J. Hart, K. Kanai, D. G. Thomas, T.-K. Yang, *J. Org. Chem.* **1983**, *48*, 289–294.
402 G. Cainelli, M. Panunzio, D. Giacomini, G. Martelli, G. Spunta, *J. Am. Chem. Soc.* **1988**, *110*, 6879–6880.
403 C. K. Murray, B. P. Warner, V. Dragisich, W. D. Wulff, R. D. Rogers, *Organometallics* **1990**, *9*, 3142–3151.
404 P. Bayard, L. Ghosez, *Tetrahedron Lett.* **1988**, *29*, 6115–6118.
405 B. Alcaide, J. Rodríguez-López, *J. Chem. Soc., Perkin Trans. 1* **1990**, 2451–2457.
406 G. I. Georg, G. C. Harriman, M. Hepperle, J. S. Clowers, D. G. V. Velde, R. H. Himes, *J. Org. Chem.* **1996**, *61*, 2664–2676.
407 G. Cainelli, M. Panunzio, E. Bandini, G. Martelli, G. Spunta, M. D. Col, *Tetrahedron* **1995**, *51*, 5067–5072.
408 G. Cainelli, D. Giacomini, M. Panunzio, G. Martelli, G. Spunta, *Tetrahedron Lett.* **1987**, *28*, 5369–5372.
409 G. Cainelli, D. Giacomini, P. Galletti, *Synthesis* **1997**, 886–890.
410 Y. Matsuda, S. Tanimoto, T. Okamoto, S. M. Ali, *J. Chem. Soc., Perkin Trans. 1* **1989**, 279–281.
411 B. Alcaide, J. Rodríguez-López, *J. Chem. Soc., Perkin Trans. 1* **1989**, 279–281.
412 E. W. Colvin, D. G. McGarry, *J. Chem. Soc., Chem. Commun.* **1985**, 539–541.
413 G. Cainelli, D. Giacomini, P. Galletti, A. Gaiba, *Synlett* **1996**, 657–658.
414 F. A. Davis, R. E. Reddy, J. M. Szewczyk, G. V. Reddy, P. S.

Portonovo, H. Zhang, D. Fanelli, R. T. Reddy, P. Zhou, P. J. Carroll, *J. Org. Chem.* **1997**, *62*, 2555–2563.

415 M. Regitz, O. J. Scherer (Eds.), *Multiple Bonds and Low Coordination in Phosphorus Chemistry*, Georg Thieme Verlag, Stuttgart, **1990**.

416 M. Yoshifuji, K. Toyota, *The Chemistry of Organosilicon Compounds*, Vol. 3 (Eds.: Z. Rappoport, Y. Apeloig), Wiley, Chichester, **1998**, Chapter 8, 491–539.

417 M. Yoshifuji, K. Toyota, N. Inamoto, *Chem. Lett.* **1985**, 1727–1730.

418 M. Yoshifuji, K. Toyota, I. Matsuda, T. Niitsu, N. Inamoto, K. Hirotsu, T. Higuchi, *Tetrahedron* **1988**, *44*, 1363–1367.

419 G. Märkl, K. M. Raab, *Tetrahedron Lett.* **1989**, *30*, 1077–1080.

420 E. P. O. Fuchs, H. Heydt, M. Regitz, W. W. Schoeller, T. Busch, *Tetrahedron Lett.* **1989**, *30*, 5111–5114.

421 E. Fuchs, B. Breit, H. Heydt, W. Schoeller, T. Busch, C. Krüger, P. Betz, M. Regitz, *Chem. Ber.* **1991**, *124*, 2843–2855.

422 S. J. Goede, L. de Vries, F. Bickelhaupt, *Bull. Soc. Chim. Fr.* **1993**, *130*, 185–188.

423 J. Escudia, H. Ranaivonjatovo, L. Rigon, *Chem. Rev.* **2000**, *100*, 3639–3696.

424 M. Yoshifuji, K. Toyota, K. Shibayama, N. Inamoto, *Tetrahedron Lett.* **1984**, *25*, 1809–1812.

425 G. Märkl, P. Kreitmeier, *Angew. Chem. Int. Ed. Engl.* **1988**, *27*, 1360–1361.

426 T. Niitsu, N. Inamoto, K. Toyota, M. Yoshifuji, *Bull. Chem. Soc. Jpn.* **1990**, *63*, 2736–2738.

427 R. Appel, C. Behnke, *Z. Anorg. Allg. Chem.* **1987**, *555*, 23–35.

428 R. Appel, P. Fölling, B. Josten, M. Siray, V. Winkhaus, F. Knoch, *Angew. Chem. Int. Ed. Engl.* **1984**, *23*, 619–620.

429 M. Yoshifuji, S. Sasaki, N. Inamoto, *Tetrahedron Lett.* **1989**, *30*, 839–842.

430 G. Raabe, J. Michl, *The Chemistry of Organosilicon Compounds* (Eds.: S. Patai, Z. Rappoport), Wiley, Chichester, **1989**, Chapter 17, 1015–1142.

431 T. Muller, W. Ziche, N. Auner, *The Chemistry of Organosilicon Compounds*, Vol. 2 (Eds.: S. Patai, Z. Rappoport), Wiley, Chichester, **1998**, Chapter 16, 857–1062.

432 A. G. Brook, M. A. Brook, *Adv. Organomet. Chem.* **1995**, *39*, 71–158.

433 D. Bravo-Zhivotovskii, V. Braude, A. Stanger, M. Kapon, Y. Apeloig, *Organometallics* **1992**, *11*, 2326–2328.

434 C. Krempner, D. Hoffmann, H. Oehme, R. Kempe, *Organometallics* **1997**, *16*, 1828–1832.

435 D. Hoffmann, T. Gross, R. Kempe, H. Oehme, *J. Organomet. Chem.* **2000**, *598*, 395–402.

436 K. Schmohl, M. Blach, H. Reinke, R. Kempe, H. Oehme, *Eur. J. Inorg. Chem.* **1998**, 1667–1672.

437 J. Ohshita, Y. Masaoka, M. Ishikawa, *Organometallics* **1991**, *10*, 3775–3776.

438 J. Ohshita, Y. Masaoka, M. Ishikawa, T. Takeuchi, *Organometallics* **1993**, *12*, 876–879.

439 Y. Apeloig, M. Bendikov, M. Yuzefovich, M. Nakash, D. Bravo-Zhivotovskii, *J. Am. Chem. Soc.* **1996**, *118*, 12228–12229.

440 K. Sakamoto, J. Ogasawara, Y. Kon, T. Sunagawa, C. Kabuto, M. Kira, *Angew. Chem. Int. Ed. Engl.* **2002**, *41*, 1402–1404.

441 T. Gross, R. Kempe, H. Oehme, *Eur. J. Inorg. Chem.* **1999**, 21–26.

442 D. Bravo-Zhivotovskii, I. Zharov, M. Kapon, Y. Apeloig, *J. Chem. Soc., Chem. Commun.* **1995**, 1625–1626.

443 M. van der Leij, P. A. T. W. Porskamp, B. H. M. Lammerink, B. Zwanenburg, *Tetrahedron Lett.* **1978**, 811–814.

444 M. van der Leij, B. Zwanenburg, *Tetrahedron Lett.* **1978**, 3383–3386.

445 P. A. T. W. Porskamp, B. H. M. Lammerink, B. Zwanenburg, *J. Org. Chem.* **1984**, *49*, 263–268.

446 P. A. T. W. Porskamp, A. M. van de Wijdeven, B. Zwanenburg, *Recl. Trav. Chim. Pays-Bas* **1983**, *102*, 506–510.

447 P. A. T. W. Porskamp, M. van der Leij, B. H. M. Lammerink, B. Zwanenburg, *Recl. Trav. Chim. Pays-Bas* **1983**, *102*, 400–404.

3
The Julia Reaction

*Raphaël Dumeunier and István E. Markó**

3.1
Introduction

The formation of C–C double bonds is of paramount importance in organic chemistry, because of both the ubiquitous presence of alkenes in a wide range of biologically active natural products and the numerous subsequent functionalizations that may be performed on the olefinic linkage.

The large number of synthetic methodologies that have been devised over the years to efficiently accomplish this key synthetic transformation stands as a clear testimony to its cardinal position among C–C bond-forming processes.

Among the various methods, the Julia reaction [1–4] is a connective sequence that usually involves two separate steps: a condensation and a reductive elimination. In its original form, the process consisted of the formation of a carbon–carbon double bond through the coupling of an anion α to a sulfone residue **2** and a carbonyl compound **3**, thereby generating the β-hydroxysulfone **4**, followed by a reductive elimination to afford the alkene **1** (Scheme 3.1).

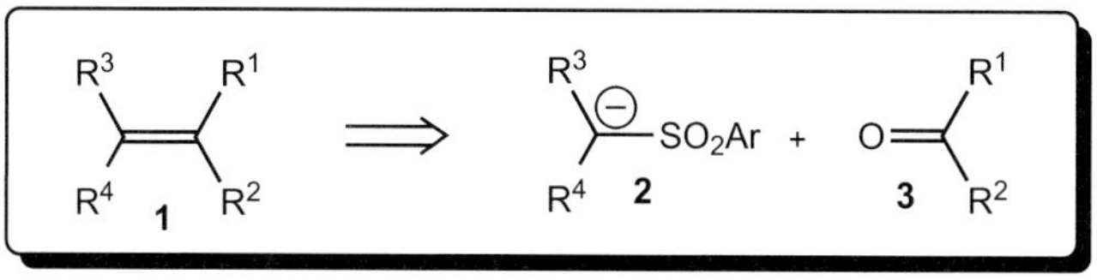

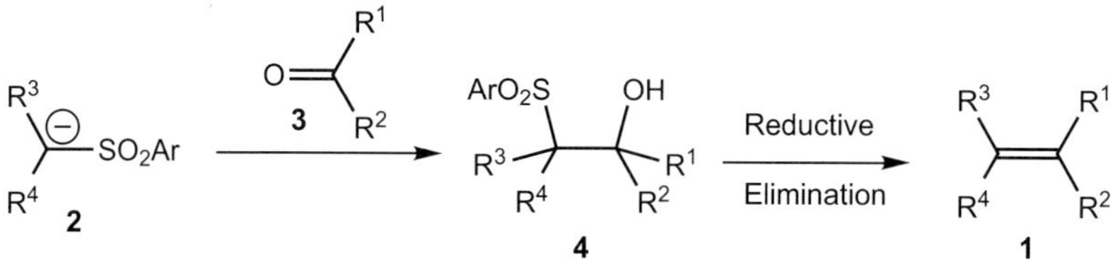

Scheme 3.1. The Julia sequence.

Modern Carbonyl Olefination. Edited by Takeshi Takeda
Copyright © 2004 WILEY-VCH Verlag GmbH & Co. KGaA, Weinheim
ISBN: 3-527-30634-X

After the initial condensation, derivatization of the β-hydroxysulfone **4** can lead to useful products. For instance, functionalization or activation of the hydroxyl group of **4** often facilitates the reductive elimination to **1**. Moreover, when the sulfone is either primary ($R^4 = R^3 = H$) or secondary (R^4 or $R^3 = H$), elimination to provide the useful vinyl sulfones **6** can easily be induced from **4** or **5** (Scheme 3.2). Although other variations have been reported, these are the two most important.

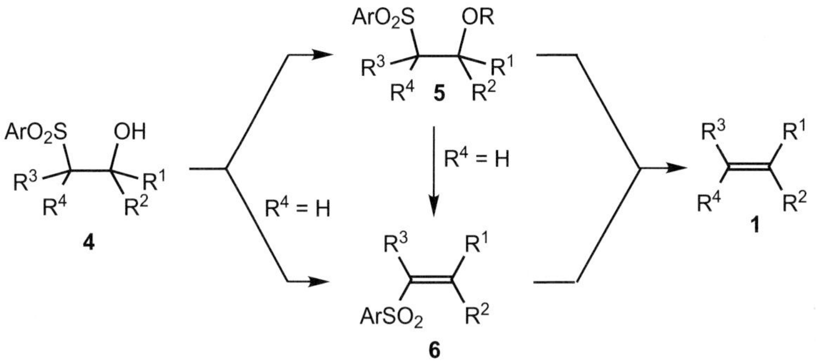

Scheme 3.2. Variations of the Julia sequence.

3.2
Historical Background

All the elements constituting the Julia reaction were known well before the assembling of this sequence by Julia and his group. In 1937, the first coupling between an aldehyde and an α-sulfonyl carbanion was performed [5] (reaction 3.1; Scheme 3.3). In reaction 3.2, a subsequent dehydration of the alcohol **12** under acidic conditions leads to the corresponding vinyl sulfone **13** [6]. An alternative method had been developed earlier by Culvenor et al. to synthesize β-hydroxy sulfones from epoxides rather than aldehydes [7] (reaction 3.3).

Since reductive elimination of aliphatic sulfones had already been studied in the early 1940s [8–10], all the pieces of the Julia sequence were at hand within the scientific community and assembling them to generate a powerful alkenylation methodology was simultaneously achieved by two different research groups. In fact, Umani-Ronchi et al. [11] published what can objectively be considered as the first Julia olefination sequence some two months before Julia [12] (Scheme 3.4).

Besides the Julia reaction, similar connective sequences had been published a little earlier, using sulfur(II) or sulfur(IV) derivatives instead of sulfones (Scheme 3.5). Whereas Durst's and Kuwajima's olefinations [13, 14] do not employ a reductive elimination step (reactions 3.5 and 3.6), Coates' method [15] employs Li in NH_3 to generate the desired olefinic linkage from **21**, and is consequently closely related to the Julia reaction (reaction 3.4).

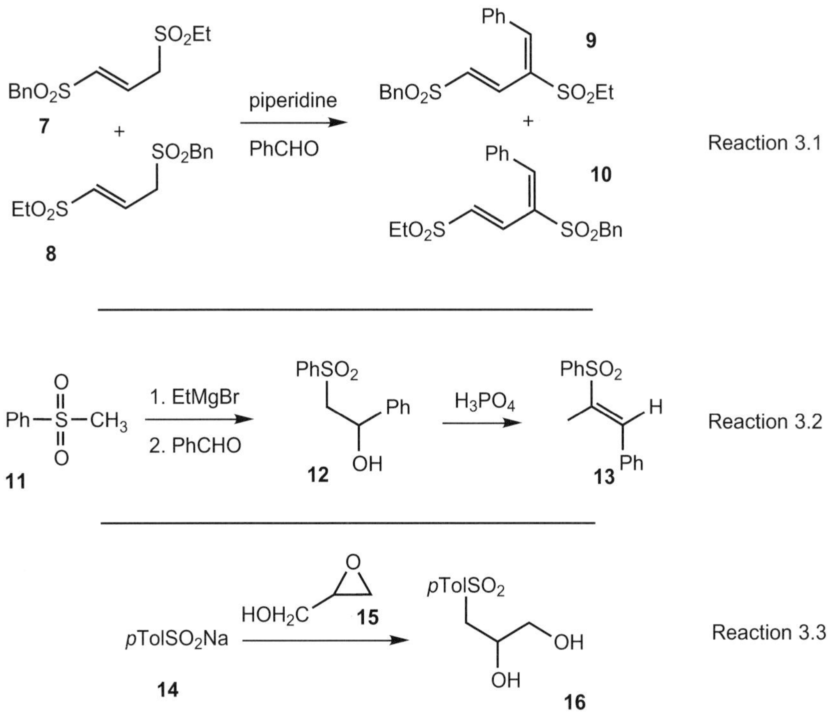

Scheme 3.3. First syntheses of hydroxy and vinyl sulfones.

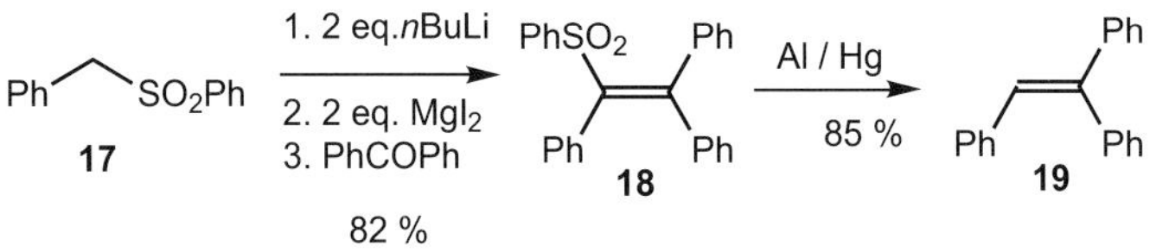

Scheme 3.4. The first reported Julia sequence.

3.3
Coupling Between the Two Precursors of the Julia Reaction

The disconnection of a C=C double bond into two fragments by a retro-Julia reaction offers two complementary possibilities. Either of the two fragments can be chosen as the carbonyl compound or as the sulfone derivative, or vice versa (Scheme 3.6). The choice is often made after considering several aspects of the connective step in order to avoid potential problems.

Typical factors that should be taken into account are the nature and number of substituents on the sulfone-bearing fragment, the nature of the counter ion, and the nature and reactivity of the carbonyl compound – aldehyde or ketone – generated by this disconnection. Some of the main problems that can usually be

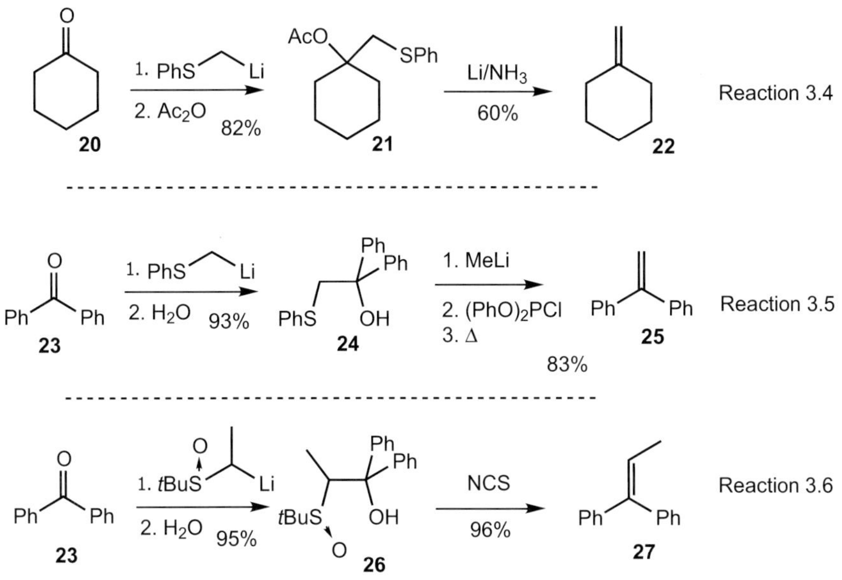

3.3 Coupling Between the Two Precursors of the Julia Reaction

Scheme 3.5. Anterior similar olefin syntheses.

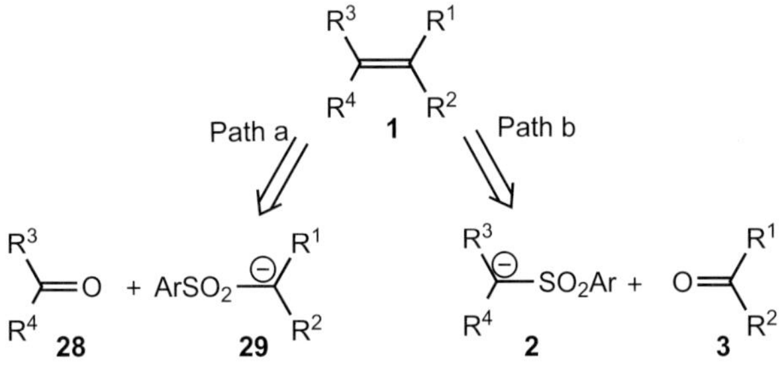

Scheme 3.6. Two possible disconnections.

avoided by a careful choice include: a difficult metallation of the sulfone derivative, a lack of reactivity of one or both partner(s), an unfavorable position of the equilibrium involving the addition reaction, a competitive elimination of a leaving group located β to the sulfone or the carbonyl compound, and a competitive reduction of the carbonyl group by the anion of the sulfone. Similar considerations should be borne in mind when a good control of the *E*- or *Z*-geometry of the alkene product is desired (*vide infra*).

3.3.1
Synthesis of Terminal Olefins

Since one or the other of the C-1 fragments generated by both disconnections are usually commercially available (i.e. MeSO$_2$Ph or HCHO), the choice of path a or path b depends much more on the accessibility of the other synthetic partner and has to be assessed on a case-by-case basis. To the best of our knowledge, no particular problems have been reported concerning these kinds of couplings. For example, primary and secondary α-metalated sulfones add smoothly to formaldehyde [16, 17] (Scheme 3.7). Although this approach is quite successful, few examples have hitherto been reported [16–23].

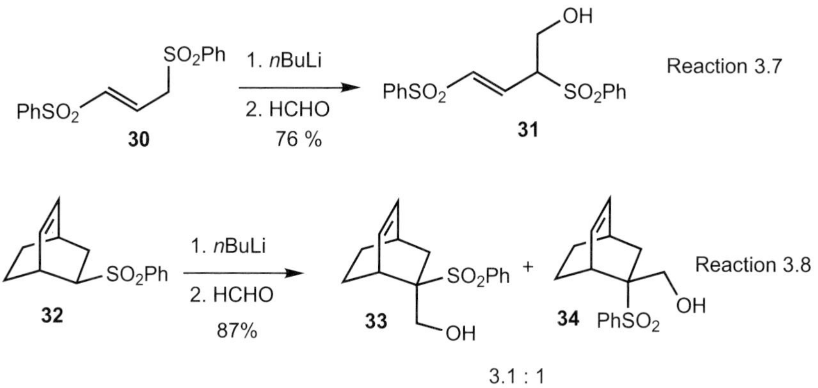

Scheme 3.7. Coupling step involving HCHO.

More widely used is the addition of MCH$_2$SO$_2$Ph to various aldehydes or ketones (Scheme 3.8). From reaction 3.9 [24], it is evident that the higher reactivity of aldehydes, coupled with the judicious choice of the metalating agent, can lead to addition products in a completely chemoselective manner.

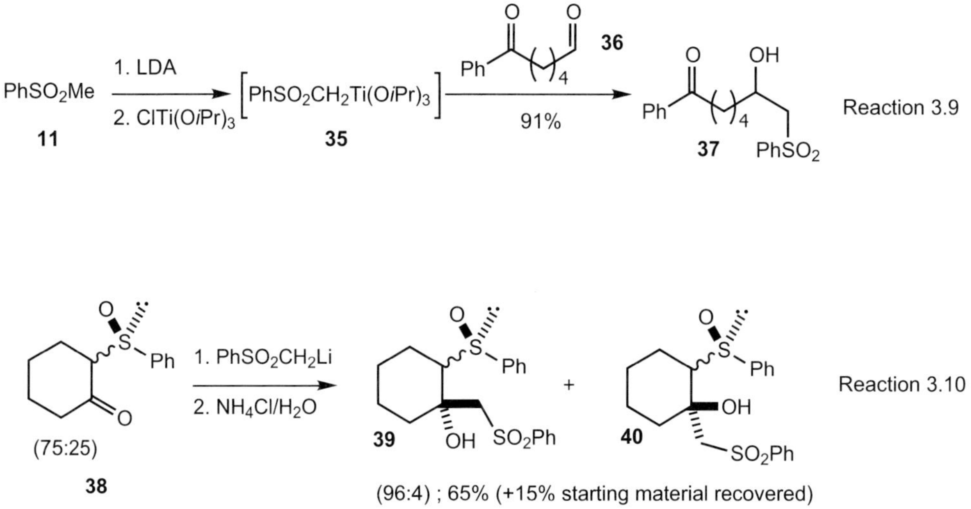

Scheme 3.8. Coupling step involving MCH$_2$SO$_2$Ph.

Condensation of sulfone-derived anions with ketones – including enolizable ketones – can also be brought to fruition. It must be noted, however, that in reaction 3.10 [25], some 15% of starting material **38** is recovered, probably originating from competitive deprotonation of the highly acidic α-keto hydrogen. Good 1,2-diastereocontrol is also exercised in this case.

3.3.2
Preparation of 1,2-Disubstituted Olefins

At the outset, it should be noted that the original Julia olefination protocol is mostly useful in the synthesis of (*E*)-olefins. Depending upon the conditions of the reductive elimination, the *E*/*Z* selectivity can be modulated, but in general (*E*)-alkenes are usually easier to obtain than their (*Z*)-counterparts. This stereochemical control – which also arises in the preparation of tri- and tetrasubstituted olefins – will be discussed in greater detail in the section on reductive elimination (Section 3.4).

Regardless of this specific consideration, 1,2-disubstituted alkenes are easily generated using the Julia sequence. In general, the sulfone derivative is unhindered and the carbonyl compound, usually an aldehyde, is often reactive enough to minimize retrograde fragmentation to the starting materials [26–28]. This retro-aldol-type reaction is commonly encountered under the basic conditions required for the initial addition of the sulfone anion to the carbonyl partner (Scheme 3.9).

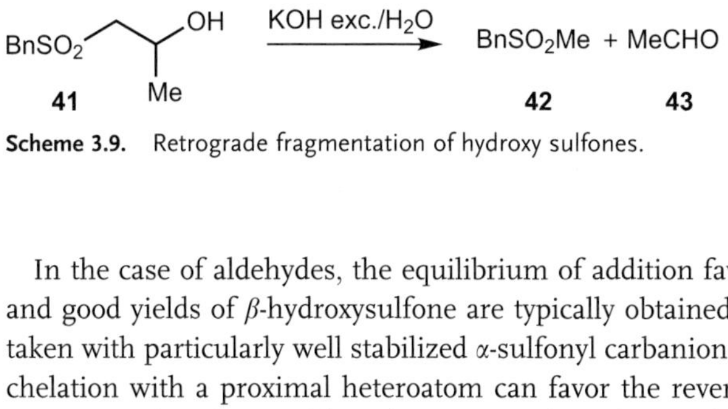

Scheme 3.9. Retrograde fragmentation of hydroxy sulfones.

In the case of aldehydes, the equilibrium of addition favors the β-alkoxy sulfone and good yields of β-hydroxysulfone are typically obtained. However, care must be taken with particularly well stabilized α-sulfonyl carbanions. Indeed, conjugation or chelation with a proximal heteroatom can favor the reverse reaction. Hannessian encountered such a problem during a total synthesis of (+)-Avermectin B1α [29] (Scheme 3.10).

As already discussed in some excellent previous reviews, several solutions have been devised to solve this problem. For instance, varying the nature of the counter ion can efficiently shift this unfavorable equilibrium. Thus, replacing lithium by magnesium [30–37] or the use of a lithiated sulfone/boron trifluoride combination [38–43] have proved to be effective. In Scheme 3.11, reaction 3.11 shows the beneficial use of magnesium salts in the Julia coupling employed as a key step in the synthesis of 24-epi-26,26,26,27,27,27-hexafluoro-1α,25-dihydroxyvitamin D2 by Kobayashi et al. [44]. In contrast, BF$_3$·OEt$_2$ was preferred by White et al. in their synthesis of the spiroketal segment of the Rutamycins [1].

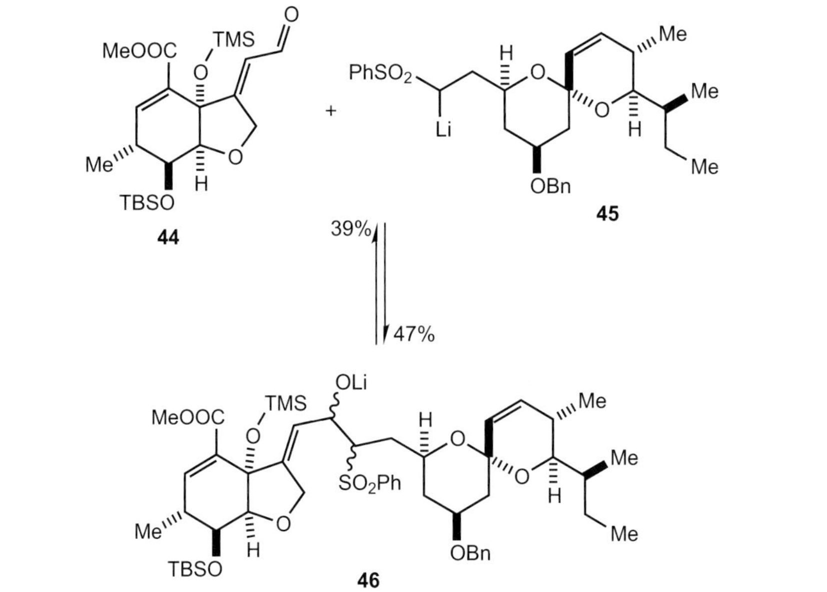

Scheme 3.10. Retrograde fragmentation during the synthesis of (+)-Avermectin B1α.

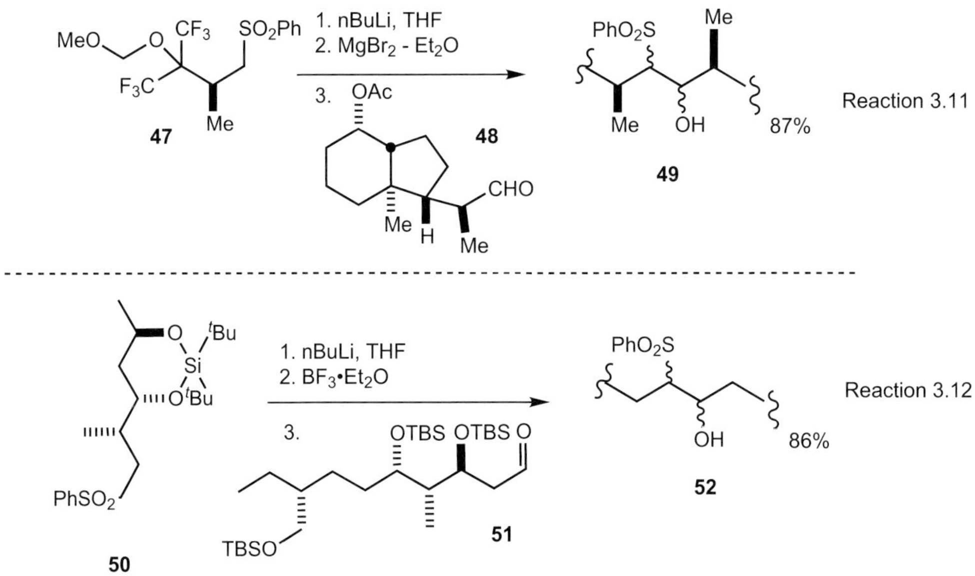

Reaction 3.11

Reaction 3.12

Scheme 3.11. Use of BF$_3$·OEt$_2$ and MgBr$_2$ in the coupling step.

Furthermore, trapping of the in situ generated alkoxide – usually with Ac$_2$O, BzCl, MsCl or TMSCl – can be a method of choice to shift the equilibrium in favor of the desired addition product. Interestingly, in many cases derivatization has been shown to also facilitate the reductive elimination step. Finally, the α-sulfonyl carbanion can be added to an ester and the resulting ketone reduced to the β-hydroxysulfone, thereby circumventing the alkoxide pitfall [41, 46–50]. Scheme 3.12 depicts such a sequence performed within the framework of a synthesis of a portion of Amphidinolide-A by O'Connor et al. in 1989 [52].

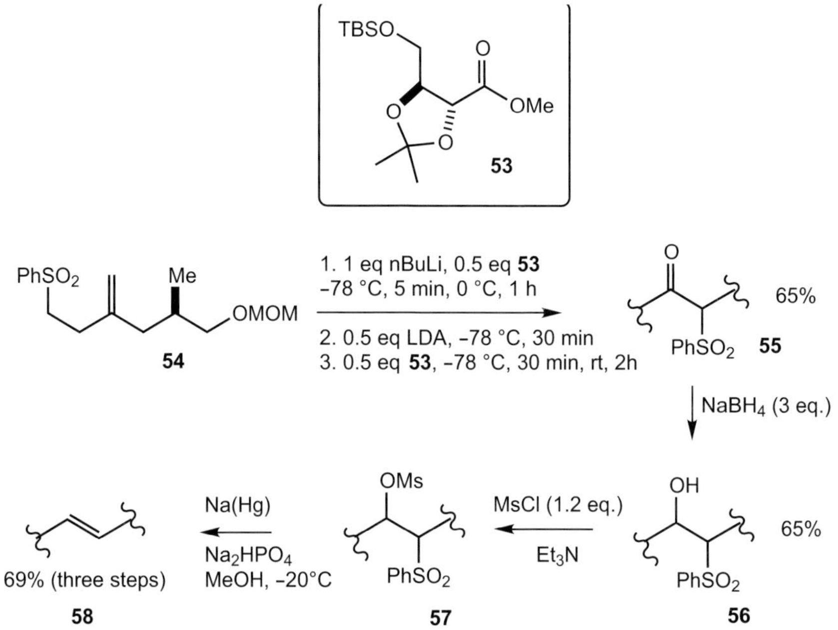

Scheme 3.12. Alternative sequence as a solution to retrograde fragmentation.

Another serious problem that can be encountered when reacting an α-sulfonyl anion with an aldehyde is competitive enolization of the carbonyl derivative, which can sometimes lead to large amounts of recovered starting material. It has been shown that judicious choice of the appropriate solvent may be a key factor for success [51]. As indicated in Table 3.1, using DME instead of THF suppresses the undesirable enolization almost completely. Whilst DME appears to be especially effective in the case of linear, aliphatic aldehydes (entries 1 and 2), THF seems to be preferred when α-substituted aldehydes are employed as coupling partners (entry 3).

Another breakthrough concerning the condensation step was reported by the group of Satoh in 1998 [52, 53]. They reasoned that the greater reactivity of a carbanion α to a sulfoxide group, as compared to a sulfone residue, would lead to im-

Tab. 3.1. Effect of DME in the coupling step.

$$nC_7H_{15}\text{-}SO_2Ph \;(59) \xrightarrow[\substack{\text{2. RCHO} \\ \text{3. NH}_4\text{Cl/H}_2\text{O}}]{\text{1. }n\text{BuLi / Solvent}} \text{HO-CHR-CH(SO}_2\text{Ph)-}nC_7H_{15}\;(60)$$

Entry	Aldehyde	Solvent	Yield (%)[a]	% Recovered
1	$nC_6H_{13}CHO$	DME	95	0
		THF	45*	45*
2	C_2H_5CHO	DME	90	3
		THF	56*	38*
3	$cC_6H_{11}CHO$	DME	89	6
		THF	94	0

[a] Isolated yields with the exception of entries marked with an asterisk, in which cases yields are based upon integration of 1H-NMR spectra of a mixture of reactant and product.

proved yields in the condensation with aldehydes. Furthermore, the higher energy of these anions should strongly disfavor retroaldolization (Table 3.2). It should be borne in mind that Durst [13], in 1973, had already obtained good yields in the additions of α-sulfoxide carbanions to ketones (Scheme 3.5; reaction 3.6).

3.3.3
Towards Trisubstituted Olefins

When the number of substituents differs on the two carbons of the double bond, a correct choice of the coupling partners becomes essential for a successful Julia olefination. In the case of trisubstituted olefins, both disconnections suffer from some shortcomings. For example, addition of an aldehyde **63** to a secondary

Tab. 3.2. Use of sulfoxide derivatives.

$$R^1\text{-SOPh}\;(61) \xrightarrow[\text{THF, -55°C}]{\text{LDA / R}^2\text{CHO}} \text{HO-CHR}^2\text{-CH(SOPh)-R}^1\;(62)$$

Entry	R^1	R^2	Yield (%)[a]
1	Ph–	$PhCH_2CH_2-$	92
2	naphthyl	benzodioxole	99
3	MeO–C$_6$H$_4$–	benzodioxole	99

[a] Isolated yields after silica gel column chromatography.

3.3 Coupling Between the Two Precursors of the Julia Reaction

α-sulfonyl carbanion **29** (path a) may lead to undesirable enolization, whilst condensation between a more reactive primary sulfone **67** and a ketone **3** (path b) is usually plagued by retroaldolization (Scheme 3.13).

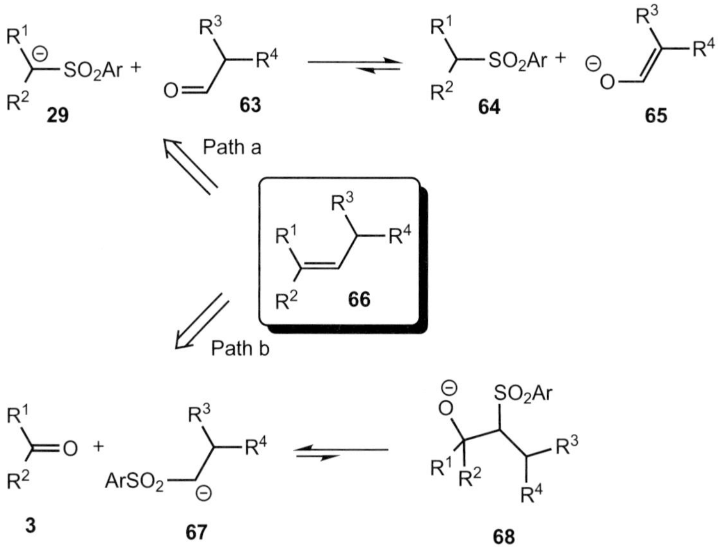

Scheme 3.13. Shortcomings in the disconnections of trisubstituted olefins.

Although path a is more frequently encountered in the literature, ketones (path b) can also be successfully employed in these coupling reactions. However, due to the unfavorable equilibrium between **68** and **3/67**, care must be taken in these cases, and the trapping protocols mentioned in Section 3.3.2 are highly recommended (Tables 3.3 [54] and 3.4 [53]).

Tab. 3.3. In situ quenching of the alkoxide.

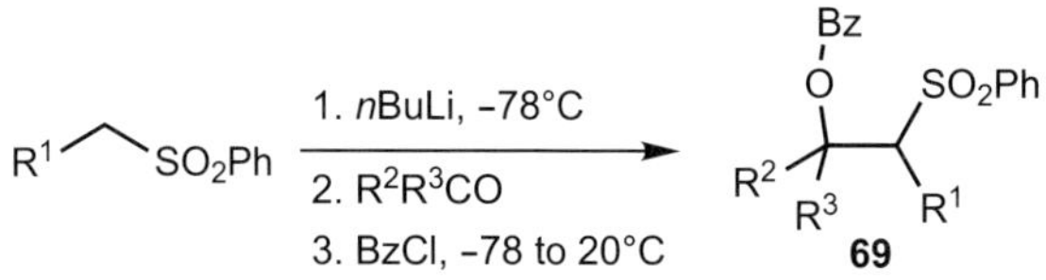

$R^1\text{–}SO_2Ph \xrightarrow{\text{1. }n\text{BuLi, }-78°C;\ \text{2. }R^2R^3CO;\ \text{3. BzCl, }-78\text{ to }20°C}$ **69**

Entry	R^1	R^2	R^3	Yield (%)[a]
1	nC_6H_{13}	$PhCH_2CH_2$	CH_3	81
2	CH_3	$PhCH_2CH_2$	CH_3	82
3	CH_3	nC_4H_9	CH_3	93

[a] All yields are for pure, fully characterized products.

Tab. 3.4. Use of sulfoxide derivatives.

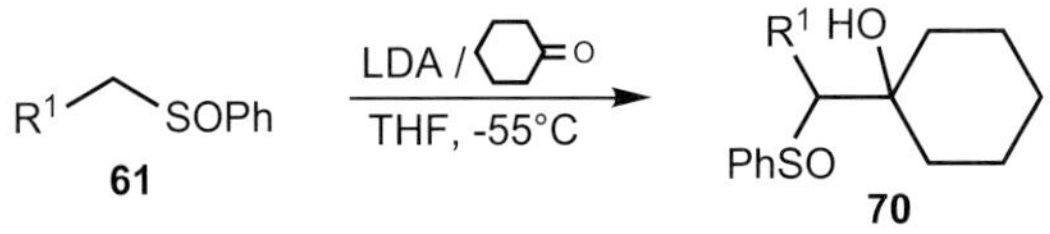

R¹–SOPh (**61**) + cyclohexanone → R¹(HO)C(cyclohexyl)(SOPh) (**70**), via LDA / THF, −55 °C

Entry	R¹	Yield (%)[a]
1	PhCH₂CH₂	87
2	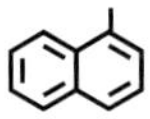 2-naphthyl	98
3	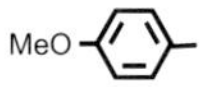 MeO-C₆H₄-	95

[a] All yields are for pure, fully characterized products.

An alternative generation of β-hydroxy sulfones was discovered by Falck and Mioskowsky, during some research pertaining to the reductive lithiation of bis(phenyl)sulfones. The authors initially used lithium naphthalenide as the reducing agent to cleave the C–S bond and generate the α-sulfonyl carbanion [55]. Subsequently, they improved their methodology by employing samarium diiodide instead of lithium naphthalenide [56] (Scheme 3.14).

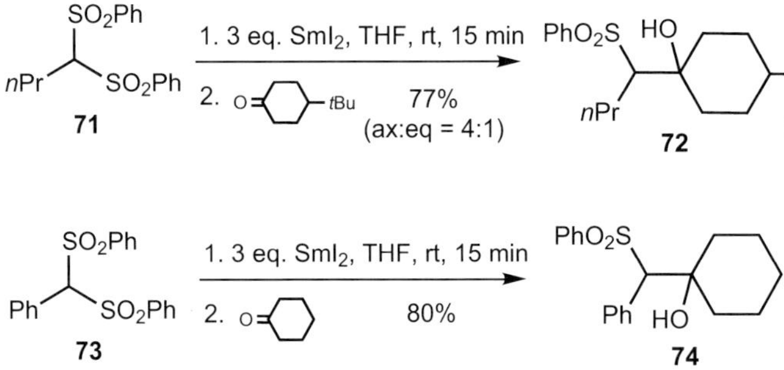

Reaction 3.13: **71** (nPr-C(SO₂Ph)₂) → **72** (nPr-C(SO₂Ph)(cyclohexyl-tBu)-OH), via 1. 3 eq. SmI₂, THF, rt, 15 min; 2. 4-tBu-cyclohexanone. 77% (ax:eq = 4:1)

Reaction 3.14: **73** (Ph-C(SO₂Ph)₂) → **74** (Ph-C(SO₂Ph)(cyclohexyl)-OH), via 1. 3 eq. SmI₂, THF, rt, 15 min; 2. cyclohexanone. 80%

Scheme 3.14. Alternative generation of hydroxy sulfones (trisubstituted olefins).

3.3.4
Towards Tetrasubstituted Olefins

As might be expected from the results discussed in the previous paragraphs, coupling reactions between secondary α-sulfonyl carbanions and ketones are rarely encountered in the literature [12, 51, 57]. Even the use of sulfoxides proved to be of no avail [52, 53]. It is thus even more remarkable that the method reported by Falck and Mioskowsky affords tetrasubstituted β-hydro-sulfones in excellent yields [56]. It is quite possible that the unique properties of the Sm^{III} salt, which acts as a strong Lewis acid (activation of the ketone partner) and forms a tightly bound cationic complex with the initially generated alkoxide (no retrograde aldol reaction), are responsible for the success of this protocol in these extremely challenging condensations (Scheme 3.15).

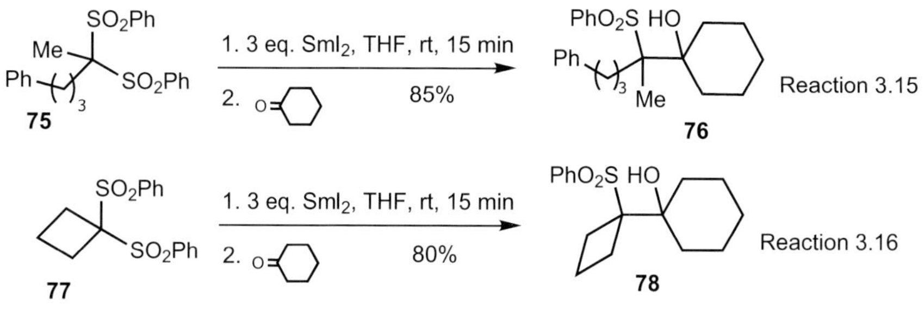

Scheme 3.15. Alternative generation of hydroxy sulfones (tetrasubstituted olefins).

In this context, it is worthy of note that a modification of the Julia sequence, introduced in 1982, allows the preparation of tetrasubstituted olefins from vinyl sulfones [58]. This modification, which concerns mainly the reductive elimination step, will be discussed in detail in the relevant section (Section 3.4.5).

3.3.5
Specific Considerations

3.3.5.1 Conjugated Olefins

In order to prepare conjugated olefins, either an allylsulfone or an α,β-unsaturated carbonyl compound is needed. In the case of allylic sulfones, the delocalized carbanion can undergo either α- or γ-addition, and the use of enals or enones can result in 1,2- or 1,4-addition products. In general, allylsulfonyl anions, derived from non-stabilized sulfones, undergo additions to carbonyls with excellent α-selectivities (reaction 3.17 [59]; Scheme 3.16). The situation is more complex with α,β-unsaturated compounds, but conditions have been defined that lead with high preference to adducts resulting from 1,2-addition (reaction 3.18) [60].

Examination of the literature reveals that Michael additions are usually facilitated when the carbanion is made softer by delocalization into a strongly electron-withdrawing group [61–68], when the enolate resulting from 1,4-addition is particularly stable [69], when the addition reaction is performed in the presence of HMPA [70–72], or when the 1,2-addition is disfavored by conformational effects [73, 74]. A few examples of such sulfonyl anions are collected in Scheme 3.17.

3.3.5.2 Leaving Groups

When the alkene product bears a potential leaving group on a carbon adjoining the double bond, as in **93**, special care should be taken in selecting the appropriate

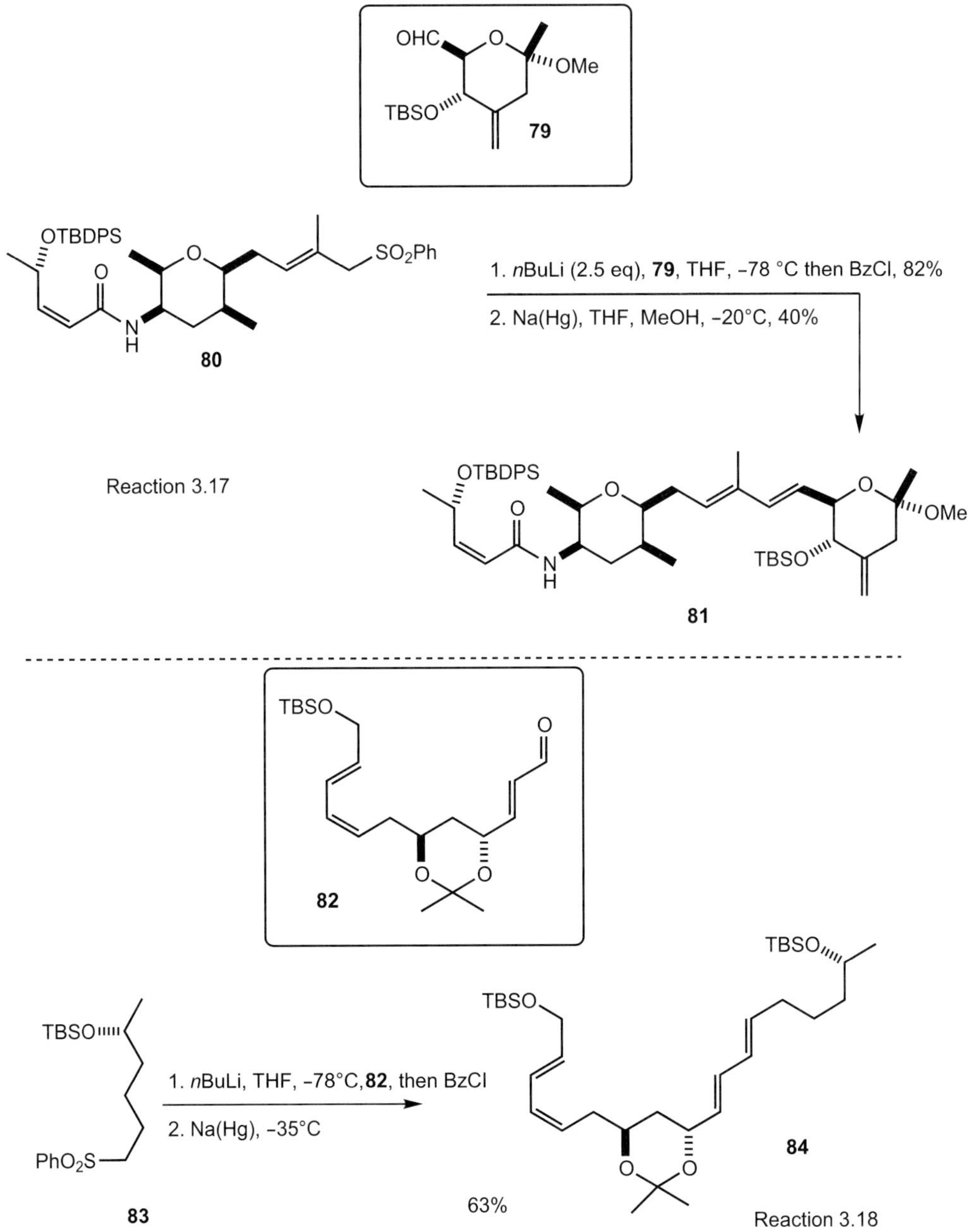

Reaction 3.17

Scheme 3.16. Synthesis of conjugated olefins.

Reaction 3.18

3.3 Coupling Between the Two Precursors of the Julia Reaction

α-Sulfonyl anions that lead to 1,4-additions

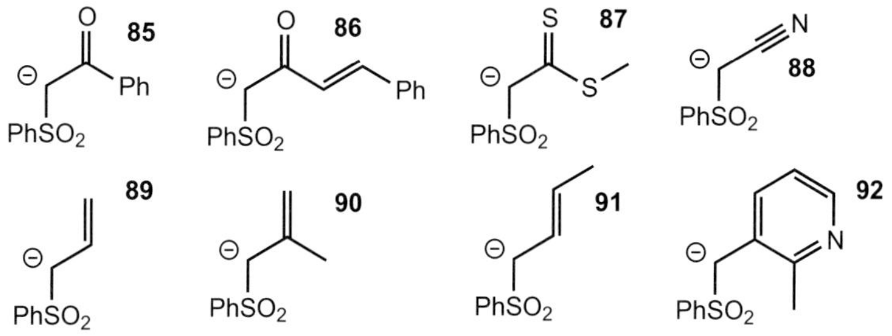

Scheme 3.17. α-Sulfonyl anions that lead to 1,4-additions.

disconnection. Indeed, the leaving group does not generally tolerate the presence of an adjacent negative charge, as in **94**, and therefore should be preferentially located on the carbonyl fragment **96** (Scheme 3.18).

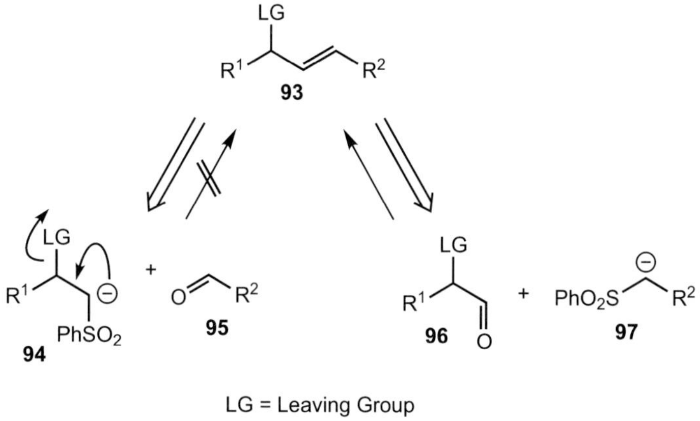

LG = Leaving Group

Scheme 3.18. Alkenes bearing an adjacent leaving group.

As depicted in Scheme 3.19, Giese et al. had to face such a problem during the course of the total synthesis of Sopharen $A_{1\alpha}$ [75]. Deprotonation of **99** led rapidly, by competitive elimination, to vinyl sulfone **100**, and no coupling between **98** and **99** was observed (reaction 3.19). Reversing the sequence (reaction 3.20) solved the problem and the expected adduct **103** was obtained in 35% yield.

In some cases, however, the elimination is sufficiently sluggish and the desired β-alkoxy sulfones can be obtained in moderate to good yields, as illustrated in reactions 3.21 [76] and 3.22 [77].

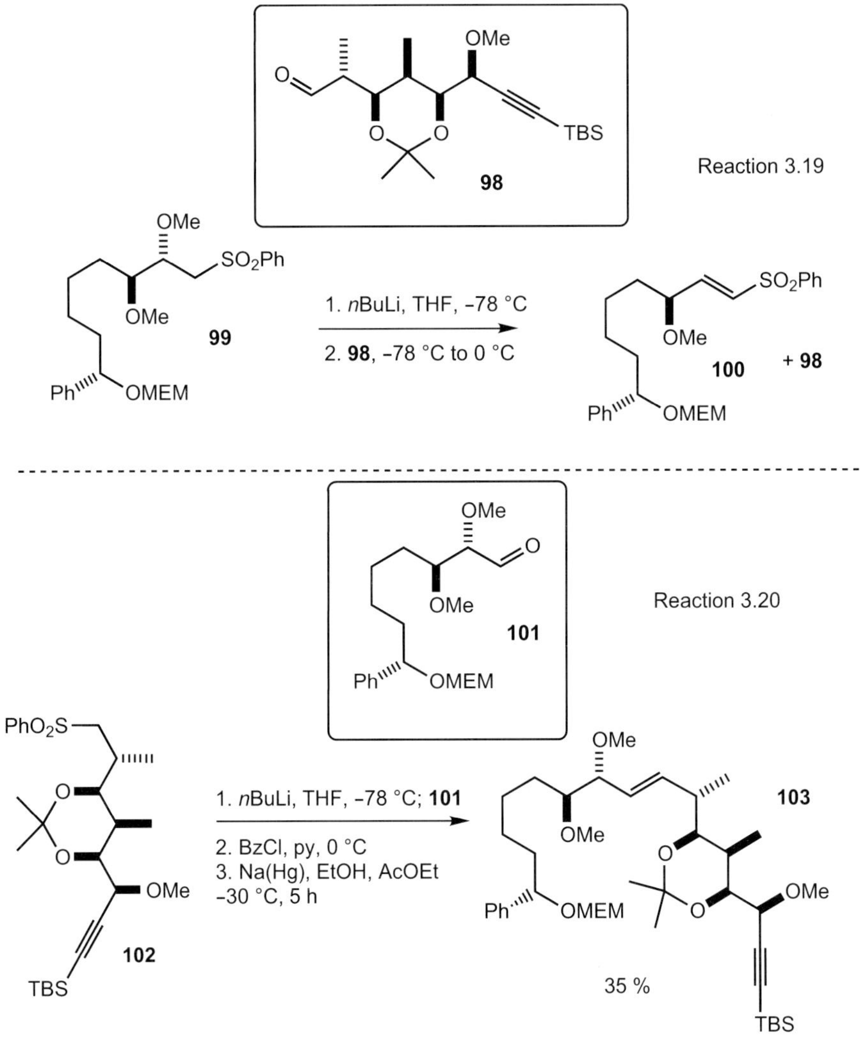

Scheme 3.19. Competitive elimination during the synthesis of Sopharen A$_{1\alpha}$.

3.3.5.3 Competitive Metallation on the Aromatic Ring of the Sulfone

As shown in Scheme 3.21, treatment of allyl phenyl sulfone with an excess of nBuLi affords a mixture of 1,o-dilithiated allyl phenyl sulfone **109** and the 1,1-dilithiated compound **110** [78]. Quenching this mixture with benzaldehyde leads to a variety of products, including **112** and **113**. However, it is interesting to note that, at 50 °C, the 1,o-dilithiated species **109** is converted into the thermodynamically more stable 1,1-dilithiated allyl phenyl sulfone **110**, which can undergo twofold reaction with benzaldehyde. This double lithiation coupling methodology has since been applied to various bis-electrophiles [79]. Competitive *ortho*-lithiation was al-

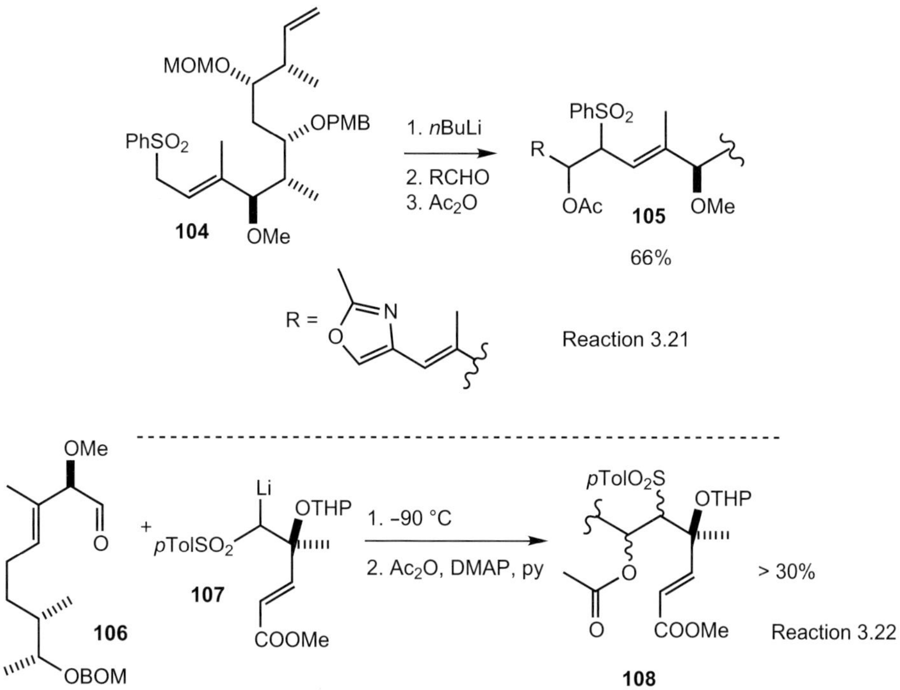

3.3 Coupling Between the Two Precursors of the Julia Reaction

Reaction 3.21

Scheme 3.20. Alternative sequence circumventing the competitive elimination.

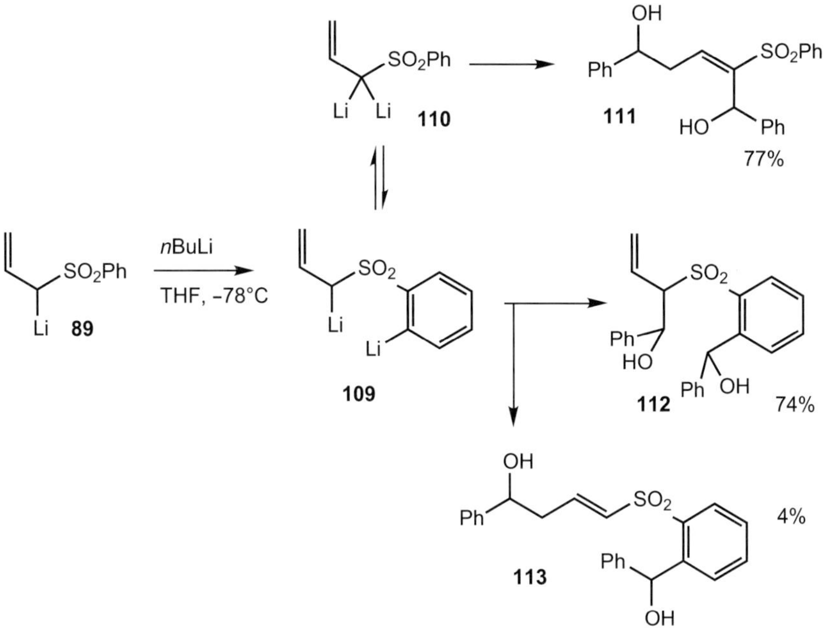

Reaction 3.22

Scheme 3.21. Competitive metallation on the aromatic ring of the sulfone.

ready mentioned by Kociensky in 1991 [3], who suggested the use of LDA instead of nBuLi in order to avoid it.

A similar problem, involving an imidazole group, arose during the total synthesis of Bryostatin 2, as reported by Evans in 1999 [80] (Scheme 3.22). Due to the high degree of steric hindrance, deprotonation and alkylation of sulfone **116** took place on the imidazole ring and product **117** was obtained (reaction 3.23). By reversing the sequence, Evans *et al.* expected a higher reactivity of the sulfone fragment **118**. Indeed, when they performed reaction 3.24, no competitive detrimental aromatic lithiation was observed. However, repeating the reaction with the correctly functionalized aldehyde **115** led to none of the expected β-hydroxy sulfone. Instead, diene **121**, resulting from a double-elimination process, was isolated (reaction 3.25). The lack of reactivity of aldehyde **115** could be due to overwhelming steric crowding around the carbonyl function. Eventually, success was achieved by replacing the imidazole group by a phenyl substituent and employing the less encumbered aldehyde **114**. The long sought after adduct **123** was thus isolated in more than 64% yield.

3.4
Reductive Elimination

Since 1973, most of the improvements concerning the Julia sequence have been directed towards the reductive elimination step. Before 1990, this reaction was mostly effected using Na(Hg) amalgam. However, examples involving other reducing systems had already been reported, such as RMgX/Pd, Fe or Ni catalyst [58, 81–84]; Bu_3SnH [21, 85–87]; Li naphthalenide [88–90]; Li or Na in liquid ammonia or in ethylamine [91, 92]; $Na_2S_2O_4$ [93]; Raney nickel [94]; potassium graphite [95]; electroreductive reactions [96]; Te/$NaBH_4$ [97]; Al(Hg) amalgam and $LiAlH_4$, with or without $CuCl_2$ [11, 98–101]. In 1990, Kende successfully employed samarium diiodide for the reductive elimination of β-hydroxy imidazolyl sulfones [102]. In conjunction with contributions from Inanaga [103] and Künzer [104], this work served to highlight the crucial role of HMPA in reductive desulfonylation. Since then, the combination SmI_2/HMPA or SmI_2 and various additives [105] has continued to prove its usefulness [106–108]. In addition, other reductive methods are regularly reported, such as the use of Mg/$HgCl_2$ [109, 110], NaTeH [111], and sulfoxide/alkyllithium [52, 53] combinations.

Moreover, two important variations of the Julia reaction were reported during the 1990s. As both procedures afford the final alkene product directly from the sulfone and carbonyl components, they shall be discussed in greater detail in a subsequent part of this chapter (Section 3.5). In this section, attention is focused on the reductive desulfonylation of various β-oxygenated sulfones and of vinyl sulfones. Insights into the advantages and disadvantages of the various protocols, some mechanistic considerations, and the stereochemical outcome of the reductive elimination step are provided.

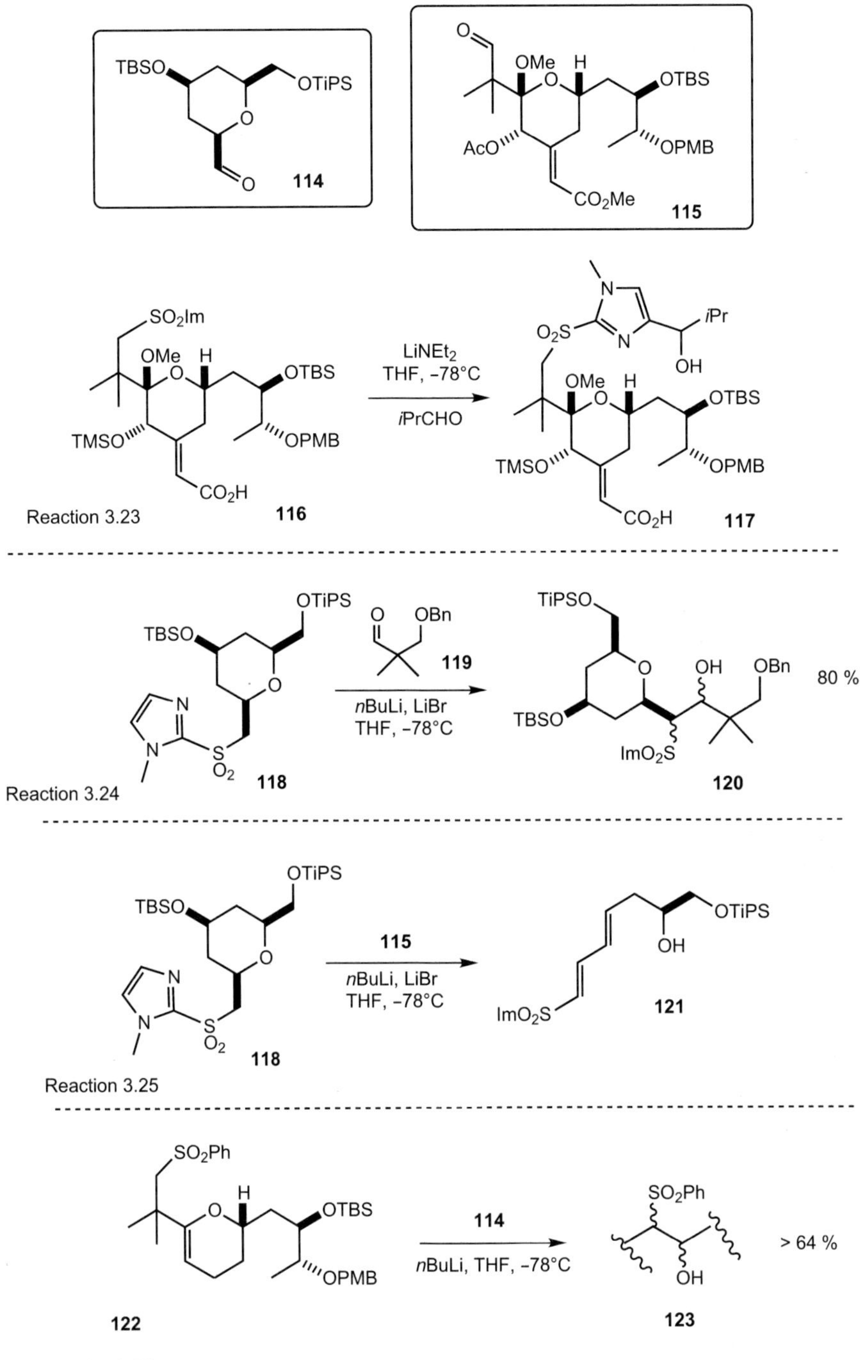

3.4 Reductive Elimination | 121

Reaction 3.23

Reaction 3.24

Reaction 3.25

Reaction 3.26

Scheme 3.22. Competitive metallation during the synthesis of Bryostatin 2.

3.4.1
Sulfones Bearing Vicinal Hydroxyl Groups

In his first contribution to the reaction that today bears his name, Julia already explored the reductive elimination of a wide range of functionalities [12]. Eliminations from vinyl sulfones as well as vicinal hydroxyl-, acetyl-, mesyl-, and tosyl sulfones to afford the corresponding alkenes were carried out using Na(Hg). He observed that the yields of the desired alkenes were invariably higher starting from any other group than a free β-hydroxyl (Scheme 3.23).

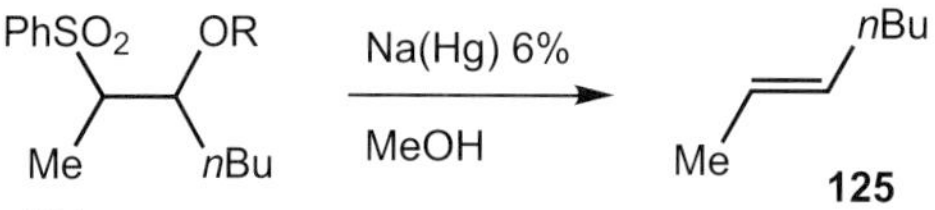

$PhSO_2$ OR → Na(Hg) 6% / MeOH → nBu / Me **125**

124

R = H - 63 % (and competitive simple desulfonation)
R = Ms - 80 %
R = Ac - 79 %
R = Ts - 64 %

Scheme 3.23. Na(Hg)-promoted reductions.

In all likelihood, conditions such as Na(Hg)/MeOH would generate a certain amount of alkoxide **128**, and an equilibrium involving retrograde fragmentation leading to **126** and **95** could then take place [26–28]. Considering the accepted mechanism outlined in Scheme 3.24, other reasons can be invoked to explain the low efficiency of the reduction.

When the reduction is performed using sodium amalgam in methanol, it is generally acknowledged that sodium – being a one-electron donor – transfers an electron to the sulfone residue. The *in situ* generated radical anion then collapses, liberating $PhSO_2^-$ and generating the radical species **129**. This radical can be over-reduced by another equivalent of sodium, and the resulting anion **130**, now vicinal to the alkoxide, can afford the olefin by elimination of Na_2O and/or NaOH. This mechanism would require a *trans* coplanar relationship between the arylsulfonyl substituent and the leaving group. Several experiments have served to confirm this mechanistic hypothesis [3, 112–117] (Scheme 3.25). Thus, reductive elimination from **133** [118], in which the anion derived from the phenylsulfonyl group and hydroxyl substituents can adopt an antiperiplanar orientation, proceeds smoothly to afford the expected alkene **134**. However, when epimer **135**, in which such a co-planar arrangement cannot be achieved, is submitted to the same conditions, only alcohol **136** is produced.

Since Na_2O can be considered as a rather poor leaving group, the anionic intermediates **130** may be quenched by a proton source before elimination occurs [118–122]. During a total synthesis of neodolabellenol in 1995, Williams et al. [123] took advantage of these considerations and intentionally performed such a desulfonation to complete their synthesis (Scheme 3.26).

The radical species **129**, generated under reductive conditions, can sometimes be a potential source of problems. For instance, during the course of the total syn-

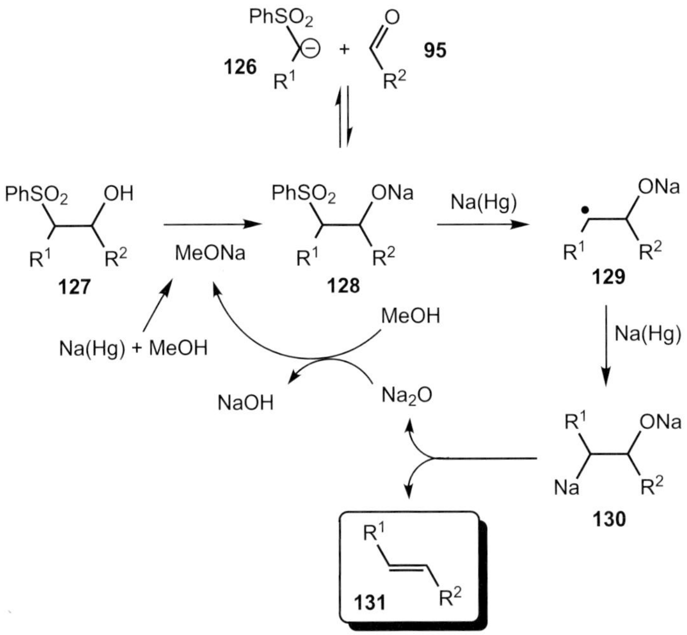

Scheme 3.24. Mechanism of Na(Hg)-promoted reductions (I).

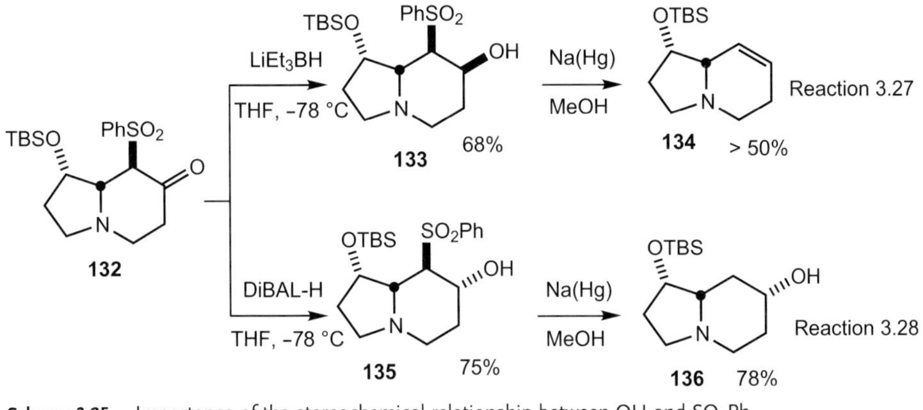

Scheme 3.25. Importance of the stereochemical relationship between OH and SO_2Ph.

thesis of FK-506, Danishefsky et al. [88] attempted the reductive elimination of **139** under a variety of conditions (Na/NH$_3$, Li-naphthalenide). An unexpected product was obtained, in which the vinyl group of the proximal allyl moiety of **139** had been lost (Scheme 3.27). The absence of this vinyl group was explained by invoking an intramolecular cyclization of a radical species arising from single-electron reduction of the phenylsulfonyl group (path a). Danishefsky et al. reasoned that if the leaving group propensity of the β-alkoxy substituent was increased, β-elimination

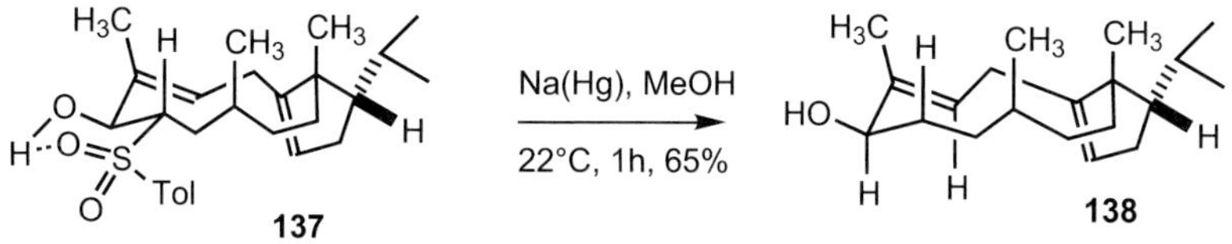

137

138

β-Neodolabellenol

Scheme 3.26. Simple desulfonation in the synthesis of neodolabellenol.

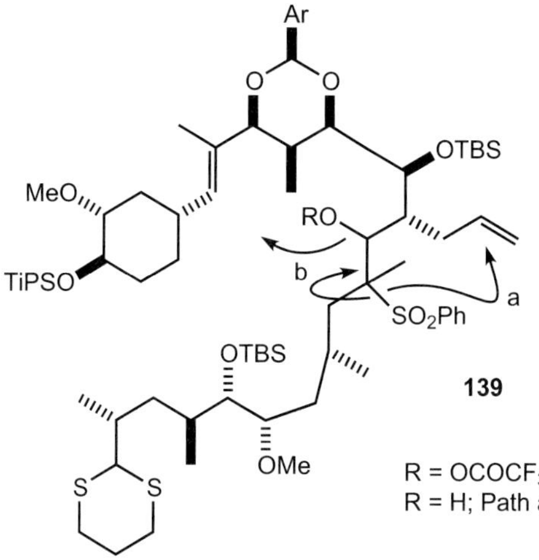

139

R = $OCOCF_3$; Path b, elimination, 65%
R = H; Path a, competitive reaction

Scheme 3.27. Quenching of a radical intermediate in the synthesis of FK-506.

might become competitive with the postulated cyclization of the radical species on the vinyl substituent. After failing to benzoylate or acetylate the secondary alcohol function, they successfully used trifluoroacetic anhydride and, after treatment with lithium naphthalenide, eventually obtained the long sought after olefin in a yield of 65% (path b).

From the viewpoint of the stereochemical outcome of the elimination step, one can reasonably assume that discrete anionic species generated under these conditions are neither configurationally nor conformationally stable, and that a dynamic equilibrium is rapidly established between **142** and **143**. The initially formed radical intermediate **141** also equilibrates rapidly, adopting an overall planar – or slightly pyramidalized – structure, and the elimination thus becomes an egregious example of Hammond's postulate. In general, 1,2-disubstituted olefins produced under these conditions predominantly have the *E*-configuration (Scheme 3.28) [127]. More highly substituted olefins are usually obtained as mixtures of both the *E*- and *Z*-geometric isomers.

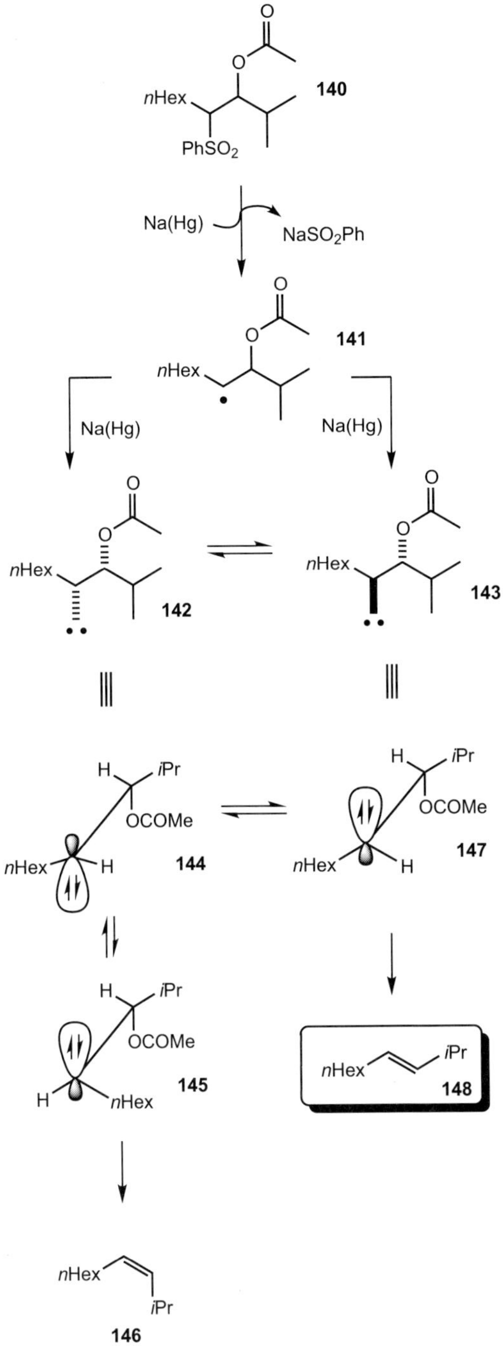

Scheme 3.28. Na(Hg) elimination favors the synthesis of (E)-olefins.

3 The Julia Reaction

The alternative use of SmI$_2$ as a non-basic reductant is quite noteworthy. When Kende et al. employed SmI$_2$ in the reductive elimination of β-hydroxy imidazolyl sulfones [102], they obtained higher yields than those usually observed using Na(Hg) (Scheme 3.29).

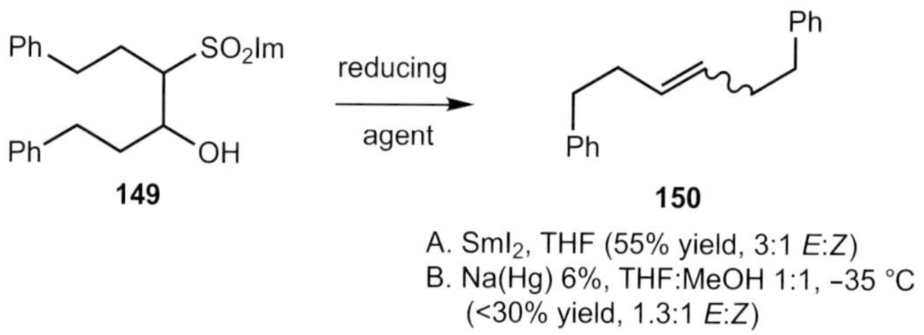

Scheme 3.29. First SmI$_2$-promoted reduction of hydroxy sulfones.

A. SmI$_2$, THF (55% yield, 3:1 *E:Z*)
B. Na(Hg) 6%, THF:MeOH 1:1, −35 °C (<30% yield, 1.3:1 *E:Z*)

The greater efficiency of the SmI$_2$-based protocol can be rationalized in terms of a change in the nature of the leaving group (presumably HOSmI$_2$) and the absence of retro-aldolization; β-alkoxide intermediates are no longer generated. However, in special cases, the radical species formed upon homolytic cleavage of the C–S bond can still lead to some competitive reactions such as additions to proximal double bonds [102] (Scheme 3.30).

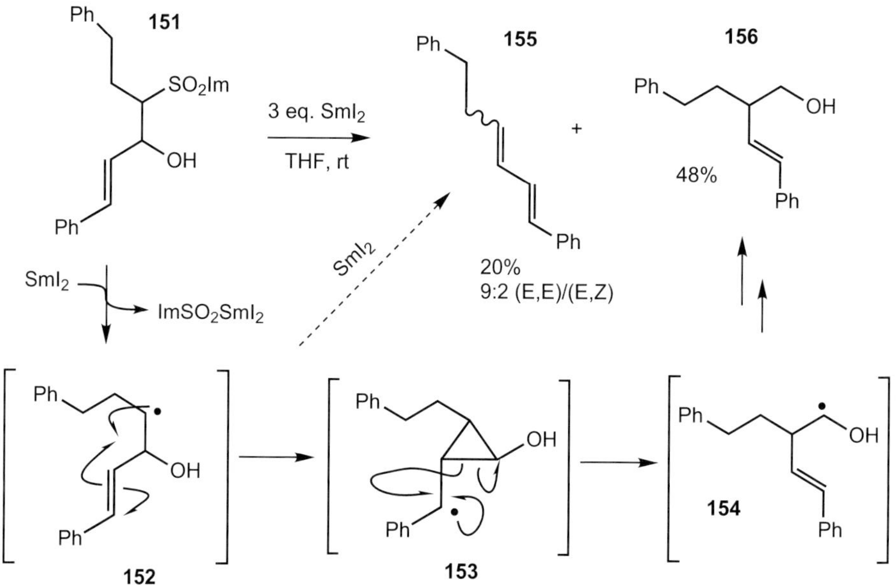

Scheme 3.30. Competitive reaction due to radical species.

3.4.2
Sulfones Bearing Vicinal Leaving Groups

In general, derivatization of the vicinal hydroxyl group is always beneficial with regard to the efficiency of the elimination step. Trapping of the alkoxide after the condensation between the two partners **2** and **3** is not only useful to counterbalance a potentially undesirable equilibrium, but the absence of an acidic hydrogen inhibits retrograde aldol reaction. Moreover, increasing the leaving group's propensity to leave should, *a priori*, result in little time for the *in situ* generated reactive intermediates to afford competitive reaction by-products such as simple reduction (**158**) or radical capture adducts (**159**) (Scheme 3.31). Due to this easier elimination, both yields and rates are often improved.

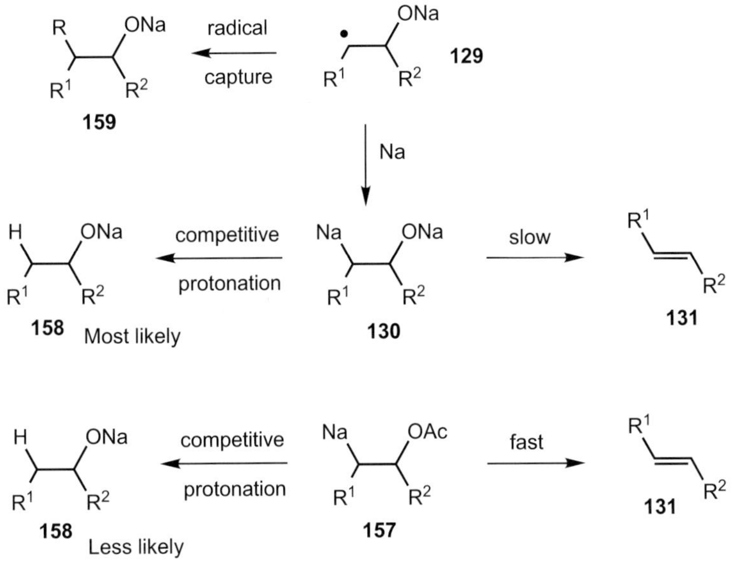

Scheme 3.31. Beneficial derivatization of the vicinal hydroxyl group.

From a mechanistic viewpoint, the reduction of vinyl sulfones and the reductive elimination of β-acetoxy sulfones with Na(Hg) or SmI$_2$ are the only procedures to have been investigated to any great extent. Recent work published by Keck et al. [105] has shown that different mechanisms are operative when Na(Hg)/MeOH or SmI$_2$/HMPA are employed. They set out to study the reductive desulfonylation of several vinyl sulfones under two sets of reaction conditions, specifically 8 equiv. SmI$_2$, *N,N*-dimethylpropyleneurea (DMPU), MeOD, 60 min. and 5% Na(Hg), Na$_2$HPO$_4$, THF/MeOD (4:1), 0 °C, 60 min (Table 3.5).

In all cases, excellent yields of alkenes, high degrees of deuterium incorporation, and virtually complete *E*-selectivities were observed. These results suggest that a vinyl radical and PhSO$_2^-$ are formed rapidly after an initial electron transfer, and

Tab. 3.5. Deuterium labeling of vinyl sulfones.

Entry	Substrate	Conditions	Deut. incorp. (%)	Yield (%)
1	160	a	98	93
2	160	b	90	91
3	162	b	80	97

a = SmI$_2$ 8 eq., THF, DMPU, D$_2$O, 1 h.
b = Na(Hg), Na$_2$HPO$_4$, THF/MeOD (4:1), 0 °C, 1 h.

that reduction of the radical to a discrete anionic species then ensues. Protonation of this vinyl anion by methanol then furnishes the corresponding olefin. However, this is still an oversimplified picture. Indeed, when [D$_4$]methanol was added to the reaction mixture after various periods of time, the deuterium incorporation and the olefin stereoselectivity dropped concomitantly as a function of time, even though the overall yield of alkene **163** remained constant (Table 3.6). It is thus quite probable that, in the absence of a proton source, another as yet unidentified reaction pathway is operating.

After these preliminary deuterium labeling studies performed on vinyl sulfones, Keck et al. focused their attention on the reductive elimination of β-acetoxy sulfones (Table 3.7). Besides the observation that the E/Z ratios obtained from SmI$_2$-promoted reductions were considerably different from those obtained using the Na(Hg) procedure, they discovered a surprisingly high degree of deuterium incor-

Tab. 3.6. Deuterium labeling of vinyl sulfones – Time dependence.

Entry	Conditions[a]	E:Z	Deut. incorp. (%)	Yield (%)
1	t = 0 min	5:1	90	91
2	t = 5 min	2:1	60	90
3	t = 60 min	1.5:1	40	90
4	t = 18 h	1.2:1	30	94

[a]SmI$_2$ 8 eq., THF, DMPU 5 eq., then 10 eq. of CD$_3$OD was added after time (t).

Tab. 3.7. Deuterium labeling of β-acetoxy sulfones.

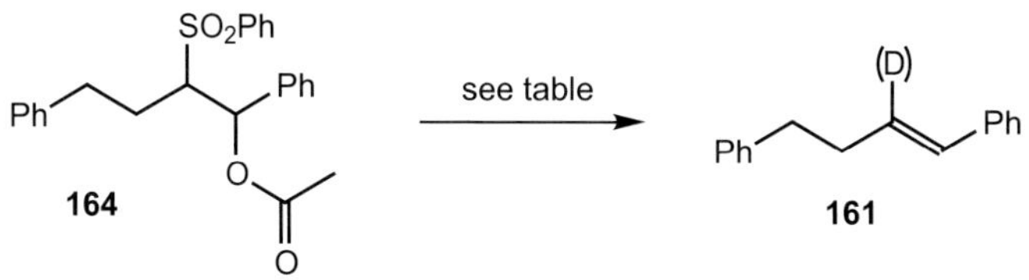

Entry	Conditions	E:Z	Deut. incorp. (%)	Yield (%)
1	a	1:1.3	–	87
2	b	1:1.4	0	88
3	c	9.3:1	91	83
4	d	9.3:1	47	85

a = SmI$_2$ 8 eq., THF, DMPU, CH$_3$OH, 1 h.
b = SmI$_2$ 8 eq., THF, DMPU, CD$_3$OD, 1 h.
c = Na(Hg), Na$_2$HPO$_4$, THF/CD$_3$OD (4:1), 0 °C, 1 h.
d = Na(Hg), Na$_2$HPO$_4$, THF, 0 °C, 5 min., then CD$_3$OD.

poration in the alkene product when the Na(Hg)-based protocol was employed. These results were at variance with the originally proposed mechanism (Schemes 3.24 and 3.28) that appears to be operative in the case of SmI$_2$-mediated reductive elimination and for which no deuterium incorporation could be observed (entries 1–4).

According to Keck, the high E-selectivities and significant deuterium incorporation obtained under the Na(Hg) reduction conditions can be rationalized in terms of the mechanism outlined in Scheme 3.32. Deprotonation of the β-acetoxy sulfone

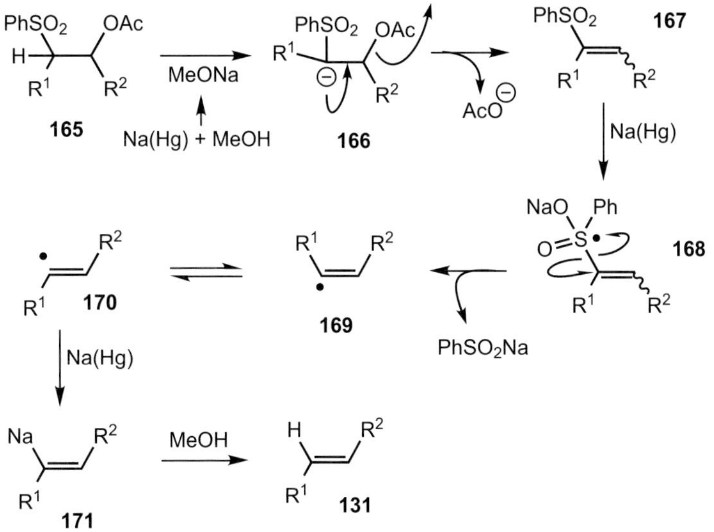

Scheme 3.32. Mechanism of Na(Hg)-promoted reductions (II).

165 by NaOMe, which is generated even under buffered conditions, leads to the vinyl sulfone **167**, which is then reduced to the vinyl radical **169**.

Rapid equilibration between the *E*- and *Z*-forms **169** and **170** then ensues, with the *E*-isomer being greatly favored; this is then further reduced to the configurationally stable vinyl anion **171**. Finally, **171** is protonated, leading to the olefin **131** as the major isomer. This mechanism nicely rationalizes both the incorporation of a deuterium atom and the observed stereoselectivity.

However, whether this reaction manifold is exclusively followed or whether it is in competition with the direct reductive elimination previously postulated is difficult to ascertain as it appears to depend upon the structure of the starting *β*-acetoxy sulfone. Indeed, when substitution prohibits elimination towards the corresponding vinyl sulfones [122, 125, 126], as in the case of substrate **172** [127], the reduction promoted by Na(Hg)/MeOH still remains highly efficient (Scheme 3.33).

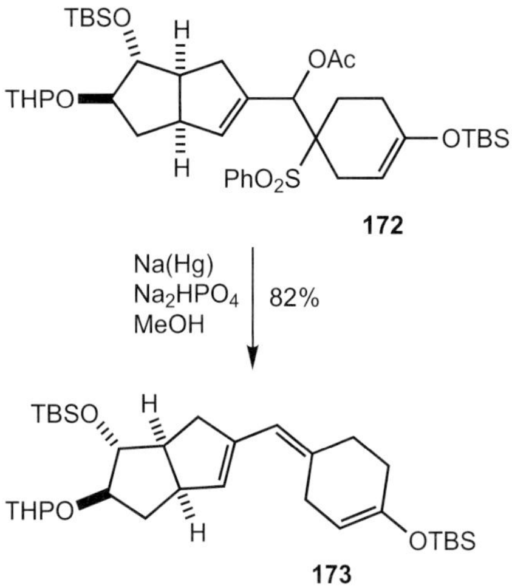

Scheme 3.33. Vinyl sulfone intermediate impossible.

3.4.3
Reverse Reductions

Up to this point, the reductive elimination of *β*-alkoxy sulfones **174** always implied the initial fragmentation of the C–S bond promoted by single electron transfer to the sulfone moiety. It was reasoned that, by suitable modification of the *β*-hydroxyl substituent, an inverse process could be induced, namely inceptive cleavage of the C–O bond to generate radical **175** followed by the elimination of either the aryl sulfonyl radical or anion (Scheme 3.34).

To the best of our knowledge, only a few groups have developed such methodologies. Lythgoe and Waterhouse, in 1977, were the first to use the Barton–McCombie radical deoxygenation approach to induce the elimination of the aryl

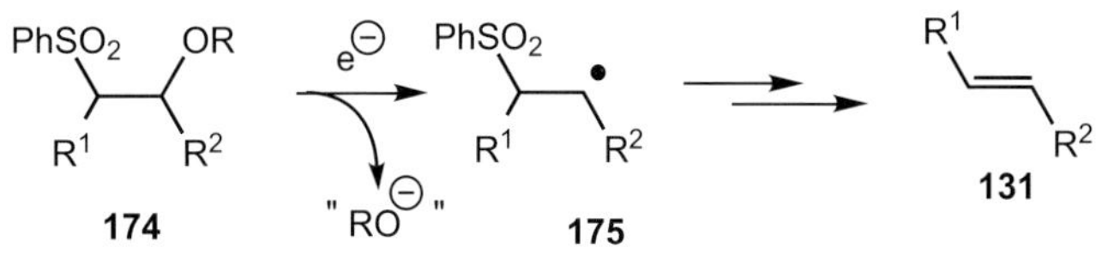

Scheme 3.34. Reverse reductions in the Julia sequence.

sulfonyl radical [85]. Later, during a total synthesis of (+)-Pseudomonic acid, Williams et al. reported high yields for olefination employing the methyl xanthate derivatives of β-hydroxy sulfones [128]. Another variation has also been successfully applied by Barrish et al. [129], while Barton et al. developed a modified procedure that circumvented the use of tin hydride [130] (Scheme 3.35).

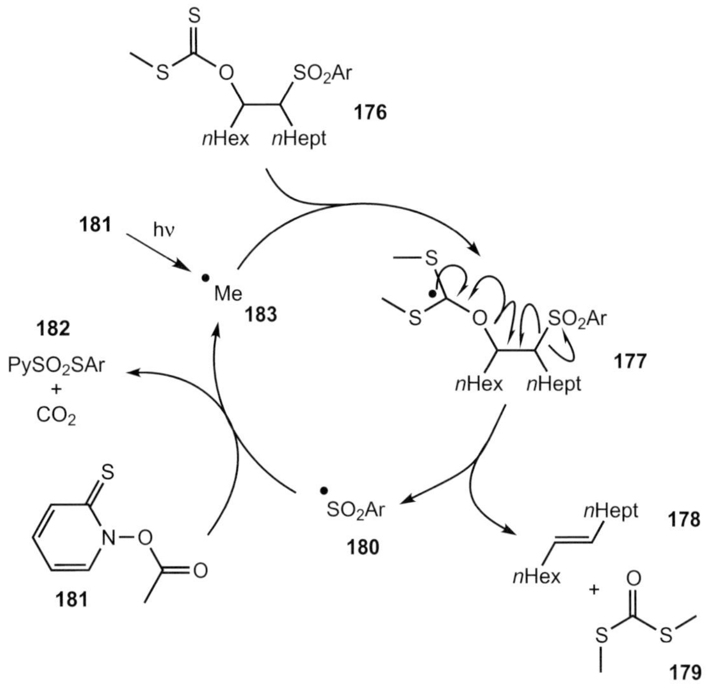

Scheme 3.35. Reverse reductions circumventing the use of Bu_3SnH.

In 1996, Markó et al. reported a reverse Julia olefination of β-sulfoxybenzoates through the use of SmI_2/HMPA [54]. Based upon the enormous differences in rate for the formation of the same alkene starting either from the β-hydroxy sulfone **188** or from the corresponding β-sulfoxybenzoate **184**, they proposed the following mechanism (Scheme 3.36). Reaction of one equivalent of SmI_2 with **184** generates a radical anion, which is believed to be located on the benzoate moiety of **185**. De-

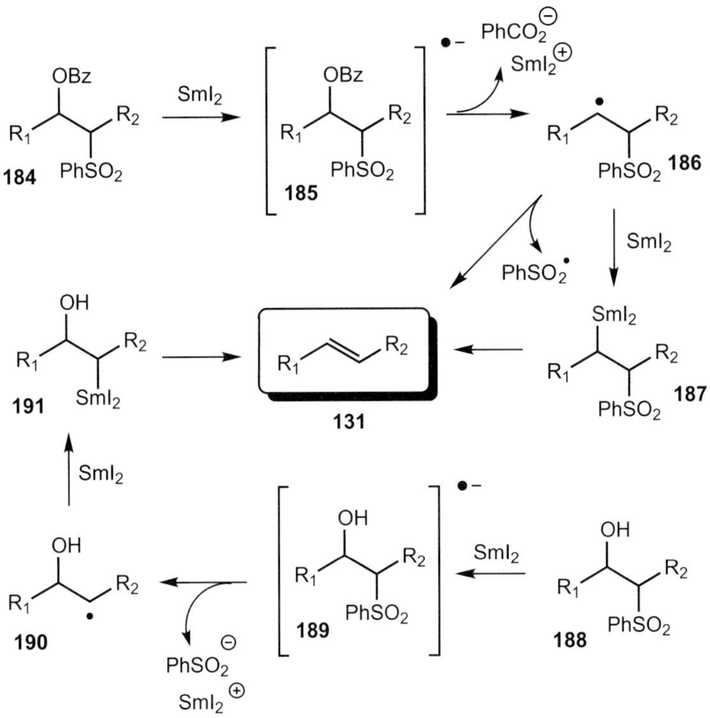

Scheme 3.36. Mechanism of SmI$_2$-promoted reverse reductions.

composition of **185** produces PhCO$_2^-$ and the radical species **186**. This radical is then over-reduced by another equivalent of SmI$_2$, leading to the organosamarium intermediate **187**, which would undergo elimination to ultimately afford olefin **131**. Alternatively, direct elimination of PhSO$_2$· can occur, producing **131** directly from **186**. In the case of β-hydroxy sulfone **188**, it is commonly acknowledged that the radical anion is located on the sulfone substituent and that its decomposition generates radical **190** and PhSO$_2^-$. The same olefin **131** is then generated from the samarium species **191**. However, it is worthy of note that elimination from **191** occurs in the opposite sense to that from **187**.

This mechanistic hypothesis has since been corroborated by several experiments (Scheme 3.37). For example, reaction of substrate **192**, possessing both a β-hydroxy sulfone and a β-benzoyloxy sulfone, with SmI$_2$ at −78 °C, affords exclusively the olefin **193** originating from the elimination of the sulfonylbenzoate. An even more remarkable example is provided in reaction 3.30, in which exclusive formation of triene **195** from substrate **194** takes place. Despite the closer proximity of the OH and phenylsulfone substituents, only the elimination involving the benzoate is observed. This reverse elimination appears to be specific to benzoate derivatives and opens new vistas in the development of mild conditions for the Julia olefination of sensitive polyfunctional molecules.

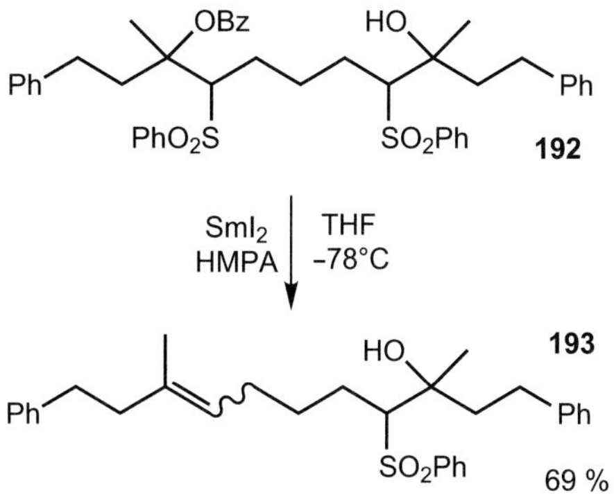

3.4 Reductive Elimination

Reaction 3.29

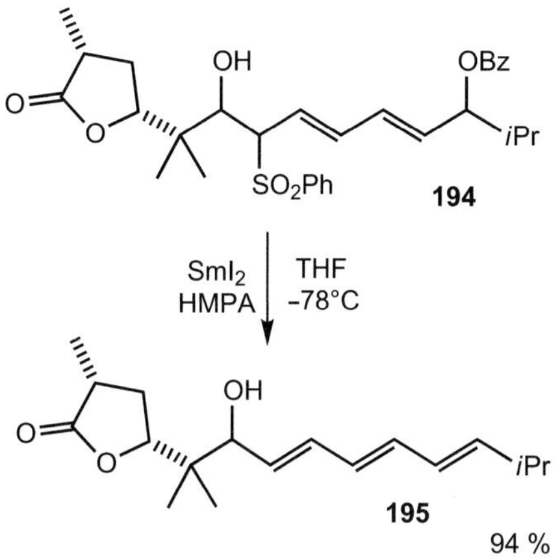

Reaction 3.30

Scheme 3.37. Examples of reverse reductions using SmI$_2$ as reducing agent.

3.4.4
Reductions of Vicinal Oxygenated Sulfoxides

The reductive elimination of β-hydroxy sulfoxides reported by Satoh et al. deserves special attention [52, 53]. The reaction of β-mesyloxy (or acetoxy) sulfoxides **196** with alkyl metals (nBuLi, tBuLi or EtMgBr) at low temperature affords the corresponding olefins in good to excellent yields (Table 3.8).

Two different mechanisms can, a priori, be postulated to explain this transformation (Scheme 3.38). The first one involves direct β-elimination of the sulfinyl and mesyloxy groups through a sulfoxide–metal exchange (path a), whilst the second one would be a two-step mechanism proceeding through the initial formation of a vinyl sulfoxide **199**, generated by base-catalyzed elimination of the mesylate leaving group. Subsequent sulfoxide–metal exchange would then lead to the vinyl anion **200**, which is protonated to afford the desired alkene **198** (path b).

Tab. 3.8. Reduction of sulfoxide derivatives.

Entry	R^1/R^2	R^3/R^4	R^5	Alk. Met.	Yield (%)
1	Ph/H	PhCH$_2$CH$_2$/H	Ms	tBuLi	92 (E:Z = 10:7)
2	PhCH$_2$CH$_2$/H	(cyclohexyl)	Ac	tBuLi	93
3	(cyclohexyl)	PhCH$_2$CH$_2$/H	Ac	tBuLi	63, 79[a]

[a] Variable with the diastereoisomers.

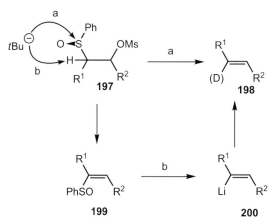

Scheme 3.38. Mechanism of the reduction of vicinal oxygenated sulfoxides (I).

To distinguish between these two possibilities, deuterium-labeling studies were performed. No isotope incorporation was observed, from which Satoh et al. concluded that the mechanism of the vicinal elimination most likely involves a direct β-elimination by sulfoxide–metal exchange of **197** (path a). It is also interesting to note that either the (E)- or the (Z)-olefin can be preferentially obtained using this protocol. Thus, treatment of *anti*-mesyloxy sulfoxide **201a** with nBuLi generates **150a** and **150b** in a 4:1 ratio, whilst reaction of **201b** affords mostly the (Z)-alkene **150b** (Scheme 3.39). Similar selectivities was observed for the reduction of both diastereoisomers of **202**. These observations reinforce the postulated mechanism implying a direct elimination of the β-sulfoxy-mesylate (path a). Unfortunately, the selectivity of this elimination depends upon the structure of the starting substrate and can be difficult to predict.

Finally, when β-hydroxy sulfoxides **204**, derived from aromatic aldehydes, are treated with mesyl chloride in the presence of triethylamine, immediate elimination takes place, directly affording the corresponding (E)-olefins **207** in good yields

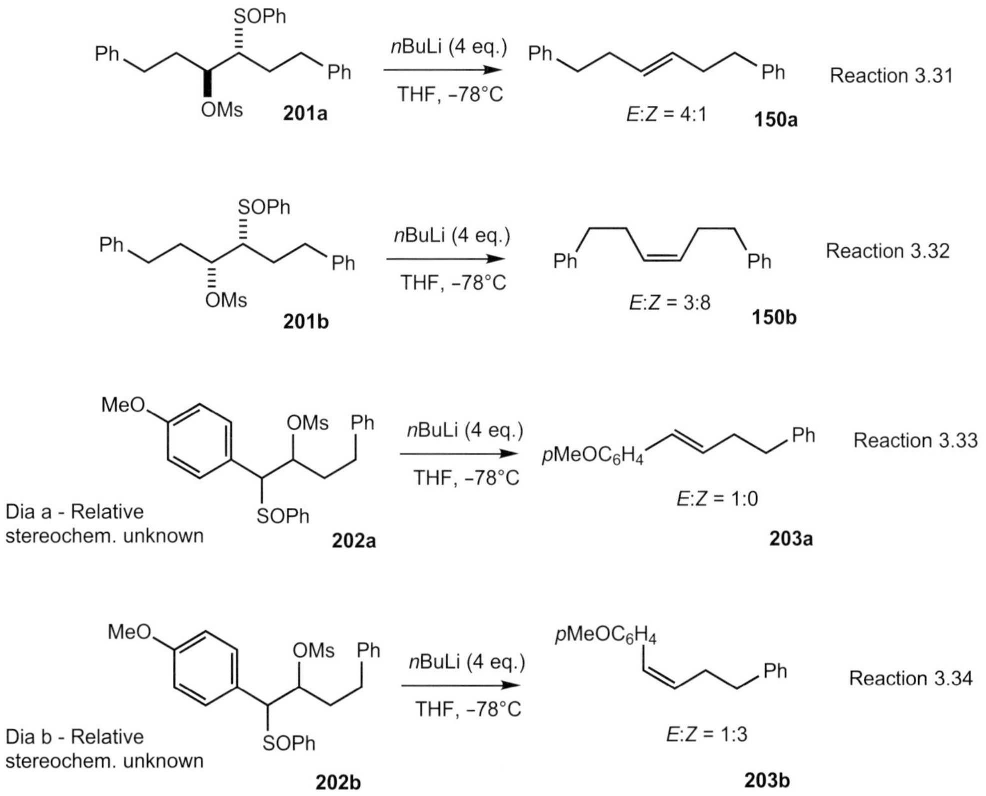

3.4 Reductive Elimination

Scheme 3.39. Stereoselective reductions of vicinal oxygenated sulfoxides.

(Scheme 3.40). According to Satoh, the mechanism is presumed to involve the benzylic-type carbocation **206**, generated by spontaneous solvolysis of the intermediate mesylate **205**. Addition of a nucleophile to the sulfoxide group then triggers its elimination.

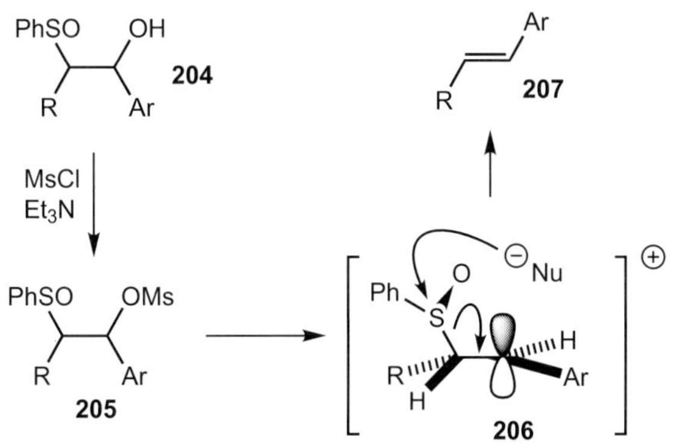

Scheme 3.40. Mechanism of the reduction of vicinal oxygenated sulfoxides (II).

Tab. 3.9. Reduction of vinyl sulfones.

Ratio		Ratio	Yield (%)
100:0		100:0	80
7:93		10:90	60

It is interesting to note that both diastereoisomers of **204** afford solely the (E)-olefin **207** in a convergent process. Of the various possible conformers of cation **206**, the one possessing the R and aryl substituents in a *trans* relationship is clearly the most stable, and the reaction is thus channelled through this intermediate.

3.4.5
Reduction of Vinyl Sulfones

Both (E)- and (Z)-vinyl sulfones can be readily prepared by several methods, and it is therefore little wonder that their transformation to stereodefined alkenes has been investigated. Both isomers of vinyl sulfones can be obtained by elimination from *erythro*- or *threo*-β-tosyl sulfones [46], generated either by controlled reduction of β-keto sufones [46, 118] or by varying the base in the coupling step [131]. The reduction of disubstituted vinyl sulfones by $Na_2S_2O_4$, discovered by Julia et al. in 1982, is particularly efficient and highly stereoselective [93] (Table 3.9).

This approach has since been extended by Julia et al. to the synthesis of trisubstituted olefins by replacing the sulfone moiety with an alkyl group. Some examples are collected in Table 3.10 [58].

3.5
Second Generation Julia Reactions

In 1991, Sylvestre Julia reported a direct synthesis of olefins from carbonyl compounds and lithiated heterocyclic sulfones [132–134]. After the initial coupling

Tab. 3.10. Replacement of the sulfone group.

Entry	R^1	R^2	R	X	Cat.	Yield (%)	E:Z
1	CH_3	CH_3	C_6H_5	Br	$Ni(acac)_2$	68	100:0
2	CH_3	CH_3	C_6H_5	Br	$Fe(acac)_2$	60	100:0
3	nPr	nBu	CH_3	Cl	$Ni(acac)_2$	64	4:96

3.5 Second Generation Julia Reactions

between the two fragments **213** and **95**, the intermediate β-alkoxy sulfone **214** underwent a sequence of consecutive reactions that led ultimately to the stereocontrolled preparation of substituted alkenes **131** (Scheme 3.41). This cascade involves the addition of the alkoxide anion to the sulfone residue to generate the cyclic species **215**, which breaks down to afford the sulfenate anion **216**. Finally, elimination of SO_2 and the alkoxide moiety completes the sequence.

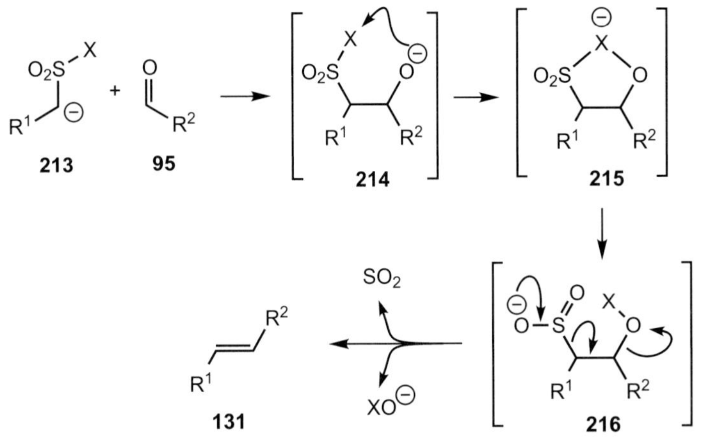

Scheme 3.41. "One-step" Julia reaction.

At the start of this work, the dual nucleophilic-electrophilic character of the α-sulfonyl carbanion **213** created some problems since this anion could react not only with the carbonyl derivative (aldehyde or ketone) but also with itself, generating via this competitive pathway some homocoupling by-products. However, a wise selection of the sulfone electrophilic substituent and the use of "Barbier-type" conditions considerably increased the chemoselectivity of the addition reaction and directed the addition onto the carbonyl function. Benzothiazole derivatives (hereafter denoted BT) were the first substrates to be utilized in this process. After some fine tuning, they proved to be suitable candidates, affording the desired alkenes with a reasonable efficiency (Table 3.11).

As was already observed in the reductive elimination of β-chloro sulfinates and β-arylsulfonyl sulfinates, the major *syn*-diastereoisomer of the β-alkoxy-BT-sulfones **218b** underwent an antiperiplanar elimination affording exclusively (Z)-olefin **219b** (Scheme 3.42). Similarly, elimination of *anti*-isomer **218a** was also highly stereoselective, producing (E)-alkene **219a** with >99% geometric purity.

Sylvestre Julia and co-workers also found that the two diastereoisomers of the β-hydroxy-BT-sulfones underwent elimination at different rates. The *syn*-isomer afforded the (Z)-alkene much faster than the *anti*-diastereoisomer produced the corresponding (E)-alkene. As can be seen in species **220** and **221**, R^1 and R^2 are in an *anti* relationship during the transfer of the BT moiety of **218b**. Further rotation about the central C–C bond is required to position the -OBT and -SO_2^- groups

Tab. 3.11. "Single-pot" Julia coupling.

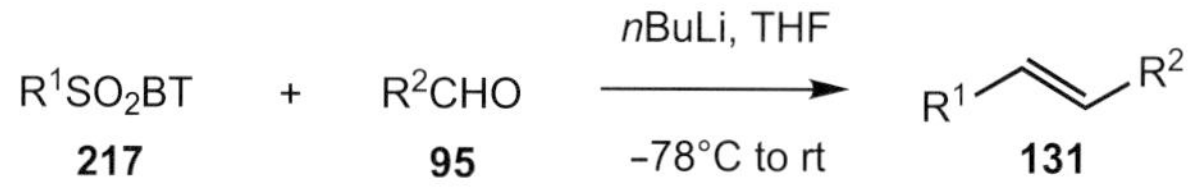

$$R^1SO_2BT \;+\; R^2CHO \xrightarrow[-78°C \text{ to rt}]{n\text{BuLi, THF}} R^1\diagup\!\!\!\diagdown R^2$$
$$\mathbf{217} \qquad\qquad \mathbf{95} \qquad\qquad\qquad\qquad \mathbf{131}$$

Entry	R^1	R^2	Yield (%)	E:Z
1	CH_2Cl	nC_8H_{17}	92	48:52
2	CH_2Cl	$p\text{MeOC}_6H_4$	95	83:17
3	nBu	$p\text{MeOC}_6H_4$	95	99:1
4	nBu	C_6H_5	68	94:6

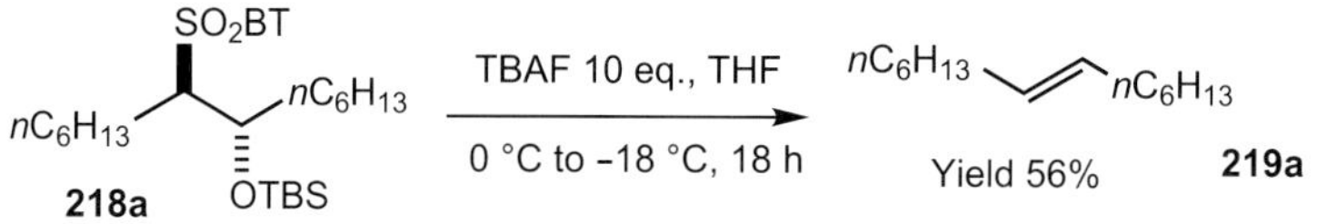

Reaction 3.35: Compound **218a** (with SO$_2$BT, nC_6H_{13}, OTBS substituents) treated with TBAF 10 eq., THF, 0 °C to −18 °C, 18 h, gives (E)-olefin **219a** (nC_6H_{13}–CH=CH–nC_6H_{13}), Yield 56%.

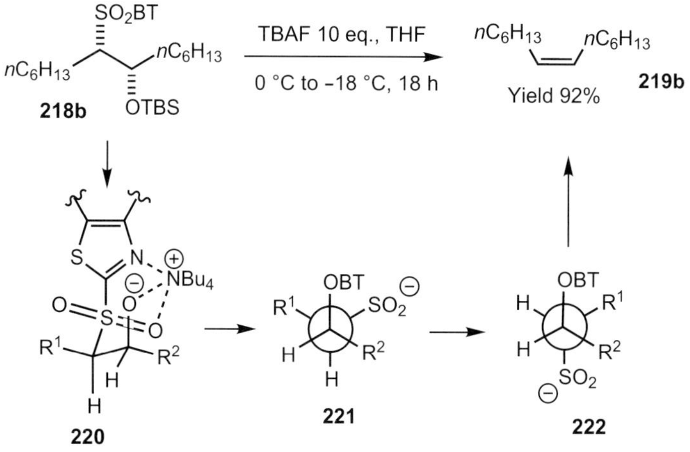

Reaction 3.36: Compound **218b** treated with TBAF 10 eq., THF, 0 °C to −18 °C, 18 h, gives (Z)-olefin **219b**, Yield 92%.

Scheme 3.42. Mechanism of the second-generation Julia reactions. (Intermediates **220**, **221**, **222**.)

antiperiplanar to each other (conformer **222**) and to bring about the elimination leading to the (Z)-olefin **219b**. Starting from *anti*-**218**, the corresponding conformers of **220** and **221** will have R^1 and R^2 in a *syn*-relationship. Due to the resultant destabilizing *gauche* interactions, these conformers will be higher in energy and the overall reaction thus proceeds more slowly. This observation nicely rationalizes both the lower rate and yield of reaction 3.35 as compared to reaction 3.36.

Julia et al. also noticed that the use of prenyl- or benzyl-BT-sulfones afforded the olefin products with reduced stereocontrol. This lower selectivity was rationalized

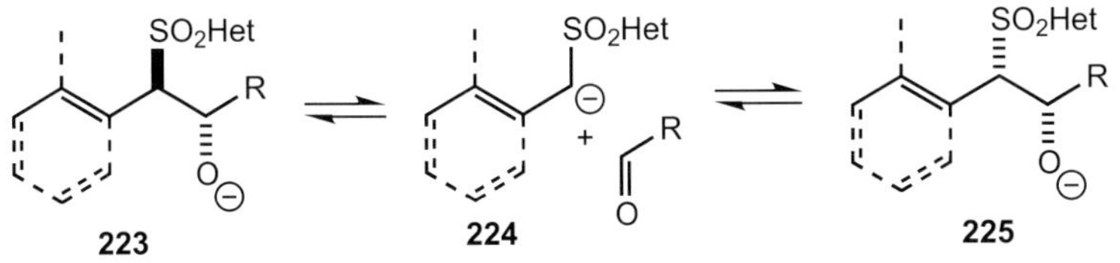

3.5 Second Generation Julia Reactions

Scheme 3.43. Equilibrium between syn- and anti-β-alkoxy sulfones.

by invoking a possible equilibrium between the syn- and anti-β-alkoxy sulfone intermediates **223** and **225** and the reactants (Scheme 3.43). Under the reaction conditions, the syn-β-hydroxy-BT-sulfones might then isomerize to the more stable anti diastereoisomer, leading to the observed erosion in the selectivity of the olefin geometry.

Although these explanations were based upon reasonable assumptions, some observations can still not be accounted for, and Julia admitted that other reaction pathways might be followed in specific cases, such as in the coupling of BT-sulfones with arylaldehydes. To rationalize these anomalous results, Sylvestre Julia invoked the participation of benzyl-type carbocations followed by E1-type elimination. Finally, S. Julia et al. studied the elimination of β-hydroxy pyridinic sulfones and observed results analogous to those reported for the BT-sulfone derivatives (Scheme 3.44).

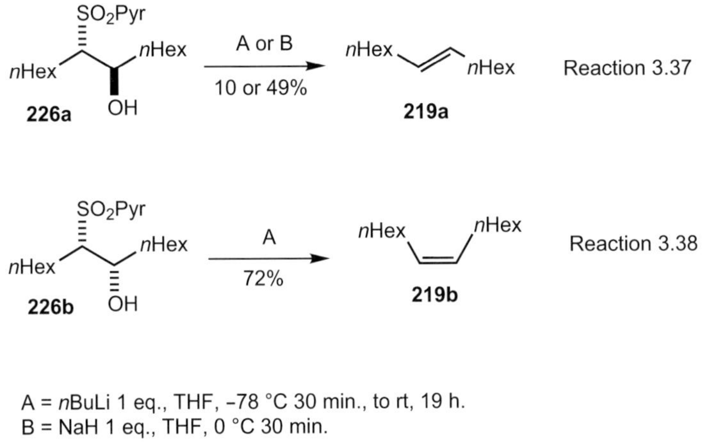

A = nBuLi 1 eq., THF, −78 °C 30 min., to rt, 19 h.
B = NaH 1 eq., THF, 0 °C 30 min.

Scheme 3.44. Elimination of β-hydroxy pyridinic sulfones.

Five years later, in 1996, Kociensky et al. successfully applied this methodology in their synthetic approach to Rapamycin [135]. They took advantage of the aforementioned observations that stabilized BT-sulfonyl α-carbanions favored E-selective olefination. However, when they performed the coupling of enal **227** with the meta-

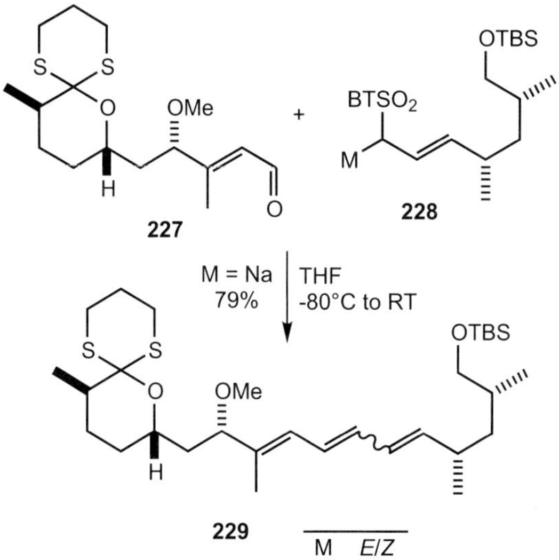

Scheme 3.45. Importance of the counter ion in the synthesis of Rapamycin.

M	E/Z
Li	1:2.4
Na	1:1.3
K	1:4.6

lated allylic BT-sulfone **228**, they observed a Z-preference, which varied slightly with the nature of the counterion (Scheme 3.45).

Several months earlier, Charette *et al.* had already reported on the critical importance of the counterion in the control of the olefinic *E:Z* ratio [136]. They also emphasized the importance of the solvent and the temperature (Table 3.12).

In 1998, Kociensky *et al.* proposed a variation of this method that obviated the use of "Barbier-type" conditions and made this sequence more compatible with complex aldehydic substrates [137]. They also extended the range of sulfone heterocycles to 1-isoquinolinoyl, 1-methyl-2-imidazoyl, 4-methyl-2-imidazoyl, 4-methyl-1,2,4-triazol-3-yl, and 1-phenyl-1*H*-tetrazol-5-yl (hereafter denoted PT), with the latter system being the most promising. Among the various trends that they observed from 96 comparative experiments, it was deduced that the *n*-alkyl PT-sulfones gave better yields than their BT-substituted counterparts (suggesting that PT sulfones are less prone to self-condensation). Moreover, careful combination of the solvent and the counter ion proved to have a profound effect on the yield and the stereochemical outcome of the elimination.

These second-generation Julia reactions have since been applied to the synthesis of natural products with varying degrees of success, as, for example, in the syntheses of vitamin D2 [138], (+)-Ambruticin [139] (Table 3.13), and (−)-Callystatin A [140]. Other variations have also been reported, for example by Kocienski in 2000 [141] and by Charette in 2001 [142, 143].

Tab. 3.12. Importance of solvent and counterion.

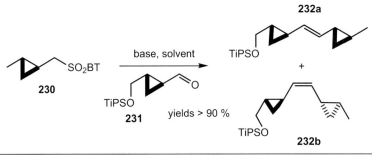

Entry	Base[a]	Solvent	E:Z[b]
1	NaHMDS	DMF	3.5:1
2	NaHMDS	THF	1.1:1
3	NaHMDS	Toluene	1:10
4	KHMDS	THF	1.2:1
5	KHMDS	Toluene	1:3.7

[a] NaHMDS 1 M in THF; KHMDS 1 M in toluene.
[b] Ratio determined by 400 MHz ^{1}H NMR.

3.6 Miscellaneous Julia Reactions

This section is dedicated to two important variations of the Julia reaction. The first one involves the replacement of the carbonyl group by a *gem*-halogeno-metal functionality, while the second employs N-analogues of sulfones. They share the common feature of having been slightly ignored by the scientific community in spite of several advantages.

3.6.1 *gem*-Halogeno-Metal Electrophiles

In 1982, Fujita *et al.* reported the one-step synthesis of 1,3-dienes by reaction of α-metalated sulfone derivatives with ICH$_2$SnBu$_3$ [144]. They also used sulfoxides or sulfides, but their yields were never higher than those obtained starting from the corresponding sulfones (Scheme 3.46; reaction 3.39). Three years later, Pearlman's group extended the methodology to moderately hindered sulfones such as **238** (reaction 3.40) [145]. It is worthy of note that similar methylations using ICH$_2$SiMe$_3$ had already been described in the literature as a two-step sequence, the elimination being triggered by fluoride ions [146].

In 1992, Julia et al. [147] employed halomethylmagnesium halides [148] instead of halomethyl stannanes. This process allows the replacement of the sulfonyl group of primary and secondary alkyl sulfones by methylene and alkylidene chains with great success (Table 3.14).

Tab. 3.13. Coupling step in the total synthesis of (+)-Ambruticin.

Entry	Base	Solvent	Temp.[a]	E:Z[b]
1	NaHMDS	THF	−78 °C	1:8
2	NaHMDS	THF	−35 °C	1:6
3	KHMDS	DMF	−60 °C	1:1
4	KHMDS	DME, 18-crown-6	−60 °C	1:3
5	LiHMDS	THF/HMPA (4:1 v/v)	−60 °C	3:1
6	LiHMDS	DMF/HMPA (4:1 v/v)	−35 °C	>30:1
7	LiHMDS	DMF/DMPU (1:1 v/v)	−35 °C	>30:1

[a] Reactions were started at this temp. and then allowed to warm to r.t.
[b] Ratios determined by ^{1}H NMR analysis of crude product mixtures.

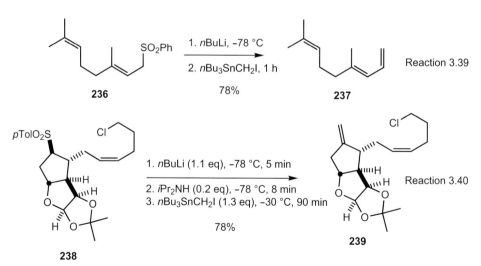

Scheme 3.46. Use of Bu_3SnCH_2I as electrophile.

3.6 Miscellaneous Julia Reactions

Tab. 3.14. Use of *gem*-haloalkyl magnesium halides.

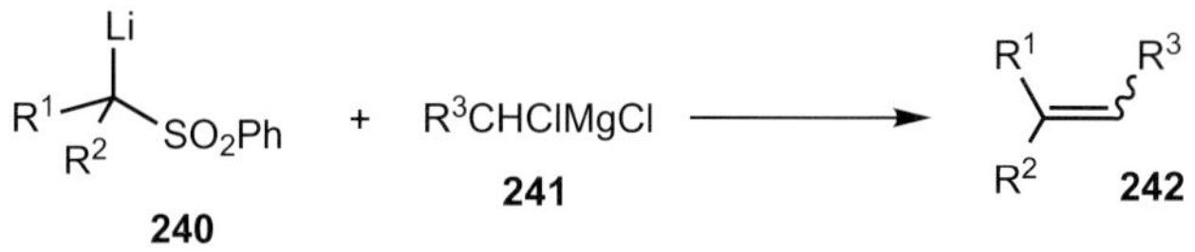

Entry	R¹	R²	R³	Yield (%)[a]	E:Z
1	Me	nHex	H	68	–
2	Me	Ph	H	76	–
3	Ph	Ph	H	65	–
4	Me	Ph	nBu	75	4:6
5	Ph	Ph	nBu	55	–

[a] Isolated yield after flash chromatography.

This methodology has since been applied by Smith et al. in their total syntheses of spongistatin [149, 150] and of phorboxazole [151] (Scheme 3.47).

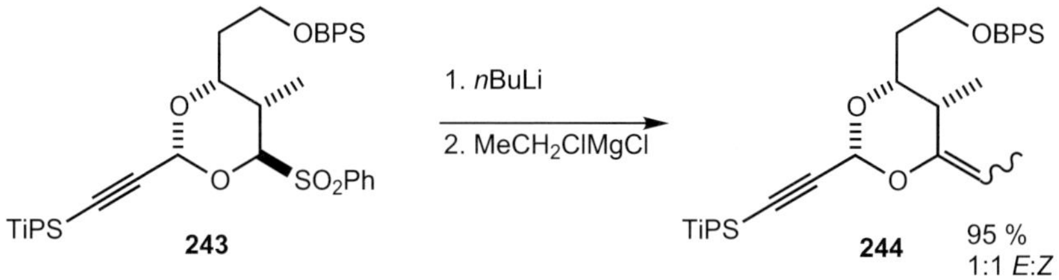

Scheme 3.47. Use of MeCH₂ClMgCl as electrophile.

3.6.2
Use of Sulfoximines

N-Analogues of sulfones can also be used in the coupling step and in the reductive elimination of the Julia sequence. They are less prone to retro-aldolization and, in certain cases, are more efficient in introducing a methylene unit in sterically demanding environments [3]. However, the use of sulfoximines for the preparation of C–C double bonds has been largely ignored by the scientific community, even though the corresponding Julia sequence was completed by Johnson et al. as long ago as 1979 [152]. Eight years earlier, Johnson's group had desulfurized β-N-methylsulfonimidoyl alcohols by using aluminum amalgam [153, 154] and thereafter focussed their efforts on the preparation of olefins. Having employed Mg/EtOH, Mg(Hg)/EtOH, Zn, Cr^{II} perchlorate, Na/NH₃, and Na(Hg)/EtOH with relative success, they explored the combination Al(Hg)/AcOH and found that the latter protocol was broadly applicable to the synthesis of mono-, di-, and trisub-

Tab. 3.15. Julia sequence applied to sulfoximines.

245 → (carbonyl compound) → 246 → 247

Entry	Carbonyl Comp.			246 Yield (%)	247 Yield (%)	E:Z[b]
	R^1	R^2	R^3			
1	H	$CH_3(CH_2)_{14}$	CH_3	90	90	–
2	H	(citronellal-type aldehyde)		85	93	–
3	Me	$CH_3(CH_2)_4$	H	75	100	60:40
4	Me	(norbornyl ketone)		75	65	78:22

stituted olefins (Table 3.15). Under these conditions, the sequence could even be performed as a single-pot operation.

As for their sulfone counterparts, double deprotonation [155] and [2,3]sigmatropic rearrangement [156, 157] of sulfoximines were studied. The most notable difference between sulfoximines and sulfones lies in the fact that the former are chiral, and therefore adducts with aldehydes or ketones may be formed in a highly diastereoselective manner, readily leading to syn- or anti-β-hydroxy-sulfoximines. Unfortunately, coupling between optically pure α-lithiated sulfoximines and aldehydes or ketones generally leads to 1:1 mixtures of syn/anti adducts [156, 158–161]. Despite this shortcoming, chiral sulfoximines have found widespread application in a number of useful synthetic transformations, including resolution [162–174].

Recently, Gais et al. succeeded in performing highly diastereoselective additions of allylic sulfoximines to aldehydes [175]. These allylic anions can lead to either α- or γ-adducts. The regio- and diastereocontrolled formation of γ-addition products has been reported by Reggelin et al. [176–179], and Gais elegantly extended this methodology to the preparation of α-hydroxy-sulfoximines.

Knowing that α-lithiated sulfoximines lead to poor diastereoselectivities and that allyltitanium species usually react via Zimmerman–Traxler transition states, Reggelin and Gais reasoned that lithium–titanium exchange might lead to enhanced stereocontrol in the coupling of these organometallics with aldehydes. Remarkably, they found that depending on the nature of the titanium ligands, α- or γ-adducts could be obtained at will. In all cases, almost total diastereoselectivity was observed (Scheme 3.48).

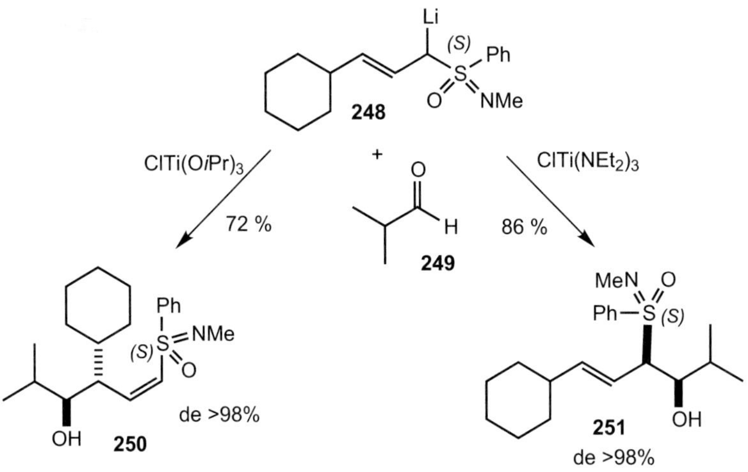

Scheme 3.48. Highly diastereoselective addition of allylic sulfoximines to aldehydes.

Unfortunately, reductive elimination of diastereoisomerically pure β-hydroxy-sulfoximines leads to mixtures of (E)- and (Z)-alkenes (Table 3.16) [152]. In general, the elimination of sulfoximines is less *trans*-selective [3, 152].

3.7 Conclusions

As can be seen from this chapter, the Julia reaction is a powerful tool for the preparation of C–C double bonds. Mono-, 1,1-di- and (E)-1,2-disubstituted olefins

Tab. 3.16. Reduction of pure diastereoisomers.

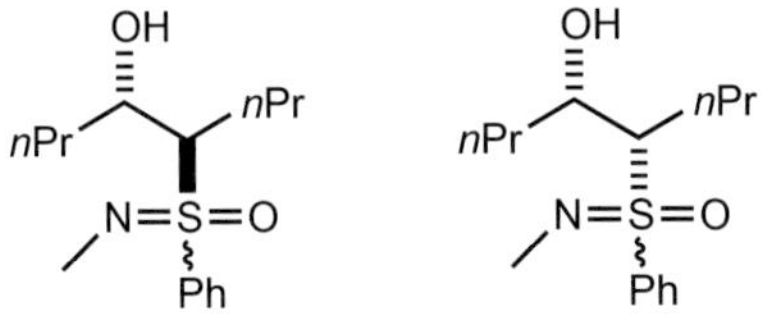

Entry	Diastereoisomer	4-octene E/Z
1	252a	60/40
2	252b	70/30
3	253a	84/16
4	253b	77/23

Conditions – Al/Hg, THF, AcOH, H$_2$O.

are easily obtained in good overall yields. It has also been shown that (Z)-1,2-disubstituted olefins and stereodefined trisubstituted olefins may be prepared by slight variations of the original sequences (Tables 3.9 and 3.10). To the best of our knowledge, no stereocontrolled tetrasubstituted olefins have yet been synthesized by means of the Julia olefination. Nevertheless, they can be synthesized by the method described in Table 3.10 [59].

Both sulfoxide- and sulfoximine-based modifications are promising alternatives for the synthesis of pure *syn* or *anti* coupling adducts and might lead to the stereocontrolled preparation of olefins. Moreover, great efforts are being made by several groups to enhance the stereoselectivity of the one-step procedures discussed in Section 3.5. It might soon be expected that any C–C double bond, no matter which isomer or how substituted it is, should be accessible by means of the Julia olefination reaction.

References

1 P. J. KOCIENSKY, *Chem. Ind.* **1981**, 548.
2 P. J. KOCIENSKY, *Phosphorus, Sulfur and Related Elements* **1985**, *24*, 97.
3 P. J. KOCIENSKY, in *Comprehensive Organic Synthesis* (Eds.: B. M. TROST, I. FLEMING), Pergamon Press, New York, **1991**, *6*, 987.
4 N. S. SIMPKINS, in *Sulfones in Organic Synthesis*, Tetrahedron Organic Chemistry series 10; Pergamon Press, London, **1993**.
5 E. ROTHSTEIN, *J. Chem. Soc.* **1937**, 309.
6 L. FIELD, *J. Am. Chem. Soc.* **1952**, 3919.
7 C. C. J. CULVENOR, W. DAVIES, W. E. SAVIGE, *J. Chem. Soc.* **1949**, 2198.
8 R. MOZINGO, D. WOLF, S. A. HARRIS, K. FOLKERS, *J. Am. Chem. Soc.* **1943**, *65*, 1013.
9 M. P. BALFE, R. E. DABBY, J. KENYON, *J. Chem. Soc.* **1951**, 385.
10 R. E. DABBY, J. KENYON, R. F. MASON, *J. Chem. Soc.* **1952**, 4881.
11 V. PASCALI, A. UMANI-RONCHI, *J. Chem. Soc., Chem. Commun.* **1973**, 351.
12 M. JULIA, J.-M. PARIS, *Tetrahedron Lett.* **1973**, *49*, 4833.
13 F. JUNG, N. K. SHARMA, T. DURST, *J. Am. Chem. Soc.* **1973**, *95*, 3420.
14 I. KUWAJIMA, S.-J. SATO, Y. KURATA, *Tetrahedron Lett.* **1972**, *9*, 737.
15 R. L. SOWERBY, R. M. COATES, *J. Am. Chem. Soc.* **1972**, 4758.
16 A. PADWA, Y. GAREAU, B. HARRISON, B. NORMAN, *J. Org. Chem.* **1991**, *56*, 2713.
17 R. V. WILLIAMS, G.-W. KELLY, J. LOEBEL, D. VAN DER HELM, P. C. B. PAGE, *J. Org. Chem.* **1990**, *55*, 3840.
18 C. N. HISAO, H. SHECHTER, *J. Org. Chem.* **1988**, *53*, 2688.
19 J.-M. BEAU, P. SINAŸ, *Tetrahedron Lett.* **1985**, *26*, 6189.
20 Y. GAONI, A. TOMAZIC, *J. Org. Chem.* **1985**, *50*, 2948.
21 Y. UENO, H. SANO, S. AOKI, M. OKAWARA, *Tetrahedron Lett.* **1981**, *22*, 2675.
22 H. KOTAKE, T. YAMAMOTO, H. KINOSHITA, *Chem. Lett.* **1982**, 1331.
23 K. INOMATA, S. IGARASHI, M. MOHRI, T. YAMAMOTO, H. KINOSHITA, H. KOTAKE, *Chem. Lett.* **1987**, 707.
24 M. T. REETZ, J. WESTERMANN, R. STEINBACK, B. WENDEROTH, R. PETER, R. OSTAREK, S. MAUS, *Chem. Ber.* **1985**, *118*, 1421.
25 J. L. G. RUANO, D. BARROS, M. C. MAESTRO, A. M. M. CASTRO, P. R. RAITHBY, *Tetrahedron: Asymmetry* **2000**, *11*, 4385.
26 E. ROTHSTEIN, *J. Chem. Soc.* **1934**, 684.
27 A. E. GUTHRIE, J. E. SEMPLE, M. M. JOULLIE, *J. Org. Chem.* **1982**, *47*, 2369.

28 H. Nemoto, K. Suzuki, M. Tsubuki, K. Minemura, K. Fukumoto, T. Kametani, H. Furuyama, *Tetrahedron* **1983**, *39*, 1123.

29 S. Hanessian, A. Ugolini, D. Dubé, P. J. Hodges, C. André, *J. Am. Chem. Soc.* **1986**, *108*, 2776.

30 P. Kociensky, B. Lythgoe, S. Ruston, *J. Chem. Soc., Perkin Trans. 1* **1978**, 829.

31 J. W. Morzycki, H. N. Schnoes, H. F. DeLuca, *J. Org. Chem.* **1984**, *49*, 2148.

32 S. J. Danishefsky, H. G. Selnick, R. E. Zelle, M. P. DeNinno, *J. Am. Chem. Soc.* **1987**, *110*, 4368.

33 T. Tagushi, R. Namba, M. Nakazawa, M. Nakajima, Y. Nakama, N. H. Kobayashi, N. Ikekawa, *Tetrahedron Lett.* **1988**, *29*, 227.

34 M. Ashwell, R. F. W. Jackson, *J. Chem. Soc., Chem. Commun.* **1988**, *10*, 645.

35 K. Hori, K. Nomura, E. Yoshii, *Heterocycles* **1989**, *29*, 663.

36 A. Talke, Ph.D. Thesis, Southampton University, 1988.

37 A. Talke, P. Kociensky, *Tetrahedron* **1990**, *46*, 4503.

38 B. Achmatowicz, E. Baranowska, A. R. Daniewski, J. Pankowski, J. Wicha, *Tetrahedron Lett.* **1985**, *26*, 5597.

39 S. L. Schreiber, H. V. Meyers, *J. Am. Chem. Soc.* **1988**, *110*, 5198.

40 B. Achmatowicz, E. Baranowska, A. R. Daniewski, J. Pankowski, J. Wicha, *Tetrahedron* **1988**, *44*, 4989.

41 J. White, Y. Ohba, W. Porter, S. Wang, *Tetrahedron Lett.* **1997**, *38*, 3167.

42 A. Barco, S. Benetti, C. De Risi, P. Marchetti, G. Pollini, V. Zanirato, *Tetrahedron Lett.* **1998**, *39*, 1973.

43 G. Zanoni, A. Porta, G. Vidari, *J. Org. Chem.* **2002**, *67*, 4346.

44 T. Iseki, S. Oishi, H. Namba, T. Tagushi, Y. Kobayashi, *Chem. & Pharm. Bull.* **1995**, *43*, 1897.

45 S. Labidalle, H. Moskowitz, M. Miocque, *Ann. Pharm. Fr.* **1981**, *39*, 545.

46 M. Julia, M. Launay, J.-P. Stacino, J.-N. Verpeaux, *Tetrahedron Lett.* **1982**, *23*, 2465.

47 E. Alvarez, T. Cuvigny, C. Herve du Penhoat, M. Julia, *Tetrahedron* **1988**, *44*, 119.

48 M. J. Ford, S. V. Ley, *Synlett* **1990**, 771.

49 D. Craig, P. S. Jones, G. J. Rowlands, *Synlett* **1977**, 1423.

50 S. J. O'Connor, P. G. Williard, *Tetrahedron Lett.* **1989**, *30*, 4637.

51 D. J. Hart, W.-L. Wu, *Tetrahedron Lett.* **1996**, *37*, 5283.

52 T. Satoh, N. Yamada, T. Asano, *Tetrahedron Lett.* **1998**, *39*, 6935.

53 T. Satoh, N. Hanaki, N. Yamada, T. Asano, *Tetrahedron* **2000**, *56*, 6223.

54 I. E. Marko, F. Murphy, C. Meerholz, S. Dolan, *Tetrahedron Lett.* **1996**, *37*, 2089.

55 J. Yu, C. Hyun-Sung, S. Chandrasekhar, J. R. Falck, C. Mioskowski, *Tetrahedron Lett.* **1994**, *35*, 5437.

56 S. Chandrasekhar, J. Yu, J. R. Falck, C. Mioskowski, *Tetrahedron Lett.* **1994**, *35*, 5441.

57 M. Pohmakotr, S. Pisutjaroenpong, *Tetrahedron Lett.* **1985**, *26*, 3613.

58 J.-L. Fabre, M. Julia, J.-N. Verpeaux, *Tetrahedron Lett.* **1982**, *23*, 2469.

59 M. Horigome, H. Motoyoshi, H. Watanabe, T. Kitahara, *Tetrahedron Lett.* **2001**, *42*, 8207.

60 J. P. Marino, M. S. McClure, D. Holub, J. V. Comasseto, F. C. Tucci, *J. Am. Chem. Soc.* **2002**, *124*, 1664.

61 V. Padmavathi, T. V. Reddy, K. V. Reddy, K. A. Reddy, D. B. Reddy, *Indian J. Chem., Sect. B: Org. Chem. including Med. Chem.* **2001**, *40B*, 667.

62 D. H. R. Barton, K. A. D. Swift, C. Tachdjian, *Tetrahedron* **1995**, *51*, 1887.

63 B. M. Trost, N. R. Schmuff, *J. Am. Chem. Soc.* **1985**, *107*, 396.

64 V. Bhat, R. C. Cookson, *J. Chem. Soc., Chem. Commun.* **1981**, 1123.

65 M. Hirama, *Tetrahedron Lett.* **1981**, *22*, 1905.

66 E. Ghera, Y. Ben David, H. Rapoport, *J. Org. Chem.* **1981**, *46*, 2059.

67 L. L. Vasil'eva, V. I. Mal'nikova, E. T. Gainullina, K. K. Pivnitskii, *Zh. Org. Khim.* **1980**, *16*, 2618.

68 B. M. Trost, N. R. Schmuff, M. J.

Miller, *J. Am. Chem. Soc.* **1980**, *102*, 5979.
69 J.-C. Spino, P. Deslongchamps, *Tetrahedron Lett.* **1990**, *31*, 3969.
70 M. R. Binns, R. K. Haynes, A. G. Katsifis, P. A. Schober, S. C. Vonwiller, *J. Org. Chem.* **1989**, *54*, 1960.
71 V. Barre, D. Uguen, *Tetrahedron Lett.* **1987**, *28*, 6045.
72 S. De Lombaert, I. Nemery, B. Roekens, J. C. Carretero, T. Kimmel, L. Ghosez, *Tetrahedron Lett.* **1986**, *27*, 5099.
73 H. Fujishima, H. Takeshita, S. Suzuki, M. Toyota, M. Ihara, *J. Chem. Soc., Perkin Trans. 1* **1999**, 2609.
74 M. Ihara, S. Suzuki, Y. Tokunaga, H. Takeshita, K. Fukumoto, *J. Chem. Soc., Chem. Commun.* **1996**, 1801.
75 S. Abel, D. Faber, O. Hüter, B. Giese, *Synthesis* **1999**, 188.
76 G. E. Keck, A. K. Savin, M. A. Weglarz, E. N. K. Cressman, *Tetrahedron Lett.* **1996**, *37*, 3291.
77 D. Tanner, P. Somfai, *Tetrahedron* **1987**, *43*, 4395.
78 J. Vollhardt, H.-J. Gais, K. L. Lukas, *Angew. Chem. Int. Ed. Engl.* **1985**, *97*, 110.
79 M. G. Cabiddu, S. Cabiddu, C. Fattuoni, C. Floris, G. S. Mellis, *Synthesis* **1993**, 41.
80 D. A. Evans, P. H. Carter, E. M. Carreira, A. B. Charette, J. A. Prunet, M. Lautens, *J. Am. Chem. Soc.* **1999**, *121*, 7540.
81 J.-L. Fabre, M. Julia, J.-N. Verpeaux, *Bull. Soc. Chim. Fr.* **1985**, 772.
82 S. Becker, Y. Fort, P. Caubere, *J. Org. Chem.* **1990**, 6194.
83 E. Alvarez, T. Cuvigny, H. C. du Penhoat, M. Julia, *Tetrahedron* **1988**, *29*, 111.
84 P. Auvray, P. Knochel, J. F. Normant, *Tetrahedron Lett.* **1986**, *27*, 5095.
85 B. Lythgoe, I. Waterhouse, *Tetrahedron Lett.* **1977**, 4223.
86 M. Ochiai, T. Ukita, E. Fujita, *J. Chem. Soc., Chem. Commun.* **1983**, 619.
87 D. R. Williams, J. L. Moore, M. Yamada, *J. Org. Chem.* **1986**, *51*, 3916.
88 A. B. Jones, A. Villalobos, R. G. Linde, S. J. Danishefsky, *J. Org. Chem.* **1990**, *55*, 2786.
89 A. Fernandez-Mayoralas, A. Marra, M. Trumtel, A. Veyrieres, P. Sinaÿ, *Tetrahedron Lett.* **1989**, *30*, 2537.
90 S. E. N. Mohamed, P. Thomas, D. A. Whiting, *J. Chem. Soc., Perkin Trans. 1* **1987**, 431.
91 G. E. Keck, D. F. Kachensky, E. J. Enholm, *J. Org. Chem.* **1985**, *50*, 4317.
92 G. Quinkert, E. Fernholz, P. Eckes, D. Neumann, G. Duerner, *Helv. Chem. Acta* **1989**, 1753.
93 J. Bremner, M. Julia, M. Launay, J.-P. Stacino, *Tetrahedron Lett.* **1982**, *23*, 3265.
94 M. Julia, D. Uguen, A. Callipolitis, *Bull. Soc. Chim. Fr.* **1976**, 519.
95 D. Savoia, C. Trombini, A. Umani-Ronchi, *J. Chem. Soc., Perkin Trans. 1* **1977**, 123.
96 T. Shono, Y. Matsumura, S. Kashimura, *J. Chem. Soc., Chem. Commun.* **1978**, 69.
97 D. H. R. Barton, J. Boivin, J. Sarma, E. Da Silva, S. Z. Zard, *Tetrahedron Lett.* **1989**, *30*, 4237.
98 S. Sawada, T. Nakayama, N. Esaki, H. Tanaka, K. Soda, R. K. Hill, *J. Org. Chem.* **1986**, *51*, 3384.
99 A. B. Smith III, K. J. Hale, L. M. Laasko, K. Chen, A. Riera, *Tetrahedron Lett.* **1989**, *30*, 6963.
100 M. Hirama, T. Nakamine, S. Ito, *Tetrahedron Lett.* **1988**, *29*, 1197.
101 L. A. Paquette, H. S. Lin, M. J. Coghlan, *Tetrahedron Lett.* **1987**, *28*, 5017.
102 A. S. Kende, J. S. Mendoza, *Tetrahedron Lett.* **1990**, *31*, 7105.
103 J. Inanaga, M. Ishikawa, M. Yamagushi, *Chem. Lett.* **1987**, 1485.
104 H. Künzer, M. Stahnke, G. Sauer, R. Wiechert, *Tetrahedron Lett.* **1991**, *32*, 1949.
105 G. E. Keck, K. A. Savin, M. A. Weglarz, *J. Org. Chem.* **1995**, *60*, 3194.
106 P. de Pouilly, A. Chénedé, J.-M. Mallet, P. Sinaÿ, *Tetrahedron Lett.* **1992**, *33*, 8065.

107 M. Ihara, S. Suzuki, T. Tanigushi, Y. Tokunaga, K. Fukumoto, *Synlett* **1994**, 859.

108 P. de Pouilly, A. Chénedé, J.-M. Mallet, P. Sinaÿ, *Bull. Soc. Chim. Fr.* **1993**, *130*, 256.

109 G. H. Lee, H. K. Lee, E. B. Choi, B. T. Kim, C. S. Pak, *Tetrahedron Lett.* **1995**, *36*, 5607.

110 G. H. Lee, E. B. Choi, E. Lee, C. S. Pak, *J. Org. Chem.* **1994**, *59*, 1248.

111 X. Huang, J. Pi, Z. Huang, *Heteroatom Chem.* **1992**, 535.

112 E. L. Grimm, S. Levack, L. A. Trimble, *Tetrahedron Lett.* **1994**, *35*, 6847.

113 A. H. Butt, B. M. Kariuki, J. M. Percy, N. S. Spencer, *J. Chem. Soc., Chem. Commun.* **2002**, 682.

114 O. Arjona, R. Menchaca, J. Plumet, *J. Org. Chem.* **2001**, *66*, 2400.

115 J. M. Bailey, D. Craig, P. T. Gallagher, *Synlett* **1999**, 132.

116 J. C. Carretero, R. Gomez Arrayas, *J. Org. Chem.* **1998**, *63*, 2993.

117 M. P. Gamble, G. M. P. Giblin, R. J. K. Taylor, *Synlett* **1995**, 779.

118 W. R. Roush, S. Russo-Rodriguez, *J. Org. Chem.* **1985**, *50*, 5465.

119 R. Carter, K. Hodgetts, J. McKenna, P. Magnus, S. Wren, *Tetrahedron* **2000**, *56*, 4367.

120 A. Barco, S. Benetti, C. De Risi, P. Marchetti, V. Pollini, *Tetrahedron Lett.* **1998**, *39*, 1973.

121 A. Kutner, M. Chodynski, S. J. Halkes, J. Brugman, *Bioorg. Chem.* **1993**, *21*, 13.

122 Y. Gaoni, A. Tomazic, *J. Org. Chem.* **1985**, *50*, 2948.

123 D. R. Williams, P. J. Coleman, *Tetrahedron Lett.* **1995**, *36*, 35.

124 R. Brückner, *Mecanismes Réactionnels en Chimie Organique*, De boek Université, Bruxelles, **1999**.

125 H. Nemoto, H. Kurobe, K. Fukumoto, T. Kametani, *J. Org. Chem.* **1986**, *51*, 5311.

126 A. G. M. Barrett, R. A. E. Carr, S. V. Attwood, G. Richardson, N. D. A. Walshe, *J. Org. Chem.* **1986**, *51*, 4840.

127 M. Sodeoka, S. Satoh, M. Shibasaki, *J. Am. Chem. Soc.* **1988**, *110*, 4823.

128 D. R. Williams, J. L. Moore, M. Yamada, *J. Org. Chem.* **1986**, *51*, 3918.

129 J. C. Barrish, H. Lin Lee, T. Mitt, G. Pizzolato, E. G. Baggiolini, M. R. Uskokovic, *J. Org. Chem.* **1988**, *53*, 4282.

130 D. H. R. Barton, C. Tachdjian, *Tetrahedron* **1992**, *48*, 7109.

131 A. Solladié-Cavallo, D. Roche, J. Fischer, A. De Cian, *J. Org. Chem.* **1996**, *61*, 2690.

132 J. B. Baudin, G. Hareau, S. A. Julia, O. Ruel, *Tetrahedron Lett.* **1991**, *32*, 1175.

133 J. B. Baudin, G. Hareau, S. A. Julia, O. Ruel, *Bull. Soc. Chim. Fr.* **1993**, *130*, 336.

134 J. B. Baudin, G. Hareau, S. A. Julia, R. Lorne, O. Ruel, *Bull. Soc. Chim. Fr.* **1993**, *130*, 856.

135 R. Bellingham, K. Jarowicki, P. Kocienski, V. Martin, *Synthesis* **1996**, 285.

136 A. B. Charette, H. Lebel, *J. Am. Chem. Soc.* **1996**, *118*, 10327.

137 P. R. Blakemore, W. J. Cole, P. J. Kocienski, A. Morley, *Synlett* **1998**, 26.

138 P. R. Blakemore, P. J. Kocienski, S. Marzcak, J. Wicha, *Synthesis* **1999**, *7*, 1209.

139 P. Liu, E. N. Jacobsen, *J. Am. Chem. Soc.* **2001**, *123*, 10772.

140 A. B. Smith III, B. M. Brandt, *Org. Lett.* **2001**, *3*, 1685.

141 P. J. Kocienski, A. Bell, P. R. Blakemore, *Synlett* **2000**, 365.

142 A. Charette, C. Berthelette, D. St-Martin, *Tetrahedron Lett.* **2001**, *42*, 5149.

143 A. Charette, C. Berthelette, D. St-Martin, *Tetrahedron Lett.* **2001**, *42*, 6619.

144 M. Ochiai, S.-I. Tada, K. Sumi, E. Fujita, *Tetrahedron Lett.* **1982**, *23*, 2205.

145 B. A. Pearlman, S. G. Putt, J. A. Fleming, *J. Org. Chem.* **1985**, *50*, 3622.

146 P. J. Kocienski, *Tetrahedron Lett.* **1979**, 2649.

147 C. De Lima, M. Julia, J.-N. Verpeaux, *Synlett* **1992**, 133.

148 J. Villieras, *Bull. Soc. Chim. Fr.* **1967**, 1511 and 1520.
149 A. B. Smith III, Q. Lin, K. Nakayama, A. M. Boldi, C. S. Brook, M. D. McBriar, W. H. Moser, M. Sobukawa, L. Zhuang, *Tetrahedron Lett.* **1997**, *38*, 8675.
150 A. B. Smith III, Q. Lin, V. A. Doughty, L. Zhuang, M. D. McBriar, J. K. Kerns, C. S. Brook, N. Murase, K. Nakayama, *Angew. Chem. Int. Ed.* **2001**, *40*, 196.
151 A. B. Smith III, K. P. Minbiole, P. R. Verhoest, T. J. Beauchamp, *Org. Lett.* **1999**, *1*, 913.
152 C. R. Johnson, R. A. Kirchhoff, *J. Am. Chem. Soc.* **1979**, *101*, 3602.
153 C. W. Schroeck, C. R. Johnson, *J. Am. Chem. Soc.* **1971**, *93*, 5305.
154 C. R. Johnson, C. W. Schroeck, J. R. Shanklin, *J. Am. Chem. Soc.* **1973**, *95*, 7424.
155 J. F. K. Muller, M. Neurburger, M. Zehnder, *Helv. Chim. Acta* **1997**, *80*, 2182.
156 M. Haramata, R. J. Claassen III, *Tetrahedron Lett.* **1991**, *32*, 6497.
157 H.-J. Gais, M. Scommoda, D. Lenz, *Tetrahedron Lett.* **1994**, *35*, 7361.
158 S. G. Pyne, *J. Org. Chem.* **1986**, *51*, 81.
159 C. R. Johnson, C. J. Stark Jr., *J. Org. Chem.* **1982**, *47*, 1193.
160 S. G. Pyne, G. Boche, *Tetrahedron* **1993**, *49*, 8449.
161 S. G. Pyne, Z. Dong, B. W. Skelton, A. H. White, *J. Org. Chem.* **1997**, *62*, 2337.
162 C. R. Johnson, J. R. Zeller, *J. Am. Chem. Soc.* **1982**, *104*, 4021.
163 C. R. Johnson, J. R. Zeller, *Tetrahedron* **1984**, *40*, 1225.
164 J. Wagner, E. Vieira, P. Vogel, *Helv. Chim. Acta* **1988**, *71*, 624.
165 L. Fitjer, H. Monzo Oltra, M. Noltemeyer, *Angew. Chem. Int. Ed. Engl.* **1991**, *30*, 1492.
166 B. J. Childs, G. L. Edwards, *Tetrahedron Lett.* **1993**, *34*, 5341.
167 A. B. Smith III, E. G. Nolen, R. Shirai, F. R. Blase, M. Oiita, N. Chida, R. A. Hartz, D. M. Fitch, W. M. Clarck, P. A. Sprengeler, *J. Org. Chem.* **1995**, *60*, 7837.
168 L. A. Paquette, S. Liang, H. L. Wang, *J. Org. Chem.* **1996**, *61*, 3268.
169 L. A. Paquette, C. F. Sturino, X. Wang, J. C. Prodger, D. Koh, *J. Am. Chem. Soc.* **1996**, *118*, 5620.
170 L. A. Paquette, L.-Q. Sun, D. Friedrich, P. B. Savage, *J. Am. Chem. Soc.* **1997**, *119*, 8438.
171 L. A. Paquette, T. M. Heidelbaugh, *Synthesis* **1998**, 495.
172 M. E. Kuehne, W. Dai, Y.-L. Li, *J. Org. Chem.* **2001**, *66*, 1560.
173 L. A. Paquette, D. R. Owen, R. T. Bibart, C. K. Seekamp, A. L. Kahane, J. C. Lanter, M. A. Corral, *J. Org. Chem.* **2001**, *66*, 2828.
174 M. A. Poupart, L. A. Paquette, *Tetrahedron Lett.* **1988**, *29*, 269.
175 J.-H. Gais, R. Hainz, H. Müller, P. R. Bruns, N. Giesen, G. Raabe, J. Runsink, S. Nienstedt, J. Decker, M. Schleusner, J. Hachtel, R. Loo, C.-W. Woo, P. Das, *Eur. J. Org. Chem.* **2000**, 3973.
176 M. Reggelin, H. Weinberger, *Angew. Chem. Int. Ed. Engl.* **1994**, *33*, 444.
177 M. Reggelin, H. Weinberger, R. Gerlach, J. Welcker, *J. Am. Chem. Soc.* **1996**, *118*, 4765.
178 M. Reggelin, T. Heinrich, *Angew. Chem. Int. Ed. Engl.* **1998**, *37*, 2883.
179 M. Reggelin, M. Gerlach, M. Vogel, *Eur. J. Org. Chem.* **1999**, 1011.

4
Carbonyl Olefination Utilizing Metal Carbene Complexes

Takeshi Takeda and Akira Tsubouchi

4.1
Introduction

In recent decades there has been an exponential increase in the application of metal-carbene complexes in organic synthesis [1]. Carbene complexes have been implicated as intermediates and reactive species in a wide range of catalytic and stoichiometric reactions. Metal carbenes, which have the general structure $M=CR^1R^2$, can be subdivided into two broad categories, namely the Fischer-type and the Schrock-type complexes. Metal carbenes having a low-valent late transition metal center, π-acceptor ligands, and π-donor substituents on the carbene carbon are called Fischer-type carbene complexes. These species are electrophilic at the carbene carbon atom. The Schrock-type carbene complexes are usually high-valent early transition metal species without π-accepting ligands, and are essentially nucleophilic in nature. The difference of reactivity of these two types of carbene complexes is of interest and extensive theoretical studies have been performed for both the Fischer- [2] and Schrock-type [3] complexes.

The Wittig-type olefination of carbonyl compounds is one of the characteristic reactions of carbene complexes. High-valent carbene complexes of early transition metals show ylide-like reactivity towards carbonyl compounds. In 1976, Schrock first demonstrated that niobium and tantalum neopentylidene complexes **1** and **2**, the typical nucleophilic Schrock-type carbene complexes, olefinate various carbonyl compounds including carboxylic acid derivatives [4].

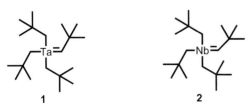

Scheme 4.1. Niobium and tantalum neopentylidene complexes.

Modern Carbonyl Olefination. Edited by Takeshi Takeda
Copyright © 2004 WILEY-VCH Verlag GmbH & Co. KGaA, Weinheim
ISBN: 3-527-30634-X

4 Carbonyl Olefination Utilizing Metal Carbene Complexes

A few years later, Tebbe and co-workers found that the methylene-bridged metallacycle **3**, which has become known as the Tebbe reagent, is useful for the methylenation of ketones and aldehydes [5]. Titanocene-methylidene **4**, the active species of this olefination, also transforms carboxylic acid derivatives into heteroatom-substituted olefins. Because the carbene complex **4** is much less basic than conventional olefination reagents such as phosphorus ylides, it can be employed for the olefination of carbonyl compounds possessing highly acidic α-protons or of highly hindered ketones, and has become an indispensable tool in organic synthesis. Various methods for the preparation of titanium-carbene complexes have since been developed for carbonyl olefination. This chapter focuses on the use of metal-carbene complexes and some related species in carbonyl olefination (Scheme 4.2).

$$R^2R^1C=ML_n \; + \; O=CR^3R^4 \; \longrightarrow \; R^1R^2C=CR^3R^4 \; + \; O=ML_n$$

Scheme 4.2. Carbonyl olefination utilizing metal-carbene complexes.

4.2
Carbonyl Olefination with Titanocene-Methylidene and Related Reagents

4.2.1
Preparation of Titanocene-Methylidene

4.2.1.1 The Tebbe Reagent

The Tebbe reagent **3** is prepared as reddish-orange crystals by the reaction of two equivalents of trimethylaluminum with titanocene dichloride [5]. The mechanism of the formation of **3** has been well investigated [6]. Theoretical studies based on ab initio calculations on a model complex suggest that the appropriate representation of the Tebbe reagent is in terms of an intramolecular complex, in which chlorine acts as the electron donor and aluminum as the acceptor [7]. Since the reagent is sensitive to air and moisture, the use of a standardized solution in benzene or toluene is recommended. Such solutions of **3** can be currently purchased from various commercial sources. The titanocene-methylidene **4** is formed by the action of a Lewis base such as pyridine or 4-dimethylaminopyridine on **3**, with the expulsion of dimethylaluminum chloride (Scheme 4.3). The carbene complex **4** itself

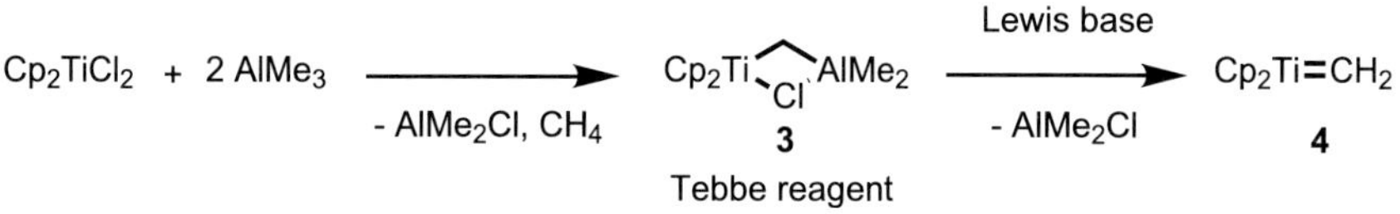

$$Cp_2TiCl_2 \; + \; 2\,AlMe_3 \; \xrightarrow{-\,AlMe_2Cl,\,CH_4} \; Cp_2Ti\underset{Cl}{\overset{}{\diagup\!\!\!\diagdown}}AlMe_2 \; \xrightarrow[-\,AlMe_2Cl]{Lewis\ base} \; Cp_2Ti=CH_2$$

$$\quad\quad\quad\quad\quad\quad\quad\quad\quad\quad\quad\quad\quad\quad\quad\quad\quad\quad\quad 3 \quad\quad\quad\quad\quad\quad\quad\quad\quad 4$$
$$\quad\quad\quad\quad\quad\quad\quad\quad\quad\quad\quad\quad\quad\quad\quad\quad Tebbe\ reagent$$

Scheme 4.3. Formation of titanocene-methylidene from the Tebbe reagent.

has not been observed, but its phosphine complexes have been prepared and characterized [8].

An improved procedure for the preparation of the Tebbe reagent **3** and the in situ preparation of **3** for large-scale reactions has been developed, requiring neither vacuum line nor Schlenk techniques [9]: the crude reaction mixture formed by the combination of titanocene dichloride and two equivalents of trimethylaluminum can be employed for further reactions [10].

The initial study by Grubbs and co-workers showed that a variety of carbonyl compounds, including carboxylic acid derivatives, are transformed into terminal olefins with the Tebbe reagent **3** in good to high yields (Scheme 4.4) [11].

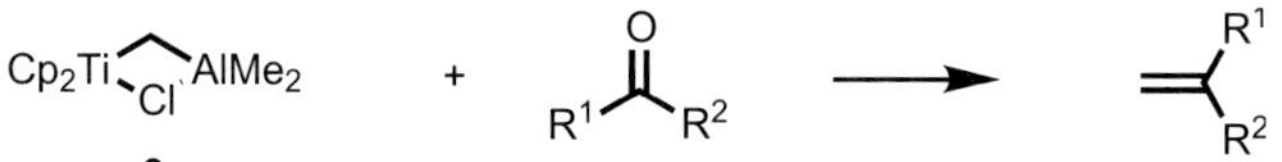

Scheme 4.4. Methylenation with the Tebbe reagent.

Grubbs and Pine proposed two pathways for the methylenation of esters with the Tebbe reagent **3**. If no Lewis base is used to generate the titanocene-methylidene **4**, the reaction of **3** with an ester is first order with respect to both **3** and the ester. On the basis of a large negative activation entropy, they suggested that the six-membered metallacycle intermediate **5** is formed. The vinyl ether is produced with the concomitant extrusion of methyltitanocene chloride **6** and the methylaluminoxy polymer **7** (Scheme 4.5). In the presence of a base such as pyridine, the reaction is much more rapid and is zero order in ester and first order in **3**, which indicates that it proceeds through the formation of oxatitanacyclobutane intermediate **8** [11b].

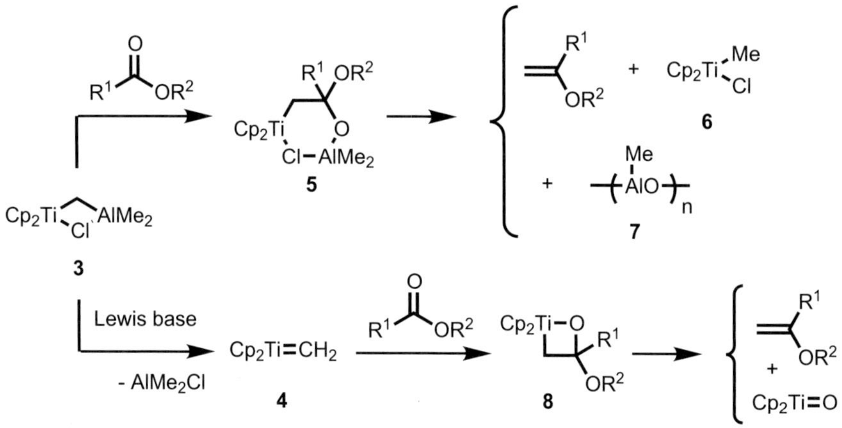

Scheme 4.5. The reaction pathways for the methylenation of esters with the Tebbe reagent.

Tab. 4.1. Methylenation of ketones with the Tebbe reagent **3**.

Entry	Ketone	Yield[a]	Ref.	Entry	Ketone	Yield[a]	Ref.
1	(2-tert-butylcyclohexanone)	96% (80%)	12	7	(bicyclic with OSiMe$_2$Bu^t, OAOM, OR groups; R = SiPh$_2$Bu^t; AOM = p-Anisoyloxymethyl)	77%	14
2	(2,4,6-trimethylacetophenone)	77% (4%)	12	8	(bicyclic N-SO$_2$CH$_2$Br ketone)	70% (−)[b]	15
3	(benzophenone)	97% (46%)	12	9	(TBDMSO, I-substituted cyclopentenone)	79%	16
4	Ph–CO–CH(Ph)	63% (38%)	12	10	(cyclopentanone with p-Tol)	56%	17
5	(2-phenylcyclohexanone)	93%	11a	11	(polyether with Et, OH groups)	60–70% (20–30%)	11a
6	(3,4-dimethoxyphenyl β-bromo ketone)	93%	13	12	(bicyclic HO-substituted ketone)	61%	18

[a] Values in parentheses are yields obtained using the Wittig reagent.
[b] Not obtained.

Methylenation of ketones: Although various reagents have been developed for the methylenation of ketones, the Tebbe reagent **3** has several advantages over the alternatives. As shown in Table 4.1, sterically hindered ketones (entries 1–3) and readily enolizable ketones (entries 4 and 5) are methylenated in better yields with **3** than with the Wittig reagent. The low basicity of the Tebbe reagent **3** enables the olefination of ketones possessing a leaving group β to carbonyl function (entries 6 and 7). In the case of the bromo ketone in entry 8, olefination with the Tebbe reagent **3** affords the terminal olefin, whereas treatment with methylenetriphenylphosphorane leads to intramolecular alkylation at the carbon α to the carbonyl group. The definite superiority of **3** over the Wittig reagent is demonstrated by the

Tab. 4.2. Methylenation of aldehydes with the Tebbe reagent **3**.

Entry	Aldehyde	Product	Yield	Ref.
1	(structure)	(structure)	35%	19
2	(structure)	(structure)	47%	20
3	(structure)	(structure)	48%[a]	21
4	(structure)	(structure)	82%	22

[a] The aldehyde prepared by oxidation of the precursor alcohol was employed in the methylenation without isolation. The yield from the alcohol is given.

methylenation of α-substituted ketones (entries 10 and 11). Treatment of ketones having a chiral center at the position α to the carbonyl group with **3** affords the terminal olefin with retaintion of the stereochemistry, whereas reaction with the Wittig reagent leads only to epimerization of the α-substituent. Another advantage of the Tebbe reagent is that it can be employed for the reactions of hydroxy ketones without any protection (entries 11 and 12).

Methylenation of aldehydes: In contrast to the methylenation of ketones, only a limited number of successful methylenations of aldehydes has been reported (Table 4.2). This is probably due to the further reaction of the resulting mono-substituted olefins with the remaining Tebbe reagent **3** to form titanacyclobutanes, which will be described later in this section. However, for the methylenation of a dialdehyde (entry 2), **3** is the reagent of choice, and other methods such as the Wittig and Peterson reactions, olefination using the $CH_2Br_2/Zn/TiCl_4$ system, and Cr(III)-mediated olefination with CH_2I_2 produce the 1,7-diene only in poor yields. All the reactions cited in Table 4.2 proceed without epimerization.

Tab. 4.3. Methylenation of esters and lactones with the Tebbe reagent **3**.

Entry	Ester	Yield	Ref.	Entry	Lactone	Yield	Ref.
1	PhO–C(=O)–Ph	94%	23	7	chroman-2-one (benzo-fused δ-lactone)	85%	23
2	Ph–CH₂–C(=O)–OEt	90%	23	8	bicyclic spiro lactone	55%[b]	25
3	cyclohex-2-enyl O–C(=O)–CH₂–Ph	96%	23	9	F₂C=C(Cl)-substituted bicyclic lactone	89%	26
4	Ph–CH=CH–C(=O)–OMe	79%	23	10	tri-OBn pyranone (BnO, BnO, OBn, OBn)	92%	27
5	phthalate di-OMe	65%[a]	24	11	vinyl tri-OBn pyranone	44%	28
6	Ph–C(=O)–O–CH₂–CH=CH₂	50%	24	12	complex bicyclic sugar lactone with OBn, HO, BnO, Ph	66%	29

[a] Both carbonyls were methylenated.
[b] The product was isolated as the *endo*-olefin due to isomerization.
[c] The methylenation product was transformed into the alcohol by hydroboration-oxidation without isolation. The yield for the two-step transformation is given.

Methylenation of carboxylic acid derivatives: The Tebbe reagent **3** is extremely valuable as a reagent for the methylenation of carboxylic acid derivatives, which is generally unsuccessful using phosphorus ylides. Esters and lactones are readily transformed into enol ethers (Table 4.3), especially when a Lewis base such as THF is present in the reaction mixture. In the methylenation of α,β-unsaturated esters, the internal olefin is not involved in the reaction, and the configuration of the double bond is maintained (entry 4). When carbonyl compounds bearing a terminal double bond are subjected to the methylenation, significantly lower yields are observed (entries 6 and 11), which may be attributable to competitive formation of a titanacycle from titanocene-methylidene **4** and the terminal olefin.

Tab. 4.4. Methylenation of amides, imides, and thiol esters with the Tebbe reagent **3**.

Entry	Carbonyl compound	Product	Yield	Ref.
1	Ph−C(=O)−N(pyrrolidine)	Ph−C(=CH₂)−N(pyrrolidine)	80%[a]	24
2	N-phenyl succinimide	mono-methylenated N-Ph imide	quant.[a]	30
3	N-Me glutarimide derivative	mono-methylenated (less hindered C=O)	65%[a]	30
4	iPr-C(=O)-SPh with OTBDMS	iPr-C(=CH₂)-SPh with OTBDMS	83%	31
5	Cp₂-dimethyl thia-titanocycle (S,O)	Cp₂-dimethyl methylene thia compound	70%	32

[a] Determined by NMR.

Similar methylenation of amides, imides, and thiol esters produces the corresponding heteroatom-substituted olefins (Table 4.4). Since the olefination of imides is greatly affected by steric hindrance, an unsymmetrical imide is monomethylenated at the sterically less hindered carbonyl group when treated with **3** (entry 3). The low basicity of **3** allows the transformation of a thiol ester to the corresponding alkenyl sulfide without racemization (entry 4).

Since esters and amides are much less reactive than aldehydes and ketones towards the Tebbe reagent **3**, chemoselective methylenation of dicarbonyl compounds has been achieved, as illustrated in Scheme 4.6 and in the examples in Table 4.2.

Synthetic application to cyclic compounds: The ability of the Tebbe reagent to transform lactone carbonyls to enol ethers provides a useful means of constructing medium-sized rings when combined with the Claisen rearrangement. In some cases, the δ-vinyl-δ-lactones are generated and rearranged in the same reaction vessel without isolation of the enol ether intermediates. The following schemes illustrate the use of this reaction sequence for the preparation of eight-membered rings that are intermediates en route to complex polycyclic molecules (Scheme 4.7).

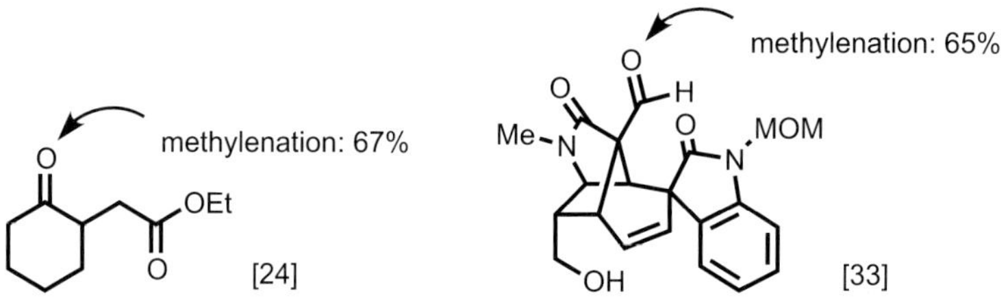

Scheme 4.6. Chemoselective methylenation with the Tebbe reagent.

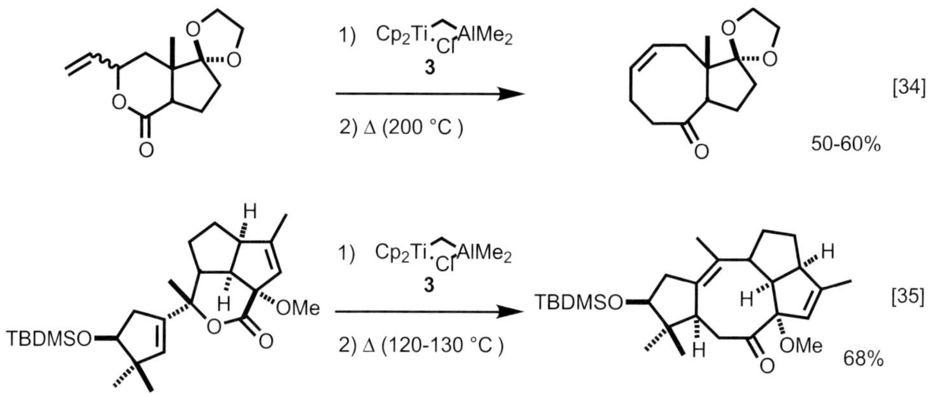

Scheme 4.7. The tandem carbonyl methylenation–Claisen rearrangement for the construction of eight-membered rings.

The combination of carbonyl olefination and ring-closing metathesis promoted by the Tebbe reagent **3** is a powerful tool for the construction of cyclic ethers (Schemes 4.8 and 4.9). This methodology was first reported by Grubbs and co-workers using the molybdenum alkylidene complex **9** as a catalyst for ring-closing olefin metathesis (Scheme 4.8) [36].

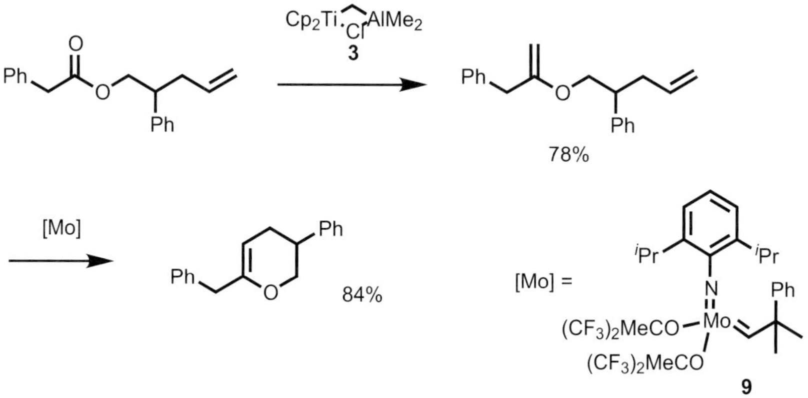

Scheme 4.8. Carbonyl methylenation with the Tebbe reagent and subsequent ring-closing metathesis using a molybdenum catalyst.

Nicolaou and co-workers employed the Tebbe reagent for both carbonyl methylenation and ring-closing metathesis (Scheme 4.9). Here, titanocene-methylidene initially reacts with the ester carbonyl group to form the vinyl ether. Subsequent olefin metathesis between titanocene methylidene and the cis-1,2-disubstituted double bond in the same molecule produces the alkylidenetitanocene **10**, ring-closing olefin metathesis of which affords the cyclic vinyl ether [37]. Similar reaction sequences have been applied for the construction of the complex cyclic polyether frameworks of maitotoxin [38] and a fragment of ciguatoxin CTX3C [39].

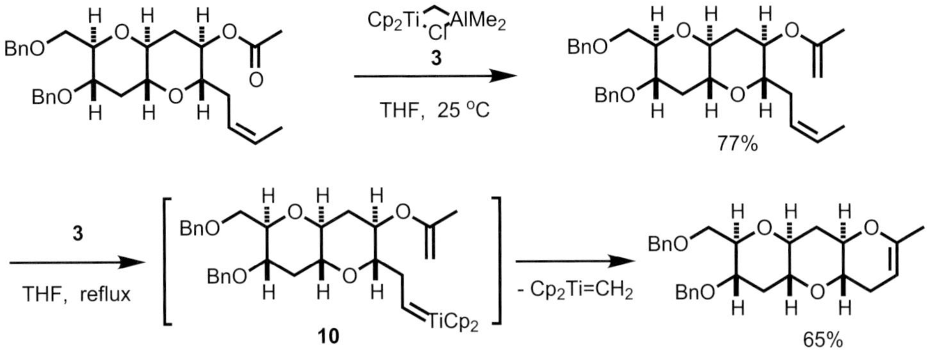

Scheme 4.9. Formation of cyclic ethers by a combination of carbonyl methylenation and ring-closing metathesis using the Tebbe reagent.

4.2.1.2 β-Substituted Titanacyclobutanes as Precursors of Titanocene-Methylidene

Alternative precursors of titanocene-methylidene **4** are titanacyclobutanes **11**, known as Grubbs' reagents, which are prepared by reaction of the Tebbe reagent **3** with appropriate olefins. The carbene complex **4** is highly reactive towards multiple bonds, such as those of alkenes, alkynes, and nitriles, as well as carbonyl functions [11, 40]. The reaction of **3** with a variety of olefins in the presence of a Lewis base such as pyridine affords four-membered metallacycles **11** [41]. Since thermolysis of these compounds effects the reverse reaction, the titanacycles **11** are employed as precursors of titanocene-methylidene **4** (Scheme 4.10). The mechanism of thermal cleavage of **11** to give a titanocene-methylidene has been extensively studied [42], and the temperature required for the thermal degradation of compounds **11** was found to be dependent on the substituents present. The crystalline metallacycles **11** can be handled in air for a reasonable period of time and are employed under various conditions (e.g. $R^1, R^2 = Me$; 5 °C; $R^1 = Me$, $R^2 = Bu$; 5 °C; $R^1 = H$,

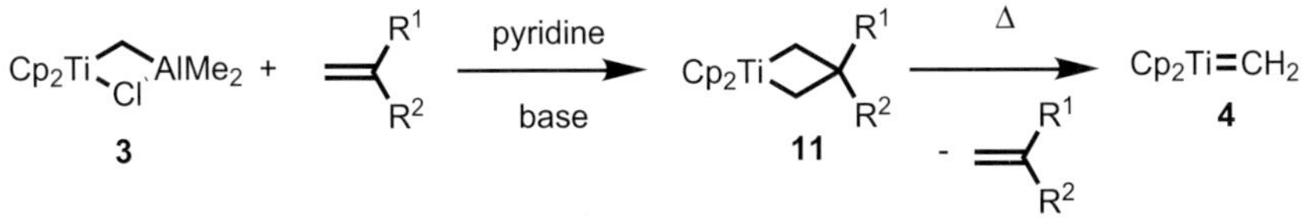

Scheme 4.10. Formation of titanacyclobutanes and regeneration of titanocene-methylidene.

Tab. 4.5. Methylenation of carbonyl compounds with the titanacyclobutanes **11**.

Entry	Carbonyl compound	Titanacyclobutane 11	Yield	Ref.
1	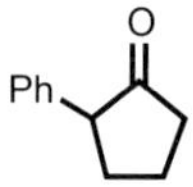 Ph-C(=O)-cyclopentyl	 Cp$_2$Ti(CH$_2$CMe$_2$CH$_2$) (with Et)	98%	43
2	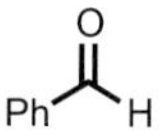 Ph-C(=O)-H	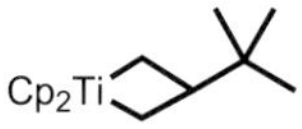 Cp$_2$Ti(CH$_2$CMe$_2$CH$_2$)	92%	11c
4	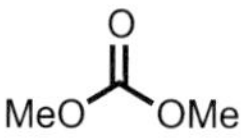 MeO-C(=O)-OMe	 Cp$_2$Ti(CH$_2$CMe$_2$CH$_2$)	60%	11a
5	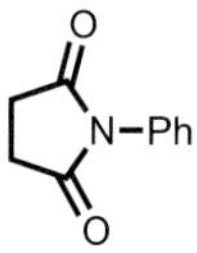 N-phenylsuccinimide	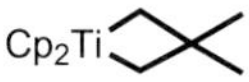 Cp$_2$Ti(CH$_2$CMe$_2$CH$_2$)	quant.[a,b]	30

[a] Determined by NMR.
[b] Both carbonyls were methylenated.

$R^2 = {}^tBu$; 60 °C) [11a]. Another advantage of these reagents is that they produce the aluminum-free titanocene-methylidene **4**, which is isolated as a phosphine complex when **11** is heated in the presence of excess phosphine [8].

As the representative examples in Table 4.5 indicate, the titanocene-methylidene precursors **11** transform various carbonyl compounds, including imides [30], in yields comparable to those obtained with the Tebbe reagent **3**. Kinetic studies indicate that the reaction of **11** with carbonyl compounds proceeds via the titanium carbene **4** [11b].

A limitation of the methylenation of carboxylic acid derivatives with Tebbe-type reagents such as **3** and **11** is that highly electrophilic acid anhydrides and acid halides cannot be transformed to the corresponding enol esters or alkenyl halides. The reaction of titanocene-methylidene generated from **11** with acid chlorides results in the formation of acylation products via titanium enolates **12** [44] rather than methylenation. Since the titanium enolates **12** are stable and do not isomerize to the more highly substituted enolates, they are useful for regioselective aldol reactions (Scheme 4.11) [44a]. Although similar titanium enolates are also produced

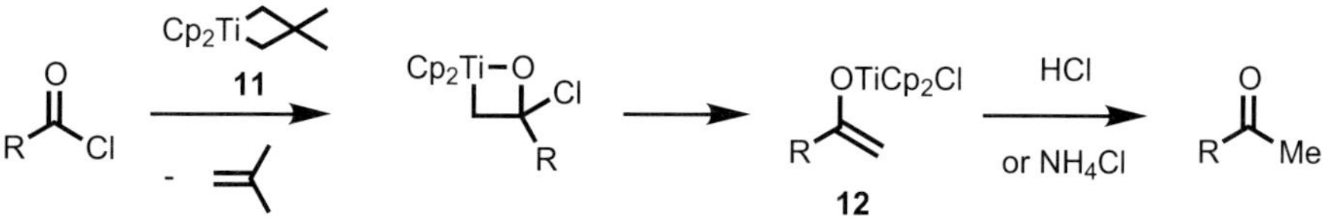

Scheme 4.11. The reaction of acid chlorides with titanacyclobutanes.

by the treatment of acid anhydrides with **11**, they fail to react with aldehydes, even at high temperatures at which the enolates decompose [30].

An alternative procedure for the preparation of titanacyclobutanes involves the treatment of titanocene dichloride with Grignard reagents prepared from 1,3-dihalides (Scheme 4.12) [45]. Although the metallacycles prepared by this method are effective methylenating agents for ketones, the corresponding zirconium and hafnium analogues are not capable of olefinating carbonyls, reflecting the reduced tendency of these species to generate metal-carbene complexes.

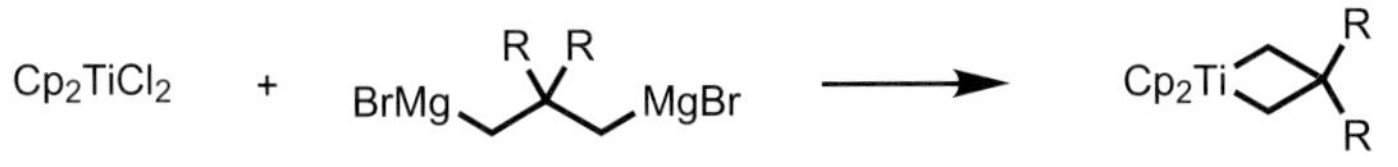

$$Cp_2TiCl_2 \quad + \quad BrMg\underset{R\ R}{\diagup\!\!\!\diagdown}MgBr \quad \longrightarrow \quad Cp_2Ti\underset{R}{\overset{R}{\diagdown\!\!\!\diagup}}$$
$$\mathbf{11}$$

Scheme 4.12. Formation of titanacyclobutanes utilizing 1,3-bis(bromomagnesio)propane.

4.2.1.3 Zinc and Magnesium Analogues of the Tebbe Reagent

Zinc and magnesium analogues of the Tebbe reagent have also been investigated and employed for the methylenation of carbonyl compounds, though their synthetic utility seems to be limited. The titanocene-methylidene compound complexed with zinc halide **13** is prepared by treating titanocene dichloride with bis-(iodozincio)methane (Scheme 4.13). The zinc complex **13** readily methylenates ketones in good to high yields [46].

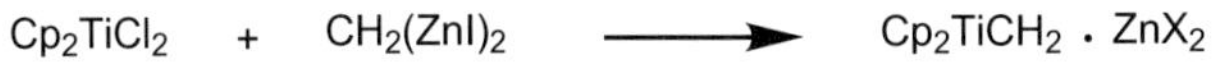

$$Cp_2TiCl_2 \quad + \quad CH_2(ZnI)_2 \quad \longrightarrow \quad Cp_2TiCH_2 \cdot ZnX_2$$
$$\mathbf{13}$$

Scheme 4.13. Formation of titanocene-methylidene zinc halide complex.

The reaction of titanocene dichloride with the methylene di-Grignard reagent affords the organotitanium species **14**, which behaves like the Tebbe reagent; **14** reacts with olefins to produce various titanacyclobutanes **11** [47]. The interesting spiro metallacycle **15**, which is formed from two equivalents of **14**, reacts with benzophenone to produce 1,1-diphenylethylene (80% yield) and forms the titanocene-methylidene complex **16** on treatment with trimethylphosphine (Scheme 4.14) [48].

4.2.2
Higher Homologues of Titanocene-Methylidene

Alkylidenation: Several attempts have been made to prepare titanium alkylidene complex **17**. The reaction of titanocene dichloride with triethylaluminum does not afford an alkylidene-bridged complex similar to the Tebbe reagent, probably due to the presence of a β-hydrogen, and produces instead the aluminotitanium hydride

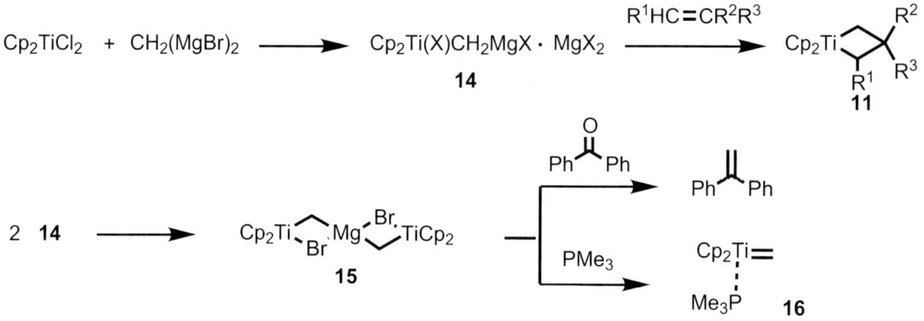

4 Carbonyl Olefination Utilizing Metal Carbene Complexes

Cp_2TiCl_2 + $CH_2(MgBr)_2$ ⟶ $Cp_2Ti(X)CH_2MgX \cdot MgX_2$ ⟶ $Cp_2Ti\begin{smallmatrix}R^2\\R^3\end{smallmatrix}$
14 **11**

with R¹HC=CR²R³ over the second arrow.

2 **14** ⟶ $Cp_2Ti\overset{Br}{\underset{Br}{\diagup}}Mg\overset{Br}{\underset{}{\diagup}}TiCp_2$ **15**

From **15**: with Ph-CO-Ph → Ph₂C=CPh₂ (Cp₂Ti=); with PMe₃ → $Cp_2Ti=$ · Me_3P **16**

Scheme 4.14. Formation and reactions of the magnesium-containing Tebbe-like reagent.

[(C₅H₅)(C₅H₄)TiHAl(C₂H₅)₂]₂ [49]. A precursor of alkylidene species **18** is obtained by the addition of an organoaluminum hydride to the double bond of the alkenyltitanium complex **19** (Scheme 4.15) [50]. The metallacycle **18** converts cyclohexanone and methyl benzoate into the corresponding trisubstituted olefin and disubstituted vinyl ether, respectively.

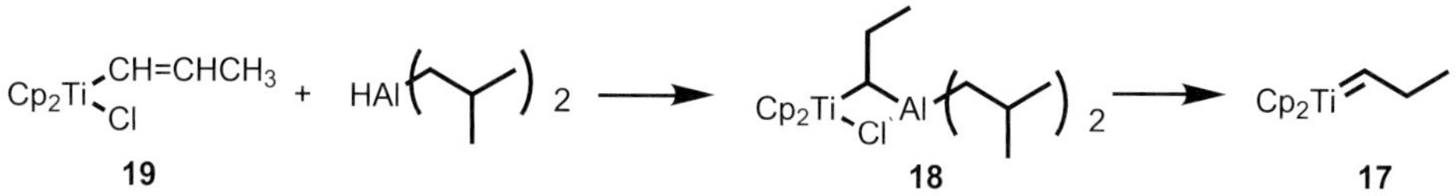

$Cp_2Ti\overset{CH=CHCH_3}{\underset{Cl}{\diagup}}$ + HAl(iBu)₂ ⟶ $Cp_2Ti\overset{}{\underset{Cl}{\diagup}}Al(iBu)_2$ ⟶ $Cp_2Ti\diagdown\diagup$
19   **18**   **17**

Scheme 4.15. The formation of higher homologues of the Tebbe reagent.

The 1,1-dimetalloalkenes **20**, prepared by the carbotitanation of 1-alkynyldimethylalanes with chloromethyltitanocene, serve as hypothetical metal-carbene species **21**, and produce allenes when treated with aldehydes and ketones (Scheme 4.16) [51].

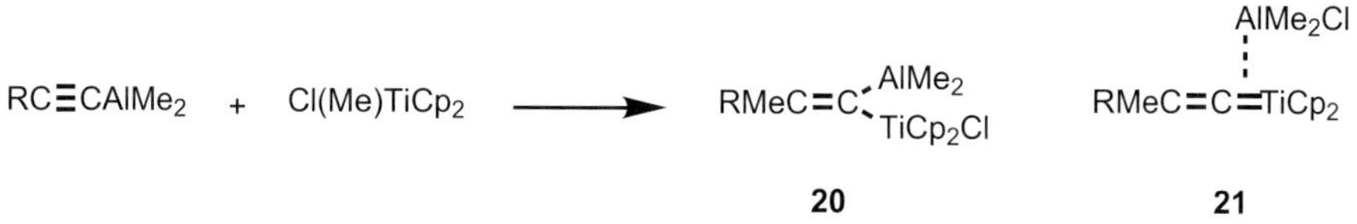

RC≡CAlMe₂ + Cl(Me)TiCp₂ ⟶ RMeC=C(AlMe₂)(TiCp₂Cl) ⟶ RMeC=C=TiCp₂ · AlMe₂Cl
                                        20   **21**

Scheme 4.16. The formation of titanocene-vinylidene precursors from alkynylalanes.

The treatment of vinylaluminum compounds with titanocene dichloride produces the organometallic species **22**, which serve as Wittig-type reagents (Scheme 4.17). The active species is suggested to be the 1,1-dimetalloalkane as the 1,1-dideuterio derivative is produced by deuterolysis with D_2O [52].

Olefin metathesis of the Tebbe reagent offers an alternative route for the preparation of titanocene-alkylidenes **17**, which can be employed for the alkylidena-

Tab. 4.6. Alkylidenation using the reagents prepared from organoaluminum compounds.

Entry	Carbonyl compound	Reagent	Product (Yield)	Ref.
1	 cyclohexanone	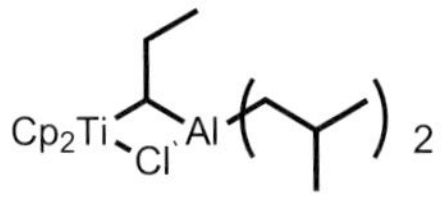 Cp$_2$Ti(Cl)Al(—)$_2$	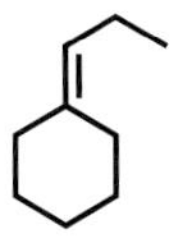 methylenecyclohexane (50%)	50
3	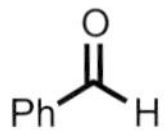 PhCHO	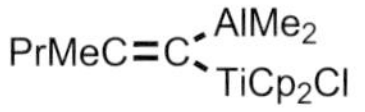 PrMeC=C(AlMe$_2$)(TiCp$_2$Cl)	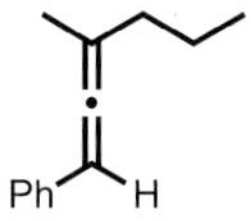 PhCH=C (67%)	51
4	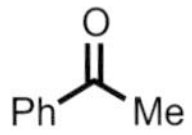 PhC(O)Me	 ~~~~ AlBu$^i_2$ / Cp$_2$TiCl$_2$	Ph(Me)C=CH~~~ (70%) (E:Z = 33:67)	52

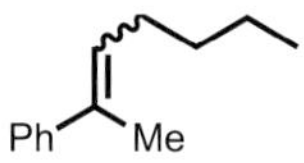

$$R^1\diagup\!\!\!\diagdown AlBu^i_2 \xrightarrow{Cp_2TiCl_2} \left[R^1 \diagup\!\!\!\diagdown \begin{matrix}ML_n\\M'L'_n\end{matrix}\right]$$

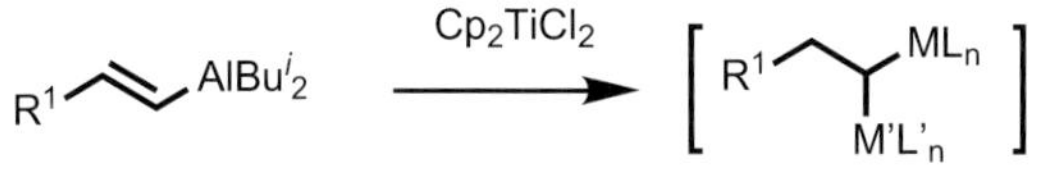

22

Scheme 4.17. The formation of 1,1-dimetalloalkanes utilizing alkenylaluminum compounds.

tion of carbonyl compounds. However, olefin metathesis between titanocene-methylidene and an olefin is not necessarily regarded as a useful synthetic procedure for the preparation of such species because the steric interaction between the substituent at the carbon α to titanium and the bulky cyclopentadienyl ligand disfavors the formation of **17** (Scheme 4.18).

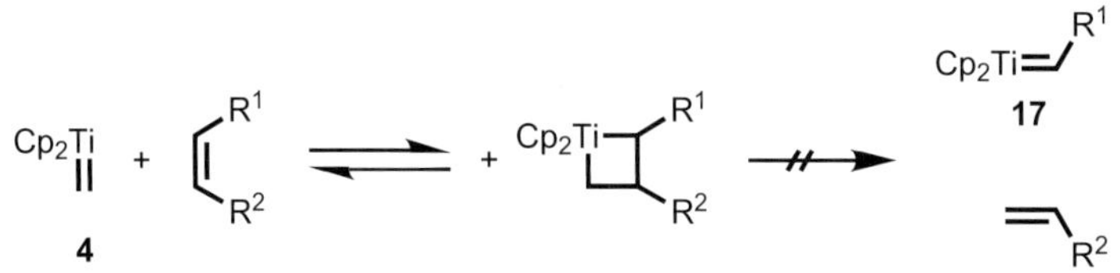

$$Cp_2Ti=CH_2 + \underset{R^2}{\overset{R^1}{\diagdown\!\!\!=}} \rightleftharpoons Cp_2Ti\underset{R^2}{\overset{R^1}{\square}} \not\!\!\rightarrow Cp_2Ti=\!\!\diagup\!\!R^1 \;\; \mathbf{17}$$

4 + olefin ⇌ titanacycle → **17** + =CHR^2

Scheme 4.18. Unfavorable formation of titanocene-alkylidenes by the metathesis between titanocene-methylidene and olefins.

If a highly strained cyclic olefin such as dimethylcyclopropene [53] or a norbornene derivative [54] is employed, however, the titanacycle is broken up to form the corresponding titanocene-alkylidene **23** or **24**, releasing the intrinsic strain energy (Scheme 4.19). The titanocene-alkylidenes thus formed react with ketones and esters in an inter- or intramolecular fashion to afford the corresponding olefins (Table 4.7). This olefin metathesis–carbonyl olefination sequence was successfully applied to the synthesis of $\Delta^{(9,12)}$-capnellene [55].

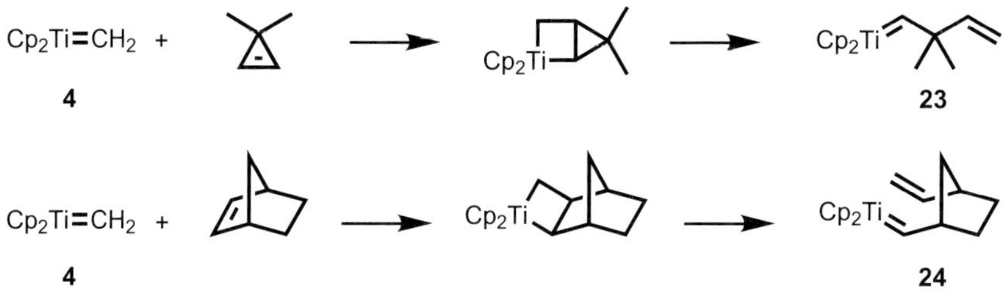

Scheme 4.19. Formation of titanocene-alkylidenes by the reaction of titanocene-methylidene with strained cyclic olefins.

Allenation: Another olefin metathesis–carbonyl olefination sequence utilizing titanacyclobutanes **11** is employed for the preparation of tri- or tetrasubstituted allenes [56]. Titanocene-methylidene **4** generated from **11** reacts with 1,1-disubstituted allenes to produce the α-alkylidenetitanacyclobutanes **25** with the liberation of an olefin. Simple treatment of **25** with ketones and aldehydes at room temperature affords substituted allenes. The vinylidene complexes **26** are formed as the active species in these transformations (Scheme 4.20).

Indirect but mechanistically interesting use of the Tebbe reagent for the olefination of carbonyl compounds is the formation of conjugated dienes via titanacyclo-

Tab. 4.7. Carbonyl olefination utilizing titanacyclobutanes **11** prepared from titanocene-methylidene and strained olefins.

Entry	Carbonyl compound	Titanacyclobutane 11	Product	Yield	Ref.
1	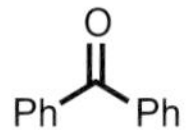 Ph-CO-Ph	 Cp₂Ti-	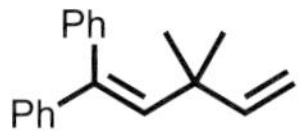 Ph₂C=CHCH₃	83%	53
2	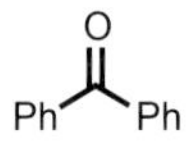 Ph-CO-Ph	 Cp₂Ti-	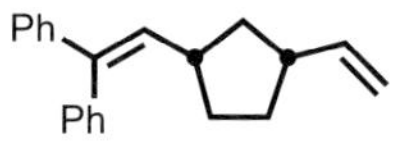 Ph₂C=...	69%	55b
3	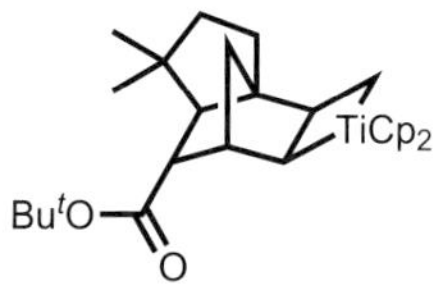 Bu^{t}O-	-TiCp₂	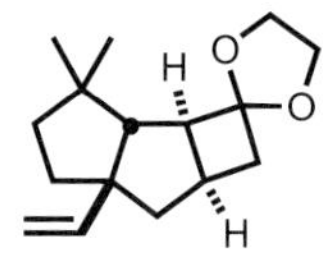	81%[a]	55b

[a] Isolated after the initial product, the vinyl ether, was transformed into the 1,3-dioxolane.

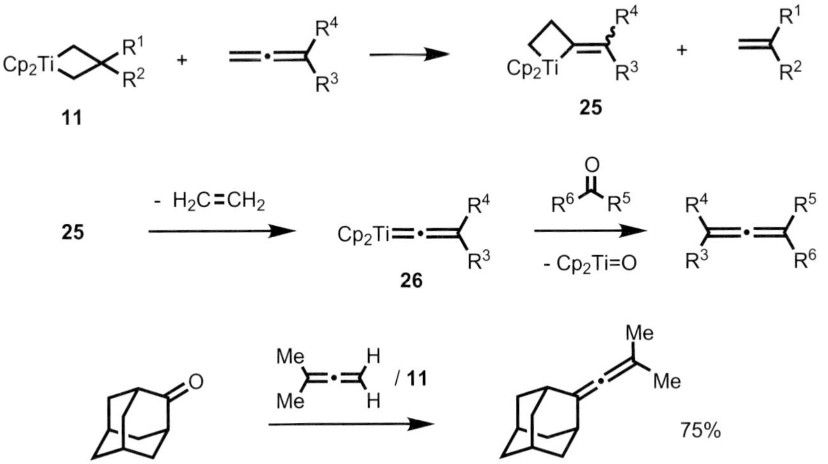

4.2 Carbonyl Olefination with Titanocene-Methylidene and Related Reagents

Scheme 4.20. The formation of titanocene-vinylidenes and their application to carbonyl allenation.

butenes **27**, which are readily generated from titanium-methylidene precursors such as the Tebbe reagent **3** [57] or titanacyclobutanes **11** with alkynes [58] (Scheme 4.21). The reaction of dialkyl-substituted titanacyclobutenes **27** with aldehydes preferentially proceeds through titanium–vinyl bond insertion to form the six-membered oxatitanacycles **28**. Thermal decomposition of the adducts **28** affords conjugated dienes with E-stereoselectivity through a concerted retro-[4+2]-cycloaddition [59]. The mode of reaction of titanacyclobutenes **27** with carbonyl com-

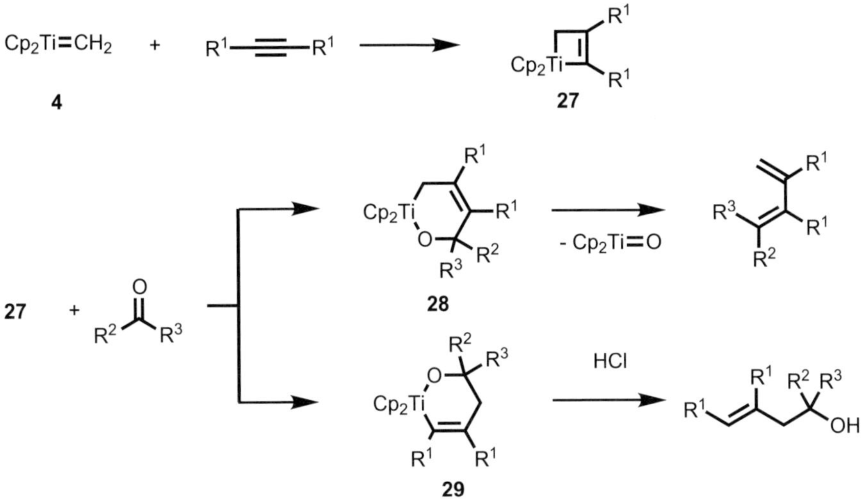

Scheme 4.21. The reaction of titanacyclobutenes with carbonyl compounds.

pounds is largely dependent on steric effects. In the case of 2,3-diphenyltitanacyclobutene, ketones and bulky aldehydes tend to insert into the titanium–alkyl bond to form the homoallyl alcohols after protonolysis of the adducts **29** [60].

4.3
Carbonyl Olefination with Dialkyltitanocenes

4.3.1
Methylenation with Dimethyltitanocene

As noted in the previous section, one drawback of the Tebbe reagent **3** is that an acidic aluminum compound is produced when it is converted into titanocene-methylidene **4**. The instability of **3** toward air and water necessitates special techniques for its preparation and is also an obstacle to its use in organic synthesis. In this sense, dimethyltitanocene **30** is a convenient precursor for the preparation of titanocene-methylidene **4** (Scheme 4.22).

$$Cp_2Ti(CH_3)_2 \xrightarrow{\Delta} Cp_2Ti=CH_2 + CH_4$$

30 **4**

Scheme 4.22. Formation of titanocene-methylidene by thermolysis of dimethyltitanocene.

McCowan and co-workers first studied the thermal decomposition of **30** and found that one of the major degradation processes was α-elimination to produce titanocene-methylidene **4** [61]. The application of this reagent in the methylenation of carbonyl compounds has been extensively investigated by Petasis and co-workers [11c–e, 62]. Dimethyltitanocene **30** is readily prepared by the reaction of titanocene dichloride with methyllithium or methylmagnesium halides [63]. The reagent **30** tolerates exposure to air and water for a short period of time, and so it can be handled and weighed in air. It is also thermally stable and can be stored in the dark as a solution in toluene or THF for several months without significant decomposition [64]. Although the reactions can be performed in either toluene or THF, greater efficiency and increased reaction rate is observed when THF is used as the solvent.

A variety of carbonyl compounds, including esters, lactones, amides, and anhydrides, are methylenated upon heating with dimethyltitanocene **30** at 60–80 °C (Scheme 4.23). Generally, at least two equivalents of dimethyltitanocene are required for complete conversion.

Methylenation of ketones and aldehydes: Similarly to the Tebbe reagent, dimethyltitanocene **30** is compatible with hydroxyl groups, and its low basicity also ex-

4.3 Carbonyl Olefination with Dialkyltitanocenes

Cp$_2$TiCl$_2$ + 2 CH$_3$M → Cp$_2$Ti(CH$_3$)$_2$ **30**

M = Li, MgX

$$Cp_2Ti(CH_3)_2 \xrightarrow{O=CR^1R^2,\ \Delta} R^1R^2C=CH_2$$

Scheme 4.23. Preparation of dimethyltitanocene and its use for carbonyl methylenation.

tends its application to base-sensitive carbonyl compounds such as easily enolizable ketones and ketones possessing an optically active center α to the carbonyl group (see Table 4.8, entry 6).

Methylenation of carboxylic acid derivatives: A variety of carboxylic acid derivatives can be transformed into heteroatom-substituted olefins (Table 4.9). The C-glycoside congeners are easily prepared by the direct methylenation of aldonolactones with **30** (entry 2) [67]. Since the generation of titanocene-methylidene **4** from **30** only produces methane as a by-product, **30** can be employed for the olefination of acid-sensitive substrates such as silyl esters and acylsilanes (entries 4 and 6). The tolerance toward acid-sensitive functional groups also allowed the application of dimethyltitanocene **30** in the conversion of a lactone bearing an acetal moiety into the corresponding cyclic vinyl ether, which was further transformed into the spiroketal ketone upon oxidation and rearrangement (Scheme 4.24) [69].

Tab. 4.8. Methylenation of aldehydes and ketones with dimethyltitanocene **30**.

Entry	Carbonyl compound	Yield	Ref.	Entry	Carbonyl compound	Yield	Ref.
1	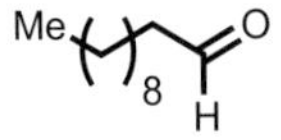 Me-(CH$_2$)$_8$-CHO	43%	64	5	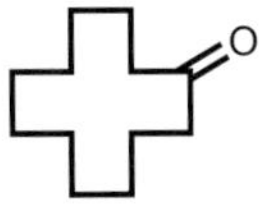 cyclobutyl ketone	83%	64
2	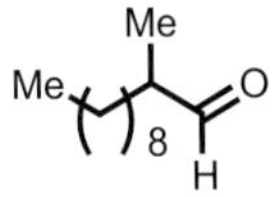 Me-CH(Me)-(CH$_2$)$_8$-CHO	62%	64	6	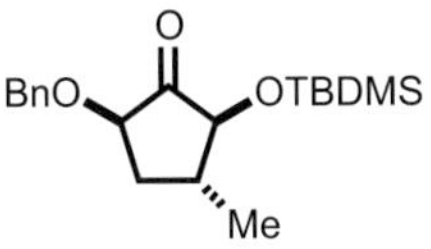 BnO-cyclopentanone-OTBDMS (Me)	77%	65
3	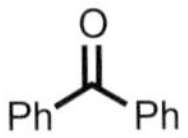 Ph-CO-Ph	90%	64	7	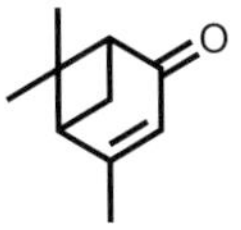 bicyclic enone	61%	64
4	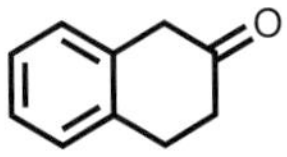 α-tetralone	60%	64	8	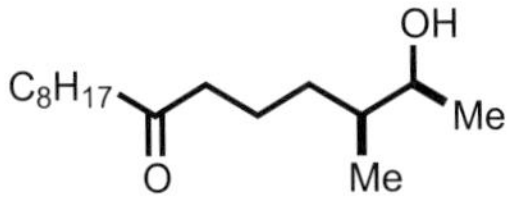 C$_8$H$_{17}$-CO-CH$_2$-C(OH)(Me)-Me	60%	66

Tab. 4.9. Methylenation of carboxylic acid derivatives with dimethyltitanocene **30**.

Entry	Carbonyl compound	Yield	Ref.	Entry	Carbonyl compound	Yield	Ref.
1	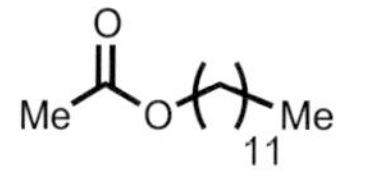 Me-C(O)-O-C(Me)₂-O-Me (11)	65%	64	5	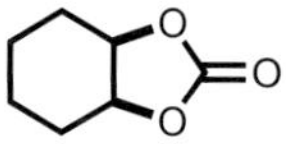 cyclohexane-fused 1,3-dioxolan-2-one	65%	68
2	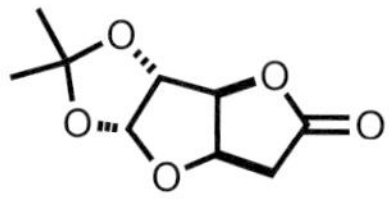 bicyclic acetal-lactone	89%	67a	6	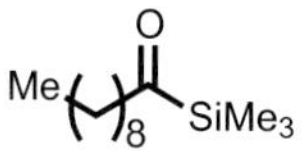 Me-(CH₂)₈-C(O)-SiMe₃	65%	68
3	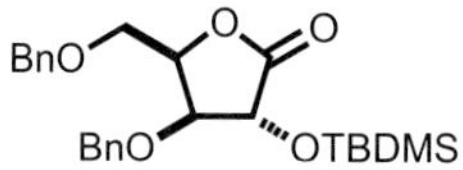 BnO, BnO, OTBDMS lactone	75%	67b	7	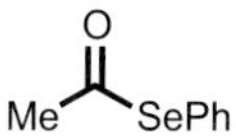 Me-C(O)-SePh	77%	68
4	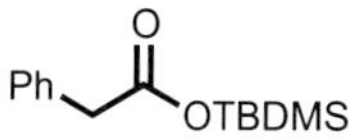 Ph-C(O)-OTBDMS	68%	68	8	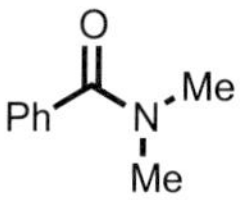 Ph-C(O)-N(Me)Me	54%	68

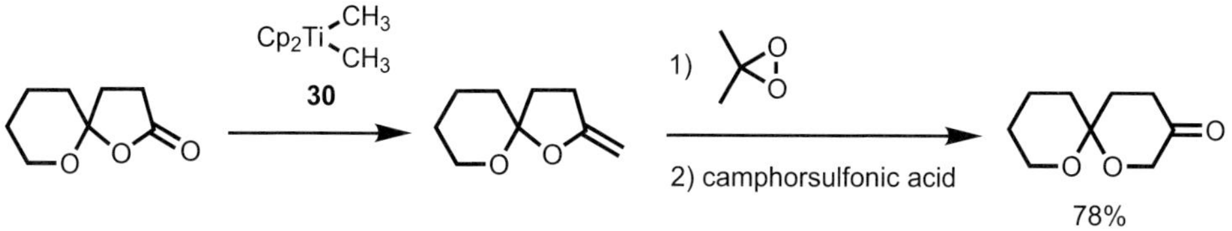

Scheme 4.24. Methylenation of acid-sensitive spiro lactones.

Stepwise methylenation of the anhydrides and thioanhydrides was achieved by using appropriate amounts of **30** [70] (Scheme 4.25). Although the same transformation is effected with the Tebbe reagent, the isomerization of dimethylene compounds to the corresponding thiophenes is essentially inevitable. Similar mono- and bis-carbonyl methylenations of imides have also been reported [68].

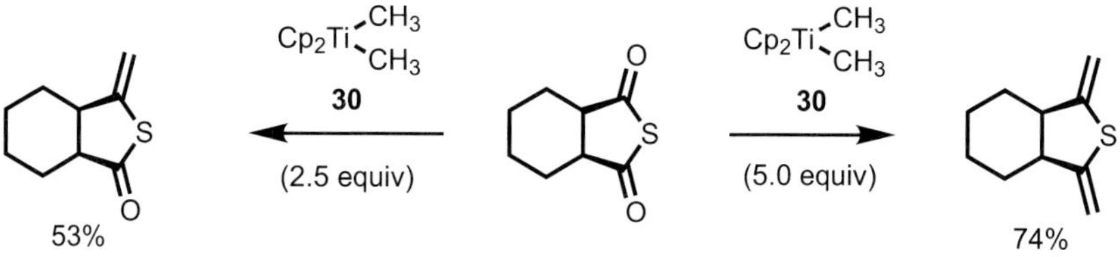

Scheme 4.25. The stepwise methylenation of thioanhydrides.

The following examples illustrate the chemoselectivity of the dimethyltitanocene-mediated methylenation (Scheme 4.26). As for the Tebbe reagent **3**, the rate of

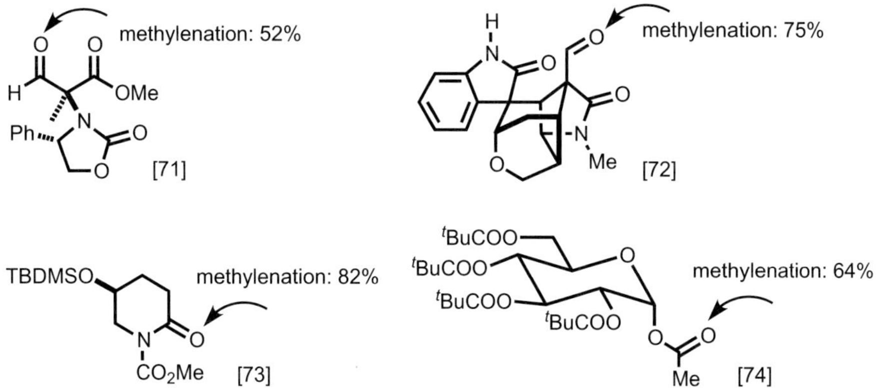

Scheme 4.26. Chemoselective methylenation utilizing dimethyltitanocene.

methylenation with dimethyltitanocene **30** is greatly affected by the electronic and steric properties of the carbonyl compounds. Ketones and aldehydes are methylenated faster than esters and amides, and sterically hindered esters such as pivalate are much less reactive than acetates.

A valuable application of dimethyltitanocene **30** is in the methylenation of small-ring diones, lactones, and lactams (Table 4.10). Selective monocarbonyl methylenation is achieved when cyclobutenediones having one alkoxy and one C-substituent are treated with dimethyltitanocene **30** (entry 1) [75]. The methylena-

Tab. 4.10. Methylenation of small-ring carbonyl compounds with dimethyltitanocene **30**.

Entry	Carbonyl compound	Product	Yield	Ref.
1	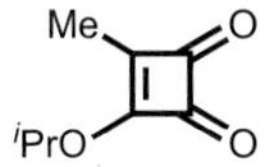 Me, ^{i}PrO-cyclobutenedione	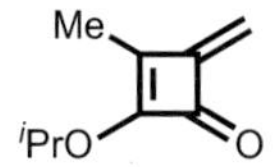 Me, ^{i}PrO-cyclobutenone (methylenated)	68%	75
2	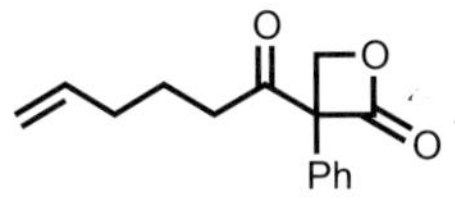 β-lactone with Ph and pentenyl chain	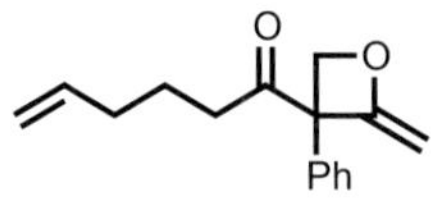 methylenated β-lactone	43%	76
3	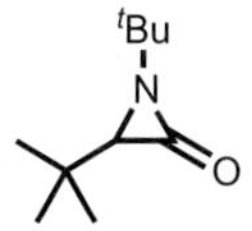 ^{t}Bu-N aziridinone	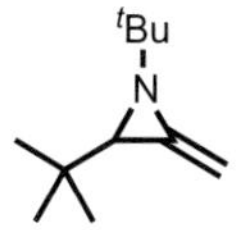 methylenated aziridine	56%	77a
4	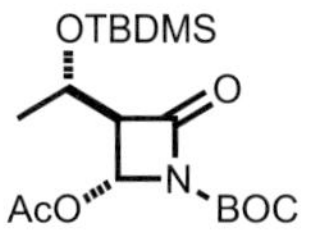 OTBDMS, AcO, N-BOC β-lactam	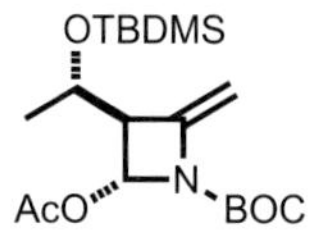 methylenated β-lactam	81%	77b

tion of a ketonic carbonyl is faster than that of a vinylogous ester carbonyl in these reactions, which is consistent with the faster reaction of **30** with ketones than with esters.

Remarkably high selectivity over various functional groups, including ester, carbamate, ketone, and alkene, is observed when β-lactones are treated with **30** (see entry 2) [76]. Similarly, the reactions of **30** with β-lactams having an ester or a carbamate moiety proceed with selective methylenation of the lactam carbonyl to afford methyleneazetidines (see entry 4) [77]. It should be noted that the chemoselectivity of these reactions is not the same as that observed in the methylenation of acyclic or larger ring compounds as described above.

The methylenation of esters or lactones with **30**, followed by further transformation, constitutes a useful synthetic protocol (Scheme 4.27). A carbonyl methylenation–Claisen rearrangement strategy using **30** is employed for the construction of medium- and large-ring ketones from lactones [78]. A phenol derivative has also been synthesized by treatment of a protected quinol ester with **30**, which involves thermal rearrangement of the intermediate vinyl ether [79].

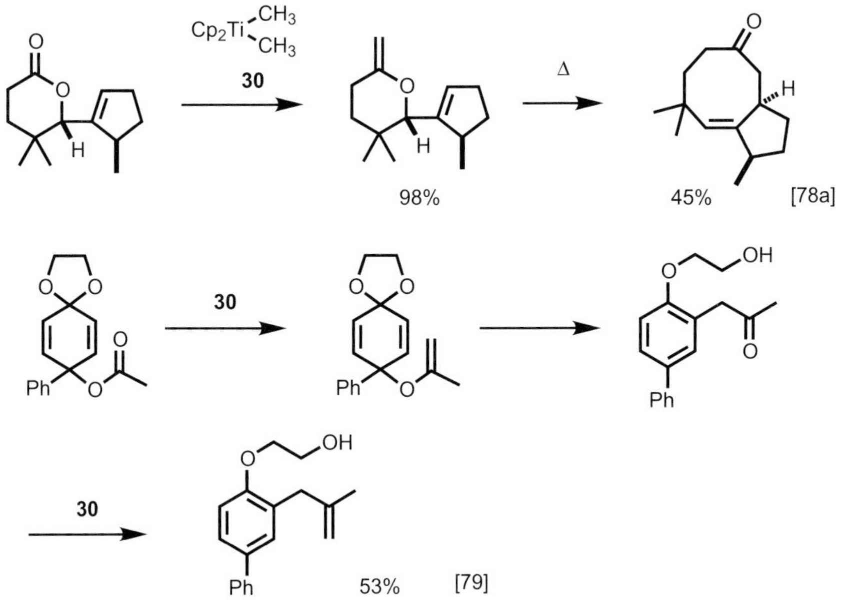

Scheme 4.27. Carbonyl methylenation–Claisen rearrangement strategy utilizing dimethyltitanocene.

As shown in Scheme 4.28, the stereocontrolled transformation of 1,3-dioxolan-4-ones into substituted tetrahydrofurans involves initial methylenation with dimethyltitanocene **30**, subsequent aluminum-mediated rearrangement of the cyclic vinyl ether, and reduction of the resulting carbonyl group [80]. A similar three-step stereocontrolled transformation of 1,3-dioxan-4-one into substituted tetrahydropyrans has been developed (Scheme 4.28) [81].

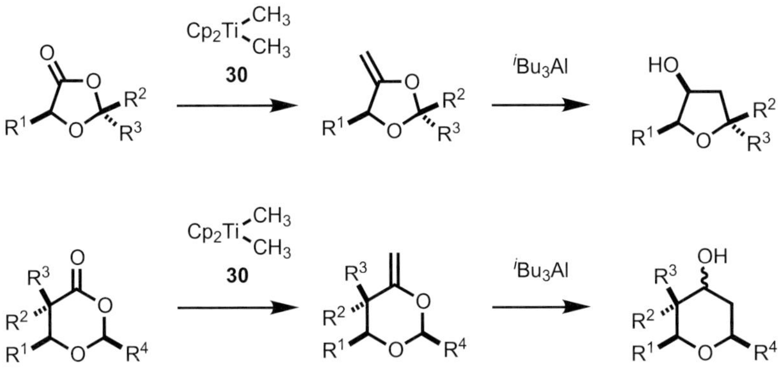

Scheme 4.28. Stereocontrolled transformation of lactones into cyclic alcohols.

The mechanism of carbonyl methylenation with dimethyltitanocene **30** is one of the major subjects of discussion in titanium-carbene chemistry. Two reaction pathways have been proposed. Based on the observation of H/D scrambling in reactions using a deuterated ester and $Cp_2Ti(CD_3)_2$, Petasis proposed that the reaction proceeds by methyl transfer to form the adduct **31** and subsequent elimination of methane and titanocene oxide (Scheme 4.29, Path A) [64]. Later, a detailed study by Hughes and co-workers using ^{13}C and D-labeled compounds showed that the methylenation of esters with **30** proceeds via a titanium carbene mechanism (Path B) [82].

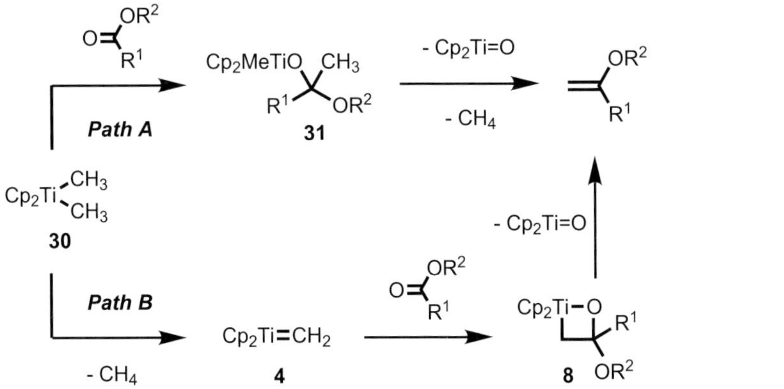

Scheme 4.29. Plausible mechanisms for carbonyl methylenation with dimethyltitanocene.

4.3.2
Alkylidenation of Carbonyl Compounds with Dialkyltitanocenes and Related Complexes

Petasis and co-workers extended the above methylenation procedure to the alkylidenation of carbonyl compounds by using dialkyltitanocenes [11e, 62]. Like methylidenation with dimethyltitanocene, the Petasis alkylidenation is believed to proceed via the formation of titanocene-alkylidenes through α-elimination of dialkyltitanocenes. This assumption is supported by the isolation of the cyclometalated product **32**, which is indicative of the intermediary formation of titanocene-benzylidene **34** by thermolysis of dibenzyltitanocene **33** bearing a *tert*-butyl group on the Cp ring (Scheme 4.30) [83].

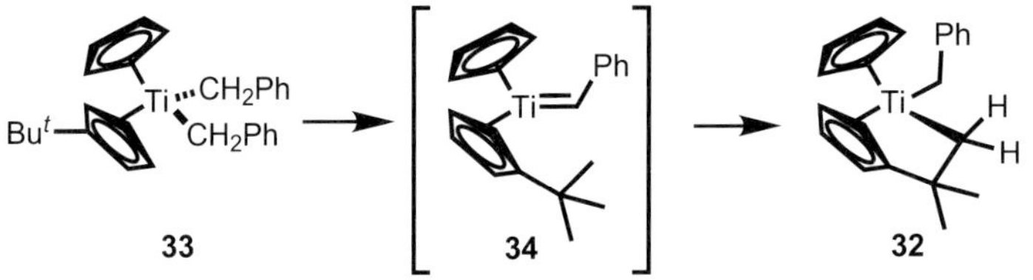

Scheme 4.30. Formation of intermediary benzylidene complex.

α-Elimination of transition metal alkyl complexes is a versatile tool for the preparation of metal carbenes. This process, however, suffers from a serious drawback in that a carbene complex generally cannot be formed from an alkyl complex possessing a β-hydrogen due to preferential β-elimination [1]. Because dialkyltitanocenes capable of facile β-elimination are thermally unstable [11e], compounds that can be used for carbonyl olefination are restricted to those without a β-hydrogen, such as dibenzyl- **35** and bis(trimethylsilylmethyl)titanocenes **36**. Bis(cyclopropyl)titanocene **37** can also be employed in the Petasis olefination because β-hydride elimination from this molecule is unfavorable due to the formation of a highly strained cyclopropene. Some related reagents, such as **38** and **39**, have also been developed for the conversion of carbonyl compounds to alkenylsilanes. These reagents are readily available from titanocene dichloride and alkyllithiums or alkylmagnesium halides (Scheme 4.31).

Benzylidenation: Dibenzyltitanocene **35** is prepared by the reaction of titanocene dichloride with benzylmagnesium chloride [84]; it reacts with carbonyl compounds upon heating at 45–55 °C to afford styrene derivatives [85] (Scheme 4.32). The benzylidenation was found to be best accomplished with 3–4 equivalents of **35** in toluene; THF and hexane proved to be less effective reaction media. Aldehydes react rather sluggishly with **35**, whereas benzylidenation proceeds well with ketones. Effective olefination also proceeds with esters, lactones, and amides giving enol ethers and enamines. In most benzylidenations with **35**, 1,2-diphenylethane is

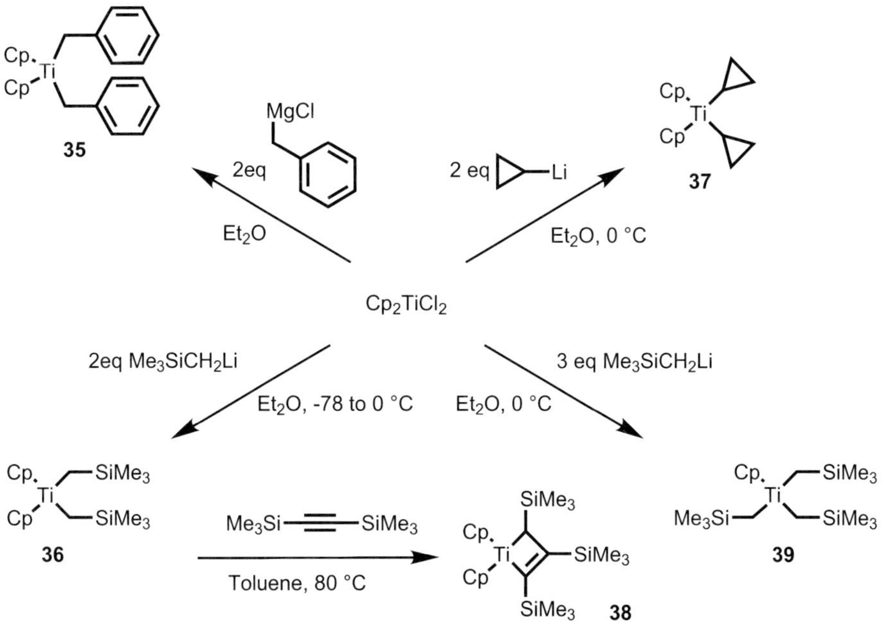

Scheme 4.31. Preparation of dialkyltitanocenes and related compounds.

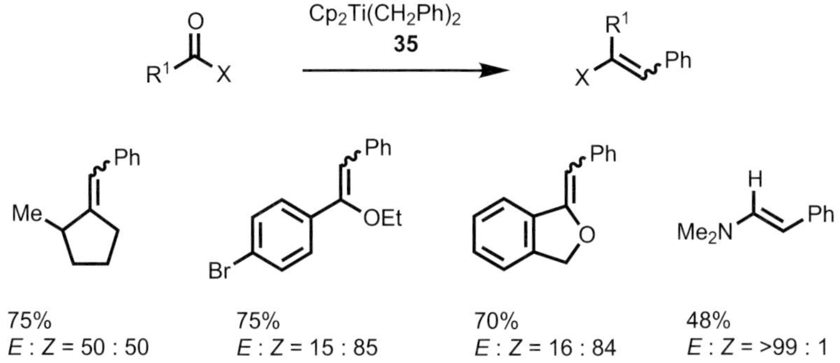

Scheme 4.32. Carbonyl olefination with dibenzyltitanocene.

formed as a by-product in variable amounts, which may be attributable to a competing reductive elimination process.

The stereoselectivity of the benzylation of isomeric esters ($R_1CO_2R_2$ and $R_2CO_2R_1$) with **35** indicates that the ester substrate should have a large acid residue and a small ester group for good Z-stereoselectivity [83]. This stereochemical outcome is explained by assuming that the benzylidenation proceeds via the four-membered transition states **A** and **B** in Scheme 4.33; the kinetic preference of their formation is dependent on the relative steric repulsion between the substituents on the ester (R^1 and R^2) and the phenyl group (Scheme 4.33).

4.3.3
Allenation of Carbonyl Compounds with Alkenyltitanocene Derivatives

Aldehydes and ketones can be converted into substituted allenes with dialkenyltitanocenes **40**, whereby vinylidene carbene complexes **41** serve as the actual olefination agents [90]. Unsymmetrical alkylalkenyltitanocenes **42** are also capable of transforming carbonyl compounds to allenes. These reagents are generated by one-pot consecutive addition of one equivalent of the alkenylmagnesium bromide to form monochlorotitanocene **43**, followed by the addition of one equivalent of the second organomagnesium bromide (Scheme 4.38).

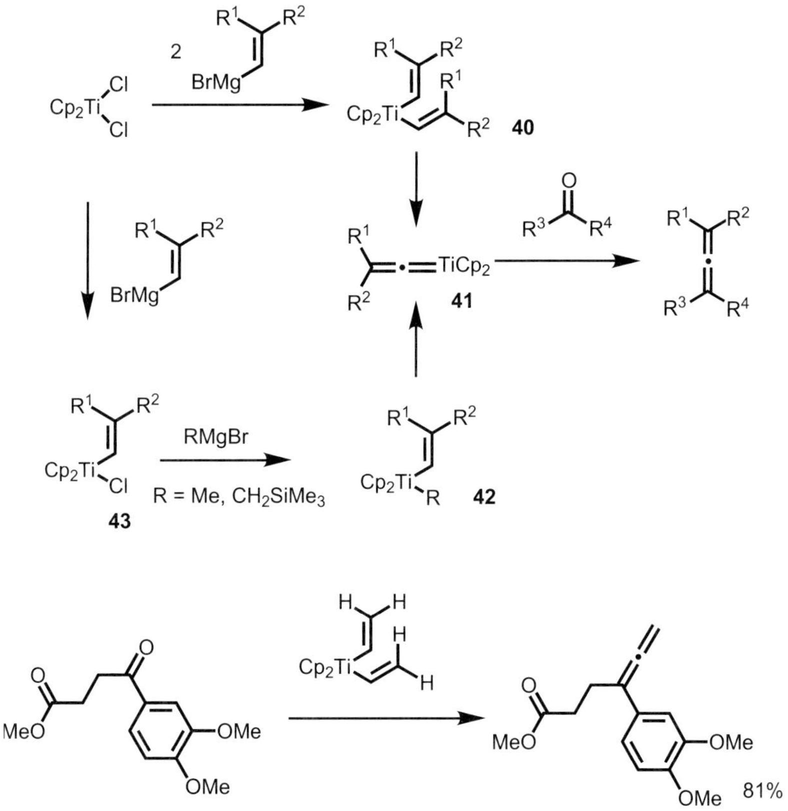

Scheme 4.38. Allenation of carbonyl compounds with alkenyltitanocenes.

4.3.4
Carbonyl Olefination Utilizing an Alkyl Halide–Titanocene(II) System

An alkyl halide–titanocene(II) system offers an operationally straightforward method for carbonyl olefination, which is believed to be mechanistically analogous to the Petasis olefination [91] (Scheme 4.39). Successive treatment of various alkyl halides possessing a substituent at the carbon β to the halogen with titanocene(II) species 44 and carbonyl compounds affords the corresponding olefins (Table 4.11).

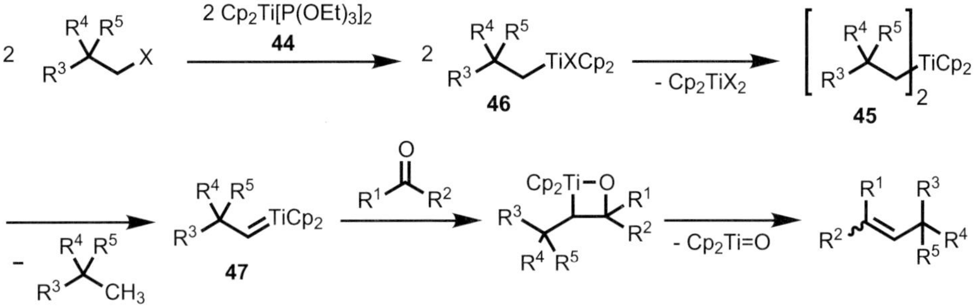

Scheme 4.39. Plausible reaction pathway for olefination utilizing an alkyl halide-titanocene(II) system.

The yield of this carbonyl olefination is largely dependent on the steric bulk of the β-substituent of the halide. When a halide lacking a β-substituent is used, no olefination product is produced. Since more than two equivalents of the alkyl halide and the titanocene(II) species are required to complete the reaction, and the dehalogenated hydrocarbon is produced as a by-product, dialkyltitanocene 45 is presumably generated as a transient intermediate in this system, as depicted in Scheme 4.39. Oxidative addition of the alkyl halide to the titanocene(II) species and subsequent disproportionation of the resulting alkyltitanium 46 affords 45. Its α-elimination gives the titanium carbene complex 47, which serves as an olefination agent. It is noteworthy that in this system carbene complexes are generated from alkyl halides even when they have a β-hydrogen.

Tab. 4.11. Carbonyl olefination utilizing an alkyl iodide-titanocene(II) system.

Entry	Carbonyl compound	Alkyl iodide	Product	Yield	E:Z
1	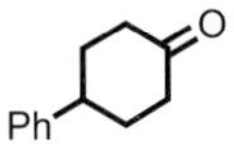	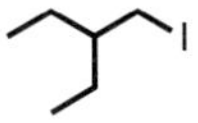	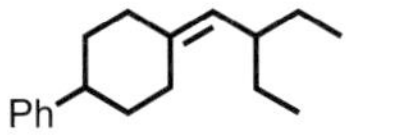	65%	
2	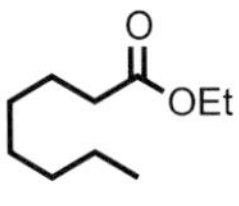	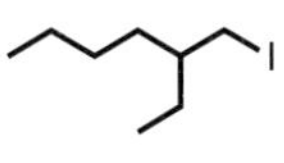	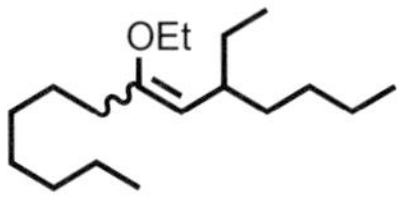	79%	24:76
3	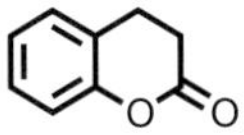	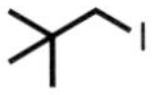	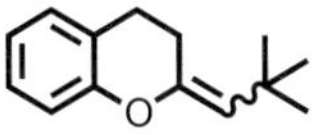	66%	34:66

4.4
Carbonyl Olefination Utilizing a Thioacetal-Titanocene(II) System

4.4.1
Formation of Titanocene-Alkylidenes from Thioacetals and Titanocene(II)

One of the most versatile procedures for the olefination of carbonyl compounds utilizes a thioacetal-titanocene(II) system [40, 92]. The organotitanium species generated by the desulfurization of thioacetals with the triethyl phosphite complex of titanocene(II) **44** react with a variety of multiple bonds, such as those of alkenes [93], alkynes [94], and nitriles [95], as well as carbonyl functions. The reactivities observed in these reactions indicate that the active species involved are titanocene-alkylidene complexes **48** (Scheme 4.40).

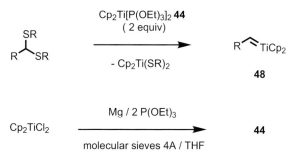

Scheme 4.40. Preparation of titanium carbene complexes from thioacetals.

As described in the previous section, α-elimination of dialkyltitanocenes is of limited use for the preparation of titanocene-alkylidenes. Since thioacetals are readily available from carbonyl compounds or through alkylation of bis(phenylthio)methane and related organosulfur compounds, a thioacetal-titanocene(II) system enables the use of different types of carbene complexes. Use of appropriate thioacetals is of crucial importance in this system; for the preparation of alkylidene complexes **48**, the corresponding diphenyl thioacetals are the starting materials of choice. No carbene complexes are formed from dialkylthioacetals. To generate vinylcarbene complexes **49**, trimethylene thioacetals of α,β-unsaturated aldehydes or 1,3-bis(phenylthio)propene derivatives are employed (Scheme 4.41). The low-valent titanium species **44** is also easily prepared by the reduction of titanocene dichloride with magnesium in the presence of triethyl phosphite at room temperature. The presence of molecular sieves 4A is essential for the reproducibility of this preparation. In relatively large-scale preparations, care should be taken to control the reaction temperature [92, 93g].

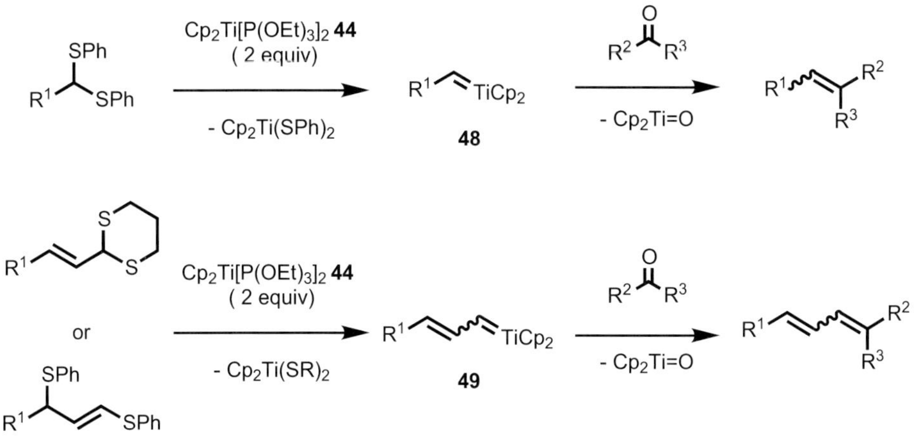

Scheme 4.41. Carbonyl olefination utilizing a thioacetal-titanocene(II) system.

4.4.2
Alkylidenation of Aldehydes, Ketones, and Carboxylic Acid Derivatives

The carbene complexes thus prepared can be employed for the olefination of a variety of carbonyl compounds. The procedure is operationally simple; treatment of the thioacetals with titanocene(II) species at ambient temperature for 5–15 min affords the organotitanium species, which are further treated with ketones and aldehydes in the same reaction vessel to give the olefins or dienes (Table 4.12) [96].

Tab. 4.12. Olefination of ketones and aldehydes with a thioacetal-titanocene(II) system.

Entry	Carbonyl compound	Thioacetal	Product	Yield	E:Z
1	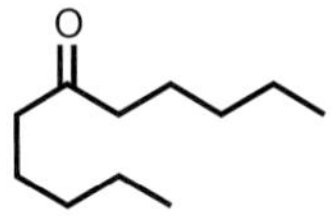	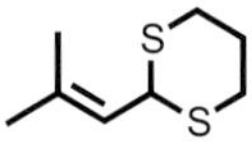	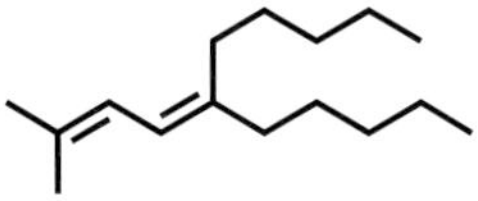	88%	
2	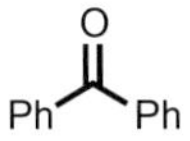	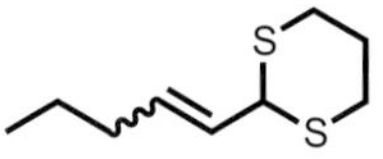 E:Z = 95:5	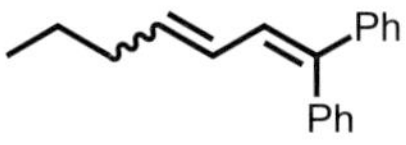	84%	79:21
3	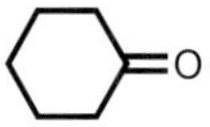	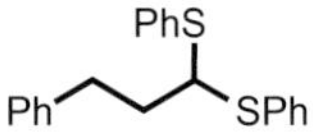	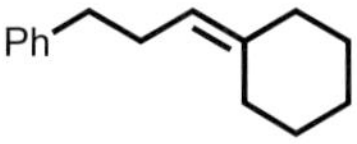	69%	
4	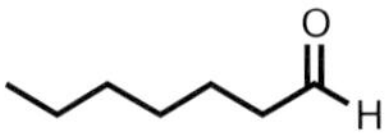	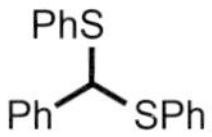		52%	56:44
5	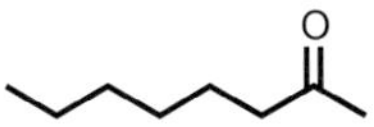	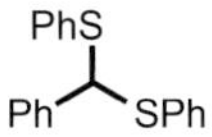	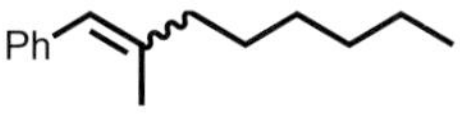	80%	54:46

180 | 4 Carbonyl Olefination Utilizing Metal Carbene Complexes

Methylenation also proceeds using bis(phenylthio)methane, although the yields of terminal olefins are lower than those obtained by other titanium-based methylenation procedures.

Titanocene-alkylidenes possessing two alkyl substituents at the carbon atom adjacent to titanium have been successfully prepared using acetone diphenyl thioacetal, and these can be employed for the isopropylidenation of carbonyl compounds [96]. The steric bulk around the thioacetal moiety seems to affect the formation of the carbene complex; olefination of 4-t-butylcyclohexanone utilizing 2,2-bis(phenylthio)butane affords the corresponding olefin only in moderate yield and no olefin is produced when thioacetals bearing two larger alkyl substituents are used [97]. This observation is consistent with the fact that alkylidenecyclobutanes are produced in good yields by olefination using cyclobutanone diphenyl thioacetals [98].

Alkylidenation with a thioacetal-titanocene(II) system is useful for the conversion of carboxylic acid derivatives into enol ethers, alkenyl sulfides, and enamines. As with other titanium-based reagents, mixtures of stereoisomers are formed in

Tab. 4.13. Olefination of carboxylic acid derivatives with a thioacetal-titanocene(II) system.

Entry	Carbonyl compound	Thioacetal	Product	Yield	E:Z	Ref.
1	Ph-C(=O)-OEt	dithiane derivative	Ph-CH=C(OEt)-	73%	23:77	96
2	Ph-CH2-C(=O)-OEt	PhS-C(cyclobutyl)-SPh	cyclobutylidene-C(OEt)=CH-Ph	80%		98
3	phthalide	PhS-CH(Ph)-SPh	3-benzylidenephthalide	75%	<1:>99	96
4	γ-phenyl-γ-butyrolactone	PhS-C(Me)2-SPh	isopropylidene lactone	64%		96
5	Ph-CH2-C(=O)-SⁱPr	PhS-C(decyl)-SPh	alkenyl sulfide with SⁱPr and Ph	78%	27:73	99
6	Ph-C(=O)-N(Me)-Ph	PhS-CH(ⁱPr)-SPh	enamine with Ph, N(Me)Ph	76%	<1:>99	100

the alkylidenation of aldehydes and ketones with this system. On the contrary, Z-stereoselectivity is observed in the alkylidenation of carboxylic acid derivatives, and in certain cases Z-isomers are produced exclusively (Table 4.13).

A significant advantage of this system is that a variety of appropriate thioacetals are readily available. Thus, allylsilanes are prepared by the olefination of carbonyl compounds utilizing β-trimethylsilyl-substituted organosulfur compounds, which are prepared by trimethylsilylmethylation of the parent compounds (Scheme 4.42) [101].

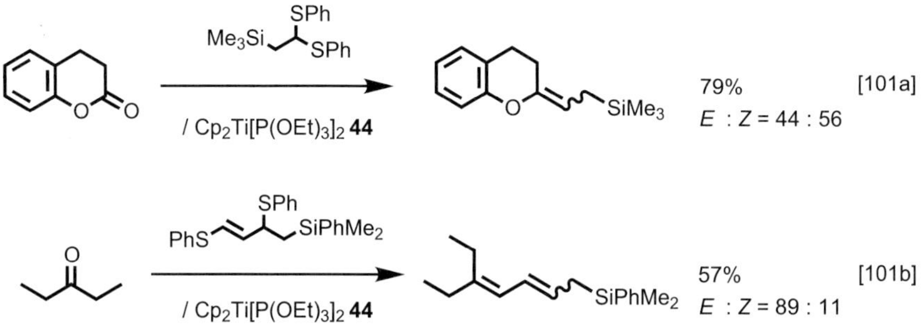

Scheme 4.42. Preparation of allylsilanes by the olefination of carboxylic acid derivatives.

Alkylidenation of esters on solid supports offers a special advantage in that the resulting enol ethers are resistant to hydrolysis (Scheme 4.43). The resin-bound ester is treated with the titanium reagent prepared from the thioacetal to produce the enol ether, which is further transformed into the benzofuran on treatment with trifluoroacetic acid [102].

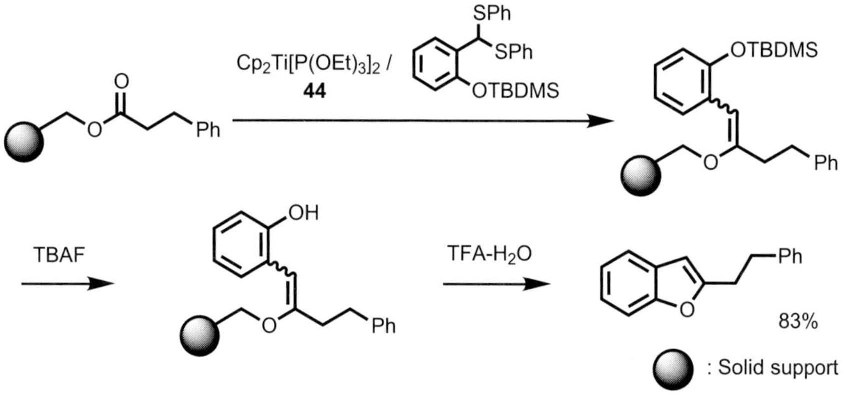

Scheme 4.43. Olefination of resin-bound esters.

4.4.3
α-Heteroatom-Substituted Carbene Complexes

By the use of di- or trithioorthoformates, heteroatom-substituted olefins are obtained. The active organotitanium species formed in these preparations are assumed to be the corresponding methoxy- and phenylthio-methylidene carbene complexes, **50** and **51**, respectively (Scheme 4.44). The success of olefinations using these species indicates that the α-heteroatom substituent does not affect the nucleophilic nature of the titanium carbene complex [103].

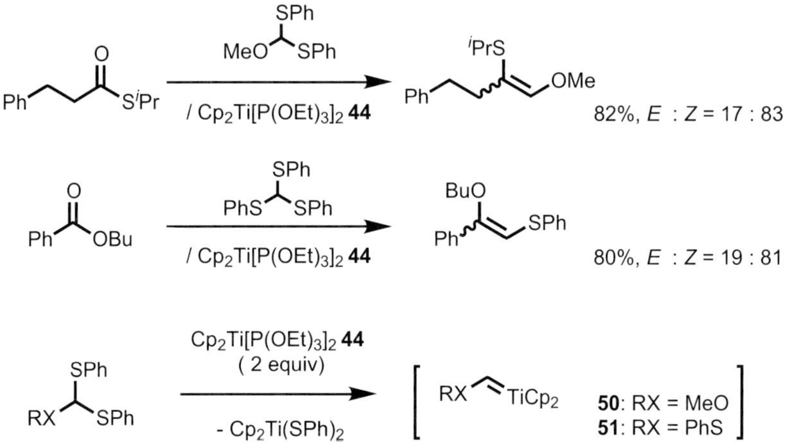

Scheme 4.44. Preparation of enol ethers and alkenyl sulfides by the use of α-heteroatom-substituted carbene complexes formed from thioacetals and titanocene(II).

4.4.4
Intramolecular Carbonyl Olefination

The capabilities of the thioacetal-titanocene(II) system have been extended to the intramolecular carbonyl olefination of esters. Titanocene(II)-promoted reaction of alkyl ω,ω-bis(phenylthio)alkanoates affords enol ethers of cyclic ketones. As summarized in Table 4.14, the treatment of esters possessing a thioacetal moiety with titanocene(II) **44** at room temperature or reflux for several hours affords the enol ethers of cyclic ketones [104]. Analogous intramolecular carbonyl olefinations using phosphorus carbanions are generally unsuccessful owing to preferential acylation; only certain five- and six-membered cyclic compounds, in which the double bond that is formed is conjugated with a carbonyl group or an aromatic ring, can be synthesized by means of these reactions [105].

The intramolecular carbonyl olefination of esters bearing a thioacetal moiety is also applicable to the synthesis of cyclic vinyl ethers. The formation of simple monocyclic vinyl ethers by the treatment of ω,ω-bis(phenylthio)alkyl alkanoates with **44** is rather complicated owing to the concomitant formation of oligomers

Tab. 4.14. Intramolecular carbonyl olefination of alkyl ω,ω-bis(phenylthio)alkanoates.

Entry	Alkyl ω,ω-bis(phenylthio)alkanoate	Product	Yield
1	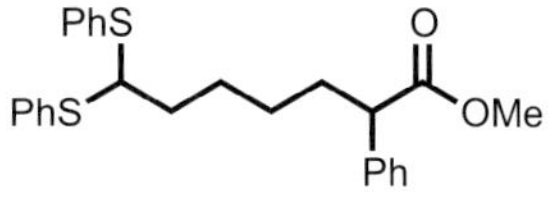 PhS, PhS, OMe, Ph (alkanoate structure)	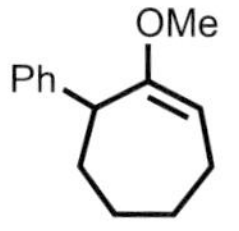 OMe, Ph (cycloheptene)	73%
2	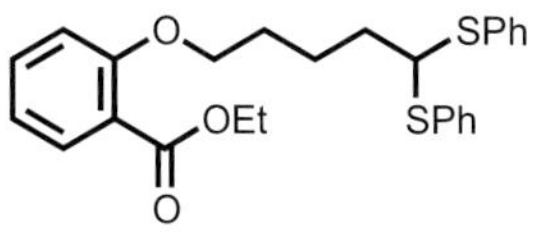 aryl-O-alkyl-SPh, SPh, OEt, C=O	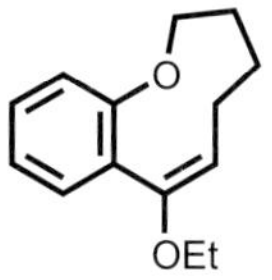 bicyclic enol ether, OEt	70%
3	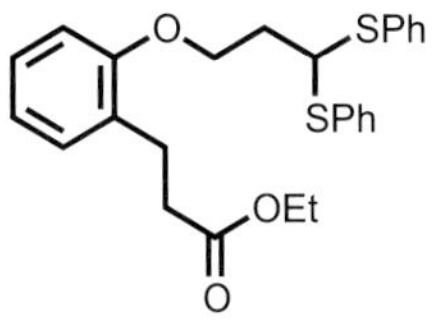 aryl-O-alkyl-SPh, SPh, OEt, C=O	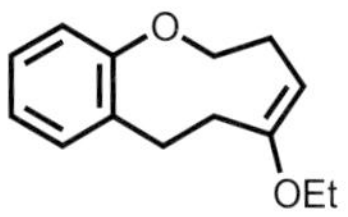 bicyclic enol ether, OEt	67%

even under high-dilution conditions. Nonetheless, this reaction is synthetically useful because hydrolysis of the cyclized products affords ω-hydroxy ketones in good yields, as illustrated in the following equation (Scheme 4.45) [106].

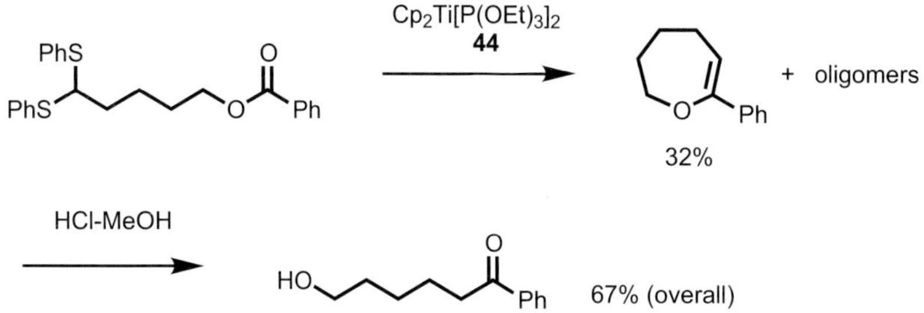

PhS-(chain)-O-C(=O)-Ph → $Cp_2Ti[P(OEt)_3]_2$ **44** → cyclic enol ether-Ph (32%) + oligomers

→ HCl-MeOH → HO-(chain)-C(=O)-Ph   67% (overall)

Scheme 4.45. Preparation of ω-hydroxy ketones by the intramolecular reaction of ω,ω-bis(phenylthio)alkyl alkanoates followed by hydrolysis.

The success of the intramolecular reaction of ω,ω-bis(phenylthio)alkyl alkanoates seems to be dependent on the conformation of the substrate; the J-ring of ciguatoxin was constructed using this methodology in good yield. The acid-labile spiroketal and methoxymethyl ether moieties remained unchanged during this transformation (Scheme 4.46) [107].

The corresponding intramolecular carbonyl olefination of thiol esters, S-[3,3-bis(phenylthio)propyl] thioalkanoates, constitutes a practical means of preparing 2,3-dihydrothiophenes (Scheme 4.47) [108].

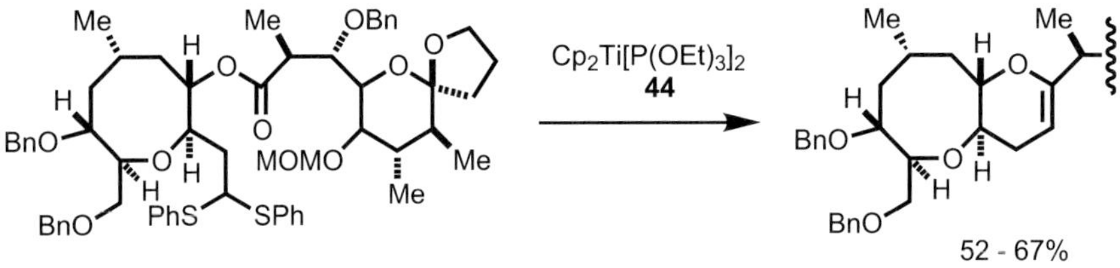

Scheme 4.46. Construction of the J-ring of ciguatoxin.

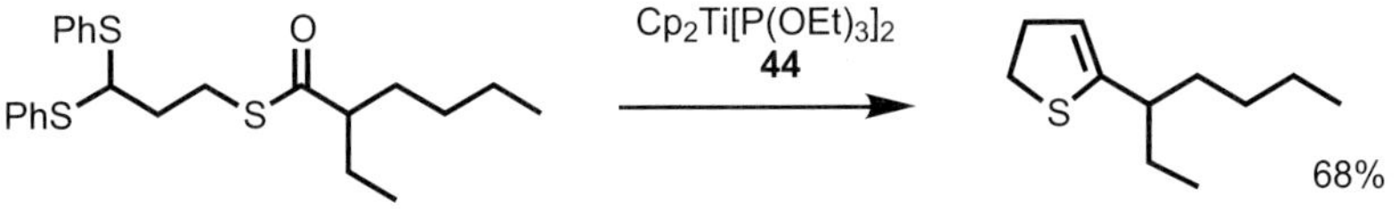

Scheme 4.47. Intramolecular carbonyl olefination of S-[3,3-bis(phenylthio)propyl]thioalkanoates.

4.4.5
Related Olefinations Utilizing *gem*-Dihalides

It is well known that the olefination of carbonyl compounds is greatly affected by steric hindrance, and that yields are generally very low in the case of tetrasubstituted olefins [109]. For example, the reactions of phosphoryl-stabilized carbanions and α-silyl carbanions proceed smoothly when only one alkyl group is present at their central carbon atoms. The preparation and reactions of these reagents having two α-substituents has been the subject of little synthetic work. Only isopropylidene- [110] and cycloalkylidene-phosphoranes [111], as well as similar α-phosphoryl [112] and α-silyl carbanions [113], whose substituents present minimal steric hindrance to the approaching carbonyl compounds, are employed for carbonyl olefination. As noted earlier in this section, carbonyl olefination utilizing a thioacetal-titanocene(II) system suffers from a similar disadvantage. However, the analogous olefination utilizing *gem*-dihalides partially eliminates this limitation. The organotitanium species generated by the reduction of *gem*-dichlorides with the titanocene(II) species **44** serve as titanocene-alkylidenes for the transformation of aldehydes, ketones, esters, and lactones into highly substituted olefins (Scheme 4.48) [97]. A variety of tetrasubstituted olefins can be prepared by this procedure, including those bearing four alkyl substituents other than a methyl group (Table 4.15). Although the structure of the organometallic species formed from the *gem*-

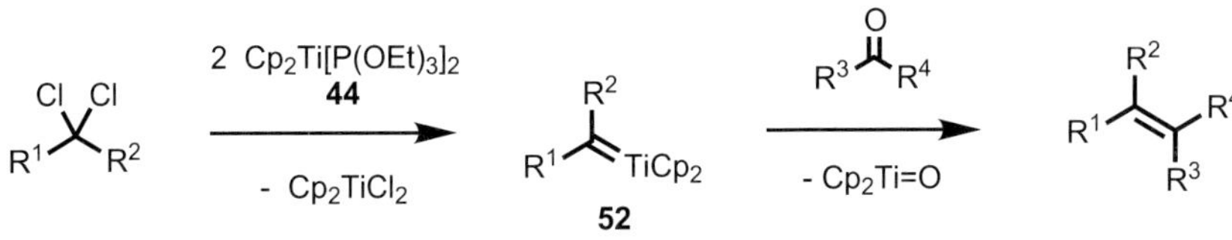

Scheme 4.48. Carbonyl olefination utilizing a *gem*-dichloride-titanocene(II) system.

Tab. 4.15. Olefination of carbonyl compounds with a gem-dichloride-titanocene(II) system.

Entry	Carbonyl Compound	gem-Dichloride	Product	Yield (Ratio of isomers)	Ref.
1	cyclohexanone	Cp₂Ti=CCl₂ cyclohexylidene dichloride	cyclohexylidene=cyclohexane	73%	97
2	Ph-C(O)-OEt	Ph-CCl₂-CH₂-CH₃	Ph-C(OEt)=CH-CH₂-Ph	74% (90:10)	97
3	γ-butyrolactone (Ph-substituted)	Ph-CCl₂-CH₂-CH₂-Ph	furanyl product with Ph groups	72%	97
4	Ph-C(O)-CH₂-Ph	CCl₄	Ph-C(CCl₂)=CH-Ph	63%	115

dichloride is unclear, the fact that the intramolecular reaction of *gem*-dichlorides with olefin affords ring-closing metathesis products in certain cases [114] suggests that the olefination proceeds via the formation of titanocene-disubstituted methylidene **52**. This procedure is applicable to the dichloromethylenation of ketones using carbon tetrachloride [115].

4.5
Carbonyl Olefination Using Zirconium, Tantalum, Niobium, Molybdenum, and Tungsten Carbene Complexes

Schrock-type carbene complexes of transition metals other than titanium are also utilized for carbonyl olefination, although their synthetic utility has not yet been fully investigated. In some cases, their reactions differ from those of titanium-carbene complexes in terms of stereo- and chemoselectivity and are complementary to olefination with titanium reagents. This section describes the use of carbene complexes of various transition metals in carbonyl olefination.

4.5.1
Zirconium Carbene Complexes

Alkylidene-bridged zirconium complexes **53**, long-chain alkylidene analogues of the Tebbe reagent **3**, are prepared by the reaction of chlorobis(cyclopentadienyl)-(vinyl)zirconium with DIBAL-H, or by addition of the Schwartz reagent, H(Cl)ZrCp₂ [116], to a solution of (vinyl)diisobutylaluminum (Scheme 4.49) [117]. Similarly to the Tebbe reagent **3**, the complex **53a** ($R^1 = {}^t$Bu) itself serves as an

4 Carbonyl Olefination Utilizing Metal Carbene Complexes

Tab. 4.16. Carbonyl olefination with zirconium carbene complexes **54**.

Entry	Carbene complex 54	Carbonyl compound	Product	Yield
1	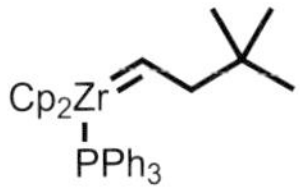 Cp$_2$Zr=CHCH$_2$CMe$_3$ · PPh$_3$	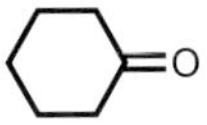 cyclohexanone	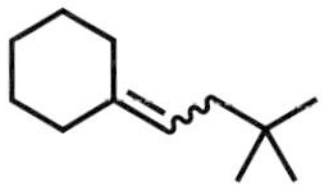 cyclohexylidene-CH-CH$_2$-CMe$_3$	quant.
2	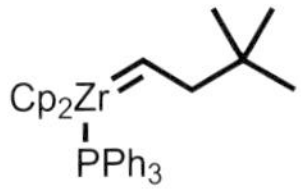 Cp$_2$Zr=CHCH$_2$CMe$_3$ · PPh$_3$	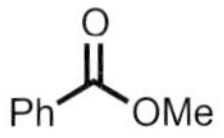 PhC(O)OMe	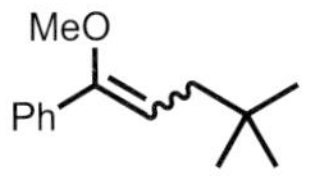 MeO-C(Ph)=CH-CH$_2$-CMe$_3$	94%
3	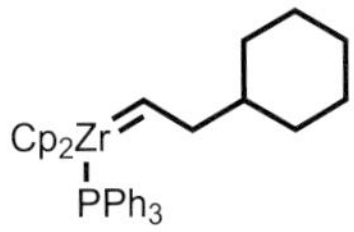 Cp$_2$Zr=CHCH$_2$Cy · PPh$_3$	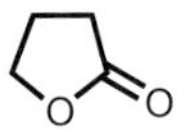 γ-butyrolactone	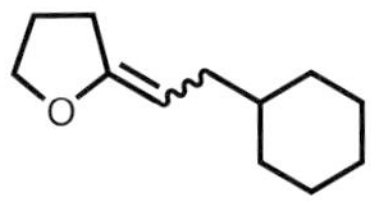 2-(cyclohexylmethylidene)tetrahydrofuran	84%

alkylidenation agent; it transforms cyclohexanone into neohexylidenecyclohexanone. Alternatively, the bimetallic complexes **53** can be converted into zirconocene-alkylidene complexes **54** by the action of HMPA (1 equiv.) in the presence of appropriate ligands such as phosphines [118]. These organozirconium species **53** and **54** have been characterized by NMR and X-ray crystal structural analysis [117b, 118]. Yields for the formation of the carbene complexes **54** are dependent on the nature of phosphine ligand. Strong σ-donor phosphines with small cone angles are favorable for the complexation and the thermal stability of the carbene complexes.

Carbene complexes **54** transform ketones to trisubstituted olefins and esters to vinyl ethers at elevated temperature (70 °C) in good to excellent yields (Scheme 4.49; Table 4.16), although no stereoselectivity is observed in these reactions.

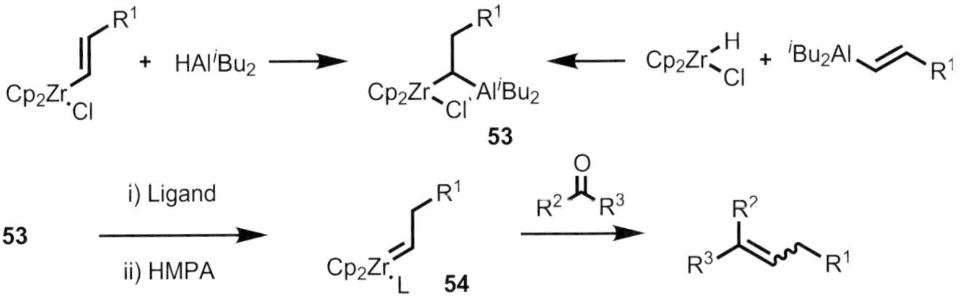

Scheme 4.49. Preparation of zirconium analogues of the Tebbe reagent and their use for carbonyl olefination.

Using the zirconium carbene complexes **54** as olefination agents, stereochemical control can be achieved by replacing a carbonyl group of the substrates with an imino group (=NR) (Scheme 4.50) [119]. The olefination of benzaldehyde with

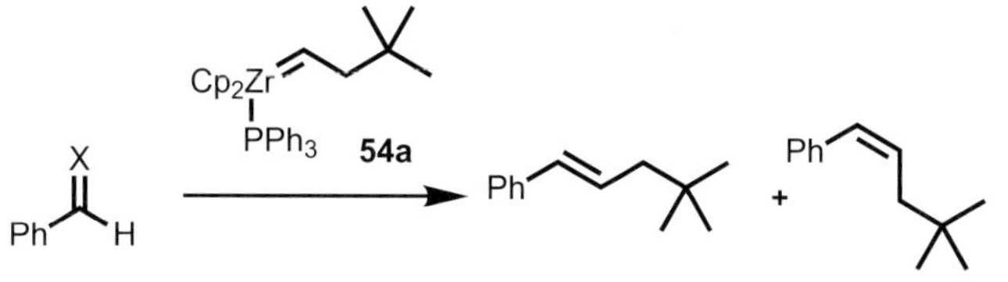

X	Yield / %, (E / Z)
O	76 (0.50)
NCH$_3$	88 (2.06)
NCH(CH$_3$)	47 (2.30)
NC(CH$_3$)	16 (4.32)

Scheme 4.50. Stereochemistry of carbonyl olefination using zirconium alkylidene complexes.

54a proceeds with preferential formation of the (Z)-styrene derivative. In contrast, the use of imines under the similar reaction conditions reverses the E/Z selectivity. The stereoselectivity is improved as the substituent on the nitrogen atom is made more bulky, although the yield of the olefination is depressed. In contrast to the olefination of lactones with **54a**, the corresponding imidates react with the zirconium-carbene complex **54a** to give enol ethers with Z selectivity.

Alternative homologues of the Tebbe reagent **3** are the substituted and highly functionalized zinc and zirconium 1,1-bimetallic reagents **55**, which are readily generated in solution by hydrozirconation of alkenylzinc halides with the Schwartz reagent. These reagents, however, are thermally unstable and decompose within 20 min at 25 °C. Therefore, carbonyl olefination with **55** is performed by adding the aldehyde or ketone immediately after the generation of **55**, giving the corresponding olefins in good yields (Scheme 4.51) [120]. Functional groups such as chloride, cyanide, and ester are inert in the formation and reaction of the bimetallic species **55**. It is noteworthy that the olefination of aldehydes proceeds with high E-selectivity. Furthermore, keto esters are chemoselectively converted into the unsaturated esters. Ketones are also transformed into tri- or tetra-substituted olefins in good yields with **55**, but the stereoselectivity is lowered compared to that for aldehydes.

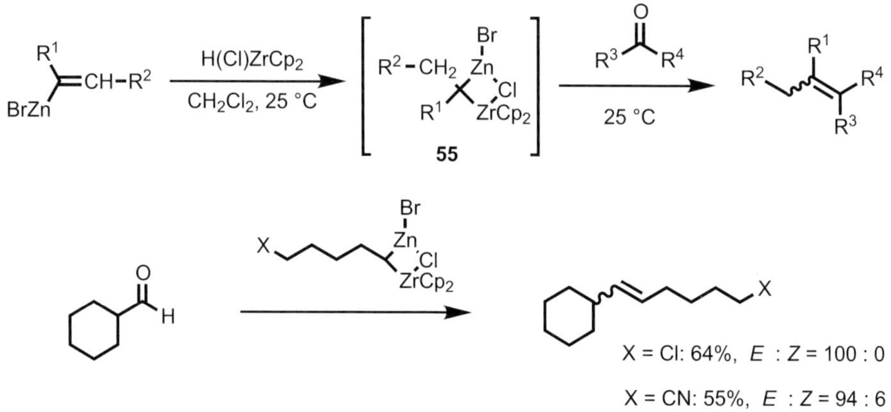

X = Cl: 64%, E : Z = 100 : 0

X = CN: 55%, E : Z = 94 : 6

Scheme 4.51. Formation of zinc and zirconium 1,1-bimetalic reagents and their reaction with aldehydes.

4.5.2
Tantalum and Niobium Carbene Complexes

Tantalum-neopentylidene complex, Ta[$CH_2C(CH_3)_3$]$_3$[$CHC(CH_3)_3$] (**1**), was the first metal-carbene complex to be isolated and used for carbonyl olefination [121]. Common carbonyl compounds such as aldehydes, ketones, esters, and amides are olefinated with **1** at 25 °C in diethyl ether, hexane, or pentane solutions (Scheme 4.52) [4]. These reactions are conveniently carried out using either the complex prepared quantitatively in situ from Ta[$CH_2C(CH_3)_3$]$_3Cl_2$ and two equivalents of LiCH$_2$C(CH$_3$)$_3$ or the isolated material. The reactions of sterically hindered carbonyl compounds bearing bulky substituents (e.g. pivalaldehyde) with **1** are, however, too slow for practical purposes. The reagent **1** also reacts with CO_2 to produce di-*tert*-butylallene. The preparation of *tert*-butyl-substituted olefins is generally difficult by conventional olefination methods due to the steric demands. Therefore, the procedure utilizing the carbene complex **1** provides a useful means of preparing *tert*-butyl-substituted olefins from carbonyl compounds.

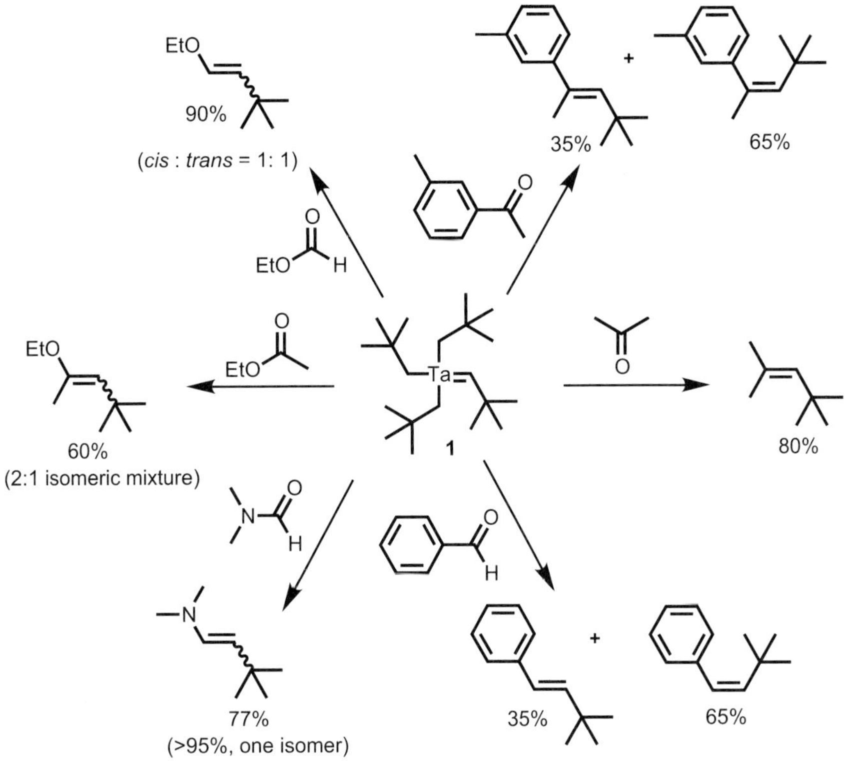

Scheme 4.52. Carbonyl olefination utilizing the tantalum carbene complex **1**.

Niobium neopentylidene complex, Nb[$CH_2C(CH_3)_3$]$_3$[$CHC(CH_3)_3$] (**2**), can also be prepared in a manner similar to that for the corresponding tantalum analogue **1**

[121]. Although the niobium complex **2** reacts with acetone to give 2,4,4-trimethyl-2-pentene in high yield, the difficulty in preparing the reagent in large amounts is a serious drawback for its practical use.

4.5.3
Molybdenum Carbene Complexes

Several molybdenum-methylidene complexes **56** are employed as olefination reagents (Scheme 4.53). Some of them differ from the conventional carbonyl olefination reagents in their acidic character and chemoselectivity. Generally, molybdenum carbene complexes are thermally unstable and attempts to isolate them have hitherto been unsuccessful. Therefore, the olefinating reagents are generated in solution in the presence of the carbonyl compounds such that they react immediately.

$Cl_3Mo=CH_2$ $Cl(O)Mo=CH_2$ $(EtO)_3Mo=CH_2$
56a   **56b**   **56c**

$(EtO)_2MeMo=CH_2$ $Cl_2(Me)Mo=CH_2$ $Cl(Me)_2Mo=CH_2$
56d   **56e**   **56f**

Scheme 4.53. Organomolybdenum compounds employed as methylenation agents.

The most easily accessible reagent **56a** is prepared by the reaction of $MoCl_5$ with two equivalents of methyllithium [122]. The reagent **56a** may possibly exist at higher temperatures in the form of the μ-CH_2 complex **56a'** (Scheme 4.54) [123].

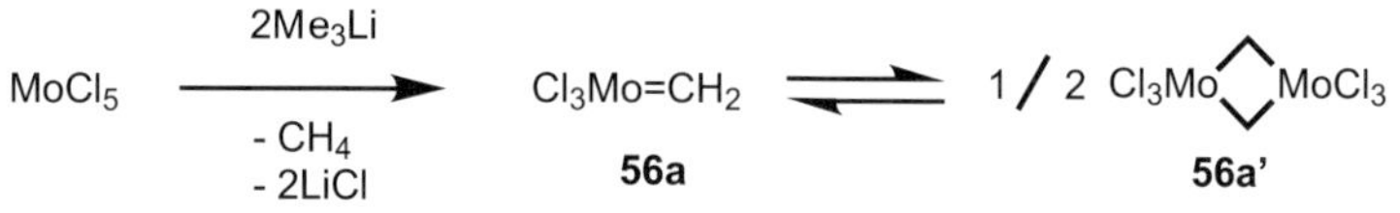

$MoCl_5 \xrightarrow[\substack{-CH_4 \\ -2LiCl}]{2Me_3Li} Cl_3Mo=CH_2 \rightleftharpoons 1/2\ Cl_3Mo(\mu\text{-}CH_2)_2MoCl_3$
                                    56a   **56a'**

Scheme 4.54. Formation of the molybdenum-methylidene complex.

Methylenation with **56a** is typically performed by adding the carbonyl compounds to a solution of the reagent prepared in situ (Table 4.17). This methylenation is of interest in that the reaction is extremely sensitive to steric hindrance. The molybdenum compound **56a** methylenates aldehydes such as heptanal and benzaldehyde. Although **56a** methylenates acetophenone, diethyl ketone and benzophenone are not transformed to the corresponding terminal olefins. Thus, a molecule containing both ketone and aldehyde functions is selectively methylenated only at the aldehyde carbonyl (entry 3).

Similar treatments of $Cl_3(O)Mo$, $(EtO)_3MoCl_2$, and $(EtO)_2MoCl_3$ with methyllithium afford the corresponding molybdenum-methylidenes **56b**, **56c**, and **56d**, which also methylenate aldehydes [122]. As regards the structure of **56b**, later investigations indicated it to be the dimeric 1,3-dioxo-1,3-dimolybdenum(V) cyclobutane complex, similar to **56a'** [124]. The reactivity order of these reagents for

Tab. 4.17. Olefination of carbonyl compounds with the molybdenum-methylidene complex **56a**.

Entry	Carbonyl compound	Product	Yield
1	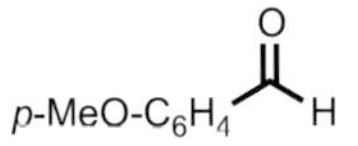 p-MeO-C$_6$H$_4$–C(O)H	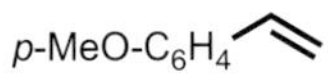 p-MeO-C$_6$H$_4$–CH=CH$_2$	84%
2	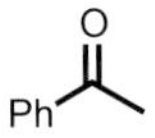 Ph–C(O)–CH$_3$	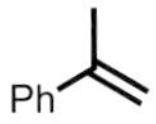 Ph–C(CH$_3$)=CH$_2$	53%
3	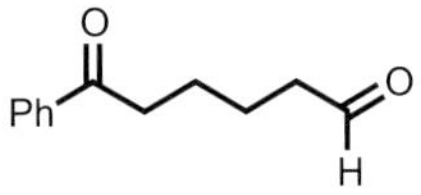 Ph–C(O)–(CH$_2$)$_3$–C(O)H	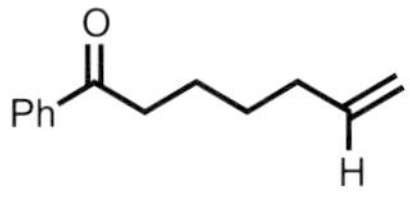 Ph–C(O)–(CH$_2$)$_3$–CH=CH$_2$	80%[a]

[a] 6-Phenyl-6-heptenal was produced in 1% yield.

methylenation is reported to be **56d** < **56c** < **56a**, **56b**; the reactions of **56c** and **56d** with acetophenone produce only a trace amount of the terminal olefin.

It is noteworthy that aldehydes and ketones can be methylenated in aqueous or ethanolic media with the molybdenum compounds **56a** and **56b** [125]. The yields of the olefins obtained by the reactions with **56b** carried out in EtOH are virtually the same as those obtained using anhydrous and alcohol-free THF as the solvent. The reagent **56a** also methylenates aldehydes in aqueous or ethanolic solvents, but the yields of the olefins are lower. These methods seem to be advantageous for the olefination of hydrophilic substrates.

The complex **56b** exhibits high selectivity in that it methylenates aldehydes in preference to ketones. Hydroxyl and amino groups in the substrates are tolerated in the methylenation using **56b**, whereby a unique chemoselectivity is observed: the carbonyl group in a β-hydroxy ketone is preferentially methylenated in competition reactions with simple ketones or β-methoxy or dimethylamino ketones (Scheme 4.55) [126].

Slow addition of three and four equivalents of MeLi to MoCl$_5$ produces Cl$_2$(Me)Mo=CH$_2$ (**56e**) and Cl(Me)$_2$Mo=CH$_2$ (**56f**), respectively. These carbene complexes react with aromatic aldehydes to afford alkenes in good yields [127].

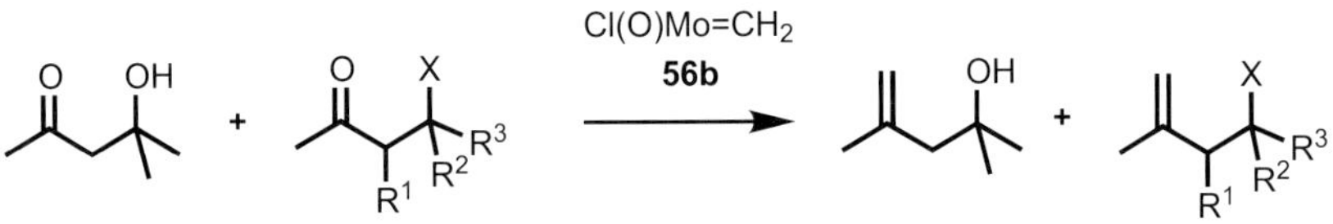

Cl(O)Mo=CH$_2$ **56b**

X = H, R^1 = H, R^2 = R^3 = CH$_3$ — 70% — 0%
X = OCH$_3$, R^1 = H, R^2 = R^3 = CH$_3$ — 78% — 12%
X = N(CH$_3$)$_2$, R^1 = CH$_3$, R^2 = R^3 = H — 75% — 8%

Scheme 4.55. Chemoselective carbonyl olefination with the molybdenum-methylidene **56b**.

Tab. 4.18. Methylidenation of aldehydes with molybdenum metallacycles **57**.

Entry	Carbonyl compound	Reagent 57	Yield
1	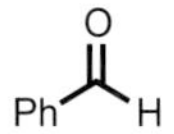 Ph-CHO	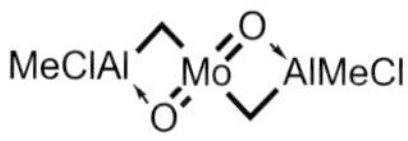 MeClAl-Mo(=O)-O-AlMeCl **57a**	98% (50%)[a]
2	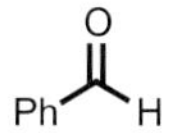 Ph-CHO	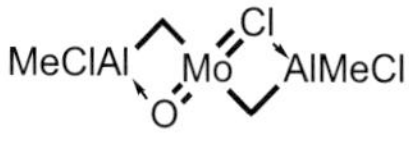 MeClAl-Mo(Cl)-O-AlMeCl **57b**	62%
3	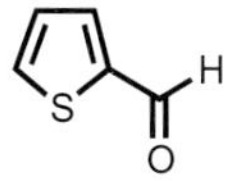 2-thienyl-CHO	**57b**	60% (31%)[a]

[a] Yield obtained using the Tebbe reagent **3**.

Treatment of MoO_2Cl_2 and $MoOCl_3$ with Me_3Al generates the metallacycles **57a** and **57b**, respectively, in solution, the structures of which are suggested to be μ-CH_2-bridged bimetallic species on the basis of their NMR data [128, 129]. In certain cases, the reagents **57** transform aldehydes to 1-alkenes in better yields than the Tebbe reagent **3** (Table 4.18).

The molybdenum-alkylidene **9** [130] promoted tandem olefin metathesis–carbonyl olefination of olefinic ketones offers an effective synthetic method for cycloalkenes (Scheme 4.56; Table 4.19). The complex **9** initially reacts with the ter-

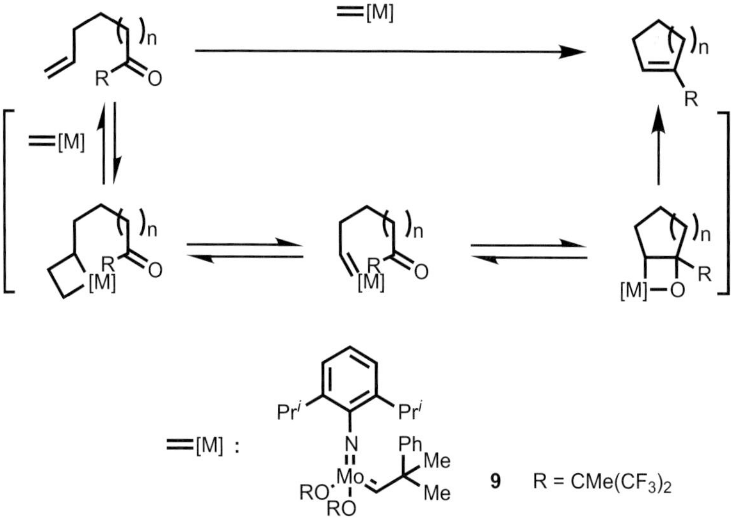

Scheme 4.56. Formation of cycloalkenes from olefinic ketones by the molybdenum-alkylidene promoted tandem olefin metathesis-carbonyl olefination.

$R = CMe(CF_3)_2$

Tab. 4.19. Tandem olefin metathesis-carbonyl olefination promoted by the molybdenum-alkylidene **9**.

Substrate	Product	Yield
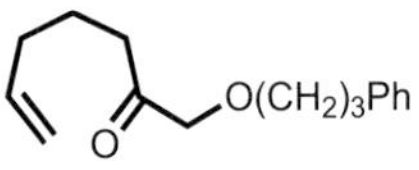 (alkene-ketone with O(CH$_2$)$_3$Ph)	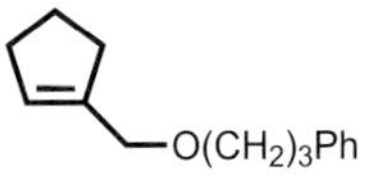 (cyclopentene with O(CH$_2$)$_3$Ph)	86%
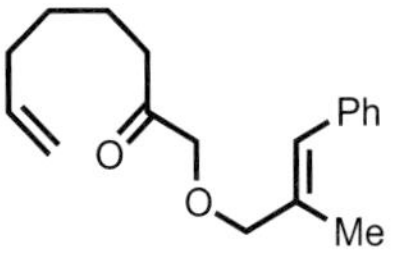 (alkene-ketone with OCH(Me)=CHPh)	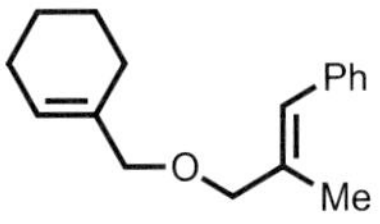 (cyclohexene with OCH(Me)=CHPh)	84%
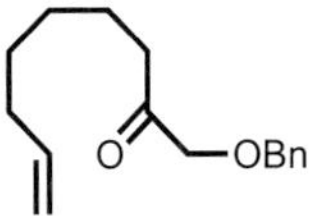 (alkene-ketone with OBn)	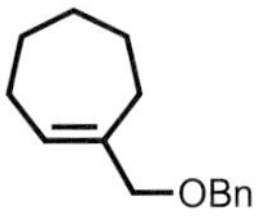 (cycloheptene with OBn)	86%

minal olefin to afford a new alkylidene bearing a pendant carbonyl group, which then undergoes intramolecular carbonyl olefination to give unsaturated five-, six-, and seven-membered rings [131].

4.5.4
Tungsten Carbene Complexes

Tungsten carbene-complexes **58** and **59** are obtained by the reaction of WOCl$_3$·(THF)$_2$ or WOCl$_4$ with two equivalents of MeLi in THF, and are more thermally stable than the corresponding molybdenum complexes (Scheme 4.57) [132].

$$\text{WOCl}_3(\text{THF})_2 \; + \; 2 \text{ MeLi} \; \xrightarrow[-\text{CH}_4]{\text{THF}} \; \text{ClW(O)=CH}_2$$
58

$$\text{WOCl}_4 \; + \; 2 \text{ MeLi} \; \xrightarrow[-\text{CH}_4]{\text{THF}} \; \text{Cl}_2\text{W(O)=CH}_2$$
59

Scheme 4.57. Preparation of tungsten-methylidenes.

Complex **58**, used in excess (3 equiv.), transforms benzaldehyde (91%), *p*-methoxybenzaldehyde (89%), and cyclohexanone (93%) into terminal olefins in high yields. Unlike molybdenum complexes, the reactivity of **58** is dramatically reduced when olefinations are carried out in aqueous or ethanolic solvents.

Alkylidene-tungsten complexes, W(=CHBu^t)(OCH$_2$Bu^t)$_2$X$_2$ (X = halogen) **60**

4.5 Carbonyl Olefination Using Zirconium, Tantalum, Niobium, Molybdenum

are prepared by the reaction of $WO(CH_2{}^tBu)_2(OCH_2{}^tBu)_2$ with AlX_3. The halide ligands of **60** can be substituted by treatment with $LiOCH_2{}^tBu$ or $AgSO_3CF_3$. Their diverse homologues, $W(=CR^1R^2)X_2Y_2$ **61**, are derived from these complexes by replacement of the neopentylidene group with terminal olefins such as hex-1-ene, styrene, or methylenecycloalkanes through olefin metathesis (Scheme 4.58) [133].

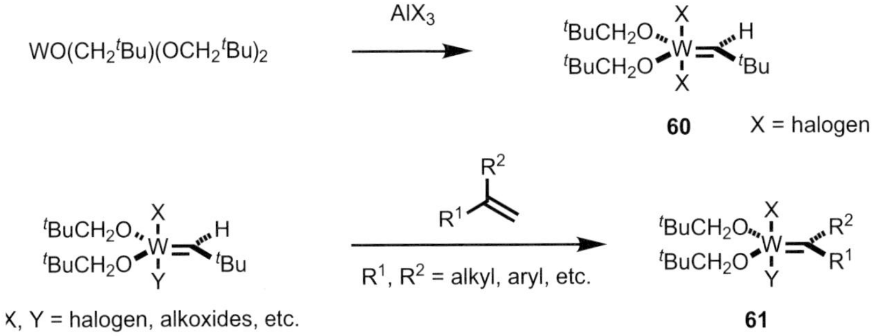

Scheme 4.58. Preparation of tungsten-alkylidenes from terminal olefins.

The alkylidene complex **61** readily converts a wide variety of carbonyl compounds into the corresponding olefins in good yields, with the following order of reactivity: aldehydes > ketones > formates > esters > amides (Table 4.20) [134]. Replacement of two neopentoxo groups by chloro, bromo, or iodo ligands reduces the reactivity of the complexes.

Besides the carbene complexes, the tungsten alkylidynes **62** [135] are also employed for carbonyl olefination. The complexes **62** react with carbonyl compounds to give oxo vinyl complexes **63**, which are hydrolyzed with 1 N NaOH to form the

Tab. 4.20. Carbonyl olefination with tungsten alkylidene complexes **60** and **61**.

Entry	Carbonyl compound	Tungsten complex	Product	Yield
1	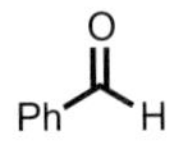 Ph-CO-H	$({}^tBuCH_2O)_4W=CH{}^tBu$	 Ph-CH=CH-${}^tBu$	95%[a]
2	 acetone	$({}^tBuCH_2O)_2I_2W=CHPh$	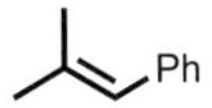 Me-C(Me)=CH-Ph	91%
3	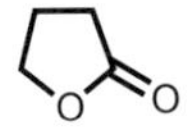 γ-butyrolactone	$({}^tBuCH_2O)_4W=CH{}^tBu$	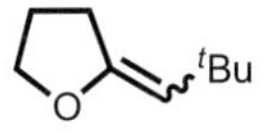 enol ether with ${}^tBu$	65%
4	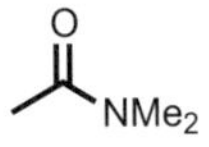 Me-CO-NMe$_2$	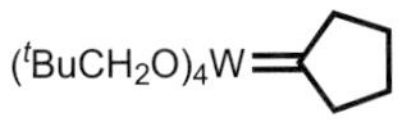 $({}^tBuCH_2O)_4W=$cyclopentylidene	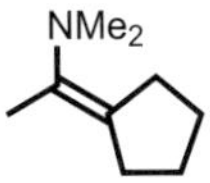 cyclopentylidene-C(Me)=NMe$_2$	76%

[a] E:Z = 80:20.

carbonyl olefination products [136]. It is noteworthy that disubstituted internal olefins are obtained with excellent Z-selectivity (Scheme 4.59).

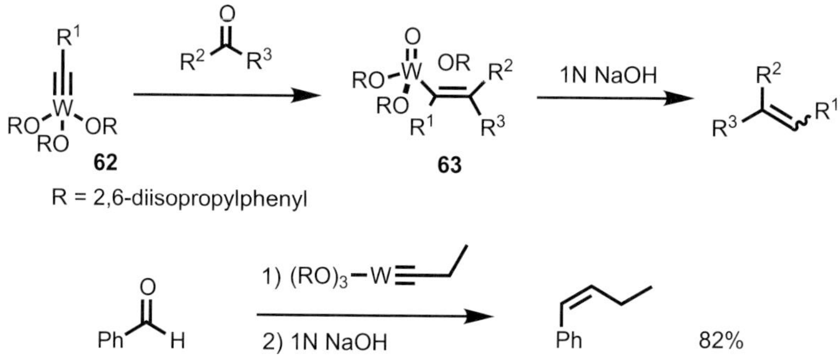

Scheme 4.59. Olefination of carbonyl compounds with tungten-alkylidyne complexes.

4.6
Conclusion

As described in this chapter, the Schrock-type metal carbene complexes and related reagents have played an important role in carbonyl olefination. Although conventional methods, such as the Wittig and Peterson reactions, have been widely used for the carbonyl olefination of aldehydes and ketones, these methods generally yield unsatisfactory results for carboxylic acid derivatives such as esters and amides. The metal carbene complexes cover this shortcoming of the conventional methods in a complementary manner, and provide a convenient means of preparing heteroatom-substituted olefins such as vinyl ethers and enamines. In addition, they enable the transformation of readily enolizable or sterically hindered carbonyl compounds into highly substituted olefins. A wide variety of reagents with different reactivities is available; therefore, by judicious choice of the reagent, stereoselective or chemoselective olefination can be achieved in some cases. Certain metal carbene complexes can even be used in protic solvents. These unique characteristics of the metal carbene complexes are sure to further extend their application in organic synthesis.

References

1 F. Z. Dörwart, *Metal Carbenes in Organic Synthesis*, Wiley-VCH, Weinheim, **1999**.
2 (a) T. E. Taylor, M. B. Hall, *J. Am. Chem. Soc.* **1984**, *106*, 1576. (b) D. G. Musaev, K. Morokuma, N. Koga, *J. Chem. Phys.* **1993**, *99*, 7859. (c) D. G. Musaev, K. Morokuma, N. Koga, K. A. Nguyen, M. S. Gordon, T. R. Cundari, *J. Phys. Chem.* **1993**, *97*, 11435. (d) H. Jacobsen, T. Ziegler, *Organometallics* **1995**, *14*, 224. (e) H. Jacobsen, T. Ziegler, *Inorg. Chem.* **1996**, *35*, 775. (f) C. C. Wang, Y.

Wang, H. J. Liu, K. J. Lin, L. K. Chou, K. S. Chan, *J. Phys. Chem. A* **1997**, *101*, 8887.

3 (a) R. J. Goddard, R. Hoffmann, E. D. Jemmis, *J. Am. Chem. Soc.* **1980**, *102*, 7667. (b) O. Eisenstein, R. Hoffmann, A. R. Rossi, *J. Am. Chem. Soc.* **1981**, *103*, 5582. (c) A. K. Rappé, W. A. Goddard, *J. Am. Chem. Soc.* **1982**, *104*, 448. (d) T. H. Upton, A. K. Rappé, *J. Am. Chem. Soc.* **1985**, *107*, 1206. (e) T. R. Cundari, M. S. Gordon, *J. Am. Chem. Soc.* **1991**, *113*, 5231. (f) A. Márquez, J. F. Sanz, *J. Am. Chem. Soc.* **1992**, *114*, 10019. (g) H. H. Fox, M. H. Schofield, R. R. Schrock, *Organometallics* **1994**, *13*, 2804. (h) L. Bencze, R. Szilágyi, *J. Organomet. Chem.* **1994**, *465*, 211. (i) Y. D. Wu, Z. H. Peng, *J. Am. Chem. Soc.* **1997**, *119*, 8043. (j) L. Luo, L. Li, T. J. Marks, *J. Am. Chem. Soc.* **1997**, *119*, 8574.

4 R. R. Schrock, *J. Am. Chem. Soc.* **1976**, *98*, 5399.

5 F. N. Tebbe, G. W. Parshall, G. S. Reddy, *J. Am. Chem. Soc.* **1978**, *100*, 3611.

6 K. C. Ott, E. J. M. de Boer, R. H. Grubbs, *Organometallics* **1984**, *3*, 223.

7 M. M. Francl, W. J. Hehre, *Organometallics* **1983**, *2*, 457.

8 J. D. Meinhart, E. V. Anslyn, R. H. Grubbs, *Organometallics* **1989**, *8*, 583.

9 J. B. Lee, K. C. Ott, R. H. Grubbs, *J. Am. Chem. Soc.* **1982**, *104*, 7491.

10 L. F. Cannizzo, R. H. Grubbs, *J. Org. Chem.* **1985**, *50*, 2386.

11 (a) K. A. Brown-Wensley, S. L. Buchwald, L. Cannizzo, L. Clawson, S. Ho, D. Meinhardt, J. R. Stille, D. Straus, R. H. Grubbs, *Pure Appl. Chem.* **1983**, *55*, 1733. (b) R. H. Grubbs, S. H. Pine, in *Comprehensive Organic Synthesis* (Ed.: B. M. Trost), Pergamon Press, New York, **1991**, Vol. 5, p. 1115. (c) S. H. Pine, *Org. React.* **1993**, *43*, 1. (d) J. R. Stille, in *Comprehensive Organometallic Chemistry II* (Eds.: E. W. Abel, F. G. A. Stone, G. Wilkinson), Pergamon Press, Oxford, **1995**, Vol. 12, p. 577. (e) N. A. Petasis, in *Transition Metals for Organic Synthesis* (Eds.: M. Beller, C. Bolm), Wiley-VCH, Weinheim, **1999**, p. 361.

12 S. H. Pine, G. S. Shen, H. Hoang, *Synthesis* **1991**, 165.

13 J. D. Winkler, C. L. Muller, R. D. Scott, *J. Am. Chem. Soc.* **1988**, *110*, 4831.

14 D. L. J. Clive, S. Sun, V. Gagliardini, M. K. Sano, *Tetrahedron Lett.* **2000**, *41*, 6259.

15 J. D. Schloss, S. M. Leit, L. A. Paquette, *J. Org. Chem.* **2000**, *65*, 7119.

16 S. Kobayashi, R. S. Reddy, Y. Sugiura, D. Sasaki, N. Miyagawa, M. Hirama, *J. Am. Chem. Soc.* **2001**, *123*, 2887.

17 J. B. Schwarz, A. I. Meyers, *J. Org. Chem.* **1995**, *60*, 6511.

18 J. E. McMurry, R. Swenson, *Tetrahedron Lett.* **1987**, *28*, 3209.

19 M. H. Ali, P. M. Collins, W. G. Overend, *Carbohydr. Res.* **1990**, *205*, 428.

20 L. Ackermann, D. E. Tom, A. Fürstner, *Tetrahedron* **2000**, *56*, 2195.

21 R. E. Ireland, R. B. Wardle, *J. Org. Chem.* **1987**, *52*, 1780.

22 (a) N. Ikemoto, S. L. Schreiber, *J. Am. Chem. Soc.* **1990**, *112*, 9657. (b) N. Ikemoto, S. L. Schreiber, *J. Am. Chem. Soc.* **1992**, *114*, 2524.

23 S. H. Pine, R. Zahler, D. A. Evans, R. H. Grubbs, *J. Am. Chem. Soc.* **1980**, *102*, 3270.

24 S. H. Pine, R. J. Pettit, G. D. Geib, S. G. Cruz, C. H. Gallego, T. Tijerina, R. D. Pine, *J. Org. Chem.* **1985**, *50*, 1212.

25 J. Adams, R. Frenette, *Tetrahedron Lett.* **1987**, *28*, 4773.

26 G. Dimartino, J. M. Percy, *Chem. Commun.* **2000**, 2339.

27 T. V. RajanBabu, G. S. Reddy, *J. Org. Chem.* **1986**, *51*, 5458.

28 W. Wang, Y. Zhang, H. Zhou, Y. Bleriot, P. Sinaÿ, *Eur. J. Org. Chem.* **2001**, 1053.

29 A. Fairbanks, P. Sinaÿ, *Tetrahedron Lett.* **1995**, *36*, 893.

30 L. F. Cannizzo, R. H. Grubbs, *J. Org. Chem.* **1985**, *50*, 2316.

31 K. Tanino, H. Shoda, T. Nakamura,

I. Kuwajima, *Tetrahedron Lett.* **1992**, *33*, 1337.

32 N. Pelloux-Leon, F. Minassian, J. Levillain, J.-L. Ripoll, Y. Vallee, *Tetrahedron Lett.* **1998**, *39*, 4813.

33 T. Fukuyama, G. Liu, *J. Am. Chem. Soc.* **1996**, *118*, 7426.

34 L. A. Paquette, T. Z. Wang, N. H. Vo, *J. Am. Chem. Soc.* **1993**, *115*, 1676.

35 S. Borrelly, L. A. Paquette, *J. Am. Chem. Soc.* **1996**, *118*, 727.

36 O. Fujimura, G. C. Fu, R. H. Grubbs, *J. Org. Chem.* **1994**, *59*, 4029.

37 K. C. Nicolaou, M. H. Postema, C. F. Claiborne, *J. Am. Chem. Soc.* **1996**, *118*, 1565.

38 K. C. Nicolaou, M. H. D. Postema, E. W. Yue, A. Nadin, *J. Am. Chem. Soc.* **1996**, *118*, 10335.

39 T. Oishi, Y. Nagumo, M. Shoji, J.-Y. L. Brazidec, H. Uehara, M. Hirama, *Chem. Commun.* **1999**, 2035.

40 T. Takeda, in *Titanium and Zirconium in Organic Synthesis* (Ed.: I. Marek), Wiley-VCH, Weinheim, **2002**, p. 475.

41 T. R. Howard, J. B. Lee, R. H. Grubbs, *J. Am. Chem. Soc.* **1980**, *102*, 6876.

42 E. V. Anslyn, R. H. Grubbs, *J. Am. Chem. Soc.* **1987**, *109*, 4880.

43 L. Clawson, S. L. Buchwald, R. H. Grubbs, *Tetrahedron Lett.* **1984**, *25*, 5733.

44 (a) J. R. Stille, R. H. Grubbs, *J. Am. Chem. Soc.* **1983**, *105*, 1664. (b) T.-S. Chou, S.-B. Huang, *Tetrahedron Lett.* **1983**, *24*, 2169.

45 J. W. F. L. Seetz, B. J. J. van de Heisteeg, G. Schat, O. S. Akkerman, F. Bickelhaupt, *J. Mol. Cat.* **1985**, *28*, 71.

46 J. J. Eisch, A. Piotrowski, *Tetrahedron Lett.* **1983**, *24*, 2043.

47 J. W. Bruin, G. Schat, O. S. Akkerman, F. Bickelhaupt, *Tetrahedron Lett.* **1983**, *24*, 3935.

48 B. J. J. van de Heisteeg, G. Schat, O. S. Akkerman, F. Bickelhaupt, *Tetrahedron Lett.* **1987**, *28*, 6493.

49 F. N. Tebbe, L. J. Guggenberger, *J. Chem. Soc., Chem. Commun.* **1973**, 227.

50 F. W. Hartner, Jr., J. Schwartz, *J. Am. Chem. Soc.* **1981**, *103*, 4979.

51 T. Yoshida, E. Negishi, *J. Am. Chem. Soc.* **1981**, *103*, 1276.

52 T. Yoshida, *Chem. Lett.* **1982**, 429.

53 L. R. Gilliom, R. H. Grubbs, *Organometallics* **1986**, *5*, 721.

54 L. R. Gilliom, R. H. Grubbs, *J. Am. Chem. Soc.* **1986**, *108*, 733.

55 (a) J. R. Stille, R. H. Grubbs, *J. Am. Chem. Soc.* **1986**, *108*, 855. (b) J. R. Stille, B. D. Santarsiero, R. H. Grubbs, *J. Org. Chem.* **1990**, *55*, 843.

56 S. L. Buchwald, R. H. Grubbs, *J. Am. Chem. Soc.* **1983**, *105*, 5490.

57 (a) F. N. Tebbe, G. W. Parshall, D. W. Ovenall, *J. Am. Chem. Soc.* **1979**, *101*, 5074. (b) F. N. Tebbe, R. L. Harlow, *J. Am. Chem. Soc.* **1980**, *102*, 6149. (c) R. J. McKinney, T. H. Tulip, D. L. Thorn, T. S. Coolbaugh, F. N. Tebbe, *J. Am. Chem. Soc.* **1981**, *103*, 5584.

58 T. R. Howard, J. B. Lee, R. H. Grubbs, *J. Am. Chem. Soc.* **1980**, *102*, 6876.

59 K. M. Doxsee, J. K. M. Mouser, *Tetrahedron Lett.* **1991**, *32*, 1687.

60 (a) J. D. Meinhart, R. H. Grubbs, *Bull. Chem. Soc. Jpn.* **1988**, *61*, 171. (b) N. A. Petasis, D.-K. Fu, *Organometallics* **1993**, *12*, 3776.

61 (a) G. J. Erskine, D. A. Wilson, J. D. McCowan, *J. Organomet. Chem.* **1976**, *114*, 119. (b) G. J. Erskine, J. Hartgerink, E. L. Weinberg, J. D. McCowan, *J. Organomet. Chem.* **1979**, *170*, 51.

62 N. A. Petasis, S.-P. Lu, E. I. Bzowej, D.-K. Fu, J. P. Staszewski, I. Akritopoulou-Zanze, M. A. Patane, Y.-H. Hu, *Pure Appl. Chem.* **1996**, *68*, 667.

63 (a) v. K. Clauss, H. Bestian, *Justus Liebigs Ann. Chem.* **1962**, 8. (b) J. F. Payack, D. L. Hughes, D. W. Cai, I. F. Cottrell, T. R. Verhoeven, *Org. Prep. Proced. Internat.* **1995**, *27*, 707.

64 N. A. Petasis, E. I. Bzowej, *J. Am. Chem. Soc.* **1990**, *112*, 6392.

65 T. Matsuura, S. Nishiyama, S. Yamamura, *Chem. Lett.* **1993**, 1503.

66 S. Ebert, N. Krause, *Eur. J. Org. Chem.* **2001**, 3831.

67 (a) R. Csuk, B. I. Glänzer, *Tetrahedron* **1991**, *47*, 1655. (b) V.

Faivre-Buet, I. Eynard, H. N. Nga, G. Descotes, A. Grouiller, *J. Carbohydr. Chem.* **1993**, *12*, 349.

68 N. A. Petasis, S.-P. Lu, *Tetrahedron Lett.* **1995**, *36*, 2393.

69 P. DeShong, P. J. Rybczynski, *J. Org. Chem.* **1991**, *56*, 3207.

70 M. J. Kates, J. H. Schauble, *J. Org. Chem.* **1994**, *59*, 494.

71 P. J. Colson, L. S. Hegedus, *J. Org. Chem.* **1993**, *58*, 5918.

72 D. Kuzmich, S. C. Wu, D.-C. Ha, C.-S. Lee, S. Ramesh, S. Atarashi, J.-K. Choi, D. J. Hart, *J. Am. Chem. Soc.* **1994**, *116*, 6943.

73 C. Herdeis, E. Heller, *Tetrahedron: Asymmetry* **1993**, *4*, 2085.

74 (a) H. K. Chenault, L. F. Chafin, *J. Org. Chem.* **1994**, *59*, 6167. (b) H. K. Chenault, A. Castro, L. F. Chafin, J. Yang, *J. Org. Chem.* **1996**, *61*, 5024.

75 N. A. Petasis, Y.-H. Hu, D.-K. Fu, *Tetrahedron Lett.* **1995**, *36*, 6001.

76 (a) L. M. Dollinger, A. R. Howell, *J. Org. Chem.* **1996**, *61*, 7248. (b) L. M. Dollinger, A. J. Ndakala, M. Hashemzadeh, G. Wang, Y. Wang, I. Martinez, J. T. Arcari, D. J. Galluzzo, A. R. Howell, *J. Org. Chem.* **1999**, *64*, 7074.

77 (a) K. A. Tehrani, N. De Kimpe, *Tetrahedron Lett.* **2000**, *41*, 1975. (b) I. Martínez, A. R. Howell, *Tetrahedron Lett.* **2000**, *41*, 5607.

78 (a) N. A. Petasis, M. A. Patane, *Tetrahedron Lett.* **1990**, *31*, 6799. (b) N. A. Petasis, E. I. Bzowej, *Tetrahedron Lett.* **1993**, *34*, 1721.

79 J. S. Swenton, D. Bradin, B. D. Gates, *J. Org. Chem.* **1991**, *56*, 6156.

80 N. A. Petasis, S.-P. Lu, *J. Am. Chem. Soc.* **1995**, *117*, 6394.

81 N. A. Petasis, S.-P. Lu, *Tetrahedron Lett.* **1996**, *37*, 141.

82 D. L. Hughes, J. F. Payack, D. Cai, T. R. Verhoeven, P. J. Reider, *Organometallics* **1996**, *15*, 663.

83 S. L. Hart, A. McCamley, P. C. Taylor, *Synlett* **1999**, 90.

84 (a) G. A. Razuvaev, V. N. Latyaeva, L. I. Vyshinskaya, *Dokl. Akad. Nauk SSSR* **1969**, *189*, 103. (b) A. Glivicky, J. D. McCowan, *Can. J. Chem.* **1973**, *51*, 2609. (c) J. Scholz, M. Schlegel, K. H. Thiele, *Chem. Ber.* **1987**, *120*, 1369.

85 N. A. Petasis, E. I. Bzowej, *J. Org. Chem.* **1992**, *57*, 1327.

86 N. A. Petasis, I. Akritopoulou, *Synlett* **1992**, 665.

87 The formation of **31** by this method is often complicated by the presence of various amounts of **33**. The preparation method had to be modified to give clean compound **31**; see: N. A. Petasis, D.-K. Fu, *J. Am. Chem. Soc.* **1993**, *115*, 7208.

88 N. A. Petasis, J. P. Staszewski, D.-K. Fu, *Tetrahedron Lett.* **1995**, *36*, 3619.

89 N. A. Petasis, E. I. Bzowej, *Tetrahedron Lett.* **1993**, *34*, 943.

90 N. A. Petasis, Y.-H. Hu, *J. Org. Chem.* **1997**, *62*, 782.

91 T. Takeda, K. Shimane, K. Ito, N. Saeki, A. Tsubouchi, *Chem. Commun.* **2002**, 1974.

92 T. Takeda, T. Fujiwara, *Rev. Heteroatom. Chem.* **1999**, *21*, 93.

93 (a) Y. Horikawa, T. Nomura, M. Watanabe, I. Miura, T. Fujiwara, T. Takeda, *Tetrahedron Lett.* **1995**, *36*, 8835. (b) Y. Horikawa, T. Nomura, M. Watanabe, T. Fujiwara, T. Takeda, *J. Org. Chem.* **1997**, *62*, 3678. (c) T. Fujiwara, M. Takamori, T. Takeda, *Chem. Commun.* **1998**, 51. (d) T. Fujiwara, T. Takeda, *Synlett* **1999**, 354. (e) T. Fujiwara, Y. Kato, T. Takeda, *Heterocycles* **2000**, *52*, 147. (f) T. Fujiwara, Y. Kato, T. Takeda, *Tetrahedron* **2000**, *56*, 4859. (g) T. Fujiwara, K. Yanai, K. Shimane, M. Takamori, T. Takeda, *Eur. J. Org. Chem.* **2001**, 155. (h) A. Tsubouchi, E. Nishio, Y. Kato, T. Fujiwara, T. Takeda, *Tetrahedron Lett.* **2002**, *43*, 5755.

94 T. Takeda, H. Shimokawa, Y. Miyachi, T. Fujiwara, *Chem. Commun.* **1997**, 1055.

95 T. Takeda, H. Taguchi, T. Fujiwara, *Tetrahedron Lett.* **2000**, *41*, 65.

96 Y. Horikawa, M. Watanabe, T. Fujiwara, T. Takeda, *J. Am. Chem. Soc.* **1997**, *119*, 1127.

97 T. Takeda, R. Sasaki, T. Fujiwara, *J. Org. Chem.* **1998**, *63*, 7286.

98 T. Fujiwara, N. Iwasaki, T. Takeda, Chem. Lett. **1998**, 741.

99 T. Takeda, M. Watanabe, N. Nozaki, T. Fujiwara, Chem. Lett. **1998**, 115.

100 T. Takeda, J. Saito, A. Tsubouchi, Tetrahedron Lett. **2003**, 44, 5571.

101 (a) T. Takeda, M. Watanabe, M. A. Rahim, T. Fujiwara, Tetrahedron Lett. **1998**, 39, 3753. (b) T. Takeda, Y. Takagi, N. Saeki, T. Fujiwara, Tetrahedron Lett. **2000**, 41, 8377.

102 E. J. Guthrie, J. Macritchie, R. C. Hartley, Tetrahedron Lett. **2000**, 41, 4987.

103 M. A. Rahim, H. Taguchi, M. Watanabe, T. Fujiwara, T. Takeda, Tetrahedron Lett. **1998**, 39, 2153.

104 M. A. Rahim, H. Sasaki, J. Saito, T. Fujiwara, T. Takeda, Chem. Commun. **2001**, 625.

105 P. J. Murphy, S. E. Lee, J. Chem. Soc., Perkin Trans. 1 **1999**, 3049.

106 M. A. Rahim, T. Fujiwara, T. Takeda, Tetrahedron **2000**, 56, 763.

107 (a) T. Oishi, H. Uehara, Y. Nagumo, M. Shoji, J.-Y. Le Brazidec, M. Kosaka, M. Hirama, Chem. Commun. **2001**, 381. (b) M. Hirama, T. Oishi, H. Uehara, M. Inoue, M. Maruyama, H. Oguri, M. Satake, Science **2001**, 294, 1904.

108 M. A. Rahim, T. Fujiwara, T. Takeda, Synlett **1999**, 1029.

109 D. H. R. Barton, B. J. Willis, J. Chem. Soc., Perkin Trans. 1 **1972**, 305.

110 For examples, see: (a) G. Wittig, D. Wittenberg, Liebigs Ann. Chem. **1957**, 606, 1. (b) U. H. M. Fagerlund, D. R. Idler, J. Am. Chem. Soc. **1957**, 79, 6473. (c) A. A. Bothner-By, C. Naar-Colin, H. Günther, J. Am. Chem. Soc. **1962**, 84, 2748. (d) G. Wittig, A. Haag, Chem. Ber. **1963**, 96, 1535.

111 For examples, see: (a) A. Mondon, Liebigs Ann. Chem. **1957**, 603, 115. (b) A. W. Johnson, J. Org. Chem. **1959**, 24, 282. (c) H. J. Bestmann, H. Z. Häberlein, Naturforsch. B: Anorg. Chem., Org. Chem. **1962**, 17, 787. (d) K. V. Scherer, Jr., R. S. Lunt, III, J. Org. Chem. **1965**, 30, 3215. (e) K. Sisido, K. Utimoto, Tetrahedron Lett. **1966**, 3267.

112 For examples, see: (a) E. J. Corey, G. T. Kwiatkowski, J. Am. Chem. Soc. **1966**, 88, 5652. (b) D. Redmore, J. Org. Chem. **1969**, 34, 1420.

113 For example, see: (a) S. Halazy, W. Dumont, A. Krief, Tetrahedron Lett. **1981**, 22, 4737. (b) T. Cohen, J. P. Sherbine, J. R. Matz, R. R. Hutchins, B. M. McHenry, P. R. Willey, J. Am. Chem. Soc. **1984**, 106, 3245. (c) B. Halton, C. J. Randall, P. J. Stang, J. Am. Chem. Soc. **1984**, 106, 6108. (d) W. H. Richardson, S. A. Thomson, J. Org. Chem. **1985**, 50, 1803. (e) D. J. Ager, J. Chem. Soc., Perkin Trans. 1 **1986**, 183.

114 T. Fujiwara, M. Odaira, T. Takeda, Tetrahedron Lett. **2001**, 42, 3369.

115 T. Takeda, Y. Endo, A. C. S. Reddy, R. Sasaki, T. Fujiwara, Tetrahedron **1999**, 55, 2475.

116 (a) D. W. Hart, J. Schwartz, J. Am. Chem. Soc. **1974**, 96, 8115. (b) L. Schwartz, J. A. Labinger, Angew. Chem. Int. Ed. Engl. **1976**, 15, 333. (c) D. W. Hart, T. F. Blackburn, J. Schwartz, J. Am. Chem. Soc. **1975**, 97, 679. (d) C. A. Bertelo, J. Schwartz, J. Am. Chem. Soc. **1976**, 98, 262.

117 (a) F. W. Hartner, Jr., J. Schwartz, J. Am. Chem. Soc. **1981**, 103, 4979. (b) F. M. Hartner, Jr., S. M. Clift, J. Schwartz, Organometallics **1987**, 6, 1346.

118 F. W. Hartner, Jr., J. Schwartz, S. M. Clift, J. Am. Chem. Soc. **1983**, 105, 640.

119 S. M. Clift, J. Schwartz, J. Am. Chem. Soc. **1984**, 106, 8300.

120 C. E. Tucker, P. Knochel, J. Am. Chem. Soc. **1991**, 113, 9888.

121 R. R. Schrock, J. Am. Chem. Soc. **1974**, 96, 6796.

122 T. Kauffmann, B. Ennen, J. Sander, R. Wieschollek, Angew. Chem. Int. Ed. Engl. **1983**, 22, 244.

123 T. Kauffmann, P. Fiegenbaum, M. Papenberg, R. Wieschollek, J. Sander, Chem. Ber. **1992**, 125, 143.

124 T. Kauffmann, P. Fiegenbaum, M. Papenberg, R. Wieschollek, D. Wingbermühle, Chem. Ber. **1993**, 126, 79.

125 T. Kauffmann, P. Fiegenbaum, R. Wieschollek, *Angew. Chem. Int. Ed. Engl.* **1984**, *23*, 531.
126 T. Kauffmann, J. Baune, P. Fiegenbaum, U. Hansmersmann, C. Neiteler, M. Papenberg, R. Wieschollek, *Chem. Ber.* **1993**, *126*, 89.
127 T. Kauffmann, G. Kieper, *Angew. Chem. Int. Ed. Engl.* **1984**, *23*, 532.
128 T. Kauffmann, M. Enk, W. Kaschube, E. Toliopoulos, D. Wingbermühle, *Angew. Chem. Int. Ed. Engl.* **1986**, *25*, 910.
129 T. Kauffmann, M. Enk, P. Fiegenbaum, U. Hansmersmann, W. Kaschube, M. Papenberg, E. Toliopoulos, S. Welke, *Chem. Ber.* **1994**, *127*, 127.
130 (a) R. R. Schrock, J. S. Murdzek, G. C. Bazan, J. Robbins, M. DiMare, M. O'Regan, *J. Am. Chem. Soc.* **1990**, *112*, 3875. (b) G. C. Bazan, J. H. Oskam, H.-N. Cho, L. Y. Park, R. R. Schrock, *J. Am. Chem. Soc.* **1991**, *113*, 6899. (c) G. C. Bazan, R. R. Schrock, H.-N. Cho, V. C. Gibson, *Macromolecules* **1991**, *24*, 4495. (d) H. H. Fox, K. B. Yap, J. Robbins, S. Cai, R. R. Schrock, *Inorg. Chem.* **1992**, *31*, 2287.
131 G. C. Fu, R. H. Grubbs, *J. Am. Chem. Soc.* **1993**, *115*, 3800.
132 T. Kauffmann, R. Abeln, S. Welke, D. Wingbermühle, *Angew. Chem. Int. Ed. Engl.* **1986**, *25*, 909.
133 A. Aguero, J. Kress, J. A. Osborn, *J. Chem. Soc., Chem. Commun.* **1985**, 793.
134 A. Aguero, J. Kress, J. A. Osborn, *J. Chem. Soc., Chem. Commun.* **1986**, 531.
135 M. R. Churchill, J. W. Ziller, J. H. Freudenberger, R. R. Schrock, *Organometallics* **1984**, *3*, 1554.
136 J. H. Freudenberger, R. R. Schrock, *Organometallics* **1986**, *5*, 398.

5
Olefination of Carbonyl Compounds by Zinc and Chromium Reagents

Seijiro Matsubara and Koichiro Oshima

5.1
Introduction

Scheme 5.1 depicts a typical representation of the Wittig reaction, which is one of the most commonly used olefin preparation reactions [1]. The reaction was developed by Georg Wittig, who was a Nobel Prize laureate in 1979. The control of diastereochemistry in the product has been studied in detail by Schlosser. Both the *E* and *Z* diastereomers can be obtained with excellent selectivity by means of Schlosser's modified method [2].

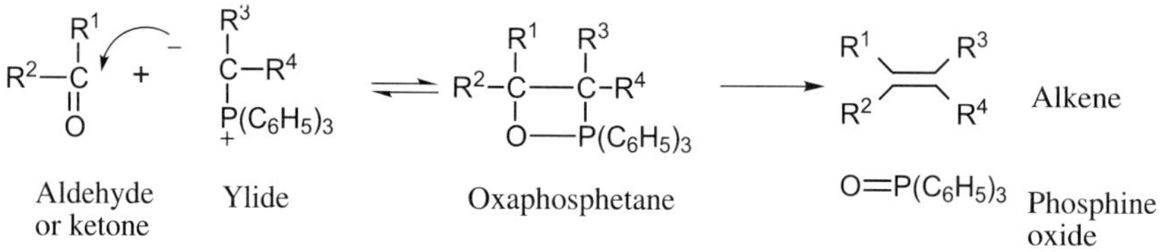

Scheme 5.1. Wittig reaction of carbonyl compounds with ylides.

This prominent name reaction looks perfect, but still has some serious drawbacks. The ylide is strongly basic, which often leads to enolization of the starting carbonyl compound. At the same time, it often suffers from a lack of nucleophilicity. For example, treatment of α-tetralone with methylenetriphenylphosphorane will not give any methylated product as the substrate is deprotonated to form the enolate. Treatment of esters with an ylide often ends with recovery of the starting ester because of insufficient nucleophilicity of the ylide. These problems, however, can be overcome by the use of a *gem*-dimetal compound. The reaction of carbonyl compounds with *gem*-dimetal compounds is outlined in Scheme 5.2.

Such compounds have two C–M bonds on the same carbon atom [3]. The first C–M bond serves to promote nucleophilic attack on the carbonyl compound, and the second facilitates the subsequent β-elimination of a metal oxide. As a result, an

Modern Carbonyl Olefination. Edited by Takeshi Takeda
Copyright © 2004 WILEY-VCH Verlag GmbH & Co. KGaA, Weinheim
ISBN: 3-527-30634-X

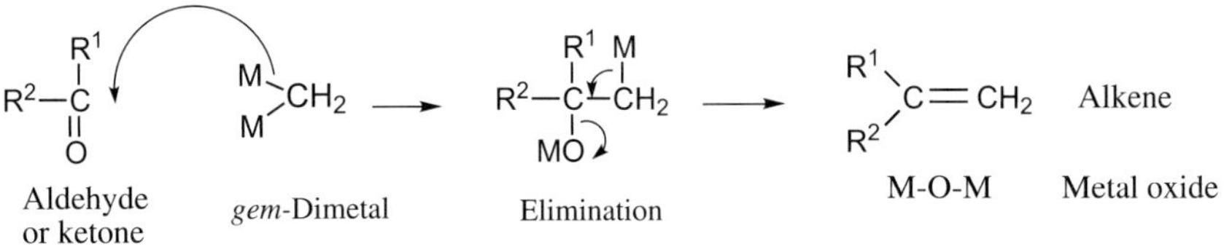

Scheme 5.2. Methylenation of carbonyl compounds with a *gem*-dimetal compound.

alkene is obtained. As the crucial carbon is substituted with a pair of electropositive metal atoms, it should be much more nucleophilic than a similar carbon in a simple organo-monometal species. In addition, careful tuning of the metal in such *gem*-dimetal species will also solve the problem of strong basicity of the reagent.

As a relative of the *gem*-dimetal compound, a metal-carbene complex should also be considered (Scheme 5.3) [4]. In this case, the reaction proceeds via a 1-metalla-2-oxacyclobutane.

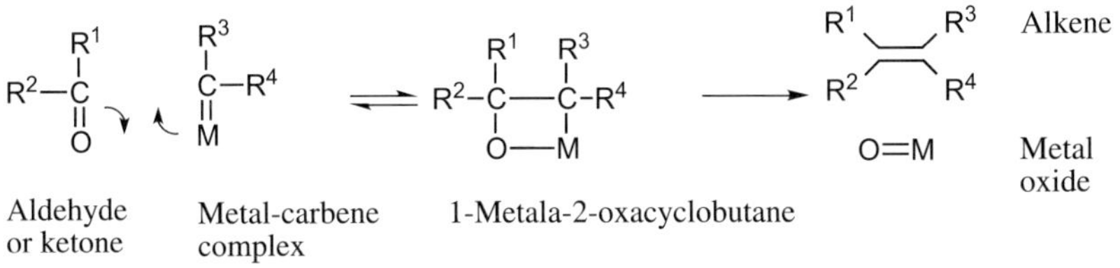

Scheme 5.3. Olefination of carbonyl compounds with metal-carbene complexes.

These organometallic reagents, however, are also associated with some problems. Compared to the Wittig reagent, they are sometimes difficult to handle because of instability, and in some cases they are prohibitively expensive. In order to obtain a desired product, these practical problems can often be ignored, but some stable and easily available *gem*-dimetal species have recently been developed that may make such considerations unnecessary. In this chapter, *gem*-dimetal reagents incorporating zinc and chromium are discussed.

5.2
Zinc Reagents

Organozinc reagents have been widely used in organic synthesis. They are not so reactive, compared to the corresponding magnesium or lithium reagents [5]. Consequently, they are reasonably stable and can be handled more easily. Although they are comparatively stable, appropriate activation such as by transmetallation and the addition of a ligand, will transform the organozinc reagent into a highly reactive and selective one. This need for "an appropriate activation" means that the reactivity of the organozinc reagents can be efficiently regulated. In this section, olefination by *gem*-dizinc species is discussed.

5.2.1
Methylenation Reactions

5.2.1.1 By Zn–CH$_2$X$_2$

The direct reduction of a dihalomethane with zinc may be considered as the easiest way to prepare a *gem*-dizinc compound, but this is also well-known as a preparation of halomethylzinc, that is, the Simmons–Smith reagent [6]. The typical procedure used to prepare the Simmons–Smith reagent involves the treatment of diiodomethane with zinc-copper couple in diethyl ether as solvent [7]. When this procedure is carried out in THF as the solvent, the *gem*-dizinc species is formed to some extent [8]. The Simmons–Smith reagent reacts with alkenes electrophilically as a carbenoid species, but does not attack carbonyl compounds nucleophilically. On the other hand, the nucleophilicity of dizinc species, formed by further reduction of the Simmons–Smith reagent, would be enhanced by the double substitution by electropositive zinc atoms. Once a dimetal species has added to a carbonyl group, the pathway that ensues may lead to Wittig-type olefination (Scheme 5.2). In 1966, Fried et al. used a *gem*-dizinc species, prepared from diiodomethane and zinc-copper couple in THF, for the methylenation of steroid derivatives (**1** → **2** in Scheme 5.4) [8, 9]. In this substrate, the hydroxyl group in the α-position of the ketone plays an important role. Thus, chelation enhances the nucleophilicity of the dizinc species [10].

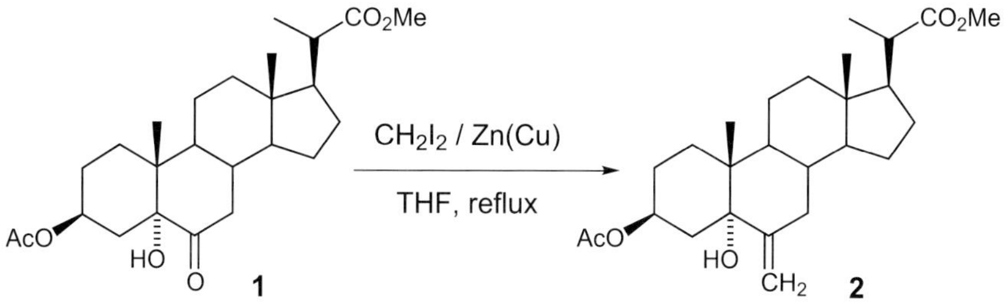

Scheme 5.4. Methylenation of an α-hydroxy ketone with CH$_2$I$_2$–Zn(Cu) in THF.

Not only zinc-copper couple, but also zinc-lead couple forms a *gem*-dimetal species starting from diiodomethane, according to Nysted's patent in 1975 [11]. He also insisted that treatment of dibromomethane with zinc-lead couple in THF at 80 °C forms a characteristic *gem*-dizinc species **3**. However, there was no direct evidence for its structure other than ^{1}H NMR data, and ^{1}H NMR spectroscopy alone was not sufficient for an unequivocal structure determination. The obtained compound was definitely a *gem*-dizinc species, but the written structure **3** (Figure 5.1) is still open to debate. The white solid **3** was obtained as a dispersion in THF, and would not dissolve in DMF or DMI. A THF dispersion is commercially available from the Aldrich Chemical Co. as "Nysted reagent". Nysted also showed that these dizinc compounds were effective for the methylenation of α-hydroxy ketone moieties in steroid derivatives.

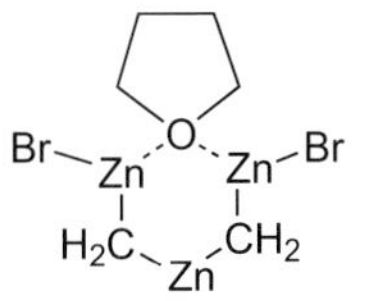

Fig. 5.1. Proposed structure of the Nysted reagent prepared from CH_2Br_2 and Zn.

5.2.1.2 By $Zn-CH_2X_2-TiCl_n$

Nozaki, Oshima, and Takai reported in 1978 that the reagent prepared from diiodomethane, zinc, and titanium(IV) chloride was effective for the methylenation of ketones [12]. In this procedure, a reagent was prepared by mixing diiodomethane (3.0 equiv.), zinc dust (9.0 equiv.), and titanium(IV) chloride (1.0 equiv.) in THF for 30 min at 25 °C. The ketone substrate (1.0 equiv.) was then added to the prepared reagent. Other metal halides were also examined (Scheme 5.5), but none of them gave better results than titanium(IV) chloride.

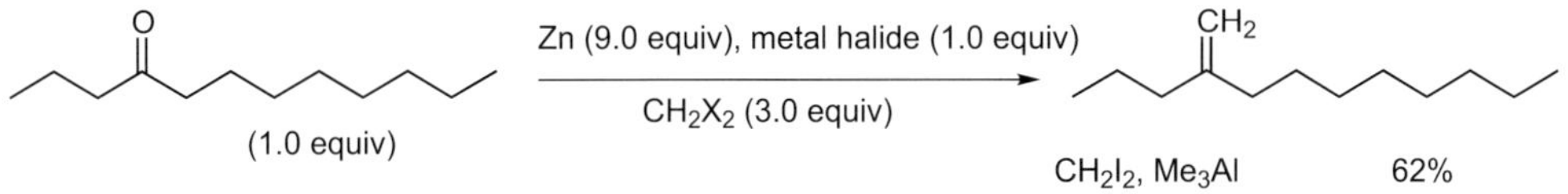

Reagents	Yield
CH_2I_2, Me_3Al	62%
CH_2I_2, $AlCl_3$	42%
CH_2I_2, VCl_4	70%
CH_2I_2, $TiCl_4$	83%
CH_2Br_2, $TiCl_4$	89%

Scheme 5.5. Methylenation of 4-dodecanone with $Zn-CH_2X_2$–metal halide systems.

The superiority of this reagent system was demonstrated by the reactions shown in Scheme 5.6. Methylenations of the highly enolizable ketones, α- and β-tetralone, were performed with the reagent system. As mentioned in the introduction, treatment of these ketones with methylenetriphenylphosphorane did not give any methylenated product [12].

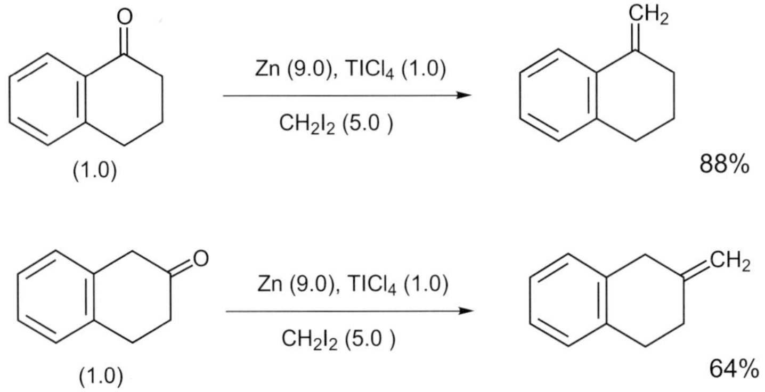

Scheme 5.6. Methylenation of highly enolizable ketones with $Zn-CH_2I_2-TiCl_4$.

The zinc dust that was used in these studies by Nozaki, Oshima, and Takai at Kyoto University was pyrometallurgy grade zinc, containing 0.04–0.07% of lead. According to Takai and Utimoto's report in 1994 [13], this lead played an important role in accelerating the further reduction of the Simmons–Smith reagent to *gem*-dizinc species. This effect was consistent with Nysted's results [11]. In other words, the ageing period required for the preparation of the dizinc reagent should be much longer when pure zinc without lead is used for this purpose.

Indeed, the reproducibility of this method was questioned by Lombardo in 1982 [14]. He attempted to apply Takai and Oshima's procedure to the methylenation of a gibberellin derivative with Zn (pure, without lead)-CH_2Br_2-$TiCl_4$, but only decomposition of the substrate was observed. Lombardo later demonstrated an improved procedure. According to his report, the requisite ageing period for the preparation of the reagent was three days (Scheme 5.7).

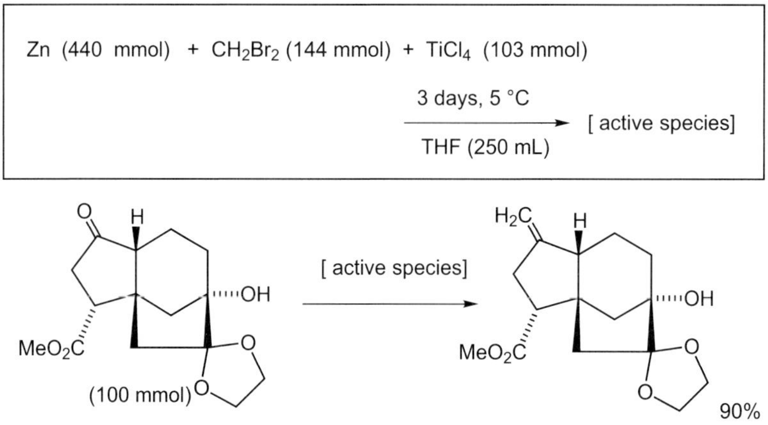

Scheme 5.7. Methylenation of a gibberellin derivative by Lombardo's method.

Methylenation of an ester group with this system did not proceed efficiently. As shown in Scheme 5.8, Takai and Utimoto used as an example the methylenation of phenyl benzoate. In this case, the addition of TMEDA was required [15]. The yield of the corresponding vinyl ether was not satisfactory. The reagent system may not have been mild enough for the acid-sensitive product. To a limited extent, it could be made to give acid-sensitive vinyl ether derivatives more efficiently by the addition of excess TMEDA to prevent the decomposition of the product (Scheme 5.9) [16].

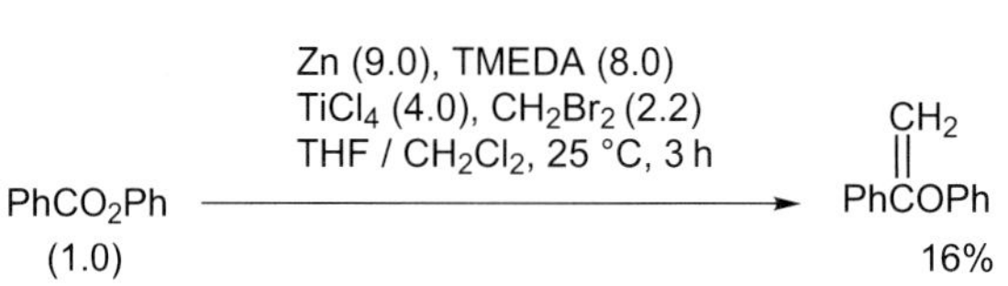

Scheme 5.8. Methylenation of an ester with Zn–CH_2I_2–$TiCl_4$–TMEDA.

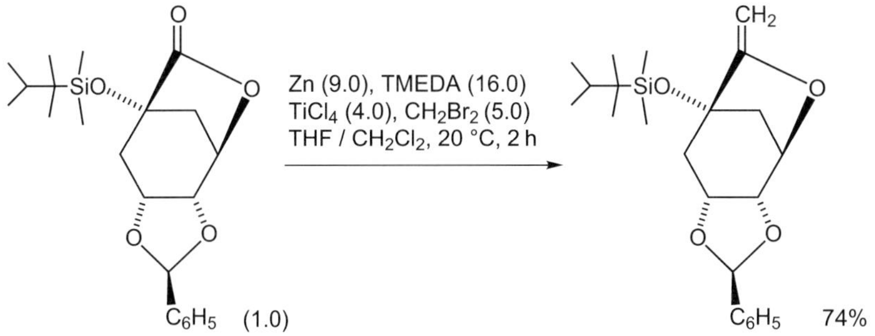

Scheme 5.9. Methylenation of a lactone with Zn–CH$_2$I$_2$–TiCl$_4$–TMEDA.

The issue of the ageing period in relation to the preparation of reactive species, as discussed by Lombardo, not only concerns the formation of the *gem*-dizinc species, although the effective role of lead is clear. It enhances the reduction rate of diiodomethane with zinc. As titanium(IV) chloride is also reduced with zinc powder [17], the titanium salt that functions as a mediator is a low-valent one. The reduction process of titanium(IV) may sometimes also lead to the problem of reproducibility. In 1998, Matsubara and Utimoto reported a general procedure for the reduction of diiodomethane with zinc powder to give a dizinc species, which was obtained as a solution in THF (Scheme 5.10) [18]. A detailed structural study of the solution by EXAFS implied that the dizinc species prepared from diiodomethane and zinc was the monomeric bis(iodozincio)methane **4** [19]. It was found that a solution of **4** in THF could be kept unchanged for at least a month in the sealed reaction vessel. Using this dizinc compound **4**, the reaction conditions as regards the titanium salts were thoroughly investigated.

Aldehydes were methylenated by bis(iodozincio)methane **4** without the mediation of a titanium salt [18]. Commercially available Nysted reagent **3** was also effective for this transformation; in this case, the addition of a catalytic amount of BF$_3$·OEt$_2$ improved the yield (Table 5.1) [20].

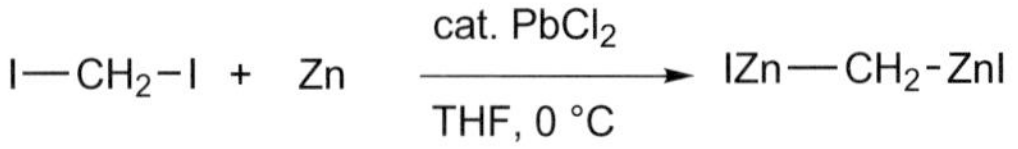

$$\text{I—CH}_2\text{—I} + \text{Zn} \xrightarrow[\text{THF, 0 °C}]{\text{cat. PbCl}_2} \text{IZn—CH}_2\text{—ZnI}$$

4 (50% yield)

Scheme 5.10. Preparation of bis(iodozincio)methane in THF.

Methylenation reactions of ketones with **4** required the addition of a titanium salt. As shown in Table 5.2, 2-dodecanone was treated with bis(iodozincio)methane **4** in the presence of various titanium salts. Titanium chloride, β-TiCl$_3$, prepared from titanium(IV) chloride and hexamethyldisilane by following Girolami's procedure [21], was shown to be the most effective. This report by Girolami corrected an earlier report by Naula and Sharma, in which it was erroneously claimed that TiCl$_2$ was formed from the reaction of titanium(IV) chloride with hexamethyldisilane

Tab. 5.1. Methylenation of aldehydes with bis(iodozincio)methane (**4**) and the Nysted reagent (**3**).[a]

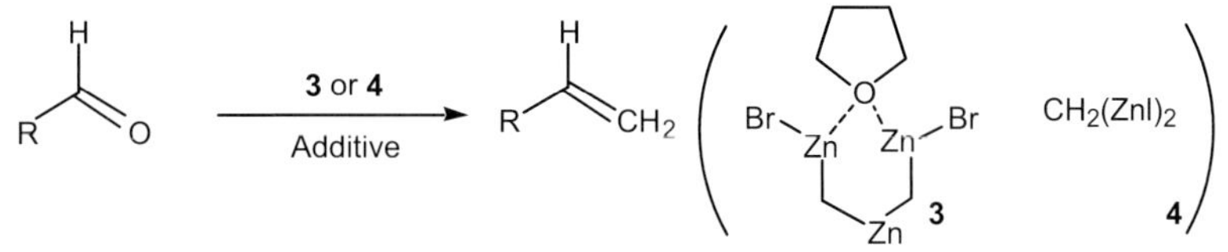

R-CHO →(3 or 4, Additive) R-CH=CH$_2$

(Br-Zn-O(CH$_2$)$_3$-Zn-Br with Zn, **3**) ; CH$_2$(ZnI)$_2$ **4**

entry	R–CHO (1.0 mmol)[b]	dizinc	Additive	Alkene (%)
1	CH$_3$(CH$_2$)$_{10}$CHO	4 (1.0 mmol)	none	74
2		4 (2.0 mmol)	none	96
3		3 (1.0 mmol)	none	68
4		3 (1.0 mmol)	BF$_3$·OEt$_2$ (0.1 mmol)	83
5	PhCH$_2$CH$_2$CHO	4 (1.0 mmol)	none	46
6	(E)-PhCH=CHCHO	4 (1.0 mmol)	none	47
7		3 (1.0 mmol)	BF$_3$·OEt$_2$ (0.1 mmol)	69
8	(S)-PhCH(CH$_3$)CHO	4 (1.0 mmol)	none	64[c]

[a] Nysted reagent was weighed according to the structure **3**, as shown in the original patent.
[b] RCHO, the dizinc compound (**3** or **4**), and additive were mixed in THF.
[c] No epimerization was observed.

(Scheme 5.11) [22]. Girolami pointed out that hexamethyldisilane cannot reduce titanium(III) chloride to titanium(II) or titanium(I) chloride regardless of the stoichiometry. Matsubara and Utimoto used the titanium chloride prepared from titanium(IV) chloride and hexamethyldisilane in their olefinations, citing Naula and Sharma's report [23]. In this series of reports, the salt was written as TiCl$_2$. It should be β-TiCl$_3$, according to Girolami's correction in 1998, although all procedures and results relating to the olefination reactions are correct [24].

TiCl$_4$ + Me$_3$SiSiMe$_3$ → TiCl$_2$ + 2 Me$_3$SiCl

Naula and Sharma (1985)

2 TiCl$_4$ + Me$_3$SiSiMe$_3$ → 2 β-TiCl$_3$ + 2 Me$_3$SiCl

β-TiCl$_3$ + Me$_3$SiSiMe$_3$ —//→ No further reduction

Hermes and Giriolami (1998)

Scheme 5.11. Reduction of TiCl$_4$ with Me$_3$Si–SiMe$_3$.

An equimolar mixture of **4** and β-TiCl$_3$ generated reactive species for the methylenation, as shown in Table 5.2. As described above, the methylenation procedure due to Takai or Lombardo employed diiodomethane, zinc, and titanium(IV) chloride. In these procedures, the titanium(IV) chloride would be reduced to titanium(0), titanium(I), titanium(II), or titanium(III). The oxidation state of tita-

Tab. 5.2. Titanium salt mediated methylenation of ketones with bis(iodozincio)methane (4).[a]

n-C$_{10}$H$_{21}$-CO-CH$_3$ (1.0 mmol) + CH$_2$(ZnI)$_2$ (**4**) / Ti salt (1.0 mmol) → n-C$_{10}$H$_{21}$-C(=CH$_2$)-CH$_3$

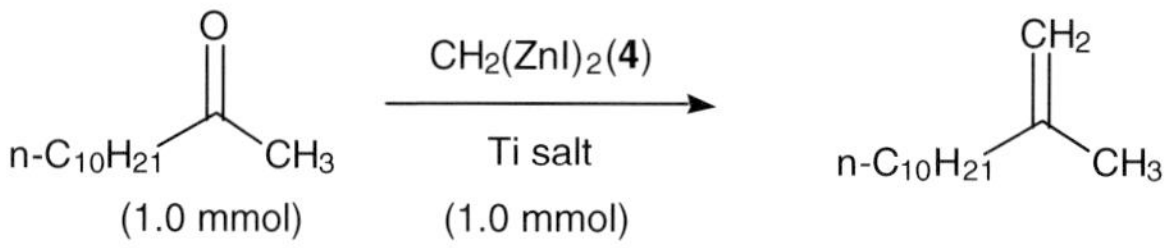

entry	Ti salt[b]	4 (mmol)	Alkene[c]	recovery (%)
1	TiCl$_4$	1.0	26%	<5
2	TiCl$_4$	2.0	78	<5
3	3TiCl$_3$·AlCl$_3$	2.0	43	33
4	α-TiCl$_3$	1.0	<5	72
5	β-TiCl$_3$	1.0	83	<5
6	β-TiCl$_3$	2.0	87	<5
7	TiCl$_2$	2.0	<5	86

[a] S. Matsubara, Y. Yokota, K. Oshima, unpublished results.
[b] 3TiCl$_3$·AlCl$_3$ (Aldrich), α-TiCl$_3$ (Aldrich), and β-TiCl$_3$.
[c] Isolated yields.

nium is also crucial for the reproducibility of the reaction. As higher valent titanium is a better Lewis acid, the use of the dizinc reagent itself and β-TiCl$_3$ from the beginning is desirable. TiCl$_4$ is also a possible mediator with two equivalents of **4**, as the dizinc reagent also reduces TiCl$_4$ to TiCl$_3$.

In Schemes 5.12 and 5.13, the results obtained using a combination of the gem-

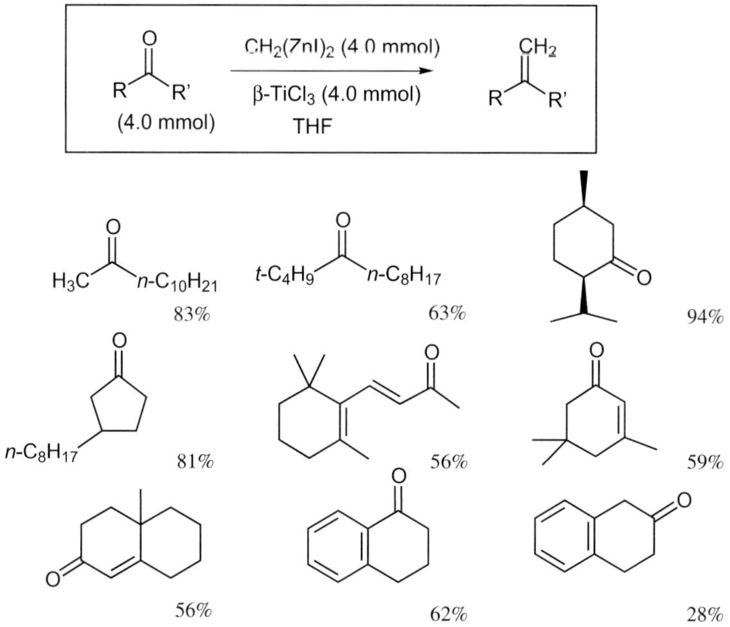

R-CO-R' (4.0 mmol) + CH$_2$(ZnI)$_2$ (4.0 mmol) / β-TiCl$_3$ (4.0 mmol) / THF → R-C(=CH$_2$)-R'

H$_3$C-CO-n-C$_{10}$H$_{21}$ 83%

t-C$_4$H$_9$-CO-n-C$_8$H$_{17}$ 63%

94%

n-C$_8$H$_{17}$ 81%

56%

59%

56%

62%

28%

Scheme 5.12. Methylenation of ketones with bis(iodozincio)methane **4** and β-TiCl$_3$.

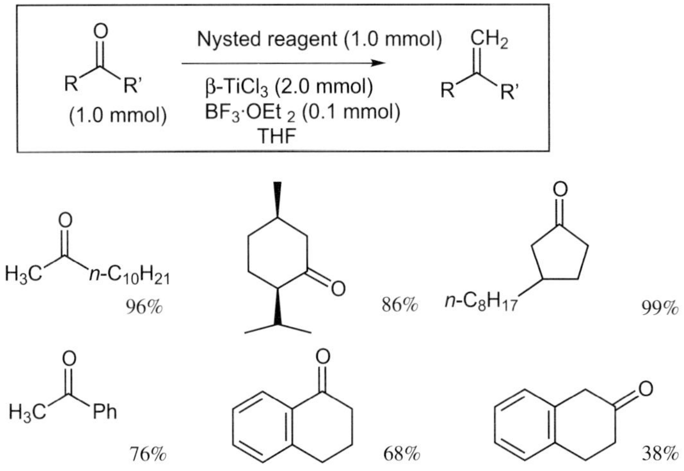

Scheme 5.13. Methylenation of ketones with the Nysted reagent **3** and β-TiCl₃.

dizinc reagent and β-TiCl₃ are shown [18, 20]. As shown in Scheme 5.13, the Nysted reagent could also be applied here. It is best used in the presence of BF₃·OEt₂. As shown in Scheme 5.14, methylenation of a polyketone was performed using bis(iodozincio)methane **4** and β-TiCl₃. With this substrate, neither the Wittig reagent, the Tebbe reagent, nor Zn–CH₂X₂–TiCl₄ gave satisfactory results. A combination of **4** and β-TiCl₃ gave the fully methylenated product without racemization (Scheme 5.14) [25].

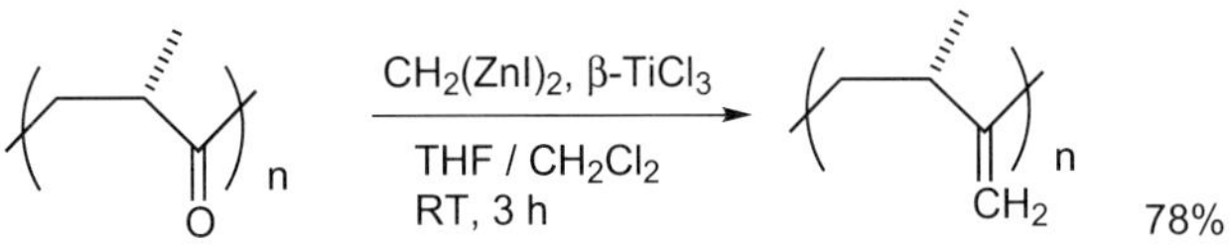

Scheme 5.14. Methylenation of optically active polyketones with bis(iodozincio)methane **4** and β-TiCl₃.

A reagent consisting of $CH_2(ZnI)_2$ (**4**), β-TiCl₃, and TMEDA was also examined for the methylenation of esters, and the results are shown in Table 5.3 [26].

5.2.2
Alkylidenation Reactions

5.2.2.1 From *gem*-Dihaloalkanes

Following the same strategy as used for methylenation, general alkylidenation of carbonyl compounds would be realized by preparation of the corresponding *gem*-dizinc species. The preparation of these *gem*-dizinc species, however, is con-

Tab. 5.3. Methylenation of esters with bis(iodozincio)methane (4) and β-TiCl₃.

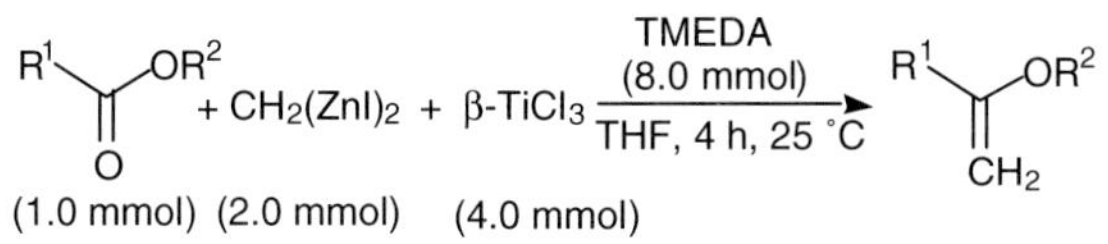

$$\underset{(1.0\text{ mmol})}{R^1\!-\!\underset{O}{\overset{\|}{C}}\!-\!OR^2} + \underset{(2.0\text{ mmol})}{CH_2(ZnI)_2} + \underset{(4.0\text{ mmol})}{\beta\text{-TiCl}_3} \xrightarrow[\text{THF, 4 h, 25 °C}]{\text{TMEDA (8.0 mmol)}} R^1\!-\!\underset{CH_2}{\overset{\|}{C}}\!-\!OR^2$$

entry	R¹	R²	β-TiCl₃	Vinyl ether[a] (%)
1	n-C₇H₁₅	Et	4.0	75
2	n-C₉H₁₉	ⁱPr	4.0	90
3	n-C₇H₁₅	ᵗBu	4.0	46
4			4.0	34[b]
5	Ph	Me	4.0	89
6	Ph	Et	4.0	51

[a] Isolated yields.
[b] The mixture was stirred for 24 h.

siderably more troublesome than the preparation of bis(iodozincio)methane. The simple reduction of a *gem*-dihaloalkane bearing a β-hydrogen will be hampered by β-elimination and thus lead to the formation of the elimination product. Moreover, depending on the substrate, the intermediary α-haloalkylzinc compound will be less stable than the α-halomethylzinc with respect to α-elimination (Scheme 5.15).

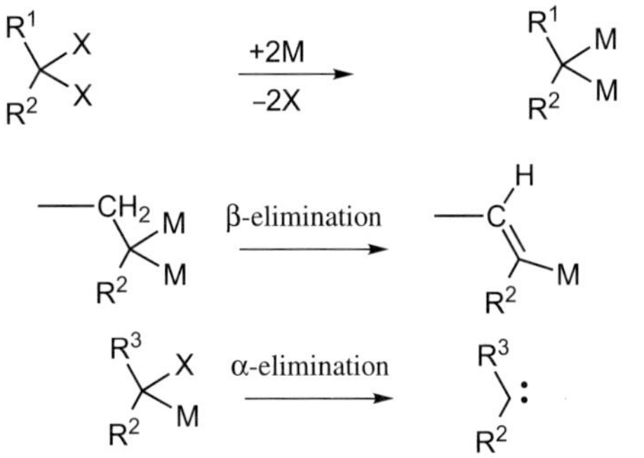

Scheme 5.15. Reduction of a *gem*-dihaloalkane and possible decomposition routes of the product.

These side reactions, however, can be suppressed by the addition of TMEDA. As shown in Scheme 5.16, several types of *gem*-dizinc compounds have been prepared. In the reaction of 1,1-diiodoethane, the addition of TMEDA proved unnecessary [18]. These dizinc species are not as stable as bis(iodozincio)methane **4**. They can be stored under Ar at 25 °C for just a few hours, and gradually decompose through β-elimination.

Using these dizinc species in conjunction with β-TiCl₃, the alkylidenation of

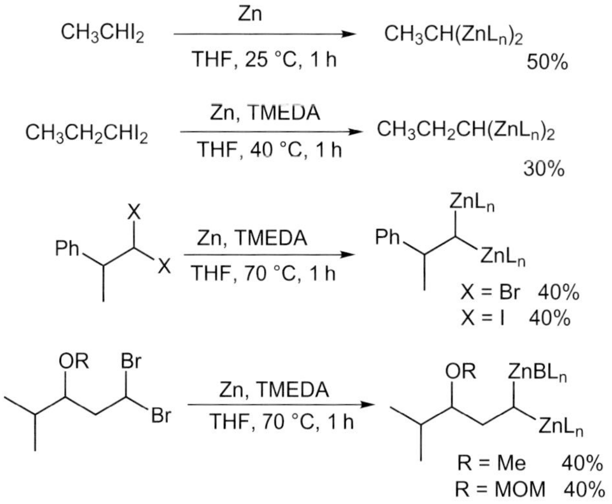

5 Olefination of Carbonyl Compounds by Zinc and Chromium Reagents

$$CH_3CHI_2 \xrightarrow[\text{THF, 25 °C, 1 h}]{\text{Zn}} CH_3CH(ZnL_n)_2 \quad 50\%$$

$$CH_3CH_2CHI_2 \xrightarrow[\text{THF, 40 °C, 1 h}]{\text{Zn, TMEDA}} CH_3CH_2CH(ZnL_n)_2 \quad 30\%$$

Ph–CH(X)–CH(CH$_3$)–X $\xrightarrow[\text{THF, 70 °C, 1 h}]{\text{Zn, TMEDA}}$ Ph–C(ZnL$_n$)=C(CH$_3$)(ZnL$_n$)

X = Br 40%
X = I   40%

(iPr)–CH(OR)–CH(Br)–CH$_2$–Br $\xrightarrow[\text{THF, 70 °C, 1 h}]{\text{Zn, TMEDA}}$ (iPr)–CH(OR)–C(ZnBL$_n$)–CH$_2$–ZnL$_n$

R = Me   40%
R = MOM 40%

Zn in these equations denotes pyrometallurgy grade zinc, containing 0.04–0.07% lead. When pure, lead-free zinc is used, a catalytic amount of PbCl$_2$ should be added.

Scheme 5.16. Preparation of dizinc species from *gem*-dihalo compounds.

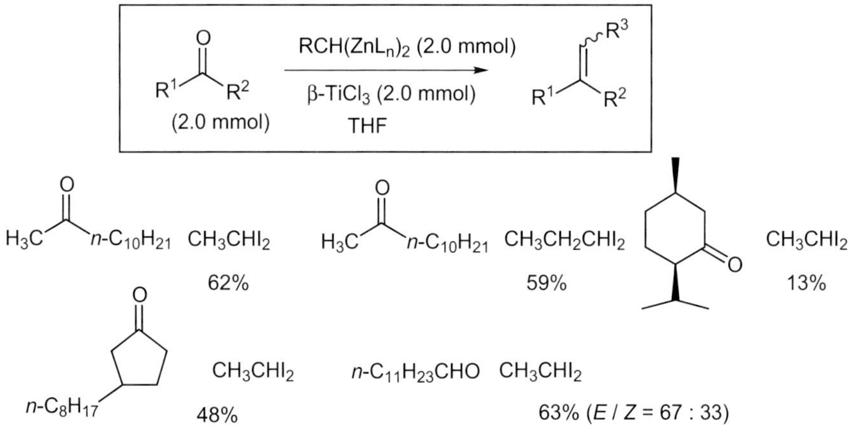

$$R^1\text{–CO–}R^2 \xrightarrow[\beta\text{-TiCl}_3 \text{ (2.0 mmol)}]{\text{RCH(ZnL}_n)_2 \text{ (2.0 mmol)}, \text{THF}} R^1R^2C=CHR^3$$
(2.0 mmol)

| H$_3$C–CO–n-C$_{10}$H$_{21}$ | CH$_3$CHI$_2$ | H$_3$C–CO–n-C$_{10}$H$_{21}$ | CH$_3$CH$_2$CHI$_2$ | (4-isopropylcyclohexanone) | CH$_3$CHI$_2$ |
| 62% | | 59% | | 13% | |

(2-n-octylcyclopentanone) CH$_3$CHI$_2$ n-C$_{11}$H$_{23}$CHO   CH$_3$CHI$_2$
n-C$_8$H$_{17}$ 48%   63% (E / Z = 67 : 33)

Scheme 5.17. Alkylidenation of carbonyl compounds with *gem*-dizinc reagents and β-TiCl$_3$.

carbonyl compounds was examined. The diastereoselectivities (E, Z) were not excellent.

On the other hand, it is not necessary to prepare such unstable *gem*-dihaloalkanes in advance. Esters can be alkylidenated by treatment with Zn–CH$_3$CHBr$_2$–TiCl$_4$–TMEDA (Scheme 5.18) [15]. This reaction requires a large excess of the reagent. It should be noted that the zinc powder used in these procedures was also pyrometallurgy grade zinc, containing about 0.04–0.07% of lead.

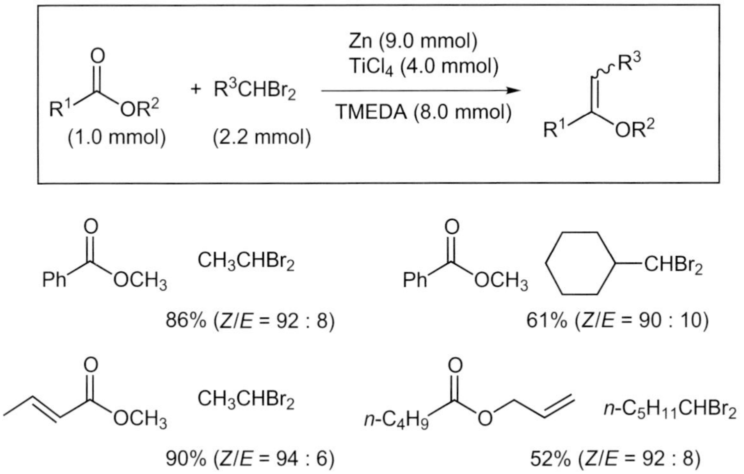

Scheme 5.18. Alkylidenation of esters with Zn–RCHBr$_2$–TiCl$_4$–TMEDA.

Diastereoselectivity was observed in this reaction, the Z-isomers being the major products.

This Zn–CH$_3$CHBr$_2$–TiCl$_4$–TMEDA reagent can also be used in the ethylidenation of thioesters and amides, as shown in Scheme 5.19 [27].

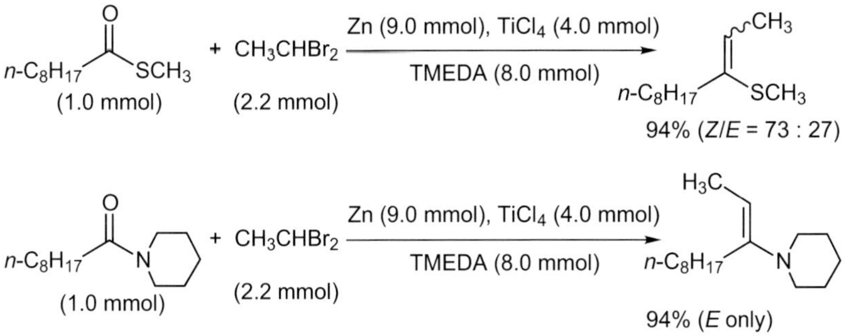

Scheme 5.19. Ethylidenation of a thioester and an amide with Zn–CH$_3$CHBr$_2$–TiCl$_4$–TMEDA.

5.2.2.2 Via Carbometallation

gem-Dimetal compounds can be prepared via an alternative route. Knochel and Normant reported the preparation of *gem*-dimetals by carbometallation (Scheme 5.20). Treatment of an alkenylmagnesium halide with an allylzinc halide afforded the *gem*-dimetal compound in good yield [28].

Treatment of aldehydes with these *gem*-dimetal compounds led to alkylidenation. In these cases, the addition of BF$_3$·OEt$_2$ as a Lewis acid was required to activate the aldehyde (Scheme 5.21).

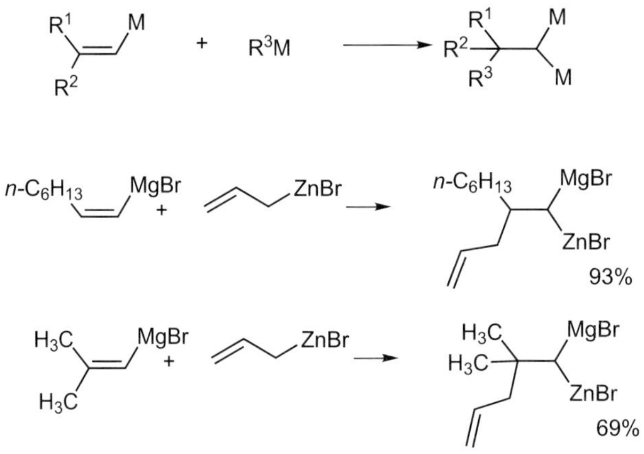

Scheme 5.20. Preparation of *gem*-dimetal compounds by carbometallation.

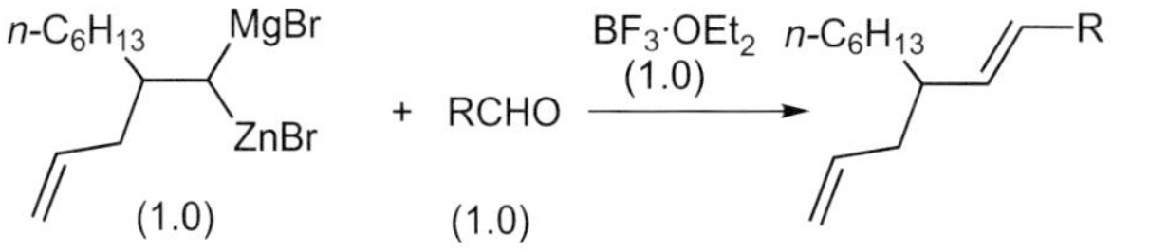

Scheme 5.21. Alkylidenation of aldehydes with *gem*-dimetal reagents prepared by carbometallation in the presence of BF$_3$·OEt$_2$.

PhCHO 78% (E / Z = 99 : 1)
CH$_3$CH$_2$CHO 61% (E / Z = 95 : 5)
t-C$_4$H$_9$CHO 58% (E / Z = 92 : 8)

Nakamura et al. reported that zincated hydrazones react with vinyl Grignard reagent to afford *gem*-dimetal compounds. These were treated with benzaldehyde to give alkenyl hydrazones (Scheme 5.22) [29].

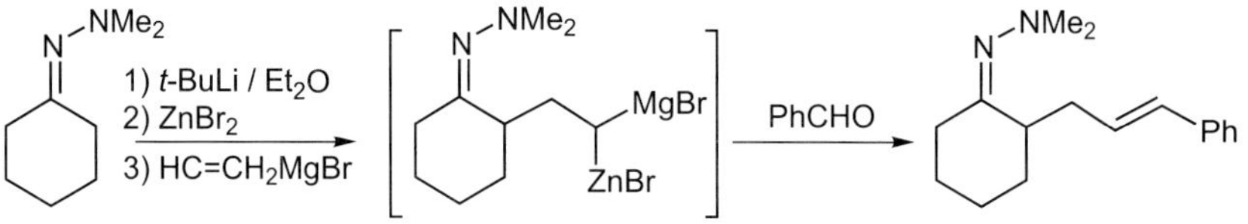

76% (E/Z = 82 : 18)

Scheme 5.22. Alkene synthesis using a hydrazone-derived *gem*-dimetal compound.

5.2.3
Alkenylsilane, -germane, and -borane Synthesis

Treatment of trimethylsilyldibromomethane with zinc in THF affords the corresponding dizinc compound in good yield [30]. Again, the zinc dust used for these preparations was pyrometallurgy grade zinc, containing about 0.04–0.07% of lead.

When pure, lead-free zinc is used, a catalytic amount of PbCl$_2$ should be added. The silyl group may promote reduction of the C–Br bond at the α-position, as it stabilizes a radical species at this α-position. The obtained *gem*-dizinc species is fairly stable as it has no β-hydrogen. In the same way, *gem*-dizinc species bearing germyl and boryl groups at the α-position have also been prepared in good yields [31, 32].

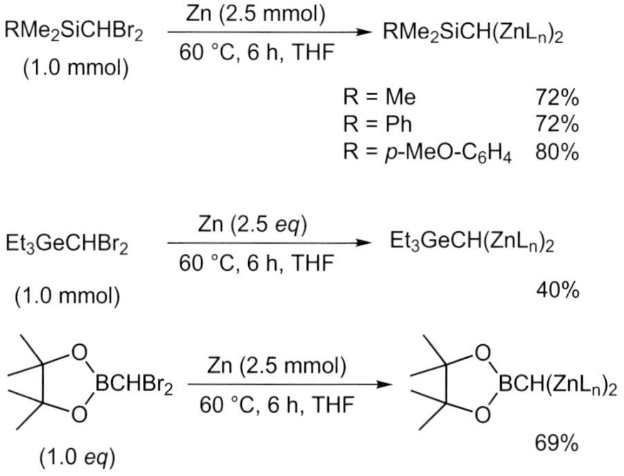

$$RMe_2SiCHBr_2 \xrightarrow[60\,°C,\,6\,h,\,THF]{Zn\,(2.5\,mmol)} RMe_2SiCH(ZnL_n)_2$$
(1.0 mmol)

R = Me   72%
R = Ph   72%
R = p-MeO-C$_6$H$_4$ 80%

$$Et_3GeCHBr_2 \xrightarrow[60\,°C,\,6\,h,\,THF]{Zn\,(2.5\,eq)} Et_3GeCH(ZnL_n)_2$$
(1.0 mmol)   40%

$$\text{(pinacolboryl)}CHBr_2 \xrightarrow[60\,°C,\,6\,h,\,THF]{Zn\,(2.5\,mmol)} \text{(pinacolboryl)}CH(ZnL_n)_2$$
(1.0 eq)   69%

Zn in these equations denotes pyrometallurgy grade zinc, containing 0.04–0.07% lead. When pure, lead-free zinc is used, a catalytic amount of PbCl$_2$ should be added.

Scheme 5.23. Preparation of silyl-, germyl-, and boryl-substituted *gem*-dizinc compounds.

Reactions of carbonyl compounds with these *gem*-dizinc species gave the corresponding heteroatom-substituted alkenes. In these reactions, the addition of a titanium salt was necessary. β-TiCl$_3$ was used, except in the case of reaction of α-boryl-substituted *gem*-dizinc, where TiCl$_4$ was used instead (Table 5.4).

Esters were also alkylidenated by treatment with Zn (pyrometallurgy grade zinc, containing lead)-Me$_3$SiCHBr$_2$–TiCl$_4$–TMEDA (Scheme 5.24) [33]. This reagent was found to convert thioesters into the vinyl sulfides.

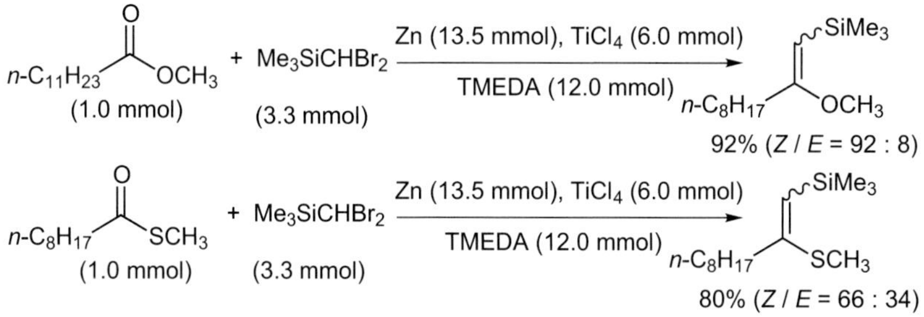

$$n\text{-}C_{11}H_{23}C(O)OCH_3 + Me_3SiCHBr_2 \xrightarrow[TMEDA\,(12.0\,mmol)]{Zn\,(13.5\,mmol),\,TiCl_4\,(6.0\,mmol)} n\text{-}C_8H_{17}C(SiMe_3)=CHOCH_3$$
(1.0 mmol)   (3.3 mmol)   92% (Z / E = 92 : 8)

$$n\text{-}C_8H_{17}C(O)SCH_3 + Me_3SiCHBr_2 \xrightarrow[TMEDA\,(12.0\,mmol)]{Zn\,(13.5\,mmol),\,TiCl_4\,(6.0\,mmol)} n\text{-}C_8H_{17}C(SiMe_3)=CHSCH_3$$
(1.0 mmol)   (3.3 mmol)   80% (Z / E = 66 : 34)

Scheme 5.24. Vinyl ether and sulfide preparation using Zn–Me$_3$SiCHBr$_2$–TiCl$_4$–TMEDA.

Tab. 5.4. Reactions of carbonyl compounds with α-heteroatom-substituted dizinc species.[a]

$$R^1CH(ZnL_n)_2 \; (1.0 \text{ mmol}) + R^2R^3C=O \; (1.0 \text{ mmol}) \xrightarrow[25\,°C]{\text{Titanium salt (1.0 mmol)}} \underset{H}{\overset{R^1}{>}}=\underset{R^3}{\overset{R^2}{<}}$$

entry	R^1	R^2	R^3	Titanium salt	Yield (%)[b]	E/Z
1	Me$_3$Si	n-C$_{11}$H$_{23}$	H	β-TiCl$_3$	92	89:11
2		PhCH$_2$CH$_2$	H	β-TiCl$_3$	78	90:10
3		(E)-PhCH=CH	H	β-TiCl$_3$	55	68:32
4		Ph	CH$_3$	β-TiCl$_3$	54	61:39
5	PhMe$_2$Si	n-C$_{11}$H$_{23}$	H	β-TiCl$_3$	86	87:13
6		PhCH$_2$CH$_2$	H	TiCl$_4$	70	74:26
7	pinacolboryl	n-C$_{11}$H$_{23}$	H	TiCl$_4$	58	74:26
8		n-C$_{10}$H$_{21}$	CH$_3$	TiCl$_4$	71	59:41
9	Et$_3$Ge	PhCH$_2$CH$_2$	H	β-TiCl$_3$	79	74:26
10		Ph	CH$_3$	β-TiCl$_3$	63	65:35

[a] These dizinc species were prepared as shown in Scheme 5.20.
[b] Isolated yields.

5.3
Chromium Compounds

Although the strong electron-donating ability of chromium(II) chloride had long been known, its practical use in organic synthesis was only initiated by Hiyama and Nozaki in 1976. They used anhydrous chromium(II) chloride in the reduction of allylic halides to obtain allylic chromium reagents [34]. Since then, useful C–C bond-forming reactions between organic halides and carbonyl compounds that are mediated by a chromium(II) salt have been developed. The most important features of these reactions are chemoselectivity and stereoselectivity. In these transformations, treatment of the organic halides with a chromium(II) salt was considered to afford the intermediary organochromium compounds, although these compounds have not been isolated. In analogy to the reductions described in the previous section, reduction of *gem*-dihalides with a chromium(II) salt as the reductant may give *gem*-dichromium species. When carbonyl compounds are exposed to these species, they should be transformed into olefins.

5.3.1
Alkylidenation

Takai and Utimoto showed that reactions of aldehydes and a *gem*-diiodoalkane with chromium(II) chloride gave Wittig-type olefination products (Scheme 5.25) [35]. The notable points of this transformation are stereoselective (E)-alkene formation and chemoselective reaction with the aldehyde. Ketones are recovered unchanged. Instead of a *gem*-diiodoalkane, α-acetoxy bromide can also be used for this transformation [36].

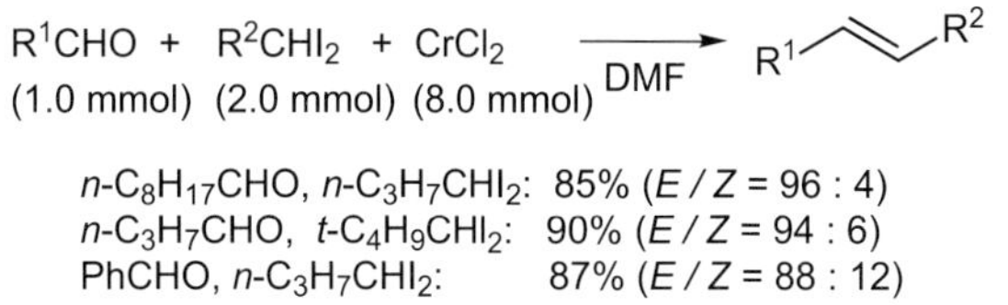

$R^1CHO + R^2CHI_2 + CrCl_2 \xrightarrow{DMF} R^1\diagup\diagdown R^2$
(1.0 mmol) (2.0 mmol) (8.0 mmol)

$n\text{-}C_8H_{17}CHO, n\text{-}C_3H_7CHI_2$: 85% (E/Z = 96 : 4)
$n\text{-}C_3H_7CHO, t\text{-}C_4H_9CHI_2$: 90% (E/Z = 94 : 6)
$PhCHO, n\text{-}C_3H_7CHI_2$: 87% (E/Z = 88 : 12)

Scheme 5.25. Alkylidenation of aldehydes with $RCHI_2$–$CrCl_2$.

The reagent prepared from a *gem*-dibromoalkane, samarium metal, and samarium(II) iodide in the presence of a catalytic amount of chromium(III) chloride was found to transform ketones into alkenes via a Wittig-type reaction. This method allows the alkylidenation of easily enolizable ketones, such as β-tetralone (Scheme 5.26) [37].

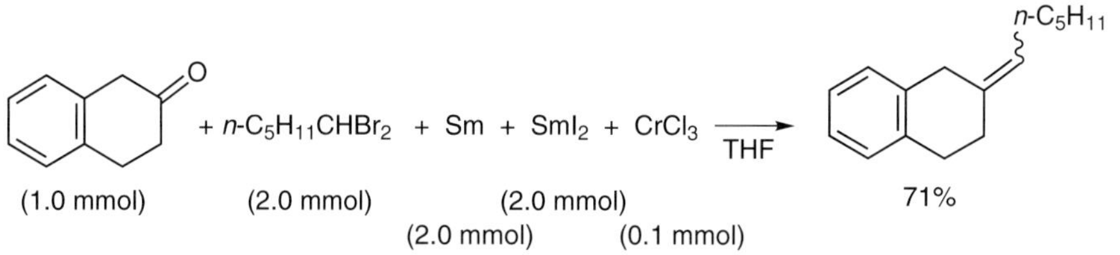

β-tetralone + $n\text{-}C_5H_{11}CHBr_2$ + Sm + SmI_2 + $CrCl_3$ $\xrightarrow{THF}$ alkene product with $n\text{-}C_5H_{11}$

(1.0 mmol)   (2.0 mmol)   (2.0 mmol)   (2.0 mmol)   (0.1 mmol)   71%

Scheme 5.26. Alkylidenation of β-tetralone with $RCHBr_2$–Sm–Cr.

5.3.2
Preparation of Alkenylboranes, -silanes, and -stannanes with E-Configuration

Chromium(II) chloride also mediates Wittig-type reactions of α-heteroatom-substituted *gem*-dihalides and aldehydes. In Scheme 5.27, representative examples of the preparation of alkenylboranes [38], -silanes [39], and -stannanes [40] are shown. In each case, high E-selectivity was observed. As these compounds are very important substrates for Suzuki, Hiyama, and Stille couplings, their stereoselective formation enhances the value of the chromium(II) chloride mediated reactions.

In the chromium(II) chloride mediated reaction, the reaction mixture is not basic, and nucleophilic attack occurs only on aldehydes. This mild and selective feature can be exploited in the reactions of polyfunctionalized substrates. Hodgson described an example involving transformation of an optically active aldehyde into an (E)-alkenylstannane without epimerization (Scheme 5.28) [40]. Aldehydes can also be transformed into 1,1-disilylalkenes by $CrCl_2$–$(Me_3Si)_2CBr_2$ [41].

5.3.3
Preparation of (E)-Haloalkenes

Treatment of an aldehyde with chromium(II) chloride and a haloform affords the (E)-haloalkene with high diastereoselectivity (Scheme 5.29) [42]. Here again, aldehydes are transformed selectively; ketones are recovered unchanged.

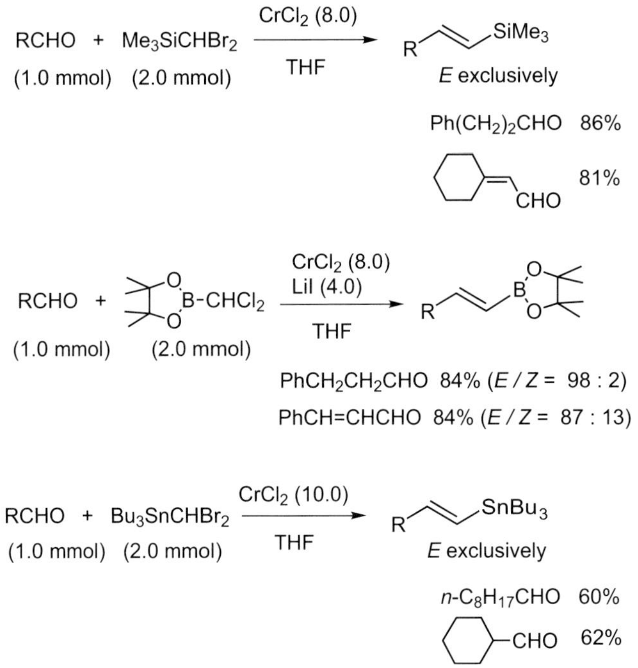

5 Olefination of Carbonyl Compounds by Zinc and Chromium Reagents

RCHO + Me$_3$SiCHBr$_2$ $\xrightarrow[\text{THF}]{\text{CrCl}_2 \text{ (8.0)}}$ R–CH=CH–SiMe$_3$

(1.0 mmol) (2.0 mmol) E exclusively

Ph(CH$_2$)$_2$CHO 86%

cyclohexyl-CHO 81%

RCHO + pinacolborane-CHCl$_2$ $\xrightarrow[\text{THF}]{\text{CrCl}_2 \text{ (8.0)}, \text{LiI (4.0)}}$ R–CH=CH–B(pin)

(1.0 mmol) (2.0 mmol)

PhCH$_2$CH$_2$CHO 84% (E/Z = 98 : 2)
PhCH=CHCHO 84% (E/Z = 87 : 13)

RCHO + Bu$_3$SnCHBr$_2$ $\xrightarrow[\text{THF}]{\text{CrCl}_2 \text{ (10.0)}}$ R–CH=CH–SnBu$_3$

(1.0 mmol) (2.0 mmol) E exclusively

n-C$_8$H$_{17}$CHO 60%

cyclohexyl-CHO 62%

Scheme 5.27. Preparation of (E)-alkenylsilanes, -stannanes, and -boranes.

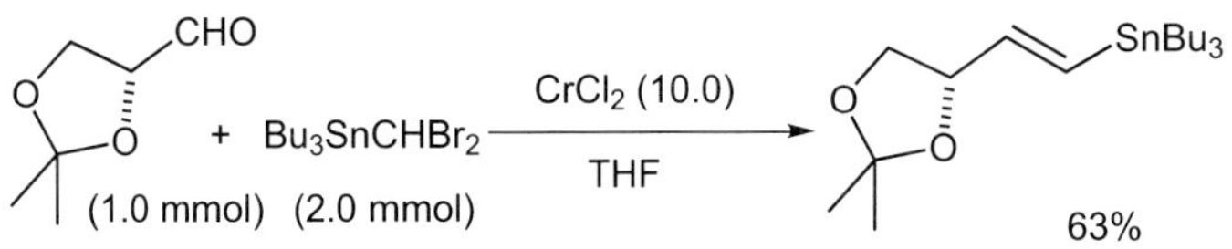

chiral acetonide aldehyde + Bu$_3$SnCHBr$_2$ $\xrightarrow[\text{THF}]{\text{CrCl}_2 \text{ (10.0)}}$ (E)-vinylstannane product

(1.0 mmol) (2.0 mmol) 63%

Scheme 5.28. Chromium(II) chloride mediated reaction of α-tributylstannyldibromomethane with a chiral aldehyde.

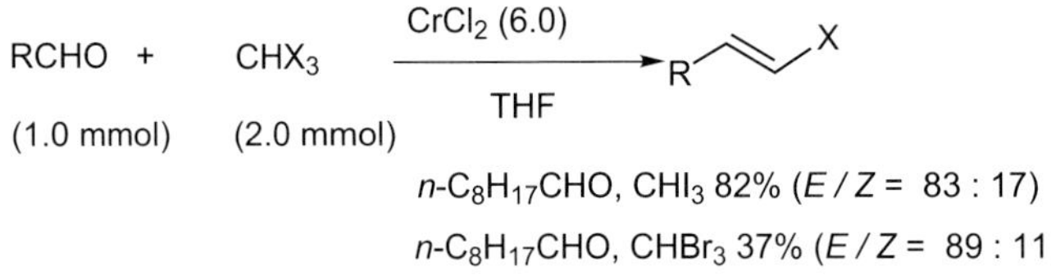

RCHO + CHX$_3$ $\xrightarrow[\text{THF}]{\text{CrCl}_2 \text{ (6.0)}}$ R–CH=CH–X

(1.0 mmol) (2.0 mmol)

n-C$_8$H$_{17}$CHO, CHI$_3$ 82% (E/Z = 83 : 17)
n-C$_8$H$_{17}$CHO, CHBr$_3$ 37% (E/Z = 89 : 11)

Scheme 5.29. (E)-Haloalkene preparation from aldehydes and CHX$_3$–CrCl$_2$.

The chemoselectivity and diastereoselectivity of this method are remarkable, so many natural product syntheses make use of this transformation. The diastereoselectivity obtained in the original report (Scheme 5.29) was not satisfactory, and so the reaction conditions had to be optimized. The optimized conditions were applied to the total synthesis of (+)-Lepicidin A by Evans, as outlined in Scheme 5.30.

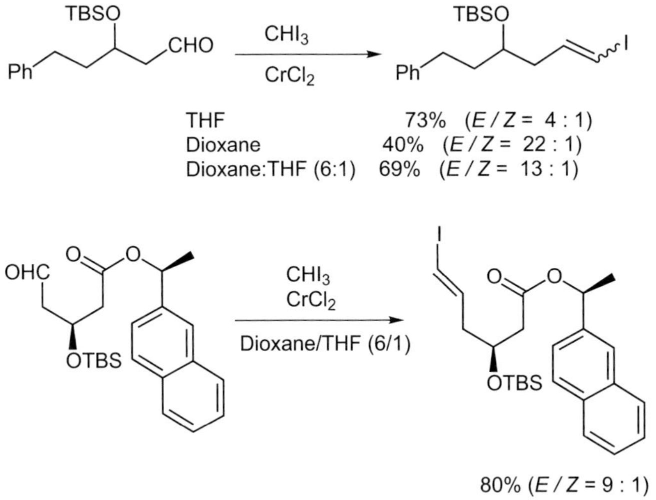

Scheme 5.30. Optimization of the solvent for (E)-iodoalkene synthesis.

Thus, a 6:1 mixture of dioxane and THF gave the best diastereoselectivity and a reasonable yield [43]. Kende also reported that α-alkoxy aldehydes were converted into iodoalkenes with extraordinary E-selectivity [44] (Scheme 5.31).

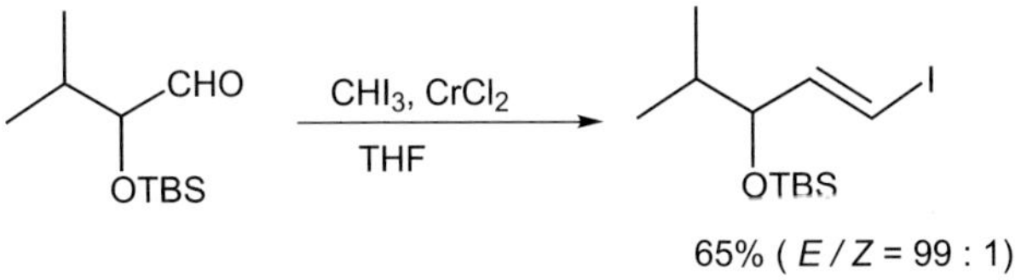

Scheme 5.31. Iodoalkene preparation from an α-alkoxy aldehyde.

More than fifteen years have elapsed since this method was developed, but it has not lost any utility in natural product synthesis. It remains one of the most utilized haloalkene preparation methods. Even in cases where a substrate possesses a lot of sensitive functional groups and chiral centers, selective transformation to the (E)-haloalkene can be accomplished with this reagent.

5.4 Applications in Natural Product Synthesis

The alkylidenation methods described in this chapter have been applied in numerous natural product syntheses. In these processes, the method should selectively target the carbonyl group without affecting other groups. Some recent examples are summarized below.

5.4.1
Zn–CH$_2$X$_2$–TiCl$_4$

Zn–CH$_2$Br$_2$–TiCl$_4$
6-*epi*-sarsolilide A
J. Zang, X. Xu, *Tetrahedron Lett.* **2000**, *41*, 941

Zn–CH$_2$Br$_2$–TiCl$_4$
88%
No racemization

(−)-necrodol
J. M. Galano, G. Audran, L. Mikolajezyk, H. Monti, *J. Org. Chem.* **2001**, *66*, 323

Zn–CH$_2$Br$_2$–TiCl$_4$
90%

amphidinolide A
G. J. Hollingworth, G. Pattenden, *Tetrahedron Lett.* **1998**, *39*, 703.

5.4.2
CHX$_3$–CrCl$_2$

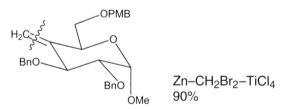

CrCl$_2$, dioxane, THF, 88% *E* / *Z*=11:1

dermostatin A
C. J. Sinz, S. D. Rychnovsky, *Tetrahedron* **2002**, *58*, 6561

muricatetrocin C
D. J. Dixon, S. V. Ley, D. J. Reynolds, *Chem. Eur. J.* **2002**, *8*, 1621

CrCl$_2$, CHI$_3$, THF, >80% E / Z = 4:1

(−)-hennoxazole A
F. Yokokawa, T. Asano, T. Shioiri, *Tetrahedron* **2001**, *57*, 6311

CrCl$_2$, CHI$_3$, THF >68%

apicularen A
K. C. Nicolaou, D. W. Kim, R. Baati, *Angew. Chem. Int. Ed.* **2002**, *41*, 3701.

CrCl$_2$, CHI$_3$, THF 91% (E / Z = 9:1)

sphingofungins E and F
B. M. Trost, C. B. Lee, *J. Am. Chem. Soc.* **2001**, *49*, 12191

CrCl$_2$, CHI$_3$ >68%

callipeltoside A
B. M. Trost, J. L. Gunzner, O. Dirat, Y. H. Rhee, *J. Am. Chem. Soc.* **2002**, *124*, 10396

CHI$_3$, CrCl$_2$
Dioxane/THF (6:1)

(+)-ambruticin
P. Liu, E. N. Jacobsen, *J. Am. Chem. Soc.* **2001**, *43*, 10772

CrCl$_2$, CHI$_3$, THF
>72%

(+)-crocacin C
L. C. Dias. L. G. de Oliveira, *Org. Lett.* **2001**, *3*, 3951

CrCl$_2$, CHI$_3$, THF
>67% (*E* >95%)

himandravine
S. Chackalamannil, R. Davies, A. T. McPhail, *Org. Lett.* **2001**, *3*, 1427

CrCl$_2$, CHI$_3$, THF 63%

(*S*)-9-methylgermacrene-B
S. Kurosawa, K. Mori, *Eur. J. Org. Chem.* **2000**, 955

CBr$_2$Me$_2$, Sm, SmI$_2$, CrCl$_3$, 60%

5.5
Conclusion

The *gem*-dizinc species described in this chapter can be handled without any special techniques. It is nearly 40 years since the *gem*-dizinc reagent was first described as a methylenation reagent, but the classical Wittig reaction is still the predominantly used method for olefination. In some cases, however, the Wittig reagent cannot achieve the desired goal. The ylides are stabilized anions, and so suffer from a lack of nucleophilicity. On the contrary, the dizinc reagents are enhanced anion equivalents (i.e. dianion equivalents). They possess high nucleophilicity and show some chemoselectivity through their coordination to substrates, but their diastereoselectivity in the alkylidenation reactions of aldehydes (i.e. E/Z selectivity) is unsatisfactory compare to that of modified Wittig reactions. As far as the methylenation reaction is concerned, their potential exceeds that of ylides.

The use of a combination of chromium(II) chloride and a *gem*-dihalo compound is not such a readily accessible reaction considering the availability of chromium(II) chloride, but shows high performance in reactions with aldehydes. High chemoselectivity and diastereoselectivity (E-selectivity) are observed. In particular, the combination of chromium(II) chloride and a haloform constitutes a very important method for the preparation of (E)-haloalkenes.

References

1. A. MAERCKER, *Org. React.* 1965, *14*, 270; B. E. MARYANOFF, A. B. REITS, *Chem. Rev.* 1989, *89*, 863.
2. M. SCHLOSSER, *Top. Stereochem.* 1970, *5*, 1; B. SCHAUB, T. JENNY, M. SCHLOSSER, *Tetrahedron Lett.* 1984, *25*, 4097.
3. I. MAREK, J.-F. NORMANT, *Chem. Rev.* 1996, *96*, 3241; S. MATSUBARA, K. OSHIMA, K. UTIMOTO, *J. Organomet. Chem.* 2001, *617–618*, 39; S. MATSUBARA, *J. Org. Synth. Soc. Jpn.* 2000, *58*, 118.
4. R. R. SCHROCK, R. T. DEPUE, J. FELDMAN, K. B. YAP, D. C. YANG, W. M. DAVIS, L. PARK, M. DIMARE, M. SCHOFIELD, J. ANHAUS, E. WALBORSKY, E. EVITT, C. KRUGER, P. BETZ, *Organometallics* 1990, *9*, 2262.
5. F. BERTTINI, P. GASSELLI, G. ZUBIANI, G. CAINELLI, *Tetrahedron* 1970, *26*, 1281; B. J. J. VAN DE HEISTEEG, M. A. SCHAT, G. TINGA, O. S. AKKERMAN, F. BICKELHAUPT, *Tetrahedron Lett.* 1986, *27*, 6123.
6. H. E. SIMMONS, R. D. SMITH, *J. Am. Chem. Soc.* 1958, *80*, 5323.
7. A. B. CHARETTE, J. F. MARCOUX, *Synlett*, 1995, 1197.
8. P. TURNBELL, K. SYORO, J. H. FRIED, *J. Am. Chem. Soc.* 1966, *88*, 4764.
9. I. T. HARRISON, R. J. RAWSON, P. TURNBULL, J. H. FRIED, *J. Org. Chem.* 1971, *36*, 3515.
10. X. CHEN, E. R. HORTELANO, E. L. ELIEL, S. V. FRYE, *J. Am. Chem. Soc.* 1992, *114*, 1778; K. UTIMOTO, A. NAKAMURA, S. MATSUBARA, *J. Am. Chem. Soc.* 1990, *112*, 8189.
11. The Nysted reagent (L. N. NYSTED, US Patent No. 3, 865 848 (1975); *Chem. Abstr.* 1975, *83*, 10406q) is commercially available from Aldrich Chemical Co.
12. K. TAKAI, Y. HOTTA, K. OSHIMA, H. NOZAKI, *Tetrahedron Lett.* 1978, *27*, 2417; J.-I. HIBINO, T. OKAZOE, K. TAKAI, H. NOZAKI, *Tetrahedron Lett.* 1986, *26*, 5579 and 5581.
13. K. TAKAI, T. KAKIUCHI, Y. KATAOKA,

K. Utimoto, *J. Org. Chem.* **1994**, *59*, 2668.

14 L. Lombardo, *Tetrahedron Lett.* **1982**, *23*, 4293; L. Lombardo, *Org. Synth.* **1987**, *65*, 81.

15 T. Okazoe, K. Takai, K. Oshima, K. Utimoto, *J. Org. Chem.* **1987**, *52*, 4410.

16 B. M. Johnson, K. P. C. Vollhardt, *Synlett* **1990**, 209.

17 T. Mukaiyama, T. Sato, J. Hanna, *Chem. Lett.* **1973**, 1041.

18 S. Matsubara, T. Mizuno, T. Otake, M. Kobata, K. Utimoto, K. Takai, *Synlett* **1998**, 1369.

19 S. Matsubara, Y. Yamamoto, K. Utimoto, *Synlett* **1998**, 1471.

20 S. Matsubara, M. Sugihara, K. Utimoto, *Synlett* **1998**, 313.

21 A. R. Hermes, G. S. Girolami, *Inorg. Synth.* **1998**, *32*, 309.

22 S. P. Naula, H. K. Sharma, *Inorg. Synth.* **1985**, *24*, 181.

23 K. Ukai, D. Arioka, H. Yoshino, H. Fushimi, K. Oshima, K. Utimoto, S. Matsubara, *Synlett* **2001**, 513, and references cited therein.

24 Y. Hashimoto, U. Mizuno, H. Matsuoka, T. Miyahara, M. Takakura, M. Yoshimoto, K. Oshima, K. Utimoto, S. Matsubara, *J. Am. Chem. Soc.* **2001**, *123*, 1503.

25 K. Nozaki, N. Kosaka, V. M. Graubner, T. Hiyama, *Macromolecules* **2001**, *34*, 6167.

26 S. Matsubara, K. Ukai, T. Mizuno, K. Utimoto, *Chem. Lett.* **1999**, 825.

27 K. Takai, O. Fujimura, Y. Kataoka, K. Utimoto, *Tetrahedron Lett.* **1989**, *30*, 211.

28 P. Knochel, J.-F. Normant, *Tetrahedron Lett.* **1986**, *27*, 4427 and 4431.

29 E. Nakamura, K. Kubota, G. Sakata, *J. Am. Chem. Soc.* **1997**, *119*, 5457.

30 S. Matsubara, Y. Otake, T. Morikawa, K. Utimoto, *Synlett* **1998**, 1315.

31 S. Matsubara, H. Yoshino, K. Utimoto, K. Oshima, *Synlett* **2000**, 495.

32 S. Matsubara, Y. Otake, Y. Hashimoto, K. Utimoto, *Chem. Lett.* **1999**, 747.

33 K. Takai, M. Tezuka, Y. Kataoka, K. Utimoto, *Synlett* **1989**, 27.

34 Y. Okude, S. Hirano, T. Hiyama, H. Nozaki, *J. Am. Chem. Soc.* **1977**, *99*, 3179.

35 T. Okazoe, K. Takai, K. Utimoto, *J. Am. Chem. Soc.* **1987**, *109*, 951.

36 M. Knecht, W. Boland, *Synlett* **1993**, 837.

37 S. Matsubara, M. Horiuchi, K. Takai, K. Utimoto, *Chem. Lett.* **1995**, 259.

38 K. Takai, N. Shinomiya, H. Kaihara, N. Yoshida, T. Moriwake, *Synlett* **1995**, 963.

39 K. Takai, Y. Kataoka, T. Okazoe, K. Utimoto, *Tetrahedron Lett.* **1987**, *28*, 1443.

40 D. M. Hodgson, *Tetrahedron Lett.* **1992**, *33*, 5603; D. M. Hodgson, L. T. Boulton, G. N. Maw, *Tetrahedron* **1995**, *51*, 3713.

41 D. M. Hodgson, P. J. Comina, M. G. Drew, *J. Chem. Soc., Perkin Trans. 1* **1997**, 2279.

42 K. Takai, K. Nitta, and K. Utimoto, *J. Am. Chem. Soc.* **1986**, *108*, 7408.

43 D. A. Evans, C. Black, *J. Am. Chem. Soc.* **1993**, *115*, 4497.

44 A. S. Kende, R. J. DeVita, *Tetrahedron Lett.* **1990**, *31*, 307.

6
The McMurry Coupling and Related Reactions

Michel Ephritikhine and Claude Villiers

6.1
Introduction

In 1973, two groups discovered that carbonyl compounds could be reductively coupled to produce alkenes upon treatment with low-valent titanium reagents. Tyrlik and Wolochowicz described the coupling of aliphatic and aromatic aldehydes and ketones by means of the $TiCl_3$-Mg system [1], and suggested that tetramethylethylene was obtained via the carbene species Me_2C, itself originating from the deoxygenation of acetone (Scheme 6.1). On the other hand, Mukaiyama et al. used the $TiCl_4$-Zn combination in the reductive coupling of aromatic aldehydes and ketones [2]; they observed that benzaldehyde and acetophenone were selectively transformed into the corresponding pinacols and alkenes, respectively, when the reactions were performed in THF at low temperature or in refluxing dioxane, and proposed the intermediacy of a metallopinacolate species. The latter was envisaged as being formed either by dimerization of ketyl radicals resulting from one-electron transfer from the low-valent metal species to the carbonyl and/or, in the case of the more easily reducible and reactive aromatic ketones, by nucleophilic attack of a ketone dianion on the C=O bond. One year later, McMurry and Fleming reported on the reductive coupling of carbonyls to form olefins using $TiCl_3$ and

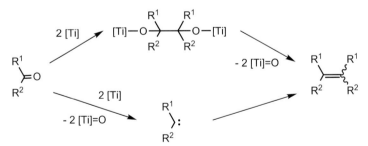

Scheme 6.1. The two mechanisms of the McMurry reaction proposed by Mukaiyama et al. (top) and by Tyrlik and Wolochowicz (bottom).

Modern Carbonyl Olefination. Edited by Takeshi Takeda
Copyright © 2004 WILEY-VCH Verlag GmbH & Co. KGaA, Weinheim
ISBN: 3-527-30634-X

LiAlH$_4$ [3]; they also proposed that metallopinacols were intermediates in this reaction, since pinacols could be isolated as by-products in many cases.

These initial studies heralded a spectacular development, due to the huge interest of the reaction in organic synthesis. In view of the leading role played by McMurry in establishing its reputation, the reductive coupling of carbonyls to give pinacols and alkenes by means of low-valent titanium compounds was called the McMurry reaction. The very broad scope of the McMurry reaction, which has been outlined in several reviews [4–11], will be summarized in the first part of this chapter, focusing on the most recent aspects. Today, a large array of ketones, aldehydes, acylsilanes, keto esters, and oxoamides can be efficiently coupled, in an inter- or intramolecular fashion, to produce a variety of products, including strained or sterically hindered olefins, macrocycles, polymers, and heterocycles. An impressive number of natural product syntheses involve an intramolecular McMurry coupling as their key step. The use of the McMurry reaction thus covers great domains of organic and organometallic chemistry, from biochemistry to materials science.

Such considerable improvements were possible after the design of more reactive low-valent titanium species and the use of new synthetic methods and procedures; these will be presented in the second part of this chapter. In comparison with its synthetic applications, the mechanism of the McMurry reaction has received much less attention, undoubtedly because chemists were firmly convinced of the involvement of pinacolate intermediates and rejected, in spite of some evidence, the alternative pathway via carbenoid species. The last part of this chapter will describe the recent studies on the structure and mode of action of low-valent titanium species, as well as some accurate analyses of the reaction products, which have allowed new insights into the mechanism, revealing its dual nature.

6.2
Scope of the McMurry Reaction

McMurry reactions will be presented in the following order: intermolecular, intramolecular, mixed (tandem) couplings of aldehydes and ketones, and finally keto ester, oxoamide, and acetal couplings. All the compounds which serve as illustrations are listed in Tables 6.1–6.10, along with the titanium reagents and solvents used for their preparation and the yields of isolated products; where not specified, the reactions were performed at solvent reflux temperature.

6.2.1
Intermolecular Coupling Reactions

6.2.1.1 Intermolecular Coupling of Aldehydes and Ketones
The intermolecular coupling of aldehydes and ketones has been used to prepare unusual molecules, in particular sterically hindered and/or strained alkenes that exhibit specific physico-chemical properties and are of theoretical interest (Figures

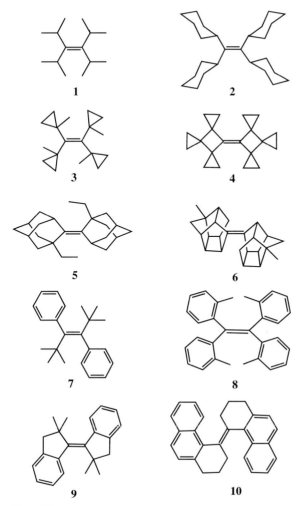

Figure 6.1.

6.1 and 6.2; Table 6.1). The driving force of the reaction is the formation of strong titanium–oxygen bonds.

These molecules were obtained from aliphatic and aromatic substrates in both the acyclic and cyclic series. In general, a mixture of E and Z isomers of the olefin is formed; the stereochemistry depends on the choice of the titanium reagent [12]. With the $TiCl_4$-Zn system, the E/Z ratio in the coupling of aliphatic methyl alkyl ketones RCOMe varies with the steric bulk of the R group, ranging from 3:1 for R = nPr to more than 200:1 for R = tBu [12]. While the synthesis of **3** represents the coupling of a "tied-back" di-*tert*-butyl ketone [13], the preparation of the long sought-after tetra-*tert*-butylethylene has hitherto remained unsuccessful. The rotational barriers of the isopropyl and cyclohexyl groups in **1** and **2** are equal to 16 and

Tab. 6.1. Syntheses of alkenes and polyenes by intermolecular McMurry reactions of saturated and unsaturated aldehydes and ketones.

Product	TiCl$_x$	Reducing Agent	Solvent	Yield (%)	Ref.
Saturated aldehydes and ketones					
1	TiCl$_3$	Zn(Cu)	DME	87	14
2	TiCl$_3$	K	THF	20	15
3	TiCl$_3$	LiAlH$_4$	THF	13	13
4	TiCl$_3$	LiAlH$_4$	THF[a]	8	16
5	TiCl$_4$	Zn	THF[b]	40	18
6	TiCl$_4$	Zn	THF[b]	73	19
7	TiCl$_4$	Zn	THF[b]	48	18
8	TiCl$_3$	LiAlH$_4$	THF	15	21
9	TiCl$_4$	Zn	THF[b]	42	22
10	TiCl$_3$	LiAlH$_4$	THF	–	23
Unsaturated aldehydes and ketones					
11	TiCl$_4$	Zn	THF/CH$_2$Cl$_2$[c]	67	27
12	TiCl$_4$	Zn	THF	56	28
13	TiCl$_4$	Zn	THF[a,b]	–	29
14	TiCl$_3$	Zn(Cu)	DME	31	30
15	TiCl$_3$	Zn(Cu)	DME[d]	12	31
16	TiCl$_3$	Zn(Cu)	DME	78	32

[a] At 20 °C; [b] In the presence of pyridine; [c] At −78 °C; [d] At 40 °C.

18.7 kcal mol^{-1}, respectively [14, 15], and it has been estimated that olefins with a strain energy lower than 19 kcal mol^{-1} may be synthesized by means of the McMurry reaction. In spirocyclopropanated bicyclobutylidenes such as **4**, each α- and β-spirocyclopropane group causes a bathochromic shift of the π–π* band compared to that of the parent bicyclobutylidene, by about 9 and 4 nm, respectively [16, 17]. Cage-functionalized alkenes such as **5** and **6** have also been synthesized [18, 19]. While the preference for the formation of Z-isomers of uncrowded stilbenes was rationalized in terms of complexation of the two aromatic groups by titanium in the pinacolate intermediate [20], *tert*-butyl phenyl ketone yielded exclusively the crowded (E)-stilbene **7**. Coupling of di-*o*-tolyl ketone gave **8** along with small amounts of the ethane derivative [21]. Coupling of 2,2-dimethyl-1-indanone gave **9**, but only the conversion of the carbonyl into a methylene group was observed with the more substituted 2,2,4,7-tetramethyl-1-indanone [22]. Compound **10** and its *cis* isomer represent the first instance in which *dl-trans* and *dl-cis* forms of one inherently dissymmetric olefin have been isolated and resolved into enantiomers [23].

6.2.1.2 Intermolecular Coupling of Unsaturated Aldehydes and Ketones

McMurry coupling of α,β-unsaturated aldehydes and ketones constitutes an easy route to conjugated polyenes (Figure 6.2), which exhibit interesting electronic properties. After β-carotene [3], a series of minicarotenes such as **11** has been syn-

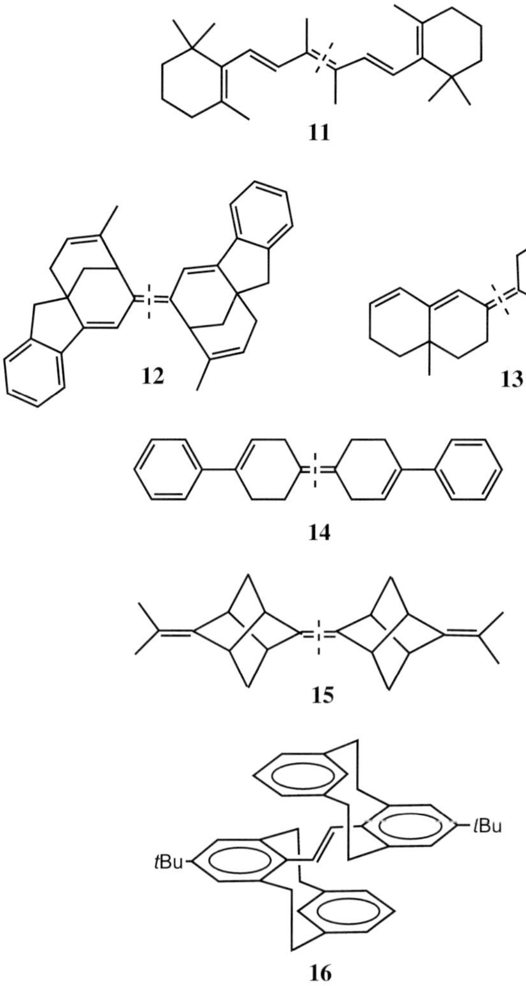

Figure 6.2.

thesized [24–27]. The rigidification in compound **12** led to a decrease in the HOMO–LUMO gap and a considerable enhancement of fluorescence yield, as compared to the open-chain analogue [28]. The pentaene **13** was prepared in order to study the kinetic and thermodynamic parameters of the *anti-syn* isomerization [29].

Non-conjugated enones have also been coupled by low-valent titanium reagents, as illustrated by the synthesis of **14** [30]. The X-ray crystal structure of the racemic distellene **15** showed that the central C–C double bond is twisted by 10 and 14.5° for the two molecules in the unit cell [31]. Compound **16** should also be considered here since it permitted study of the electronic interaction between two [2.2]metacyclophane systems connected with a C=C bond [32].

Tab. 6.2. Syntheses of alkenes by intermolecular McMurry reactions of aldehydes and ketones with functional heteroatom groups.

Product	TiCl$_x$	Reducing Agent	Solvent	Yield (%)	Ref.
Halides					
17	TiCl$_3$	LiAlH$_4$	THF	51	35
18	TiCl$_3$	LiAlH$_4$	THF	78	35
19	TiCl$_4$	Zn	THF	24	36
20	TiCl$_4$	Zn	THF	44	37
21	TiCl$_4$	Zn	DME	50	38
22	TiCl$_4$	Zn	DME	79	38
Acetals, toluenesulfonates, esters, and ethers					
23	TiCl$_3$	K	[a]	80	4
24	TiCl$_3$	LiAlH$_4$	THF	98	39
25	TiCl$_3$	Zn(Cu)	THF	80	40
26	TiCl$_3$	Zn(Cu)	THF	87	40
27	TiCl$_4$	Zn(Cu)	THF	43	42
28	TiCl$_4$	LiAlH$_4$	THF	28	43
Sulfides					
29	TiCl$_4$	Zn	THF	86	44
30	TiCl$_4$	Zn	THF	89	45
31	TiCl$_4$	Zn	THF	71	46
32	TiCl$_3$	LiAlH$_4$	THF[b]	20	47
33	TiCl$_4$	Zn	THF[b]	99	48

[a] Not specified; [b] In the presence of pyridine.

6.2.1.3 Intermolecular Coupling of Aldehydes and Ketones with Functional Heteroatom Groups

It is clear that some functional groups are not compatible with the carbonyl coupling reaction, being readily reduced by the low-valent titanium reagent. These include allylic and benzylic alcohols, pinacols, epoxides, enediones, quinones, halohydrins, α-halo ketones, nitro compounds, oximes, and sulfoxides. However, McMurry reactions can be carried out with carbonyl substrates bearing a variety of functional heteroatom groups (Figures 6.3 to 6.10; Tables 6.2 and 6.3), even if some of these (esters, toluenesulfonates) slowly react with the low-valent titanium reagent.

Halides: McMurry couplings of 2-bromobenzaldehyde [33] and 3-fluorobenzaldehyde [34] give the corresponding *E*-configured stilbenes. However, coupling of 4-bromoacetophenone gives the *cis* isomer **17** as the major product (*cis/trans* = 9); this unexpected result has been confirmed by X-ray crystallography and resolves an inconsistency in earlier literature assignments [35]. Coupling of 4-bromobenzophenone affords **18**, the *trans* isomer being very slightly favored [35] (Figure 6.3).

A series of dibromides including **19** served as monomers in the synthesis of poly(arylenevinylene)s, which exhibit high luminescence quantum yields [36],

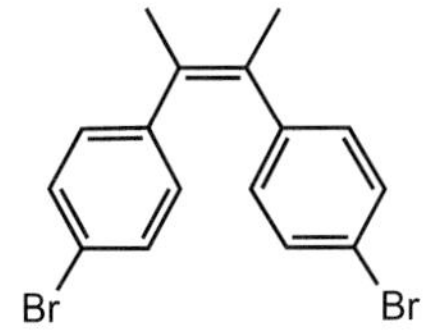

Figure 6.3.

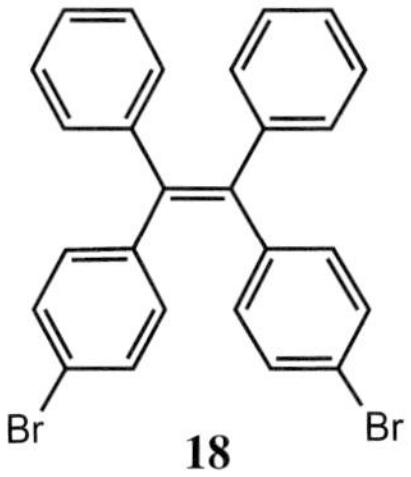

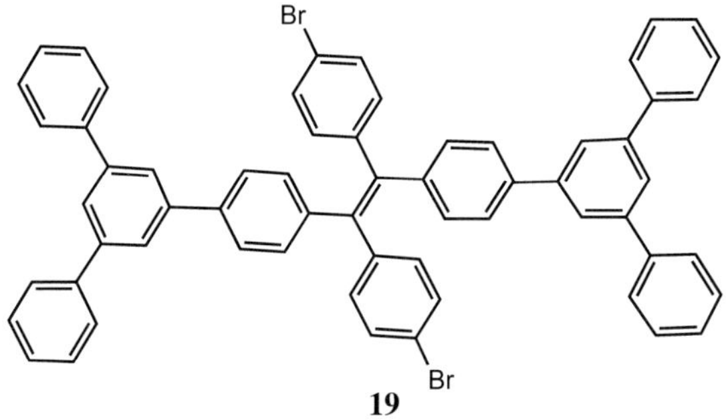

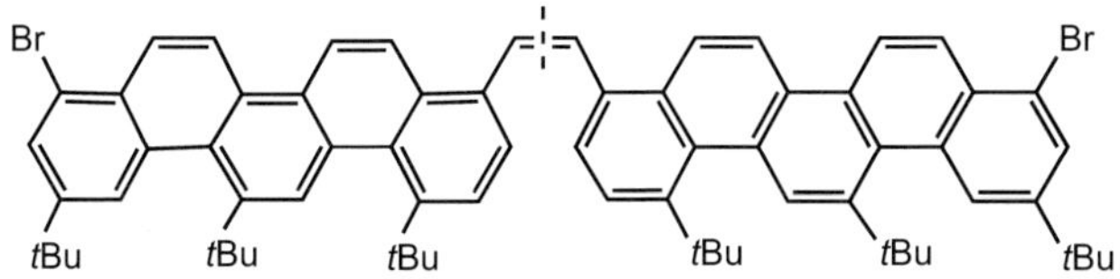

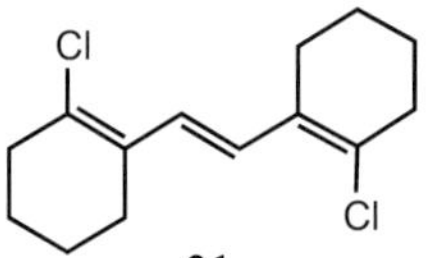

while oxidative photocyclization of diarylethylenes such as **20** led to [*n*]phenacenes, a new family of graphite ribbons [37]. Several 1,6-dihalo-1,3,5-hexatrienes, such as **21** and **22**, have been obtained with high stereoselectivity from β-haloacrylaldehydes [38].

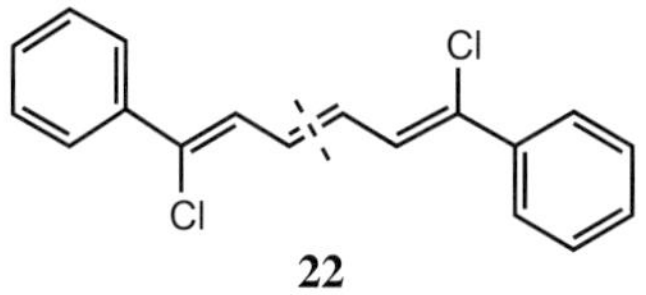

Acetals, toluenesulfonates, esters, and ethers: The compatibility of these groups with the McMurry reaction is illustrated by the coupling products of substituted cycloalkanones, **23** [4] and **24** [39], and benzaldehydes, **25** and **26** [40, 41] (Figure 6.4). However, benzylidene acetals were found to be cleaved in a modified McMurry reaction (see Section 6.2.6).

ESR and ENDOR investigations on the radical cation and radical anion of **27**, the most expanded fulvalene reported to date, indicate that the free electron is delo-

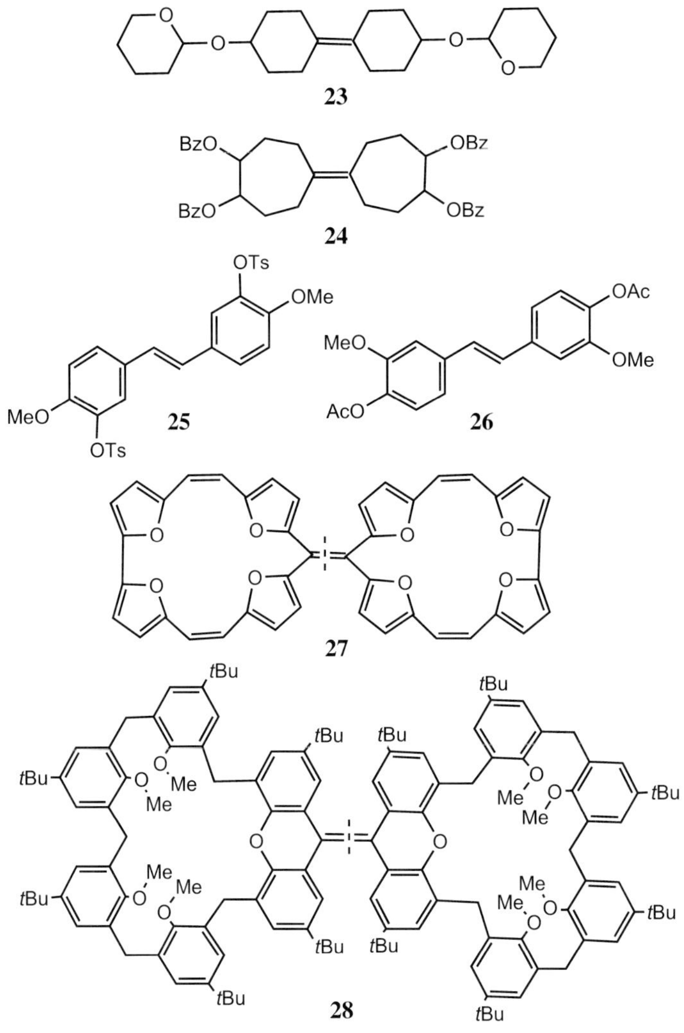

Figure 6.4.

calized over the entire molecule [42]. The dixanthylene **28** was obtained by coupling of the corresponding calixanthone [43].

Sulfides: A series of bridged dithienylethylenes such as **29**, synthesized by McMurry dimerization of cyclopenta[*b*]thiophen-6-ones, are precursors of small bandgap electrogenerated conjugated polymers [44], while bridged 1,6-dithienylhexa-1,3,5-trienes, represented by **30**, are highly photoluminescent and stable thiophene-based π-conjugated systems [45] (Figure 6.5). A family of thienylenevinylene oligomers containing **31** permitted study of the effect of chain extension on the electrochemical and electronic properties [46].

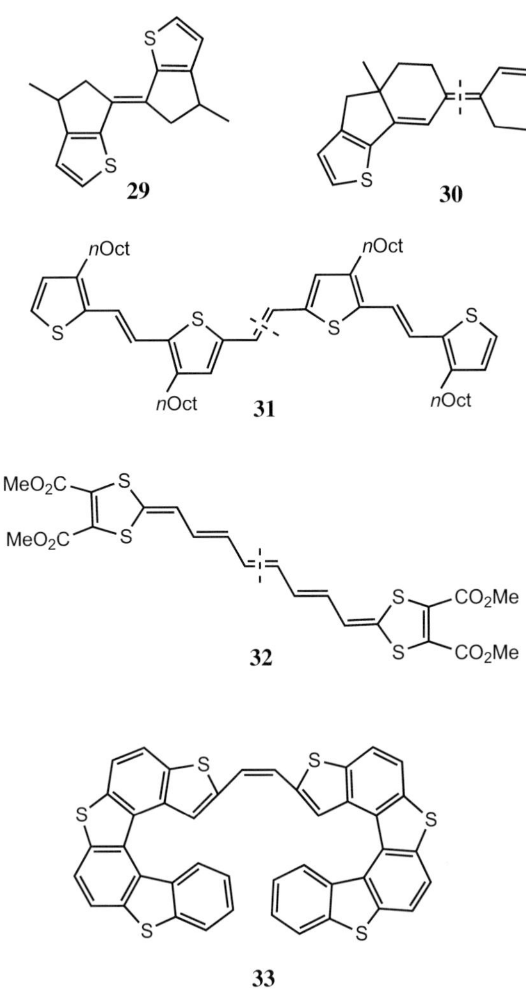

Figure 6.5.

The length of the unsaturated chain of a polyenic analogue of tetrathiafulvalene, such as **32**, has a marked influence on its effective conjugation and π-donating ability [47]. A series of precursors to thiahelicenes, as exemplified by **33**, have been obtained by McMurry coupling of benzothiophene-2-carbaldehydes or acetyl compounds [48, 49].

Amines: While 3-pyrrolylaldehydes or ketones and 2-pyridyl ketones are transformed into rearranged and/or reduced McMurry products [50, 51], 2-pyrrolylaldehydes or ketones and 3-benzoylpyridine can be coupled by low-valent titanium reagents, as shown by the synthesis of compounds **34**, or **35** and **36**, respectively [50–53] (Figure 6.6). Coupling of 3-pyridine carbaldehydes gives the dipyridyl

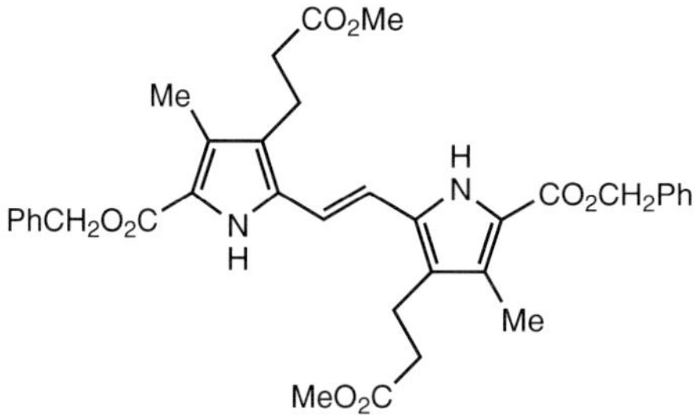

Figure 6.6.

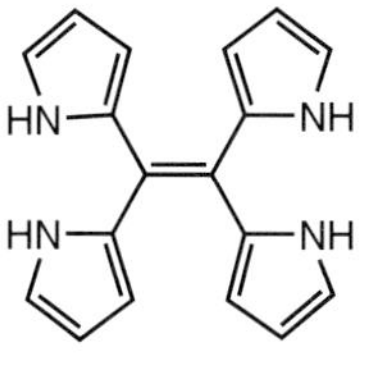

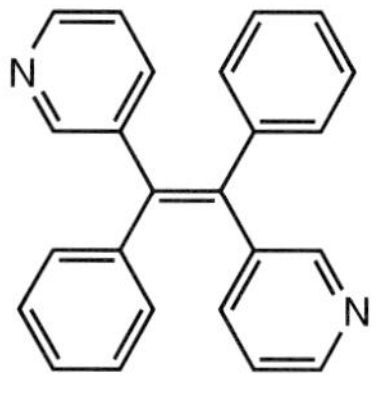

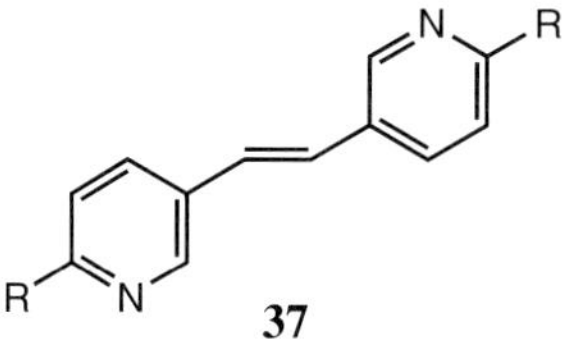

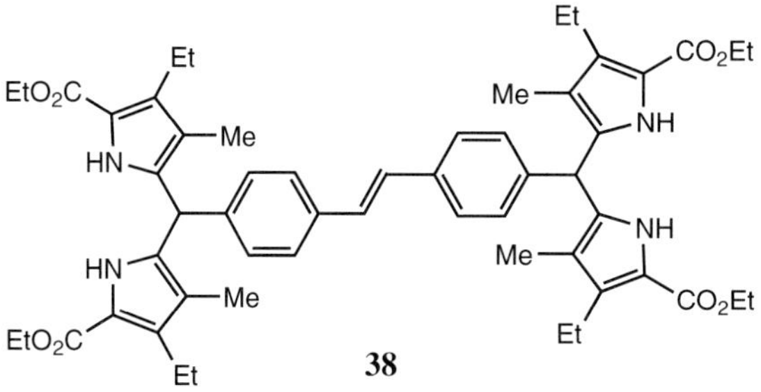

ethene compounds **37** (R = OMe or SMe), which might be useful precursors for molecular wires and π-conjugated-type polymers [54].

Compound **38** has been used as a precursor in the synthesis of a stilbene-linked bisporphyrin [55]. Bisporphyrins and bischlorins such as **39** [55] and **40** [56] can be found with both *cis* and *trans* configurations about the C–C double bond, and in some cases the *cis* product is formed preferentially [55–58] (Figure 6.7). Thus, the X-ray crystal structure of **40** shows a *cis*-ethene linkage and a cofacial arrangement of the two macrocycles; this likely reflects a strong π–π pre-association of the sub-

Tab. 6.3. Syntheses of alkenes by intermolecular McMurry reactions of aldehydes and ketones with functional heteroatom groups, and of organometallic aldehydes and ketones.

Product	TiCl$_x$	Reducing Agent	Solvent	Yield (%)	Ref.
Amines					
34	TiCl$_3$	Zn(Cu)	DME	46	52
34	TiCl$_4$	Zn	THF	86	50
35	TiCl$_3$	Zn(Cu)	DME[a]	56	53
36	TiCl$_3$	LiAlH$_4$	THF	13	51
37 (R = OMe)	TiCl$_4$	Zn	Dioxane[a]	40	54
37 (R = SMe)	TiCl$_4$	Zn	Dioxane[a]	30	54
38	TiCl$_3$	Zn(Cu)	DME	76	55
39	TiCl$_3$	Zn(Cu)	DME	53	55
40	TiCl$_3$	Zn(Cu)	DME	90	56
41 (R = Me)	TiCl$_3$	Zn(Cu)	DME[b]	44	58
41 (R = Ph)	TiCl$_3$	Zn(Cu)	DME[b]	44	58
Phospholes, alkyl- and acylsilanes, and ketones					
42	TiCl$_4$	Zn	THF	71	60
43	TiCl$_3$	Li	DME	38	61
44	TiCl$_4$	Zn	THF[a]	41	62
45 (R = OMe)	TiCl$_3$	Na(Al$_2$O$_3$)	THF	56	63
46	Ti[c]		DME	84	64
Organometallic aldehydes and ketones					
47	TiCl$_3$	LiAlH$_4$	THF[d]	25	65
48	TiCl$_3$	Li[e]	DME	80	66
49	TiCl$_3$	Zn	DME[a]	70	67
50	TiCl$_4$	LiAlH$_4$	THF	36	65
51	TiCl$_3$	LiAlH$_4$	THF	44	68
52	TiCl$_3$	Zn(Cu)	DME[a]	59	69
53	TiCl$_4$	Zn	THF	41	70

[a] In the presence of pyridine; [b] At −10 °C; [c] In the presence of Me$_3$SiCl; [d] At 20 °C; [e] With ultrasound activation.

strates before the coupling step. In the bischlorin **41**, the two chlorin subunits are coplanar and strongly electronically coupled, providing a model compound for the special pair in photosynthesis [59].

Phospholes: The only example of McMurry coupling of a carbonyl substrate bearing a phosphorus functional group is provided by the synthesis of the 1,2-bis(2-phospholyl)alkenes **42** (R = Me, Ph) [60] (Figure 6.8).

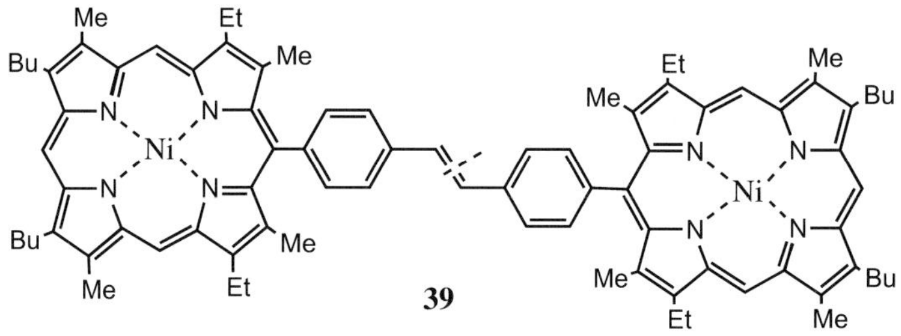

Figure 6.7.

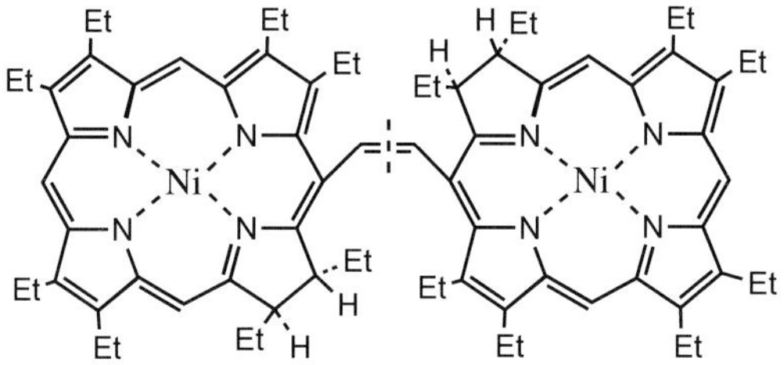

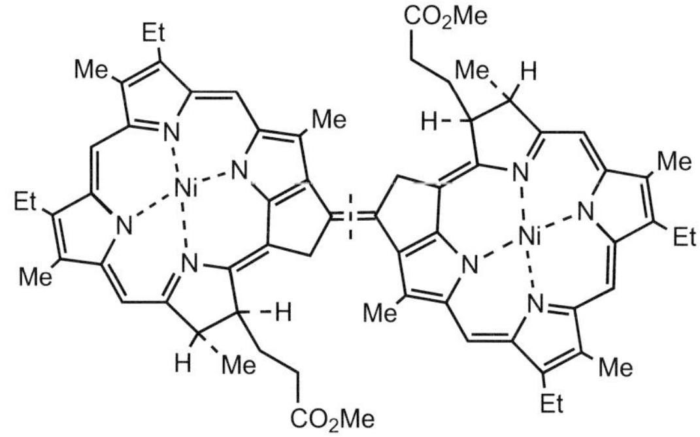

Alkyl and acylsilanes: Tetraalkylsilanes are inert to the carbonyl coupling reaction conditions, as shown by the syntheses of **43** and **44** [61, 62]. Despite their close similarity to sterically hindered ketones, McMurry reactions of acylsilanes are limited to the preparation of the 1,2-disilylated stilbenes **45** (R = H, Br or OMe); silyl enol ethers are also obtained as by-products, resulting from silatropic migration [63]. No such coupling of aliphatic acylsilanes has yet been reported.

Ketones: Examples in which unprotected ketones kinetically survive McMurry reactions are very rare. Coupling at the enone site of androsta-1,4-diene-3,17-dione

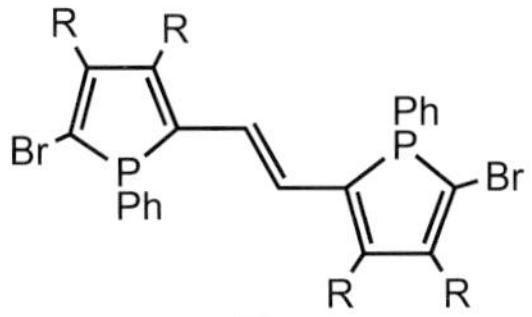

Figure 6.8.

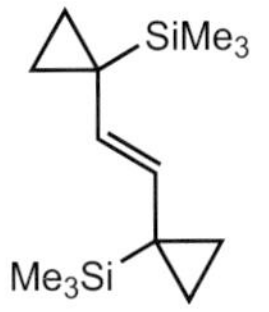

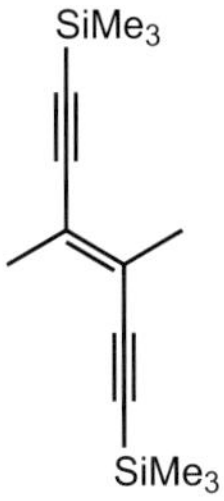

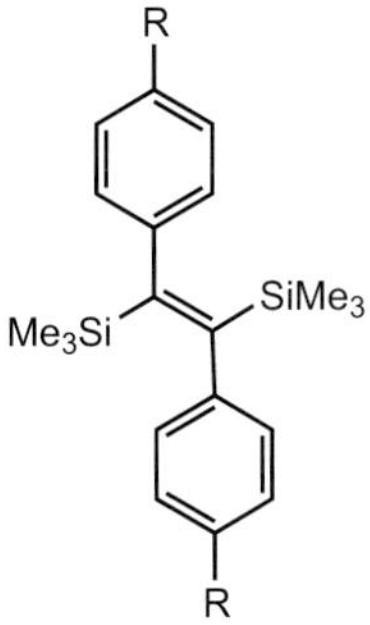

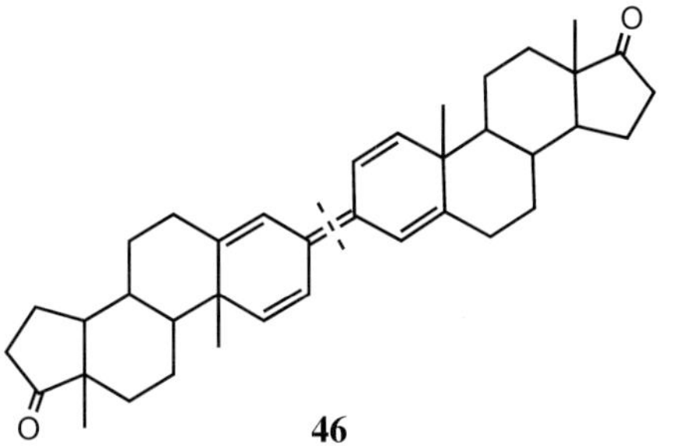

affords compound **46** [64]. In other polycarbonyl substrates, intramolecular cyclization has been found to occur so rapidly that the reaction can be stopped before competing intermolecular dimerization takes place (Sections 6.2.2.2 and 6.2.5).

6.2.1.4 Intermolecular Coupling of Organometallic Ketones and Aldehydes

McMurry coupling of formylferrocene and ferrocenylketones leads to compounds **47**, **48**, and **49** [65–67] (Figure 6.9), while the chromium and manganese complexes **50** and **51** have been obtained from the parent formylcynichroden and acylcymanthrene, respectively [65, 68] (Figure 6.10).

The phospholyl analogues **52** and **53** have been isolated from similar reactions with the corresponding formylphosphaferrocene and acylphosphacymanthrene [69, 70].

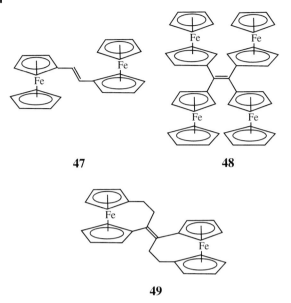

Figure 6.9.

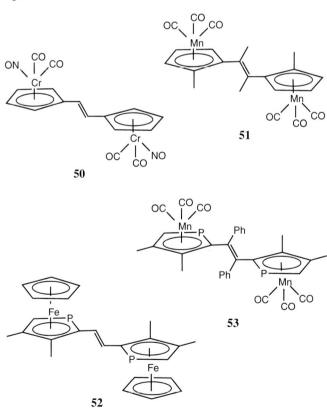

Figure 6.10.

6.2.1.5 The McMurry Reaction in Polymer Synthesis

Intermolecular coupling of dialdehydes and diketones constitutes a newly developed route to polymers [71].

Such polymerizations are successful if the spacer unit X between the two carbonyls is rigid enough to prevent competing intramolecular cyclization (Scheme 6.2; Figure 6.11). Poly(arylene vinylene)s such as **54** and poly(thienylene vinylene)s such as **55**, which have been formed with relatively high molecular weights, are materials with interesting photoluminescence and chiroptical properties [72, 73]. Organometallic polymers such as **56** have been obtained by the coupling of diformylferrocenes [74].

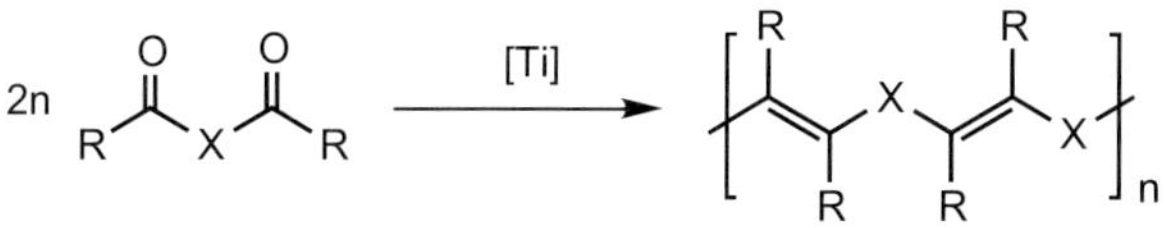

Scheme 6.2. The McMurry reaction in polymer synthesis.

6.2.1.6 Intermolecular Cross-Coupling Reactions

Intermolecular coupling of two distinct aliphatic carbonyl substrates generally affords a statistical mixture of products, unless one component is used in excess (Figures 6.12–6.14; Table 6.4). In many cases, acetone is used as the excess reagent, since this ketone and 2,3-dimethyl-2-butene are easily removed by distillation.

It is noteworthy that the R-(−) chirality was retained during the synthesis of **57** [75]. Compound **58** (R = *t*butyldimethylsilyl) was prepared by using an excess of cyclopentanone [76], and a mixture of end-functionalized polyadamantane rods **59** ($n = 1–3$) was obtained by treating adamantane-2,6-dione with 2.3 equivalents of its mono(ethylene)ketal [77]. However, **60** was obtained in 60% yield from an equimolar mixture of the two coupled carbonyls [78]. Compound **61**, which is a key intermediate in the synthesis of catechol metabolites of diethylstilbestrol and hexestrol, was synthesized by coupling of the parent mono- and dimethoxypropiophenones [79]. Higher yields are obtained from the mixed coupling reaction when one of the reactants is a diaryl ketone, because the latter undergoes a rapid two-electron reduction and the resulting dianion then nucleophilically attacks the other ketone before it can be reduced; thus, **62** was isolated in 77% yield [80]. Such a reaction constitutes a classical route to the antitumor agent tamoxifen **63** [81] and many of its derivatives [82–89], including **64** and **65**, and organometallic analogues such as **66** and **67** [90–93] (Figure 6.13).

McMurry cross-coupling reactions have also been found to be useful for the synthesis of the unsymmetrically substituted bisporphyrin **68** and the porphyrin-chlorin heterodimer **69**, which were isolated from mixtures of the corresponding homodimers [55] (Figure 6.14).

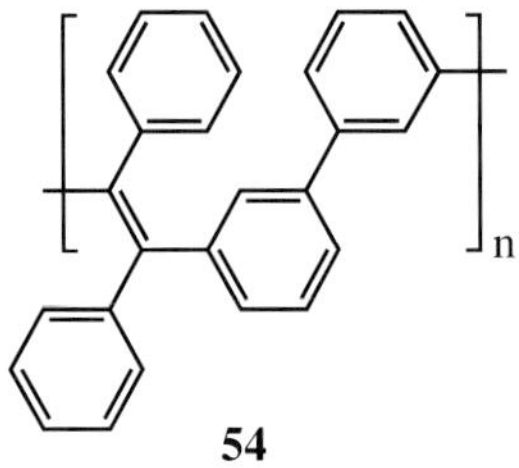

Figure 6.11.

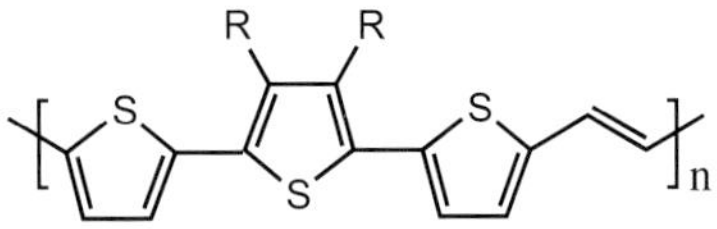

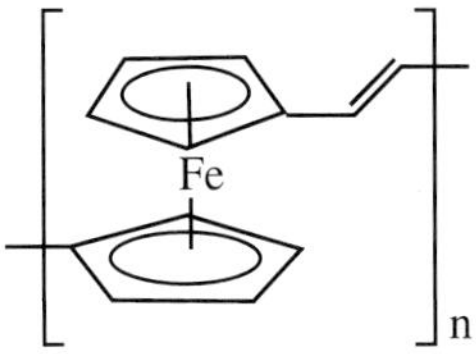

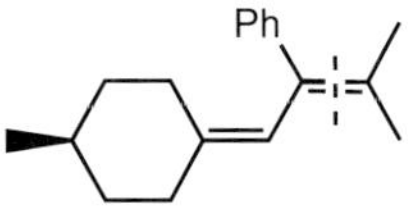

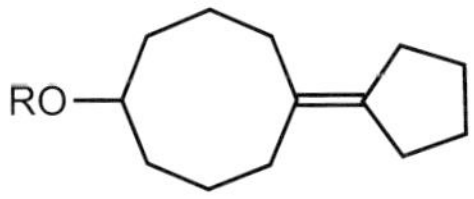

Figure 6.12.

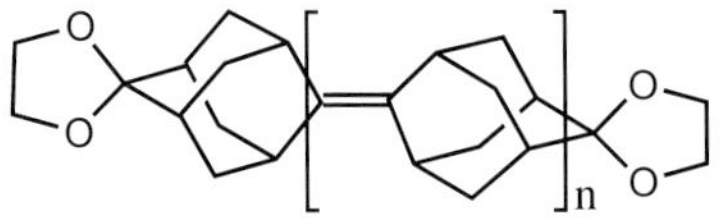

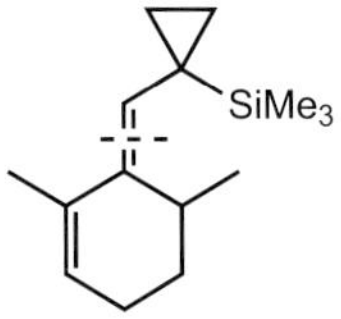

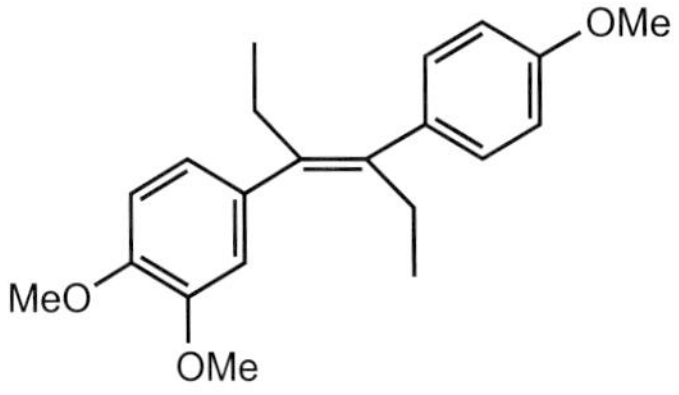

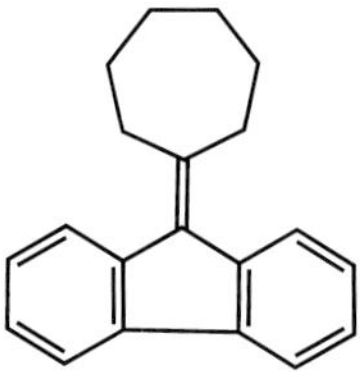

6.2 Scope of the McMurry Reaction | 239

Tab. 6.4. McMurry syntheses of alkenes by intermolecular cross-coupling reactions.

Product	TiCl$_x$	Reducing Agent	Solvent	Yield (%)	Ref.
57	TiCl$_4$	LiAlH$_4$	THF[a]	40	75
58	TiCl$_3$	Zn(Cu)	DME	95	76
59	TiCl$_3$	Na	dioxane	71	77
($n = 1$)					
60	TiCl$_3$	Zn(Cu)	DME	60	78
61	TiCl$_3$	Li	DME	55	79
62	TiCl$_3$	Li	DME	77	80
63	TiCl$_4$	Zn	DME	88	81
64	TiCl$_4$	Zn	THF	65	82
65	TiCl$_4$	Zn	THF	72	84
66	TiCl$_4$	Zn	THF	36	90
67	TiCl$_4$	Zn	THF	54	92
68	TiCl$_3$	Zn(Cu)	DME	20	55
69	TiCl$_3$	Zn(Cu)	DME	41	55

[a] At 20 °C.

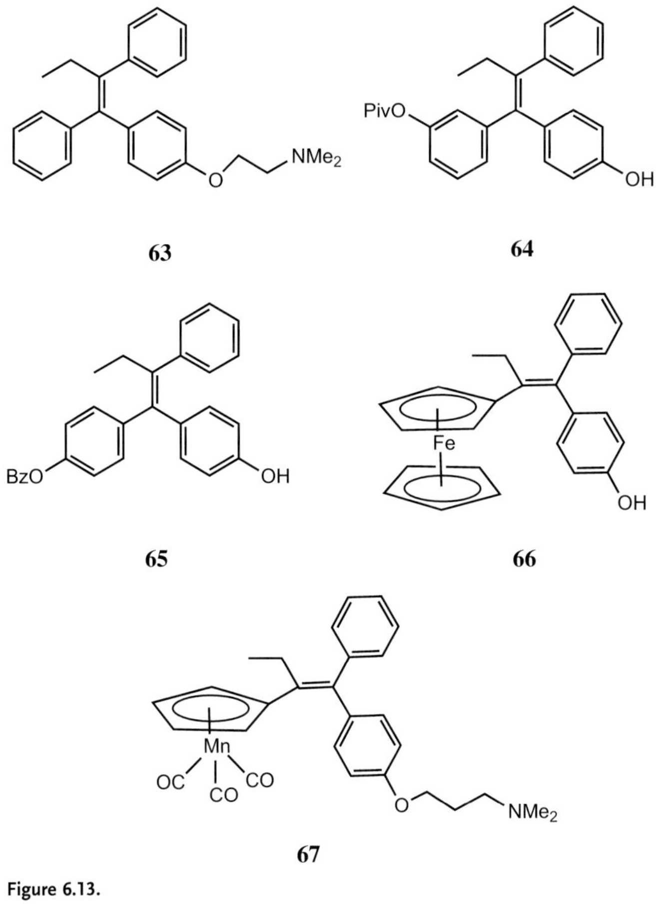

Figure 6.13.

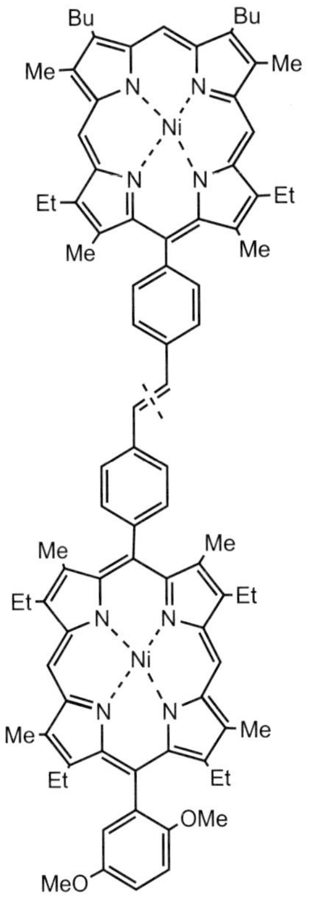

Figure 6.14.

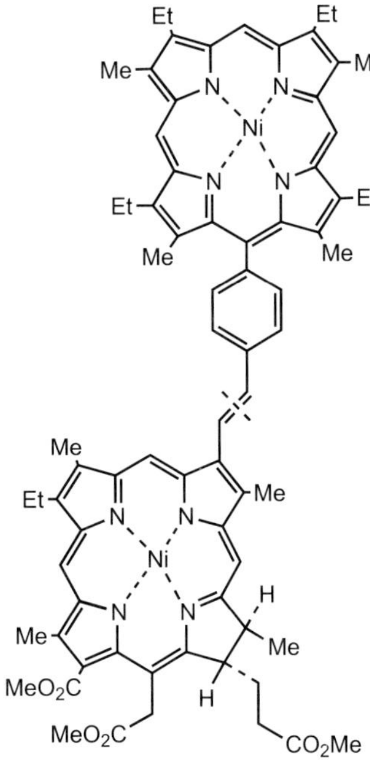

6.2.2
Intramolecular Coupling Reactions of Aldehydes and Ketones

6.2.2.1 Synthesis of Non-Natural Products

Cycloalkenes and cyclopolyenes: Cycloalkenes can be prepared in all ring sizes from three upwards in high yields, as illustrated by the syntheses of **70** [94], **71**, and **72** [95] (Figure 6.15 and Table 6.5). The McMurry reactions give much higher yields of medium-sized cycloalkenes than the acyloin, Thorpe, or Dieckmann condensations. *E*- and *Z*-isomers may be obtained for the larger ring systems, which are best prepared by using high dilution conditions to avoid intermolecular reactions. The coupling reaction has also proved to be efficient for the synthesis of the bridgehead alkene **73**, which exists in the form of the *in* isomer and served for the study of bicyclic carbocations containing a three-center two-electron C–H–C bond [96–99], and of the perfluorocyclopentene **74**, a versatile intermediate for the development of photochromic materials [100].

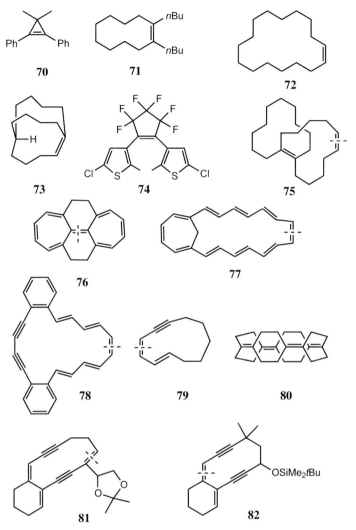

Figure 6.15.

The ability of the McMurry reaction to form medium- and large-sized rings efficiently, while at the same time building in considerable strain is also evident in the synthesis of the betweenanene **75** [101] and the heptafulvalene **76**; the latter results from a transannular carbonyl coupling reaction and was transformed into displeiadiene after dehydrogenation [102]. A variety of conjugated cyclic polyenes, such as **77** and **78** [103, 104], as well as unique examples of cyclo-1,3-dien-5-ynes **79** [105], have been prepared by the intramolecular coupling of dialdehydes. The tetraene **80**, which was investigated as a tetradentate ligand for transition metals, accommodates an Ag^+ ion within the cavity to give a unique square-planar com-

Tab. 6.5. Syntheses of cycloalkenes, cyclopolyenes, cyclophanes, and calixarenes by intramolecular McMurry reactions.

Product	TiCl$_x$	Reducing Agent	Solvent	Yield (%)	Ref.
Cycloalkenes and cyclopolyenes					
70	TiCl$_3$	LiAlH$_4$	THF	46	94
71	TiCl$_3$	Zn(Cu)	DME	75	95
72	TiCl$_3$	Zn(Cu)	DME	85	95
73	TiCl$_3$	Zn(Cu)	DME	30	96
74	TiCl$_3$	Zn	THF[a]	55	100
75	TiCl$_3$	Li	DME	44	101
76	TiCl$_4$	Zn	THF	95	102
77	TiCl$_3$	LiAlH$_4$	DME	5	103
78	TiCl$_3$	LiAlH$_4$	DME	14	104
79	TiCl$_3$	Zn(Cu)	DME	38	105
80	TiCl$_3$	Zn(Cu)	DME	90	106
81	TiCl$_3$	Zn(Cu)	DME[b]	45	107
82	TiCl$_3$	Zn(Cu)	DME[c]	41	108
Cyclophanes and calixarenes					
83	TiCl$_4$	Zn(Cu)	DME	50	112
84	TiCl$_3$	Zn(Cu)	DME	13	113
85	TiCl$_4$	Zn	THF	50	114
85	TiCl$_4$	Zn	DME	34	114
86	TiCl$_4$	Zn	THF	70	115
87	TiCl$_3$	LiAlH$_4$	THF	81	120
88	TiCl$_4$	Zn	THF[d]	79	122
89	TiCl$_4$	Mg(Hg)	THF[c]	30	123
90	TiCl$_3$	Zn	THF	25	124

[a] At 40 °C; [b] At 30 °C; [c] At 20 °C; [d] In the presence of pyridine.

plex [106]. Several polycyclic dienediynes and trienediynes, such as **81** and **82**, have been synthesized as models of the antitumor agent neocarzinostatin chromophore [107–111].

Cyclophanes and calixarenes: The McMurry reaction has recently found new applications in the synthesis of *o*-, *m*-, and *p*-cyclophane derivatives, which attract much attention in supramolecular chemistry (Figure 6.16). The naphthalenophane **83** exists exclusively in the *anti* conformation [112]. The extended ring system of compound **84** (R = C$_6$H$_{13}$) represents a discotic mesogen [113]. In many cases, as for **85** [114] and **86** [115], these compounds are formed together with the products resulting from successive inter- and intracyclization reactions (Section 6.2.3) [114–119].

Transannular coupling reactions have allowed access to the higher orthocyclophane derivative **87** [120, 121], the triply-bridged cyclophane **88** [122], as well as the highly distorted cone calix[4]arenes **89** (R = CH$_2$CH$_2$OEt) [123] and **90** (R = *n*Pr) [124].

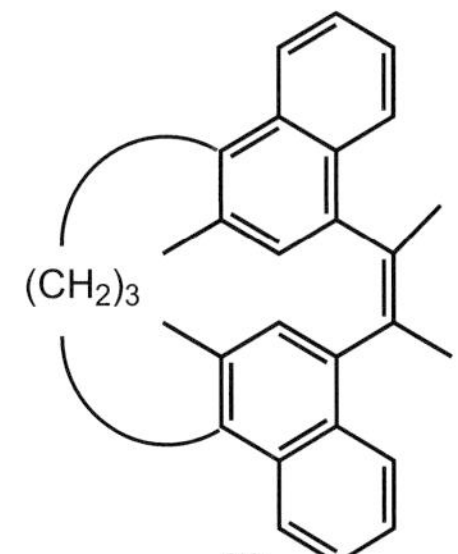

83

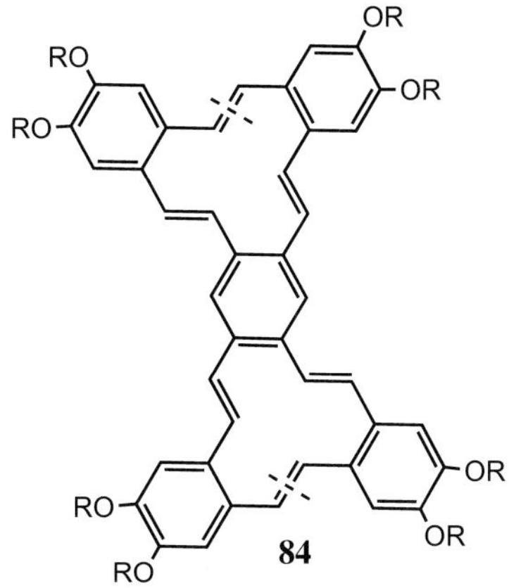

84

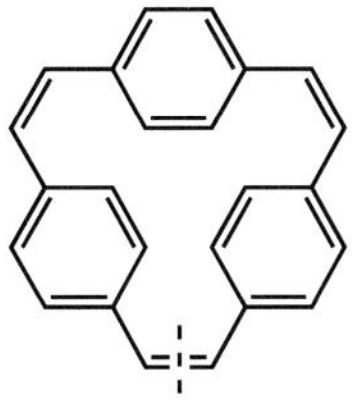

85

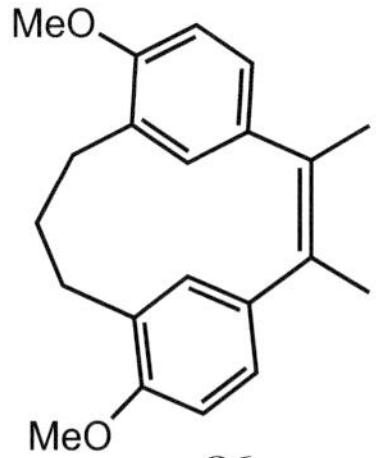

86

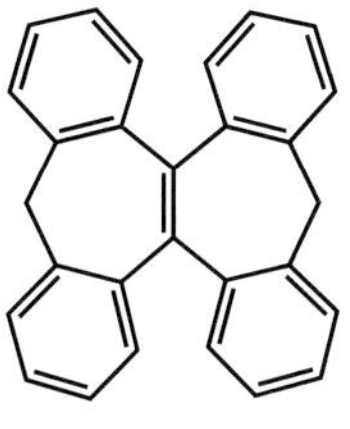

87

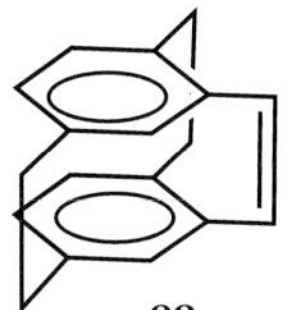

88

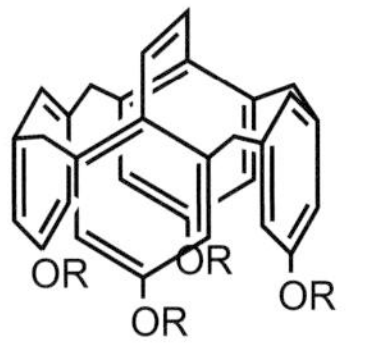

89

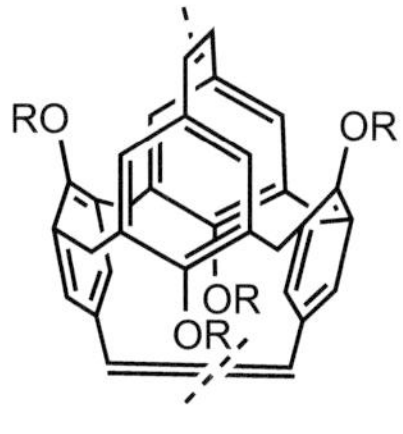

90

Figure 6.16.

Phenanthrenes, helicenes, and circulenes: A novel application of the intramolecular McMurry condensation of bis(benzaldehyde)s led to the preparation of phenanthrenes such as **91** [125], [5]helicene **92** [126], the chiral disubstituted [7]thiaheterohelicene **93** [127, 128], and [7]circulene **94** [129] (Figure 6.17; Table 6.6).

Crownophanes, crown ether-calixarenes, and polyepoxyannulenes: Comparison of the efficiencies of various low-valent titanium species showed that the metal exerts a strong template effect in the synthesis of crownophanes such as **95** [130] (Figure 6.17). Compound **96** exhibits the shortest and most rigid bridging moiety in the class of stilbenophane crown ethers [131, 132]. A series of crown ether calix[4]-

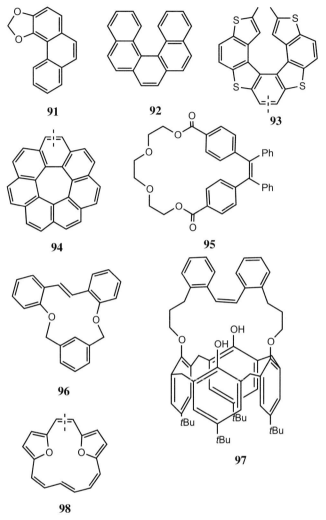

Figure 6.17.

arenes derivatives, including **97**, was synthesized for a study of their photoswitchable properties; the coupling reactions yielded both *cis-* and *trans*-stilbenes from *o-* and *m*-bis(benzaldehyde)s, while only the *cis*-stilbene was obtained from the *p*-isomer [133]. Polyepoxyannulenes such as **98** were found to be highly dynamic systems [134–136].

Thiophenes and selenophenes: While the intramolecular reductive coupling of diketo sulfides or selenides has been known since 1985, with the synthesis of a series of symmetrically and unsymmetrically substituted 2,5-dihydrothiophenes and se-

Tab. 6.6. Syntheses of phenanthrenes, helicenes, circulenes, and heterocyclic alkenes by intramolecular McMurry reactions.

Product	TiCl$_x$	Reducing Agent	Solvent	Yield (%)	Ref.
Phenanthrenes, helicenes, and circulenes					
91	TiCl$_3$	Zn(Cu)	DME	57	125
92	TiCl$_3$	Zn(Cu)	DME	21	126
93	TiCl$_3$	Zn(Cu)	DME	78	127
94	TiCl$_3$	LiAlH$_4$	DME	35	129
Crownophanes, crown ether-calixarenes, and polyepoxyannulenes					
95	TiCl$_3$	C$_8$K	THF	40	130
96	TiCl$_4$	Zn	THF/Tol	25	131
97	TiCl$_4$	Zn	THF	57	133
98	TiCl$_4$	Zn(Cu)	THF[a]	9	136
Thiophenes and selenophenes					
99	TiCl$_4$	Zn	THF	73	137
100	TiCl$_4$	Zn	THF	30	138
Pyrrolo[2,1,5-cd]indolizines and porphyrins					
101	TiCl$_4$	Zn	THF[b]	95	139
102	TiCl$_4$	Zn(Cu)	THF	23	140
103	TiCl$_4$	Zn(Cu)	THF[c]	3	141
Ferrocenophanes					
104	TiCl$_4$	Zn	THF[a]	14	142
105	TiCl$_4$	Zn	THF[a]	32	143
106	TiCl$_3$	Zn(Cu)	DME	44	144

[a] In the presence of pyridine; [b] At 20 °C; [c] At 0 °C.

lenophenes such as **99**, **100** (Figure 6.18), and other thiophenes and selenophenes obtained by dehydration of the corresponding pinacols [137, 138], this reaction has not found recent applications.

Pyrrolo[2,1,5-cd]indolizines and porphyrins: It has recently been found that McMurry coupling opens an efficient route for the preparation of pyrrolo[2,1,5-cd]indolizines **101** and other substituted derivatives, by intramolecular reductive condensation of 3,5-diacylindolizines [139] (Figure 6.18). Another application of the McMurry reaction in nitrogen heterocyclic chemistry is in the synthesis of porphyrinoids; octaethylisocorrole **102** and corphycene **103** have been synthesized by intramolecular cyclization of the corresponding tetrapyrrolic α,ω-dialdehydes [140, 141]. In fact, most of the porphyrin derivatives have been obtained by tandem reactions (Section 6.2.3).

Ferrocenophanes: Complexes **104–106** have been synthesized by intramolecular coupling of ferrocenedicarbaldehydes [142–144] (Figure 6.18). Other isomers of **105**, in which the ethylene moiety bridges the *o*- or *m*-positions, have also been

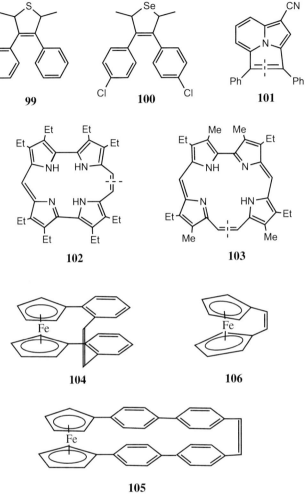

Figure 6.18.

isolated [143]. Transannular π electronic interactions between two aromatic rings of these compounds have been assessed on the basis of the NMR and electronic spectra. The *ansa*-(vinylene) ferrocene **106** is a precursor of conjugated organometallic polymers obtained by ring-opening metathesis polymerization [144].

6.2.2.2 Synthesis of Natural Products

It is primarily in its crucial and elegant use as a key step of numerous syntheses of natural products that the McMurry reaction has proved so successful. Indeed, intramolecular carbonyl coupling has proved to be one of the best methods for

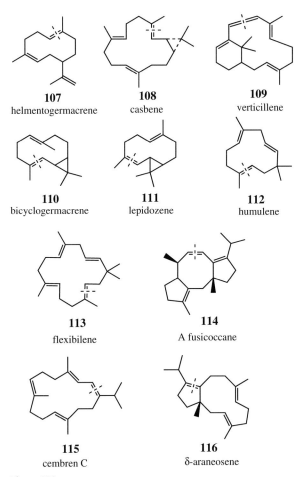

Figure 6.19.

the formation of the great variety of ring systems found in the terpene series [145–154]. Representative examples are given by compounds **107–116** (Figure 6.19; Table 6.7), many of which were synthesized by McMurry et al. in the 1980s.

The compatibility of many functional groups with the low-valent titanium reagents (see Section 6.2.1.3) allowed the syntheses of some highly substituted molecules to be envisaged, in particular poly(oxygenated) products, as exemplified by compounds **117–132** (Figures 6.20 and 6.21). The archaeal 36- and 72-membered macrocyclic ether lipids **117** and **118** are the largest rings yet obtained by means of a titanium coupling reaction [155, 156].

The McMurry reaction has also proved useful in the synthesis of a number of cembrenoid diterpenes, including crassin acetate methyl ether **119** [157], sar-

Tab. 6.7. Synthesis of natural products by intramolecular McMurry reactions.

Product	TiCl$_x$	Reducing Agent	Solvent	Yield (%)[a]	Ref.
107	TiCl$_3$	Zn(Cu)	DME	60	145
108	TiCl$_3$	Zn(Cu)	DME	75	146
109	TiCl$_3$	Zn(Cu)	DME	7	147
110	TiCl$_3$	Zn(Cu)	DME	61	148
111	TiCl$_3$	Zn(Cu)	DME	75	148
112	TiCl$_3$	Zn(Cu)	DME	60	149
113	TiCl$_3$	Zn(Cu)	DME	78	150
114	TiCl$_4$	Zn	THF[b]	84	152
115	TiCl$_4$	Zn	DME[b]	72	153
116	TiCl$_3$	Zn(Cu)	DME	90	154
117	TiCl$_3$	Zn(Cu)	DME	56	155
118	TiCl$_3$	Zn(Cu)	DME	66	156
119	TiCl$_3$	Zn(Cu)	DME	65	157
120	TiCl$_3$[c]	Zn(Cu)	THF	60	158
121	TiCl$_4$	Zn	THF[b]	58	159
122	TiCl$_3$	Zn(Cu)	DME	41	160
123	TiCl$_4$	Zn	THF	24	161
124	TiCl$_4$	Zn	DME[b]	81	163
125	TiCl$_3$	Zn(Cu)	DME	54	164
126	TiCl$_3$	Zn(Cu)	DME	11	165
127 (R = H)	TiCl$_3$	C$_8$K	DME	85	166
127 (R = Me)	TiCl$_3$	C$_8$K	DME	86	166
128	TiCl$_3$	C$_8$K	DME[d]	82	167
129	TiCl$_3$	Zn(Cu)	DME	52	168
130	TiCl$_3$	Zn(Ag)	DME	56	169
131	TiCl$_3$	Zn(Cu)	DME	32	170
132	TiCl$_3$	Zn(Cu)	DME	57	171

[a] Yield of the McMurry key step; [b] In the presence of pyridine; [c] In the presence of AlCl$_3$; [d] At 20 °C.

cophytols **120** and **121** [158, 159], and various hydroxy- and epoxy-cembrenes such as **122**–**124** [160–163]. Compounds **125** and **126** possess the ring systems of terpestacin and trinervitane, respectively [164, 165]. The synthesis of compactin and mevinolin, **127** (R = H, Me), could be performed in an efficient and reproducible way by using titanium-graphite in place of the standard TiCl$_3$-Zn(Cu) combination [166]. The compatibility of the ester group with the titanium reagents is demonstrated by the preparations of lasiodiplodin **128** and strigol **129** [167, 168].

The estrone methyl ether **130** [169], kempene-2 **131** [170], and isokhusimone **132** [171] were easily isolated following McMurry reactions of their tricarbonyl precursors as their intermolecular dimerization is much slower than the intramolecular cyclization.

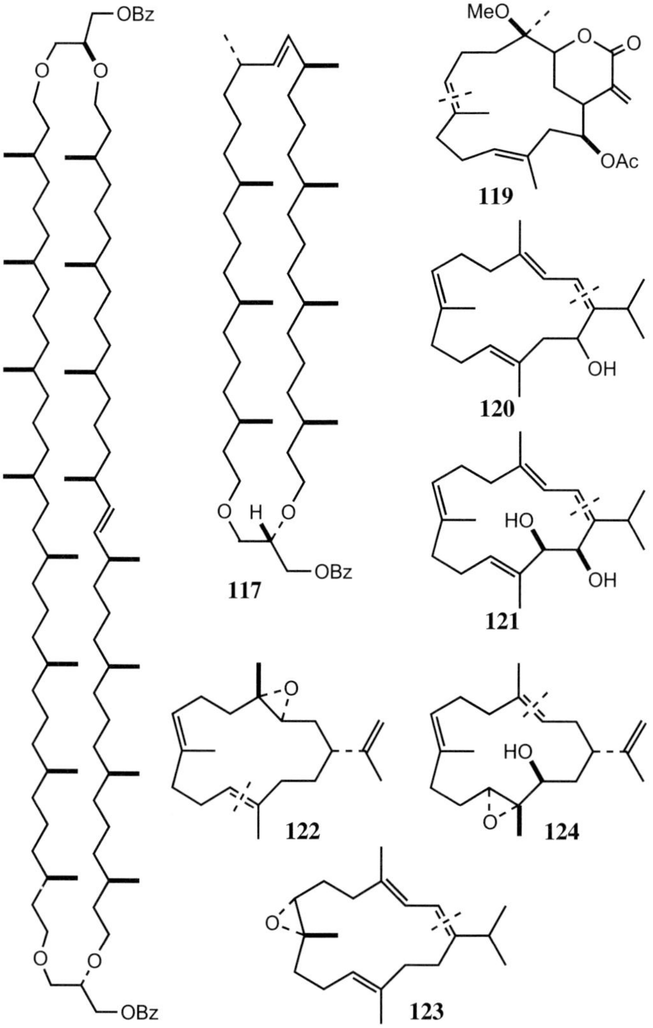

117 118 119 120 121 122 123 124

Figure 6.20.

6.2.3
Tandem Coupling Reactions

The tandem McMurry reaction of dicarbonyl compounds consists of an initial intermolecular coupling followed by an intramolecular cyclization (Scheme 6.3).

As noted for polymerization reactions (Section 6.2.1.5), the first intermolecular coupling should be favored over the cyclization. Thus, changing the propylene fragment to an ethylene or methylene group in the dicarbonyl precursor of compound **83** facilitated the tandem reaction [112]. The yields of such two-step pro-

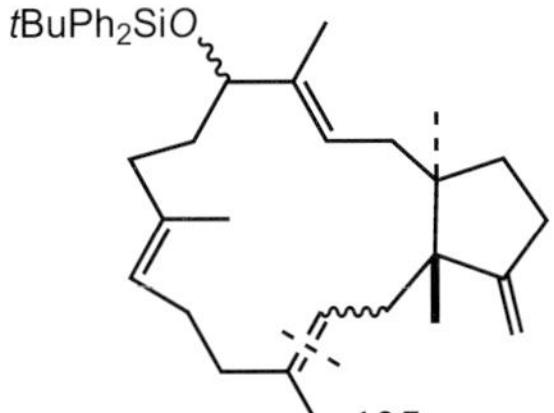

125

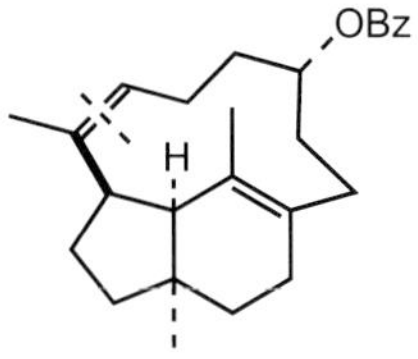

126

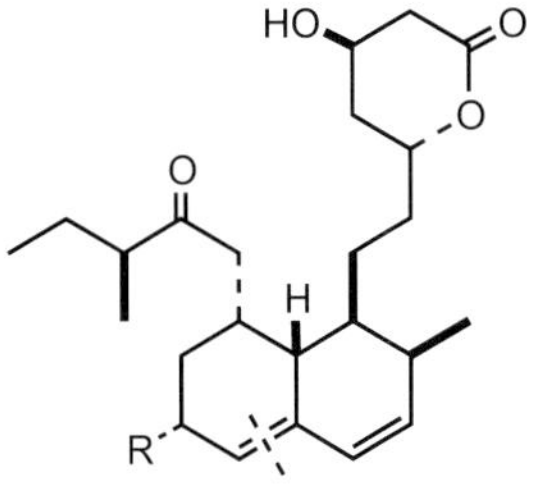

127

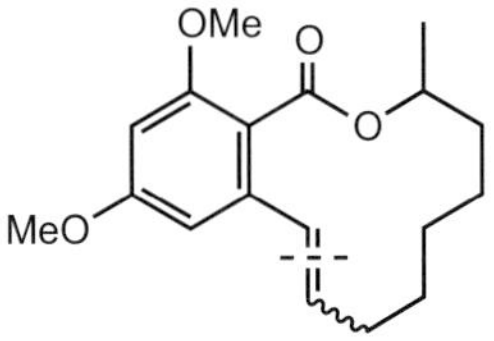

128

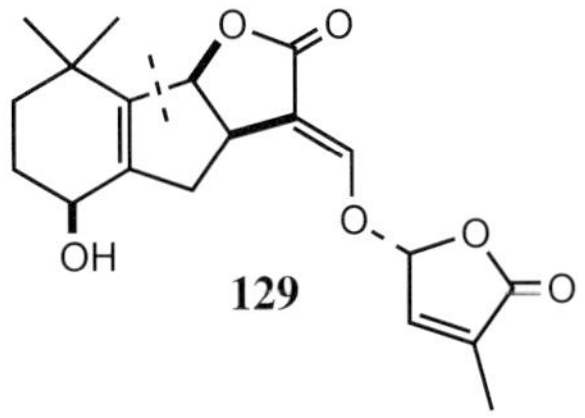

129

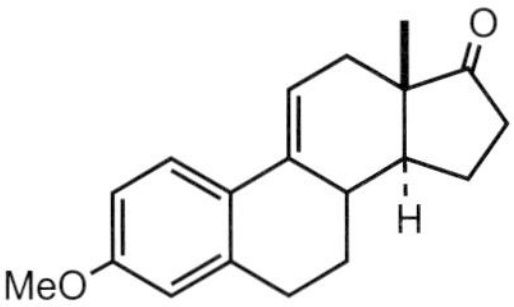

130

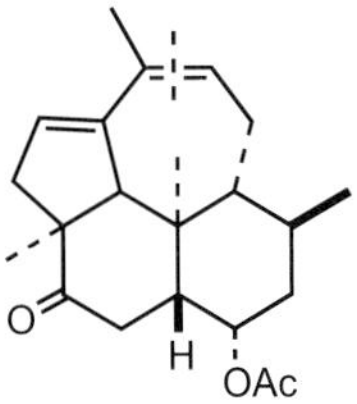

131

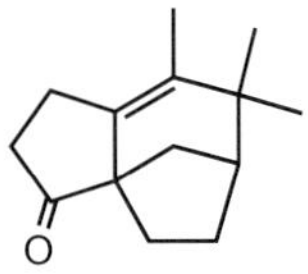

132

Figure 6.21.

cesses are usually modest; formation of the *Z*-isomer in the first step is generally required for cyclization to take place [112, 172].

The tandem McMurry reaction has been successfully used for the synthesis of cyclophanes (Figure 6.22; Table 6.8) such as **133** and **134** [114, 115, 173], and is a classical route to "expanded" porphyrin systems.

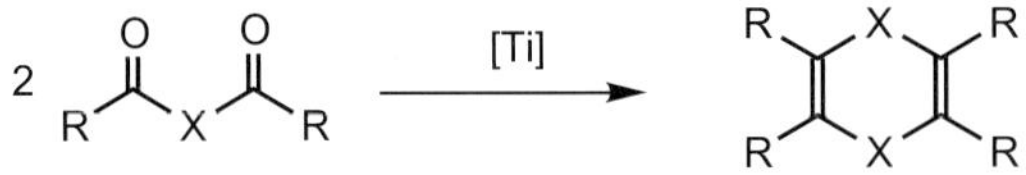

Scheme 6.3. The tandem McMurry reaction.

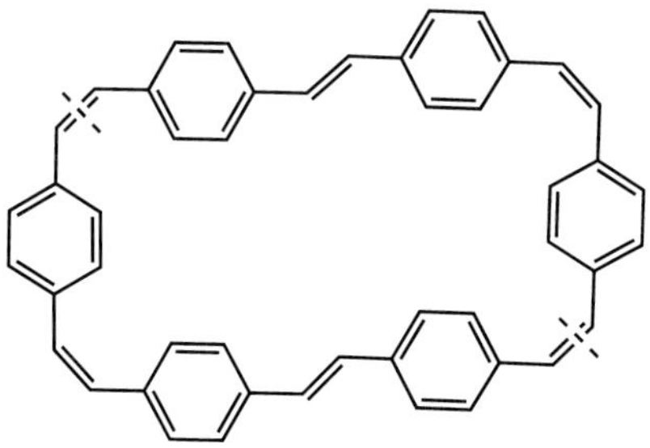

133

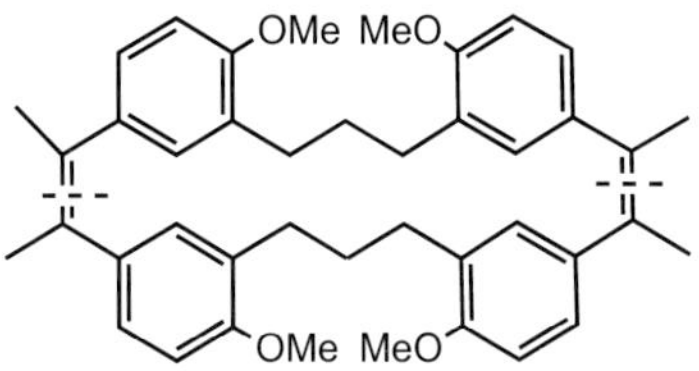

134

Figure 6.22.

After the preparation of porphycene and its substituted derivatives such as **135** [174], a variety of porphyrin analogues were isolated, in which pyrrole units were replaced with furan or thiophene rings and/or extended by additional ring systems [135, 175–185]; these are illustrated by compounds **136–142** (Figures 6.23 and

Tab. 6.8. Syntheses of cyclic alkenes and polyenes by McMurry tandem reactions.

Product	$TiCl_x$	Reducing Agent	Solvent	Yield (%)	Ref.
133	$TiCl_4$	Zn	DME/Tol	41	114
134	$TiCl_4$	Zn	THF[a]	2	115
135	$TiCl_4$	Zn	THF	15	174
136	$TiCl_4$	Zn(Cu)	THF	7	175
137	$TiCl_4$	Zn(Cu)	THF	6	135
138	$TiCl_4$	Zn	THF[a]	64	178
139	$TiCl_3$	Zn(Cu)	DME	15	179
140	$TiCl_4$	Zn	THF[a]	69	181
141	$TiCl_4$	Zn	THF[a]	7	183
142	$TiCl_4$	Zn	THF[a]	7	184
143	$TiCl_4$	Zn(Cu)	THF[a]	18	186
144	$TiCl_4$	Zn(Cu)	THF[a]	10	187
145 (R = Me)	$TiCl_4$	Zn(Cu)	DME	8	188

[a] In the presence of pyridine.

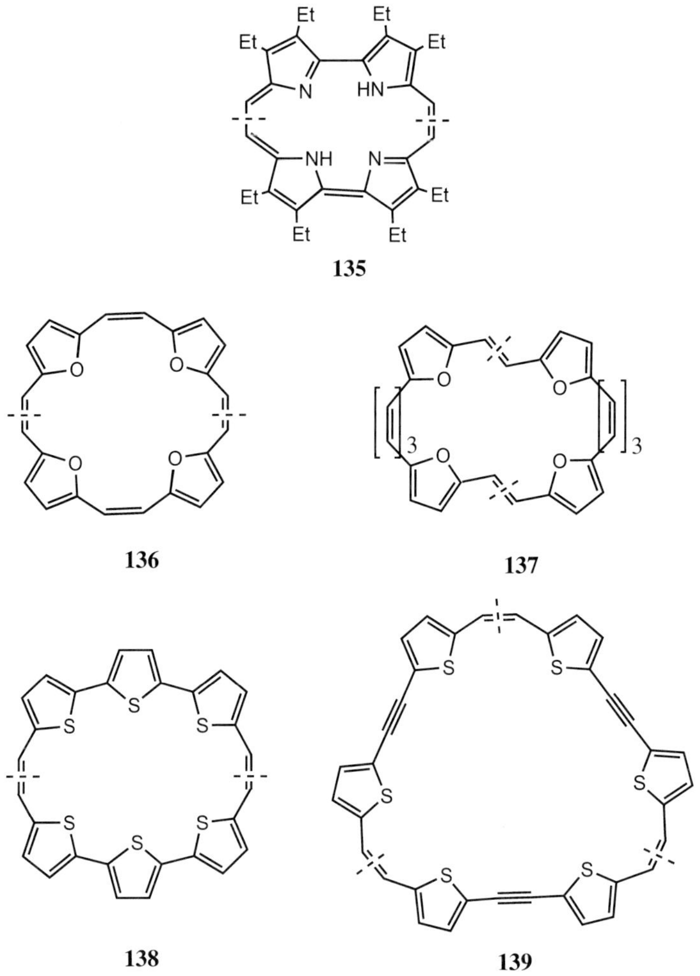

Figure 6.23.

6.24). Some of these molecules are of current interest as effective photosensitizers for biomedical applications and are employed in the photodynamic therapy of tumors and viral inhibition. Compound **142**, oxobronzaphyrin, is noteworthy since it contains three distinct heterocycles and was isolated from a cross-tandem coupling reaction between a furan dipyrraldehyde and a thiophene dipyrraldehyde [184].

More recently, novel aromatic porphycene analogues containing thiazole rings have been obtained by oxidation of **143**, which was itself prepared by a tandem reaction of the corresponding dicarbaldehydes [186] (Figure 6.25). The spectroscopic properties of the thiazole-derived annulene **144** are similar to those of **143** and do not reflect the typical paratropic behavior of the planar annulene structures of the porphycenes [187]. Finally, the tandem reaction has been elegantly used in the synthesis of the novel cage molecule **145** (R = Me or n-hexyl), with the formation of three C–C double bonds by couplings of tris(5-formyl-2-thienyl)methanes [188].

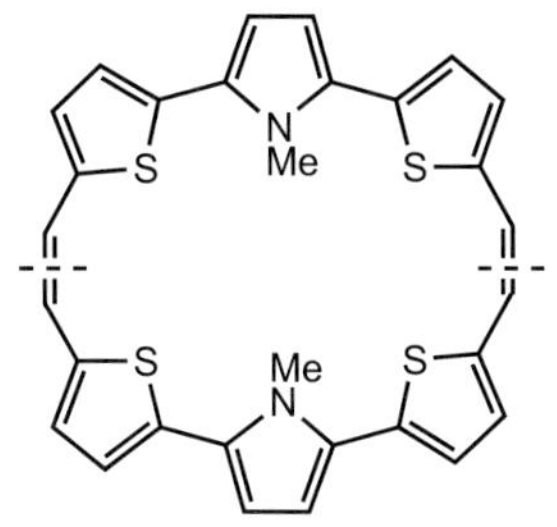

6.2 Scope of the McMurry Reaction | 253

140

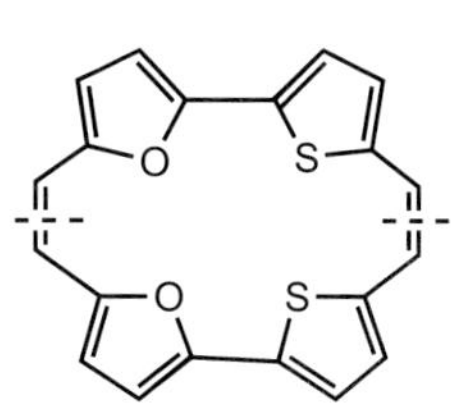

141

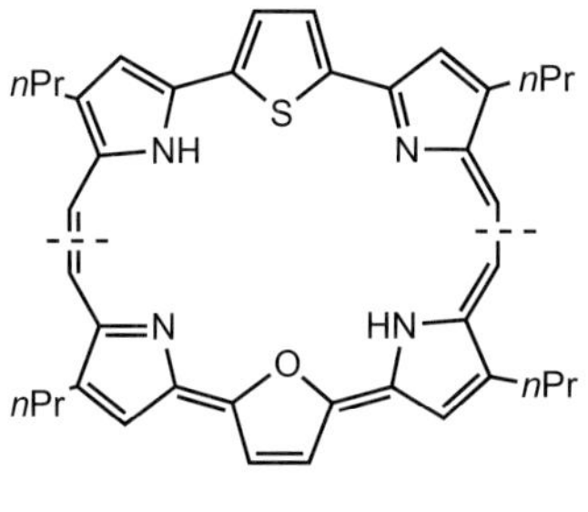

142

Figure 6.24.

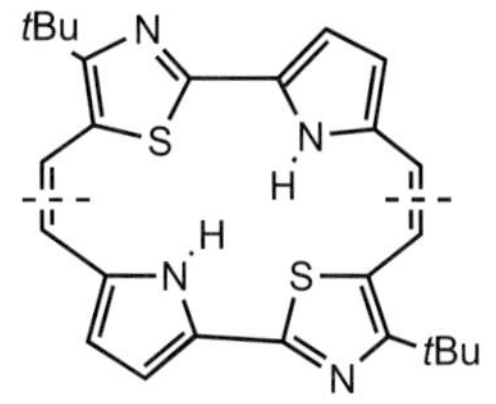

143

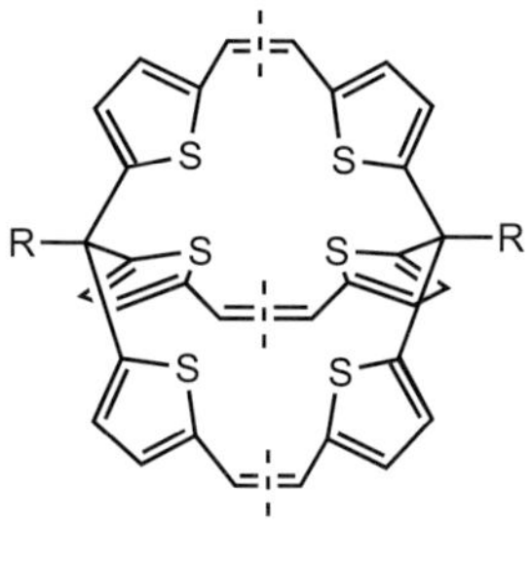

145

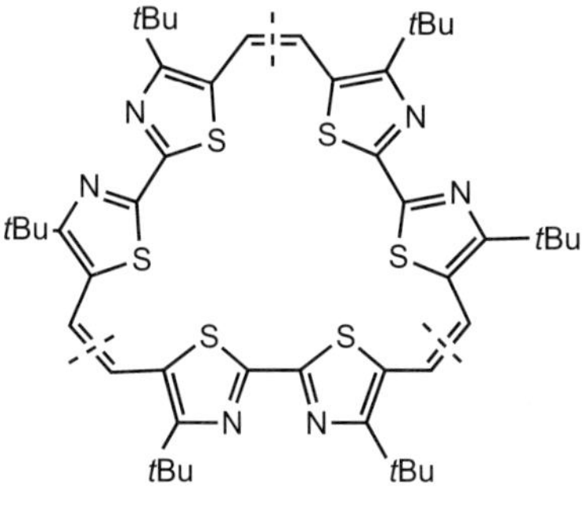

144

Figure 6.25.

6.2.4
Keto Ester Couplings

6.2.4.1 Intermolecular Keto Ester Couplings

The McMurry keto ester coupling gives enol ethers, which, after acidification, are transformed into the corresponding ketones (Scheme 6.4).

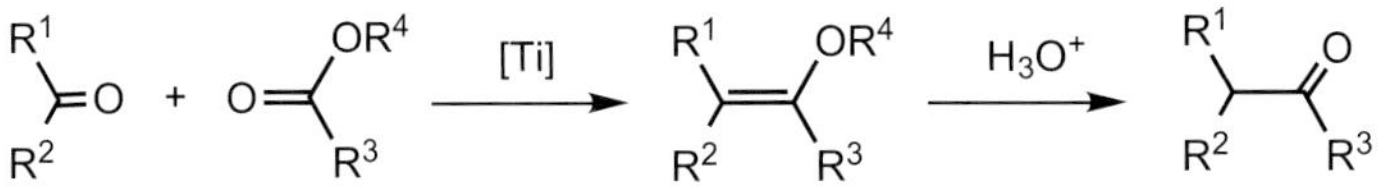

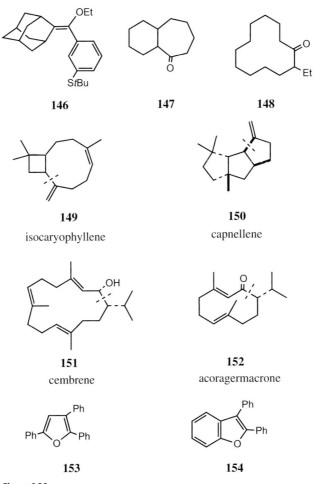

Scheme 6.4. Synthesis of ketones by McMurry keto ester coupling.

146

147

148

149
isocaryophyllene

150
capnellene

151
cembrene

152
acoragermacrone

153

154

Figure 6.26.

6.2 Scope of the McMurry Reaction | 255

Tab. 6.9. Syntheses by McMurry keto ester coupling reactions.

Product	TiCl$_x$	Reducing Agent	Solvent	Yield (%)	Ref.
146	TiCl$_3$	LiAlH$_4$	DME[a]	60	189
147	TiCl$_3$	LiAlH$_4$	DME[a]	82	190
148	TiCl$_3$	LiAlH$_4$	DME[a]	63	190
149	TiCl$_3$	LiAlH$_4$	DME[a]	38[b]	191
150	TiCl$_3$	LiAlH$_4$	DME[a]	72[b]	192
151	TiCl$_3$	LiAlH$_4$	DME[a]	81[b]	193
152	TiCl$_3$	LiAlH$_4$	DME[a]	62[b]	194
153	TiCl$_3$	C$_8$K	THF	92	196
154	TiCl$_3$	C$_8$K	THF	88	196

[a] In the presence of NEt$_3$; [b] Yield of the McMurry key step.

Very recently, the scope and limitations of an unprecedented intermolecular McMurry cross-coupling between a ketone (2-adamantanone) and an ester (3-substituted benzoate) have been studied; the products, such as **146**, were obtained in acceptable yields by using the TiCl$_3$-LiAlH$_4$-Et$_3$N combination [189] (Figure 6.26; Table 6.9).

6.2.4.2 Intramolecular Keto Ester Cyclizations; Synthesis of Cyclanones

The intramolecular McMurry coupling of keto esters proved to be more problematic than the cyclization of diketones, mainly because the esters are not very reactive. Also in contrast to conventional couplings of diketones, yields are better for smaller rings than for medium-sized rings. The best results were obtained by using the TiCl$_3$-LiAlH$_4$ reagent, sometimes in the presence of triethylamine, which inhibits protonolysis of the product enol ether, or the TiCl$_3$-C$_8$K system. This process has found application in the synthesis of substituted cyclanones, such as **147** and **148** [190], and of some natural products, **149–152** [191–195] (Figure 6.26).

6.2.4.3 Intramolecular Cyclizations of Acyloxycarbonyl Compounds; Synthesis of Furans

McMurry-type reactions of acyloxycarbonyl compounds (Scheme 6.5) are carried out with the TiCl$_3$-C$_8$K system, giving furans and benzo[b]furans in good yields; compounds **153** and **154** are representative examples [196] (Figure 6.26).

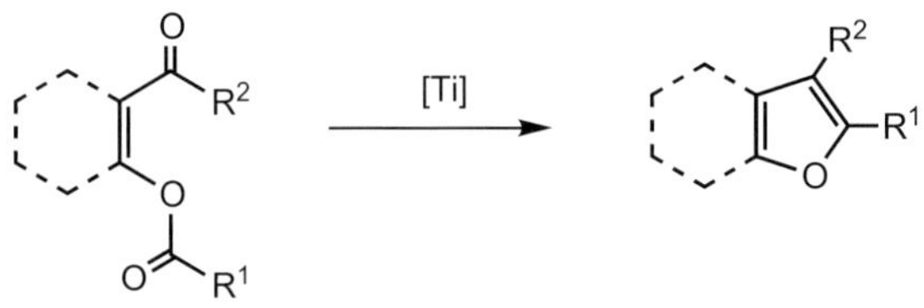

Scheme 6.5. Synthesis of furans by McMurry ketone ester cyclization.

6.2.5
Intramolecular Couplings of Acylamidocarbonyl Compounds; Synthesis of Pyrroles and Indoles

The intramolecular coupling of oxoamides (Scheme 6.6), which opens an efficient route to pyrroles [197] and indoles [64, 198–204], represents a remarkable and very useful extension of the McMurry reaction.

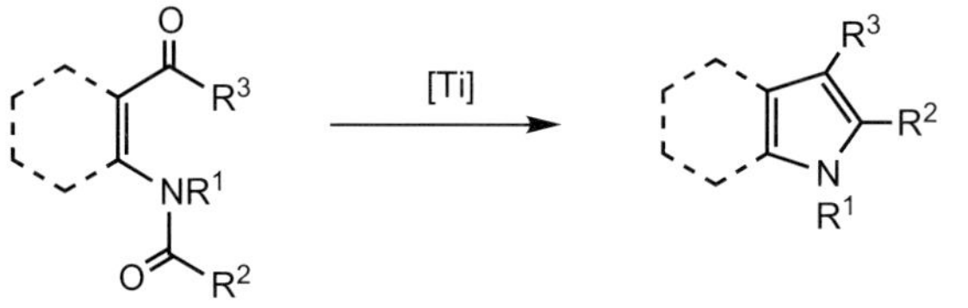

Scheme 6.6. Synthesis of pyrroles and indoles by McMurry ketone amide cyclization.

This cyclization reaction was discovered and developed by Fürstner et al., who mainly used the $TiCl_3$-C_8K combination as the coupling agent, as well as the

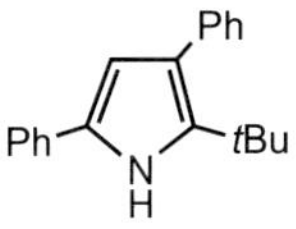

155

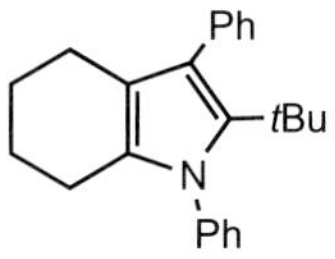

156

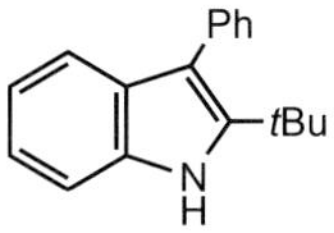

157

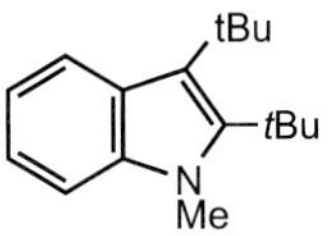

158

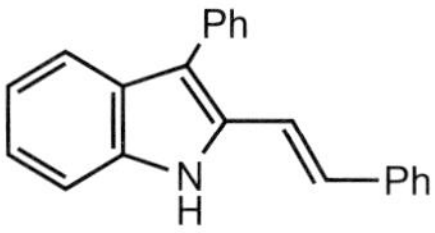

159

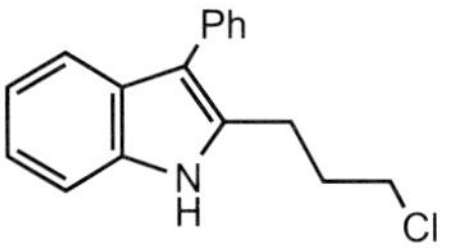

160

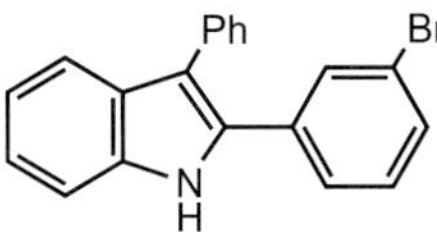

161

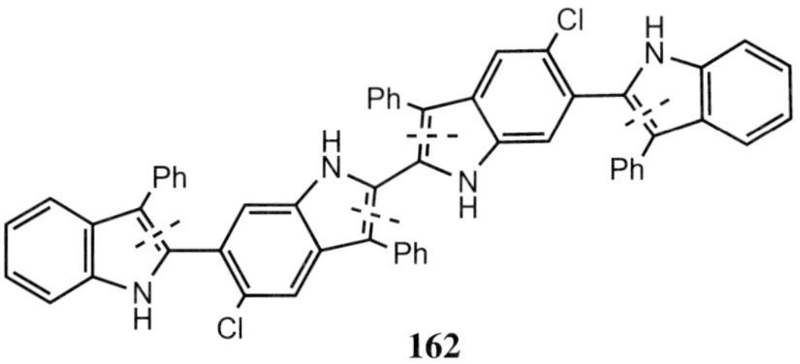

162

Figure 6.27.

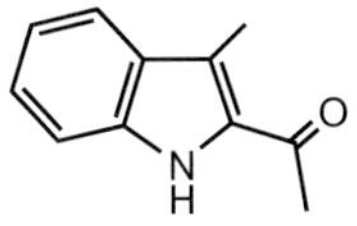

163

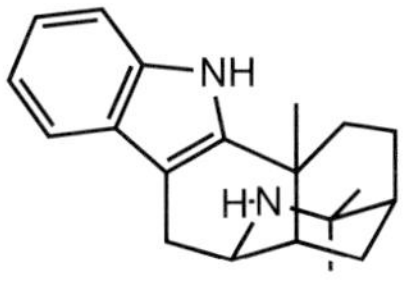

164

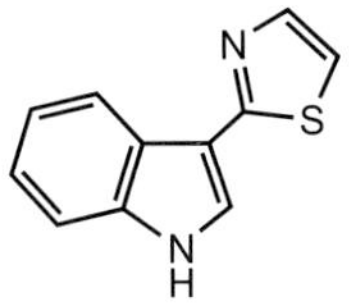

165

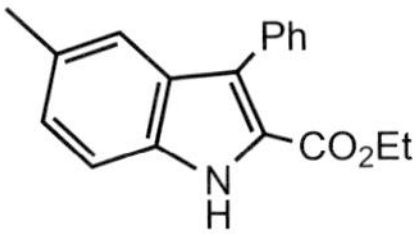

166

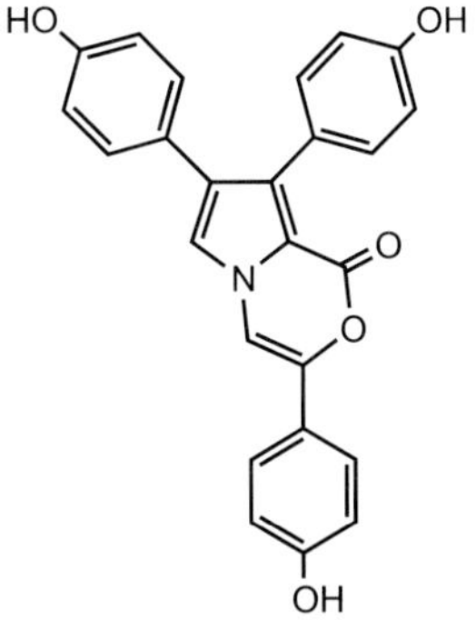

167

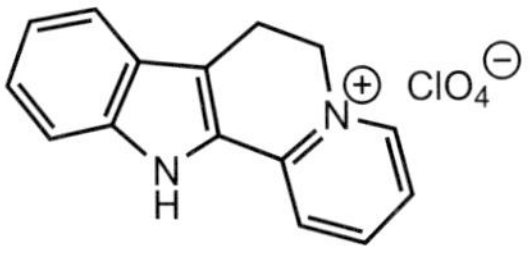

168

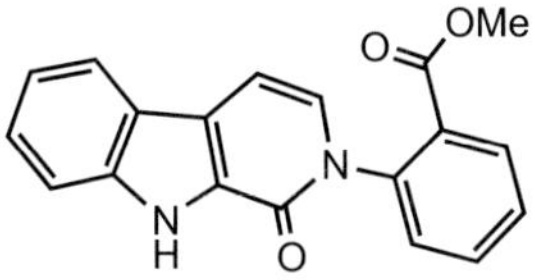

169

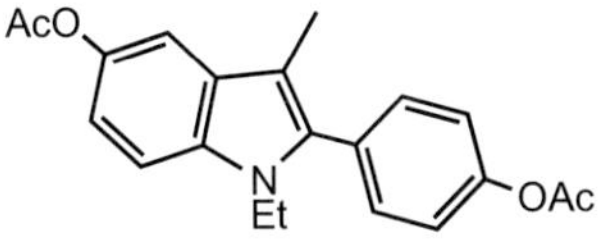

170

Figure 6.28.

$TiCl_3$-Zn system in the "instant method" (Section 6.3.1). Examples **155–170** show that the reaction is highly flexible with respect to the substitution pattern in the enamine region of the molecule, especially in relation to the substituent at C2 on the heteroarene ring (Figures 6.27 and 6.28; Table 6.10).

The method is suitable for the synthesis of sterically encumbered indoles, such as **158** [199], and is compatible with a variety of functional groups, including alkenes (e.g. **159** [196]) and alkyl or aryl halides (e.g. **160** and **161** [196, 199]). Most strikingly, the formation of the pyrrole or indole ring is highly favored over inter- or intramolecular McMurry couplings of polyfunctional substrates. This is illustrated by the synthesis of oligoindoles such as **162**, which were obtained in yields greater than 80%; neither McMurry coupling of the keto groups, nor formation of 2-quinolones upon reaction of the ketones with the distal amides, nor intermolecular

Tab. 6.10. Syntheses of pyrroles and indoles by McMurry reactions.

Product	TiCl$_x$	Reducing Agent	Solvent	Yield (%)	Ref.
155	TiCl$_3$	Zn	DME	60	197
156	TiCl$_3$	Zn	THF	58	197
157	TiCl$_3$	C$_8$K	DME	84	196
158	TiCl$_3$	C$_8$K	DME	35	199
159	TiCl$_3$	C$_8$K	THF	68	196
160	TiCl$_3$	C$_8$K	THF	90	199
161	TiCl$_3$	C$_8$K	THF	81	196
162	TiCl$_3$	Zn	DME	80	200
163	TiCl$_3$	C$_8$K	DME	60	201
164	TiCl$_3$	Zn	THF	75	199
165	TiCl$_3$	Zn	DME	71	202
166	TiCl$_3$	C$_8$K	DME	93	201
167	TiCl$_3$	C$_8$K	DME	52	197
168	TiCl$_3$	C$_8$K	THF	22	202
169	TiCl$_3$	C$_8$K	THF	60	203
170	TiCl$_3$	C$_8$K	DME	71	204

coupling processes were found to interfere with the domino two-by-two ketoamide cyclizations [200].

The regio- and chemoselectivity of this titanium-induced synthesis of indoles have led to elegant applications to natural products and biologically active compounds, including salvadoricine **163** [201], (+)-aristoteline **164** [199], camalexin **165** [202], and **166**, a known precursor of diazepam [201]. It is important to note that the stereogenic centers in the substrate are not racemized under the reaction conditions. The ketoamide cyclization has also been the key step in the syntheses of other pyrrolo- and indolo-alkaloids, such as those of lukianol A **167** [197], indolopyridocoline **168** [202], secofascaplysin **169** [203], and the anticancer agent zadoxifen **170** [204].

Most notably, the oxoamide cyclization could be rendered catalytic in titanium when the reaction was carried out in the presence of a chlorosilane [64]. The method is based upon the in situ generation of the active low-valent titanium species by reduction of TiCl$_3$ with Zn in the presence of the substrate, followed by reconversion of the formed titanium oxychlorides into titanium chloride through ligand-exchange reaction with the chlorosilane (Scheme 6.7). The quantity of TiCl$_3$ necessary for good yields depends on the chlorosilane: 5–10 mol% with Me$_3$SiCl, 1–2 mol% with ClMe$_2$SiCH$_2$CH$_2$SiMe$_2$Cl. The synthesis of 3-alkylated indole derivatives is favored by using ClMe$_2$Si(CH$_2$)$_3$CN as the additive.

6.2.6
Reductive Coupling of Benzylidene Acetals

The reductive cleavage of benzylidene acetals using the TiCl$_3$-Li combination was found to give aryl alkanes and stilbenes (22–45% yield). Aliphatic acetals, however,

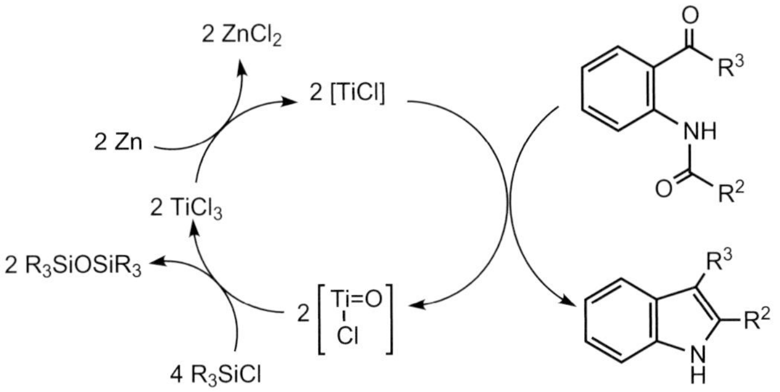

Scheme 6.7. Titanium-catalyzed indole synthesis.

remain unaffected. This cleavage offers an attractive, alternative stereoselective route to stilbenes in a modified McMurry reaction [205].

6.3
Procedures and Reagents Used in the McMurry Reactions

6.3.1
Procedures

When they discovered the reductive coupling of carbonyl compounds by means of low-valent titanium reagents, Tyrlik and Mukaiyama used two distinct procedures. In the Tyrlik experiment, $TiCl_3$ was reduced with Mg in THF and the carbonyl substrate was then added to the slurry thus obtained [1]. This two-step protocol was also followed by McMurry, who used $TiCl_3$ in combination with a variety of reducing agents (Section 6.3.2.1), and is still that commonly employed in most of the coupling reactions. Such a two-step sequence is imperative when the carbonyl substrate is not inert towards the reducing agent, and also when high dilution techniques are required in intramolecular cyclization reactions. However, when the reducing agent, zinc dust or iron powder, is not strong enough to affect the carbonyl group, McMurry reactions can be carried out in a one-pot procedure, by mixing and heating all the components together. This one-step protocol was that used by Mukaiyama with the $TiCl_4$-Zn system [2] and has been reintroduced by Fürstner et al. with the so called "instant method" [199]. In fact, a two-step procedure with this system is not possible since $TiCl_3$ can be reduced by Zn only if its redox potential has been lowered by coordination to the carbonyl; the active titanium species is then produced regioselectively at the reaction site. The one-step and two-step procedures, with the $TiCl_3$-Zn and $TiCl_3$-C_8K systems, respectively, were found to be equally effective in the reductive coupling of oxoamides to indoles, but the former is much easier to perform as all hazardous reagents are avoided. One-pot cyclizations, which are compatible with many functional groups, can also be carried out in non-ethereal solvents such as dimethylformamide (DMF), ethyl acetate, and acetonitrile [199].

It is well established that temperature has a strong influence on the course of the McMurry reaction, higher temperatures favoring the deoxygenation of the pinacolate intermediates. Although not studied systematically, it is clear that the solvent also plays a major role during the coupling process, affecting the yield, the nature, and the stereochemistry of the products [89, 206]. For example, treatment of acetophenone with the TiCl$_3$-Li system in tetrahydrofuran (THF) or dimethoxyethane (DME) affords the corresponding alkene in excellent yield but with E:Z ratios equal to 0.4 and 9, respectively [206]. The tandem reaction leading to the cyclophane **85** was found to be favored over the intramolecular cyclization of the dialdehyde precursor when a mixed solvent of DME and toluene was used in place of DME or THF alone [114].

6.3.2
Reagents

With a few exceptions, the active low-valent titanium species in McMurry reactions are obtained by reduction of TiCl$_4$ or TiCl$_3$; the choice of the reducing agent is of great importance.

6.3.2.1 The TiCl$_4$- and TiCl$_3$-Reducing Agent Systems

After the discovery of the TiCl$_3$-LiAlH$_4$ system [3], McMurry made a great effort to prepare more efficient low-valent titanium species by reduction of TiCl$_3$ with alkali metals, according to the method of Riecke [207–209], and with zinc-copper couple [210]. The best results were obtained with the TiCl$_3$(DME)$_{1.5}$-Zn(Cu) combination, which gave the highest yields in a reproducible manner [14]. The efficacy of this reagent is illustrated by the coupling of iPr$_2$C=O, which afforded iPr$_2$C=CiPr$_2$ in 87% yield, instead of 17% by using TiCl$_3$-LiAlH$_4$ and 37% by using TiCl$_3$-K. The sky-blue solvate complex of TiCl$_3$ and DME is more stable and less prone to hydrolysis than the free salt, and DME has been strongly recommended over THF as the solvent for the coupling reaction because THF, being a good donor of H·, would quench a proportion of the intermediate ketyl radicals before they could dimerize. However, the TiCl$_3$-LiAlH$_4$ system remains the most efficient for the keto ester coupling reaction (Section 6.2.4).

Although the TiCl$_3$-Zn(Cu) system in DME has been recommended as the standard coupling reagent for the synthesis of alkenes, examination of Tables 6.1–6.10 shows that the TiCl$_4$-Zn and TiCl$_4$-Zn-pyridine systems in refluxing THF or dioxane are also efficient and frequently used in the reductive coupling of aldehydes and ketones. However, it is noteworthy that fluorenone was not coupled to give bifluorenylidene with the TiCl$_4$-Zn-pyridine system, but was instead transformed into terfluorenyl in 71% yield [211].

While the TiCl$_3$-Na and TiCl$_3$-K reagents are no longer used as general coupling reagents for reasons of safety and efficacy, Fürstner et al. found that reduction of TiCl$_3$ with high surface area sodium gave a highly active titanium species supported on Al$_2$O$_3$, NaCl or TiO$_2$ [63, 212], and that titanium-graphite combinations could be obtained by treating TiCl$_3$ with the potassium graphite laminate C$_8$K; the most efficient reagents were formed with only two equivalents of sodium carrier or C_8K [199]. Aromatic aldehydes and ketones are coupled to form the corresponding alkenes by low-valent titanium on alumina, and the most interesting feature of this reagent is its pronounced template effect since cycloalkenes are obtained irrespective of ring size, without using high dilution techniques [212]. This reagent also proved useful in the first example of a McMurry-type coupling of an acylsilane, leading to the formation of the disilylated stilbenes **45** [63]. Titanium-graphite was found to be an efficient coupling agent for all types of substrates, in particular the oxoamides [64, 198, 199, 201]. In the synthesis of (+)-compactin **127** (R = H), it was the only one among all the titanium reagents to promote the key cyclization step [166].

Reaction of $TiCl_4$ with two equivalents of BuLi in THF gave black, analytically pure $TiCl_2$, which was separated from the LiCl by-product and was obtained as the bis(THF) adduct in toluene [213–215]. It was found that this Ti(II) reagent was capable of coupling aromatic aldehydes and ketones to give the corresponding alkenes with high *E*-stereoselectivity.

In some cases, the efficacy of the low-valent titanium species has been enhanced by ultrasonication. Under such conditions, reductive coupling of acetophenone with $TiCl_3$-Li was complete after 45 min at 30 °C in THF, as compared to 16 h at 65 °C without sonication; however, the reaction stopped at the pinacol stage when performed in DME [206]. The synthesis of tetraferrocenylethylene **48** could only be achieved when the $TiCl_3$-Li reagent was ultrasonicated [66].

6.3.2.2 Effect of Additives on the $TiCl_4$- and $TiCl_3$-Reducing Agent Systems

The effects of a number of additives on the $TiCl_4$- and $TiCl_3$-reducing agent systems have been considered with a view to enhancing the formation of the alkene products of the coupling reaction or, on the contrary, to stop it at the pinacol stage. This tuning of the activity of the low-valent titanium species has been achieved by changing the ligand environment of the metal and/or by adding chemical redox agents.

Promoted McMurry olefination reactions: The beneficial effect of pyridine on the $TiCl_3$-Zn reagent has long been recognized [216]. Thus, with this system, bis([3]-ferrocenophane-1-ylidene) **49** and its *cis* isomer could be prepared in 70% yield only after the addition of two equivalents of pyridine (py); in the absence of the aromatic base, a pinacol rearrangement product was obtained in 78% yield, which can be attributed to the Lewis acidity of the $ZnCl_2$ that is formed during the reaction [67] (Scheme 6.8).

Activation of low-valent titanium species is highly desirable in the case of McMurry alkene syntheses where the required high temperature is detrimental for compounds containing thermolabile functional groups. External addition of group I and II metal halides, iodine, or arenes to the $TiCl_3$-Li system in THF dramatically enhances its activity (Table 6.11).

Thus, coupling of acetophenone with this system in the presence of two equivalents of KCl or 0.25 equivalents of I_2 afforded the corresponding stilbene in 82

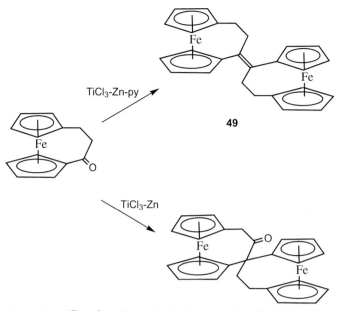

Scheme 6.8. Effect of pyridine on the McMurry coupling of a ferrocenyl ketone.

and 90% yield, respectively, after 2 h at room temperature [217, 218], whereas the same reaction without additive gave the pinacol in 89% yield. Aliphatic ketones such as cyclopentanone and cyclohexanone were also coupled to give the corresponding alkenes in good yield at 25 °C. The *E*-selectivity in the formation of stilbene was augmented on addition of KCl or I_2 to the $TiCl_3$-Li reagent at low temperature. Metal-arenes have been found to be efficient soluble organic reductants for $TiCl_3$; the $TiCl_3$-Li-naphthalene combination proved to be the best for the McMurry olefination of both aromatic and aliphatic substrates at a lower temperature and in a much reduced time as compared to the conventional procedures, and has also proved useful for the synthesis of O- and N-heterocycles through intramolecular carbonyl coupling at room temperature [219]. Subtle changes in the

Tab. 6.11. Effect of additives on the McMurry reaction of acetophenone with the $TiCl_3$-Li system.

Additive (equiv.)	Time (h); temperature (°C)	Alkene % yield (E:Z)	Pinacol % yield (dl:meso)	Ref.
none	16; 25	trace	89 (75:25)	217
none	2; reflux	87 (75:25)	0	217
KCl (2)	2; 25	82 (85:15)	0	217
I_2 (0.25)	2; 0–5	90 (80:20)	0	218
Naphthalene (0.25)	1.5; reflux	89 (56:44)	0	219
Naphthalene (0.25)	2; 25	77 (68:32)	0	219

preparation of the reagents were found to influence the stereoisomeric ratio of the alkenes.

The McMurry pinacol reaction: The challenge of the McMurry pinacol reaction is to obtain the 1,2-diol products in good chemical yields, with a high and predictable diastereoselectivity. Usually, the reaction is performed at room temperature or below, in order to prevent deoxygenation of the pinacolate intermediates and formation of the corresponding alkene; the pinacol is liberated upon basic hydrolysis. The $TiCl_3(DME)_{1.5}$-Zn(Cu) system, which has been recommended for the McMurry alkene synthesis, has also proved very useful for the preparation of cyclic 1,2-diols by carrying out the reaction at 25 °C rather than at solvent reflux temperature. By assuming that the product stereochemistry is controlled by steric interactions in the ring being formed, some prediction of the reaction diastereoselectivity could be made [220]. Various natural products have also been synthesized, namely sarcophytol B [221], crassin [222], periplanone C [223], and taxol [224–228], the latter being perhaps the most famous natural product to have been prepared with the aid of low-valent titanium compounds. The $TiCl_4$-Mg(Hg) system was found to be very efficient, at 0 °C in THF, for the intra- and intermolecular pinacol coupling of aliphatic and aromatic aldehydes and ketones; reaction of a cyclic ketone with an excess of acetone gave the unsymmetrical diol [229].

The original Mukaiyama and Tyrlik reagents, $TiCl_4$-Zn and $TiCl_3$-Mg in THF, have been modified in order to favor the intermolecular McMurry pinacol reaction and to enhance its diastereoselectivity. The effects of solvent, additives, and/or modified ligands can be illustrated by the reaction of benzaldehyde, and are summarized in Table 6.12.

Aromatic aldehydes are coupled in CH_2Cl_2 with the $TiCl_4(THF)_2$-Zn system to yield pinacols with high *threo* selectivity; the diastereoselectivity is even better after addition of DME or tetramethylethylenediamine (TMEDA) [230]. The $TiCl_4$-Bu_4NI system has the same reducing ability as low-valent titanium reagents, indicating that non-metallic species can also reduce Ti(IV) compounds [231]. The reductive dimerization of acetophenone could be stopped at the pinacol stage by the addition of 10 equivalents of pyridine to the $TiCl_3$-Mg system and, in the presence of a stoichiometric amount of a mono- or dihydroxy auxiliary, pinacols were obtained in higher yields and with better selectivities ($dl/meso \approx 4$–5). Among these additives, catechol proved to be the most useful for the complete pinacolization of aromatic carbonyl substrates, even under reflux conditions [232]. Pinacols were formed with poor stereoselectivity ($dl/meso = 1.3$) by the coupling of aromatic carbonyl compounds with aqueous $TiCl_3$ in basic media, but with $TiCl_3$ in anhydrous CH_2Cl_2 the pinacolization was highly diastereoselective ($dl/meso > 100$) [233].

Recently, it has been found that reductive coupling of aromatic and aliphatic aldehydes and ketones proceeds under mild conditions to give the corresponding pinacols in moderate to high yields and with good *dl*-diastereoselectivities by using combinations of either $TiCl_2$ and Zn, $TiBr_2$ and Cu, or TiI_4 and Cu in dichloromethane in the presence of pivalonitrile [234, 235]. Pinacol coupling of benzalde-

Tab. 6.12. Pinacol coupling of benzaldehyde induced by various low-valent titanium systems and reaction conditions.

Titanium system	Temp. (°C)	Solvent	Additive	Yield % (dl:meso)	Ref.
$TiCl_4(THF)_2$-Zn	25	THF	–	98 (3:1)	230
$TiCl_4(THF)_2$-Zn	25	CH_2Cl_2	–	57 (94:1)	230
$TiCl_4(THF)_2$-Zn	25	CH_2Cl_2	TMEDA	77 (dl only)	230
$TiCl_4$-nBu_4NI	–78	CH_2Cl_2	–	99 (90:10)	231
$TiCl_3$-Mg	80	THF	catechol	76 (1.16:1)	232
$TiCl_3$	25	THF	–	41 (100:1)	233
$TiCl_3$	25	CH_2Cl_2	–	65 (200:1)	233
$TiCl_2$-Zn	–23	CH_2Cl_2	tBuCN	91 (45:55)	234
$TiBr_2$-Cu	–23	CH_2Cl_2	tBuCN	95 (96:4)	234
TiI_4-Cu	–23	CH_2Cl_2	tBuCN	94 (98:2)	234
$TiCl_2$	25	THF	TMEDA	86 (dl only)	236
$TiCl_2$	25	THF	chiral amine	82 (82:5.7)[a]	236
$TiCl_2$	–78	THF	chiral amine	31 (81:19)[b]	238
$TiCl_3$[c]-Zn-Me_3SiCl	0	THF	tBuOH	90 (69:31)	240
$TiCl_3$[c]-Zn-Me_3SiCl	0	THF	tBuOH, DEPU	89 (88:12)	240
$TiCl_4$[c]-Mn-Me_3SiCl	25	MeCN	Schiff base	40–80 (>90:10)	241

[a] 41% ee. [b] 65% ee. [c] Catalytic amount of titanium chloride.

hyde with $TiCl_2$ in THF in the presence of TMEDA gave the dl-hydroxybenzoin with high diastereoselectivity; by using an optically active amine in place of TMEDA, the homogeneous reaction afforded the product with 40% ee [236, 237]. Similar reaction of benzaldehyde with $TiCl_2$ in the presence of two equivalents of (S)-2-methoxymethyl-pyrrolidine generates the diol with an enantiomeric excess of 65% [238].

The first pinacol coupling reactions to be catalytic in titanium were performed by using the $TiCl_4$-Li(Hg) system in the presence of $AlCl_3$; valeraldehyde was transformed into 5,6-decanediols in 78% yield with a dl:meso ratio equal to 70:30. In this reaction, transmetallation of the titanium pinacolate intermediates with $AlCl_3$ regenerated the precatalyst $TiCl_4$ and gave aluminum diolates, which were inert towards the reducing agent and not transformed into the alkene [239]. Catalytic pinacolization of benzaldehyde was achieved with 1% $TiCl_3(THF)_3$ in the presence of Zn and Me_3SiCl. This combination was not effective for the coupling reactions of less electrophilic aldehydes and the diastereoselection was very low, although addition of 5 mol% of tBuOH led to a more reactive system that catalyzed the pinacolization of aromatic and aliphatic aldehydes and aryl methyl ketones with dl:meso ratios ranging from 1.5 to 4.8. Moreover, the threoselectivity of the homocoupling of aryl aldehydes was substantially enhanced (dl:meso = 6.7 to 10.1) when 30 mol% of 1,3-diethyl-1,3-diphenylurea (DEPU) was added to the $TiCl_3(THF)_3$-tBuOH catalyst [240]. Another method for the McMurry pinacol reaction of aldehydes employs 3 mol% of a complex prepared in situ from $TiCl_4(THF)_2$ and a Schiff base in acetonitrile, the catalytic cycle being realized with

Me$_3$SiCl and Mn as reducing agent; the diastereoselectivity, which is excellent, depends on the Schiff base used [241]. Stoichiometric reactions of aromatic aldehydes with chiral titanium(IV) Schiff-base compounds in the presence of various reductants afforded the chiral diols in high yields and with enantioselectivities up to 91%; these transformations were rendered catalytic by the addition of TMSCl [242].

Pinacol coupling can be effected by organotitanium compounds, as demonstrated long ago for the arene complex (η-C$_6$Me$_6$)Ti(AlCl$_4$)$_2$ [229] and the CpTiCl$_3$-LiAlH$_4$ [229] and Cp$_2$TiCl$_2$-iPrMgCl [243] systems (Cp = η-C$_5$H$_5$). Such reactions have undergone a significant improvement in recent years. Thus, catalytic pinacolization of aliphatic aldehydes has been performed with Cp$_2$TiCl$_2$ or Cp$_2$TiPhCl in the presence of Zn and Me$_3$SiCl [244–246]. It has been reported that [Cp$_2$TiCl]$_2$ is capable of reductively coupling aromatic and α,β-unsaturated aldehydes to form 1,2-diols in both anhydrous and aqueous media; the measured diastereoselectivity was high, with dl/meso ratios greater than 91:9 [247]. Pinacol coupling of aromatic aldehydes has been catalyzed by 3 mol% of rac-ethylenebis(η-indenyltitanium) dichloride in the presence of MgBr$_2$, Me$_3$SiCl, and Zn to give the racemic 1,2-diols in good yield and with excellent diastereoselectivity (dl:meso > 96:4). These results are encouraging for further investigation of asymmetric induction by using enantiomerically pure metallocene catalysts [248]. Recent studies have revealed substantial effects of the structure of the titanocene-type catalyst on the chemo- and stereoselectivity of the pinacolization reaction [249–252]. The nature of the active species and intermediates in these pinacol coupling reactions, which are performed with the aid of rather complicated systems, is not known. It is generally proposed that the high dl stereoselectivity of benzaldehyde or acetophenone coupling is due to the dimerization of ketyl radicals oriented in a manner that minimizes steric interactions between the phenyl groups (Scheme 6.9).

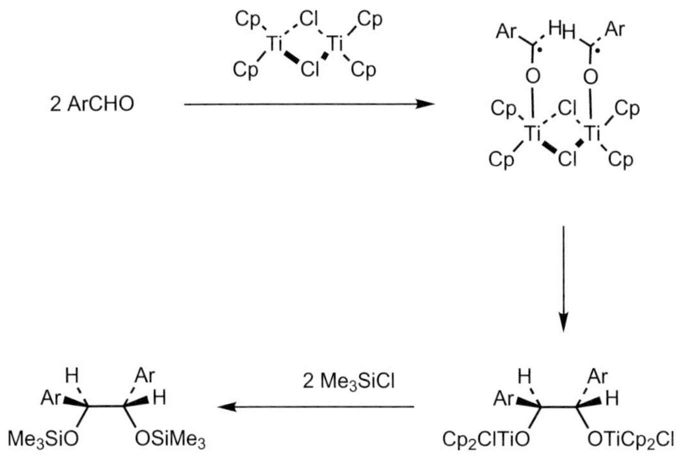

Scheme 6.9. Proposed mechanism for the diastereoselective coupling of benzaldehyde.

6.3.2.3 Other Systems for the McMurry Alkene Synthesis: Organotitanium Complexes, Titanium Oxides, Titanium Metal

Low-valent organotitanium complexes are generally much less successful in the alkene-forming coupling of carbonyl compounds than in the McMurry pinacol reaction. Aromatic aldehydes have been coupled to form pinacols and alkenes by the Ti(II) compound $Cp_2Ti(CO)_2$ [253], while reaction of paraformaldehyde with the Ti(III) complex $CpTiCl_2(THF)_{1.5}$ afforded ethylene and the oxide $(CpTiCl_2)_2O$ [254]. The titanium(0) complex $(\eta\text{-}C_6H_6)_2Ti$ was found to be highly active in the reductive coupling of both aromatic and aliphatic ketones, and no pinacols were ever obtained in these reactions; furthermore, the reactions of pinacols with this compound were much slower than the reactions of ketones [255]. These features suggest that there are important mechanistic differences between the soluble organometallic complexes and heterogeneous systems prepared from $TiCl_4$ or $TiCl_3$. The zerovalent titanium compounds $(\eta\text{-}C_6H_5Me)_2Ti$ and $Ti(biphenyl)_2$, as well as the Ti(II) complex $Cp_2Ti(PMe_3)_2$, were also found to be efficient in the oxoamide cyclization reaction [199].

Reductive coupling of aldehydes and ketones has also been carried out as a gas-solid process on the surface of reduced titania [256–259]; the reaction could be performed catalytically in the presence of hydrogen. Such reactions do not require strong reducing agents, and a reduced oxide catalyst would be cheaper and easier to handle than the usual McMurry reagents in liquid-solid slurries. Moreover, the gas-solid reaction represents a potential route for the coating of surfaces with conducting polymers, as may be envisaged by the reductive coupling of p-benzoquinone [257].

A titanium(0) colloid, prepared by the reduction of $TiCl_4$ with $KBEt_3H$, has been used in the synthesis of indoles [260]. Nanostructured titanium clusters, produced electrochemically by using a titanium sacrificial anode as the metal source and Bu_4NBr in THF as the electrolyte and stabilizer, were found to induce olefin-forming McMurry-type coupling of oxoamides, aromatic aldehydes, and ketones [261].

The titanium-catalyzed indole synthesis revealed that chlorosilanes were able to regenerate the active low-valent titanium chloride species from the formed titanium oxides (Section 6.2.5). By applying the same methodology, chlorosilanes were found to effect the activation of commercial titanium powder, which was then used as an efficient off-the-shelf reagent for the reductive coupling of aromatic and α,β-unsaturated carbonyl compounds. This reagent exerts a remarkable template effect, allowing macrocyclization reactions without recourse to high dilution. The ketone-ester and ketone-amide cyclizations were also possible with the $Ti\text{-}Me_3SiCl$ combination [64].

6.4 Mechanisms of the McMurry Reaction

As noted above, there was apparently no doubt about the involvement of pinacolate intermediates in the McMurry reaction and the main questions concerned instead the nature of the active species, and in particular its oxidation state.

6.4.1
Nature of the Active Species

It is clear that McMurry reactions can be performed with low-valent titanium species in various oxidation states. Pinacol and alkene-forming reductive coupling reactions have been achieved with organometallic complexes of Ti(0), Ti(II), and Ti(III) (see Section 6.2.2.3). While no evidence was found for the presence of Ti(0) on reduced TiO_2 surfaces active in benzaldehyde coupling, X-ray photoelectron spectroscopy showed that the active site required for gas-solid reductive coupling is an ensemble of Ti cations in the +1, +2, and +3 oxidation states that collectively effect the four-electron reduction [258].

The first studies on the $TiCl_3$-M (M = Li, K, Mg) and $TiCl_3$-$LiAlH_4$ systems, which involved measuring the amount of reducing agent in the reduction and following the reaction by ESR spectroscopy, led to the assumption that finely divided titanium particles are the active species in McMurry reactions [262]. However, more recent investigations have revealed that the reduction of $TiCl_3$ with Mg or $LiAlH_4$ does not afford metallic titanium, but gives instead low-valent titanium complexes with distinct structures and reactivities. Thus, reaction of $TiCl_3$ or $TiCl_4$ with excess magnesium in THF at room temperature proceeds in two steps, giving $TiMgCl_2(THF)_x$ **171** and $(TiMgCl)_2(THF)_x$ **172**, which are in the formal oxidation states of 0 and −2, respectively [263, 264].

On the basis of EXAFS studies, **172** would be the inorganic Grignard reagent represented in Figure 6.29 [265]. The low-valent titanium species obtained

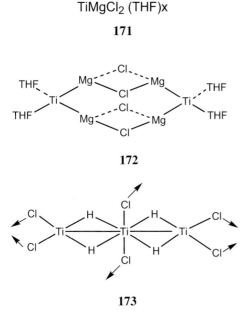

Figure 6.29.

by reduction of $TiCl_3$ with $LiAlH_4$ was shown to be the Ti(II) chlorohydride $[HTiCl(THF)_{0.5}]_x$ **173**, a possible structure of which is also shown in Figure 6.29 [266].

6.4.2
Characterization of the Pinacolate Intermediates

There was never any doubt that metallopinacols were intermediates in McMurry reactions since, for many substrates, pinacols and alkenes could be obtained by using the same titanium system, according to the experimental conditions. The course of the reductive coupling of acetophenone with $[HTiCl(THF)_{0.5}]_x$ and the $TiCl_3$-Zn(Cu) combination was determined by analysis of the products obtained upon hydrolysis of aliquots taken at intervals [267]. As expected, these reactions were shown to proceed by two consecutive steps: generation of the pinacolate intermediates, which occurred at room temperature, followed by alkene formation at solvent reflux temperature. In the proposed mechanisms shown in Scheme 6.10, the alkene would be formed by deoxygenation of the same pinacolate intermediate **174**. Because of homolytic dissociation of the hydride ligand, complex **173** behaves as a strong two-electron reductant and the alkene is produced by using a 1:1 molar ratio of **173** and ketone.

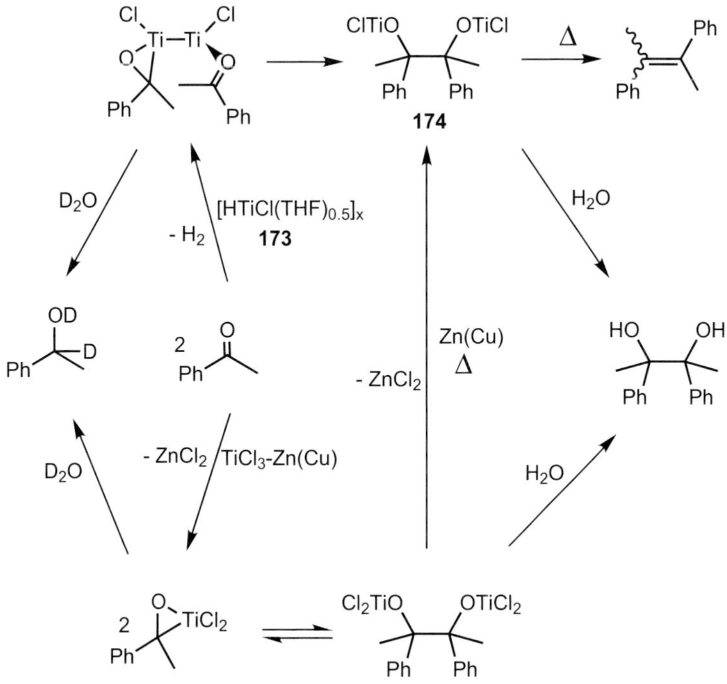

Scheme 6.10. Proposed mechanisms for the reductive coupling of acetophenone with the $TiCl_3$-$LiAlH_4$ and $TiCl_3$-Zn(Cu) systems.

With the TiCl$_3$-Zn(Cu) system, each of the two reaction steps requires a one-electron reduction of Ti(III) by Zn [267]. TiCl$_2$ [213] was also found to act as a one-electron reductant for the formation of the alkene from the pinacolate intermediate. Interestingly, deuterolysis experiments performed at an early stage of each of these reactions produced 1-phenylethanol with 98% deuterium incorporation at the 1-position. This result indicates that the precursor of the titanium pinacolate is a ketone dianion and supports a nucleophilic rather than a radical mechanism for the C–C coupling step of aromatic ketones [267]. This mechanism was supported by quantum mechanical calculations [268].

However, titanium-bound ketyl radicals have been generated through the reaction of the sterically encumbered complex Ti(silox)$_3$ (silox = tBu$_3$SiO) with ketones and aldehydes, and have been characterized by ESR spectroscopy [269]. These radicals, which cannot couple to give pinacolates for steric reasons, were trapped with Ph$_3$SnH to give the corresponding alkoxides. The ketyl complex obtained from benzophenone is in equilibrium with the dimerization product **175**, resulting from *para*-phenyl/carbonyl carbon coupling, which is reminiscent of the coupling of trityl radicals (Scheme 6.11).

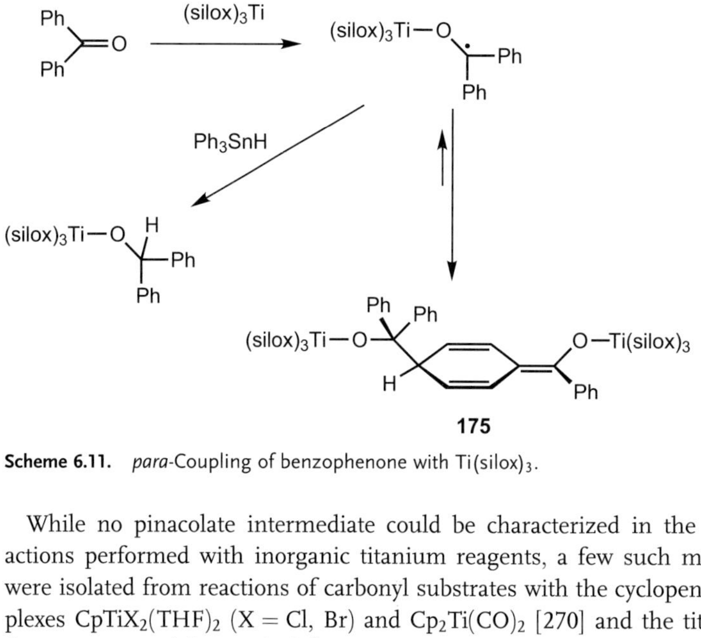

Scheme 6.11. *para*-Coupling of benzophenone with Ti(silox)$_3$.

While no pinacolate intermediate could be characterized in the McMurry reactions performed with inorganic titanium reagents, a few such metallopinacols were isolated from reactions of carbonyl substrates with the cyclopentadienyl complexes CpTiX$_2$(THF)$_2$ (X = Cl, Br) and Cp$_2$Ti(CO)$_2$ [270] and the titananorbornadiene compound (DMSC)Ti[η^6-1,2,4-(Me$_3$Si)$_3$C$_6$H$_3$] (DMSC = 1,2-alternate dimethylsilyl-bridged *p-tert*-butylcalix[4]arene) [271]. Unsymmetrical diolate complexes have been obtained by treatment of the dialkyl or diaryl complexes L$_2$TiR$_2$ (R = Me or Ph, L = N,N'-dimethylaminotroponiminate) with CO and aldehydes or ketones; a reactive titanium η^2-carbonyl compound, formed by double alkyl migration to CO, was proposed as a key intermediate in these reactions [272]. The reversible dissociation of the titanapinacolates (DMSC)Ti(OCAr$_2$CAr$_2$O) (Ar = Ph or *p*-MeC$_6$H$_4$) was demonstrated by analysis of the kinetics of their reactions with *t*BuC≡CH, which revealed that the fragmentation reactions proceed via a pre-equilibrium mechanism giving (DMSC)Ti(η^2-OCAr$_2$) and Ar$_2$CO, followed by rate-limiting reaction of (DMSC)Ti(η^2-OCAr$_2$) with the alkyne molecule [273]. In the titanium pinacolates that have been crystallographically characterized, short Ti-O bond lengths (ca. 1.86 Å) and a long pinacolic C–C distance (ca. 1.62 Å) have been observed. No alkene was obtained from these titanium pinacolates.

Metallopinacols have been synthesized by reaction of ketones with the UCl$_4$-M(Hg) systems (M = Li or Na). Uranium and titanium complexes exhibit strong analogies in terms of structure and reactivity, and uranium metal powder is effective in the reductive coupling of aromatic ketones [208, 209]. Moreover, uranium compounds have some advantages over their titanium counterparts: they can be easily detected by their very shifted paramagnetic NMR signals, and they often crystallize more readily. Therefore, the chances of isolating and characterizing the intermediates are greater. The active uranium species in the UCl$_4$-M(Hg) systems were shown to be in the +3 oxidation state; it was demonstrated by electrochemical studies that reduction of UCl$_4$ to UCl$_4^-$ was rapidly followed by a chloride ion transfer from UCl$_4^-$ to UCl$_4$, giving UCl$_3$ and the anionic U(IV) complexes U$_2$Cl$_9^-$ and UCl$_4^-$, which were then reduced at lower potentials [274].

Reaction of benzophenone with UCl$_4$ and Na(Hg) afforded successively the mono- and bis(benzopinacolate)s UCl$_2$(O$_2$C$_2$Ph$_4$) **176** and U(O$_2$C$_2$Ph$_4$)$_2$(THF)$_2$ **177**; the latter was characterized by its X-ray crystal structure [275]. These compounds gave benzopinacol upon hydrolysis and were transformed into tetraphenylethylene after reduction with Na(Hg) at 25 and 65 °C for **176** and **177**, respectively (Scheme 6.12).

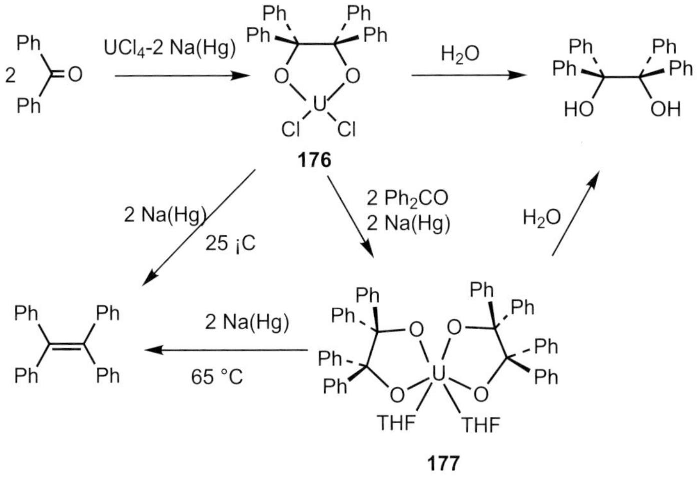

Scheme 6.12. Metallopinacols isolated from the reaction of benzophenone with the UCl$_4$-Na(Hg) system.

Similar treatment of aromatic ketones PhCOR with the UCl$_4$-Li(Hg) system afforded a mixture of the corresponding pinacol and the keto alcohol resulting from

6.4 Mechanisms of the McMurry Reaction

para coupling of the ketone; the proportion of the latter increased with the steric bulk of the R group and the steric saturation around the active metal species [276].

Several metallopinacols have been isolated from the reaction of acetone with UCl_4 and M(Hg); their structures and their eventual transformation into $Me_2C=CMe_2$ were found to be strongly influenced by the molar ratio of the reactants, the nature of the amalgam used, Li(Hg) or Na(Hg), and the salts formed, LiCl or NaCl (Scheme 6.13) [277, 278].

Whatever the amalgam, the first intermediate was **179** resulting from dimerization of the ketyl complex **178**, which was trapped with Ph_3SnH; the crystal struc-

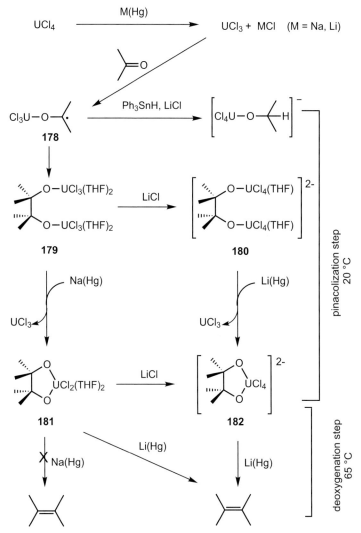

Scheme 6.13. Metallopinacols isolated from the reaction of acetone with the UCl_4-M(Hg) systems (M = Li or Na).

ture of the HMPA adduct $\{UCl_3(HMPA)_2\}_2(OCMe_2CMe_2O)$ has been determined. When Li(Hg) was used, **179** was transformed into **180** by the addition of chloride ions. Reactions of **179** and **180** with the alkali metal amalgam gave **181** and **182**, respectively, and these latter were the true precursors of tetramethylethylene. The mechanism of the rate-determining deoxygenation step was not clarified, but it was found that the reducing agent also has a great influence on this process, which could be achieved only with Li(Hg) in refluxing THF, while the LiCl salt was responsible for some side reactions leading to a lower yield of tetramethylethylene. This alkene was obtained in good yield by controlling the reaction time and heating or, more reliably, by performing the coupling and deoxygenation steps in a sequential manner. These results emphasize the complexity of an apparently simple McMurry reaction, which could only be successfully accomplished with the appropriate stoichiometry and experimental conditions. These factors are not always easily controlled in a heterogeneous medium and would constitute major sources of non-reproducibility.

6.4.3
Evidence of Carbenoid Intermediates

Reaction of diisopropyl ketone with the MCl_4-Li(Hg) systems (M = Ti, U) was found to be quite different from that of acetone since the only coupling product formed at 25 °C was $iPr_2C=CiPr_2$ [279, 280]. In fact, formation of the corresponding pinacol was never observed in any McMurry reaction of iPr_2CO. Moreover, it was demonstrated that $iPr_2C=CiPr_2$ could not be obtained by deoxygenation of the pinacolate Cl_3M-$OCiPr_2CiPr_2O$-MCl_3, which, in the case of M = Ti, was not stable with respect to rupture of the pinacolic C–C bond and was readily transformed into $TiCl_3$ and iPr_2CO. The facile cleavage of the titanium pinacolates and the absence of pinacol in the product mixtures indicated that reductive coupling of $iPr_2C=O$ would not proceed by dimerization of ketyl radicals. A more careful analysis of the products showed that the McMurry reactions of iPr_2CO afforded a significant quantity of $Me_2C=C(H)iPr$, resulting from deoxygenation of the ketone. This observation, which had hitherto been overlooked, revealed that carbenoid species were likely intermediates. These would be formed by reduction and deoxygenation of the ketyl radicals; their reaction with another molecule of ketone would give $iPr_2C=CiPr_2$, presumably via a metallaoxetane complex, and their rearrangement by H migration would afford $Me_2C=C(H)iPr$ (Scheme 6.14).

The great difference in reactivity between Me_2CO and iPr_2CO can be accounted for by steric factors; the more sterically hindered the ketyl radicals are, the more difficult they are to dimerize. Steric hindrance would also destabilize the pinacolate intermediates with respect to rupture of their C–C bond and reverse conversion to the ketyl radicals.

The involvement of carbenoid species in the McMurry reaction of sterically hindered ketones was further demonstrated by analysis of the products resulting from the reduction of di-*tert*-butyl ketone with the MCl_4-Li(Hg) systems (Scheme 6.15) [280, 281]. In this case, H migration within the $[M]$–$CtBu_2$ species gave the ex-

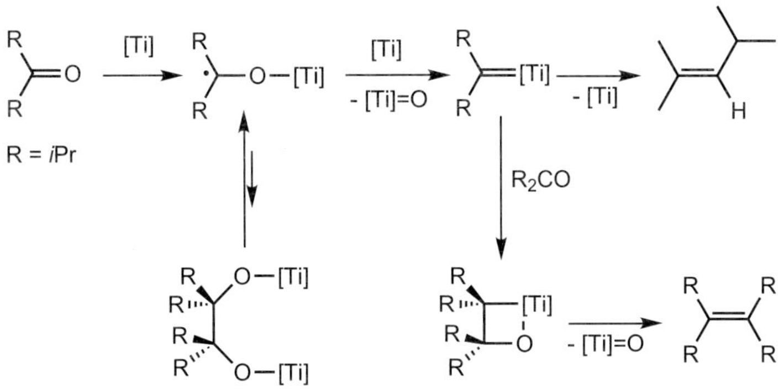

Scheme 6.14. Reductive coupling of diisopropyl ketone via a carbenoid species.

pected cyclopropane compound. The major product was tBu_2CH_2, which was liberated in 60% yield upon hydrolysis of the reaction mixture; deuterolysis experiments confirmed that this alkane was formed by successive addition of H(D) atoms to the carbenoid species [M]=$CtBu_2$. The latter were thus much more stable than [M]=$CiPr_2$ and could also be trapped with aldehydes RCHO to give the cross-coupling products tBu_2C=$C(R)H$ (R = Me, tBu).

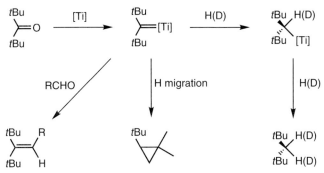

Scheme 6.15. Involvement of carbenoid species in the reaction of di-*tert*-butyl ketone with the MCl$_4$-Li(Hg) system (M = U, Ti).

6.4.4
Mechanistic Analogies Between the McMurry, Wittig, and Clemmensen Reactions

The results presented in Section 6.4.3 revealed that the McMurry reactions of sterically hindered ketones can in fact be viewed as Wittig-like olefination reactions. While these two reactions were for a long time thought to be mechanistically different, formation of carbenoid species by reduction of ketones with low-valent titanium complexes is not really surprising in view of the highly oxophilic and reducing character of titanium. Carbonyl olefinations by means of titanium carbenoid species are well documented; these are performed with a variety of reagents, as devised by Tebbe [282], Grubbs [283], Petasis [284], and Takai et al. [285]. The more recently discovered carbonyl olefination using thioacetals with the aid of $Cp_2Ti(P(OEt)_3)_2$ also involves a titanium carbene intermediate [286]. Alkylidene complexes of zirconium [287], niobium [288], and group 6 metals [289–292] are also effective in such Wittig-like olefination reactions, which are believed to proceed via metallaoxetane intermediates. All these reactions are presented and discussed in Chapters 4 and 5 of this book.

On the other hand, the formation of alkanes and alkenes by hydrogenation and deoxygenation, respectively, of sterically hindered ketones in McMurry reactions is reminiscent of the Clemmensen reaction, that is the reduction of a carbonyl to a methylene group by means of zinc and hydrochloric acid [293], or its modified version, that is the synthesis of alkenes by deoxygenation of carbonyl compounds with zinc and chlorosilane [294–296]. These reactions, which were shown to involve zinc carbenoid intermediates resulting from reduction and deoxygenation of ketyl radicals, also produce a number of side products, in particular pinacols and coupled alkenes, which are the normal products of McMurry reactions; it has been demonstrated that the deoxygenative coupling process does not proceed via pinacol-type intermediates. It is very interesting to note that the same orthocyclophane **87** was obtained in 91% yield by treating its dicarbonyl precursor with Zn(Hg) and HCl in place of the $TiCl_3$-$LiAlH_4$ system [121]. Moreover, the chlorosilane-modified Clemmensen reduction has, in some cases, proved to be a useful alternative to the McMurry alkene synthesis, in particular with electron-rich aromatic carbonyl compounds and metallocenic ketones [297]. These marked similarities between the McMurry and Clemmensen reactions may reflect the involvement of similarly reactive carbenoid species in the two reactions.

Evidence for carbenoid intermediates in some McMurry reactions can also be found in previous work. Thus, the formation of large amounts of alkanes in the McMurry reactions of sterically hindered ketones had already been noted [18, 67, 70, 142, 253]; alkenes resulting from deoxygenation of ketones had also been observed. For example, cyclohexene was detected among the products of the reductive coupling of cyclohexanone with the $TiCl_3$-K system [262], and the alkenes $RCH=CH_2$ (R = Me, Ph) were formed during the coupling of acetone and acetophenone on reduced alumina [259]; these alkenes are indicative of the occurrence of carbenes as intermediates, even if such species could not be trapped with the usual reagents. The involvement of carbenoid intermediates in the reductive coupling of diferrocenyl ketone was clearly evident from the traces of hexaferrocenylcyclopropane, a formal trimer of diferrocenylmethylidene, that was formed alongside the coupled alkene **48** [66].

6.4.5
The Different Pathways of the McMurry Reaction

The McMurry reaction of ketones RCOR' by means of the MCl_4-Li(Hg) system (M = U, Ti) will follow different pathways depending on steric factors, that is the steric bulk of the R and R' substituents, and the steric saturation of the coordination sphere of the active metal species. These routes are summarized in Scheme 6.16.

In all cases, the first step of the reaction is formation of a ketyl radical by one-electron transfer from titanium to the carbonyl. This ketyl can disproportionate or abstract a H atom from the solvent, leading, after hydrolysis, to the alcohol RR'CHOH and aldolization products. When the ketyl is not sterically hindered, it can dimerize to give a titanium pinacolate, which is readily transformed into the α-diol RR'C(OH)C(OH)RR' by hydrolysis, or subsequently deoxygenated to the corresponding alkene RR'C=CRR' in the rate-determining step. This route corresponds to the mechanism that was generally accepted for the McMurry reaction. However, the symmetric pinacol coupling of an aromatic ketone PhCOR is impeded by the presence of bulky R groups and/or by coordination of sterically demanding ligands on the metal complex. Dissymmetric dimerization of the ketyl radical via *para*-phenyl/carbonyl carbon coupling was then found to occur. Dimerization of ketyl radicals generated from aliphatic ketones is also prohibited by steric factors and, in this case, the ketyls undergo deoxygenative reduction to carbenoid species. These react further with the ketone to afford the coupled alkene RR'C=CRR' and their rearrangement by H migration gives the alkene RCH=CHR" (R' = CH$_2$R"), which can be the major product of the reaction. Finally, in the case of severely congested ketones, the ketyl radical is also reduced to the carbenoid species, but the latter is not able to react with the ketone and is preferentially converted into the alkane RCH$_2$R'.

6.5
Conclusion

The McMurry reaction continues to find great success in synthetic organic chemistry and new developments are still being made. Its field of application has been extended to the synthesis of more and more sophisticated molecules and materials having interesting physico-chemical properties. A number of conjugated systems, in particular cyclophanes, bisporphyrins and bischlorins, have been prepared. The utility of the McMurry reaction in the synthesis of cyclic compounds and especially of natural products has been demonstrated. Tandem coupling reactions have proved to be efficient in many cases. The new route to pyrroles and indoles by cyclization of acylamido carbonyl compounds constitutes a major extension of the scope of reductive carbonyl couplings.

Significant progress has clearly been made during recent years in the development of new reagents and methods, leading to further interesting applications of low-valent titanium complexes. New insights into the mechanism of the McMurry reaction have been provided with the recognition that not only pinacolates but also metallocarbenes are involved in the reductive coupling. The formation and structure of these intermediates are strongly dependent on the experimental conditions and the nature of the carbonyl substrate, the titanium compound, the reducing

276 | 6 The McMurry Coupling and Related Reactions

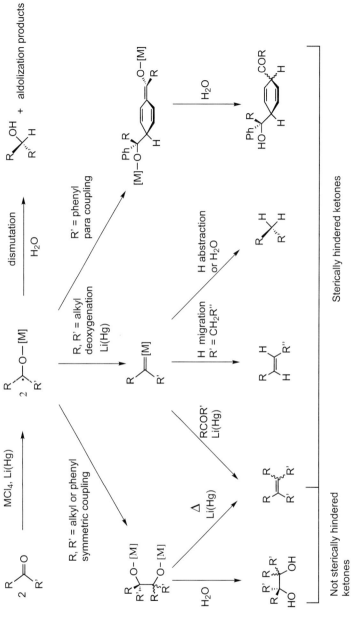

Scheme 6.16. Distinct pathways for the McMurry reactions of ketones RCOR′.

agent, and the by-products. Other aspects of the mechanism, for example the deoxygenation of the pinacolate intermediates and the role of additives in determining the stereoselectivity, remain obscure. The search for chiral ligands and auxiliaries will represent a next step in the development of enantioselective McMurry couplings.

The McMurry reaction does not cease to find new important applications, despite some pervasive problems of reproducibility; new practical reagents and simplified methods have been developed, but rather complicated systems are still necessary for obtaining higher selectivities. These contradictory features of the McMurry reaction, which leave many challenges for further study, and the dual nature of its mechanism, reinforce the fascinating character of this versatile transformation, which is irreplaceable in organic synthesis.

References

1 I. Tyrlik, I. Wolochowicz, *Bull. Soc. Chim. Fr.* **1973**, 2147–2148.
2 T. Mukaiyama, T. Sato, J. Hanna, *Chem. Lett.* **1973**, 1041–1044.
3 J. E. McMurry, M. P. Fleming, *J. Am. Chem. Soc.* **1974**, 96, 4708–4709.
4 J. E. McMurry, *Chem. Rev.* **1989**, 89, 1513–1524.
5 D. Lenoir, *Synthesis* **1989**, 883–897.
6 Y. Dang, H. J. Geise, *J. Organomet. Chem.* **1991**, 405, 1–39.
7 G. M. Robertson, in *Comprehensive Organic Synthesis*, Vol. 3 (Eds.: B. M. Trost, I. Fleming, G. Pattenden), Pergamon, Oxford, **1991**, pp. 563–611.
8 R. G. Dushin, in *Comprehensive Organic Chemistry II*, Vol. 12 (Ed.: L. S. Hegedus), Pergamon, Oxford, **1995**, pp. 1071–1085.
9 T. Leckta, in *Active Metals. Preparation, Characterization, Applications* (Ed.: A. Fürstner), VCH, Weinheim, **1996**, pp. 85–131.
10 A. Fürstner, B. Bogdanovic, *Angew. Chem. Int. Ed. Engl.* **1996**, 35, 2442–2469.
11 M. Ephritikhine, *Chem. Commun.* **1998**, 2549–2554.
12 D. Lenoir, *Chem. Ber.* **1978**, 111, 411–414.
13 G. Bohrer, R. Knorr, *Tetrahedron Lett.* **1984**, 25, 3675–3678.
14 J. E. McMurry, T. Lectka, J. G. Rico, *J. Org. Chem.* **1989**, 54, 3748–3749.
15 I. Columbus, S. E. Biali, *J. Org. Chem.* **1994**, 59, 3402–3407.
16 H. Wenck, A. de Meijeire, F. Gerson, R. Gleiter, *Angew. Chem. Int. Ed. Engl.* **1986**, 25, 335–336.
17 A. de Meijere, H. Wenck, S. Zollner, P. Merstetter, A. Arnold, F. Gerson, P. R. Schreiner, R. Boese, D. Blaser, R. Gleiter, S. I. Kozhushkov, *Chem. Eur. J.* **2001**, 7, 5382–5390.
18 D. Lenoir, H. Burghard, *J. Chem. Res. (S)* **1980**, 396–397.
19 A. P. Marchand, A. Zope, F. Zaragoza, S. G. Bott, H. L. Ammon, Z. Du, *Tetrahedron* **1994**, 50, 1687–1698.
20 J. Leimner, P. Weyerstahl, *Chem. Ber.* **1982**, 115, 3697–3705.
21 R. Willem, H. Pepermans, K. Hallenga, M. Gielen, R. Dams, H. J. Geise, *J. Org. Chem.* **1983**, 48, 1890–1898.
22 D. Lenoir, P. Lemmen, *Chem. Ber.* **1980**, 113, 3112–3119.
23 B. Feringa, H. Wynberg, *J. Am. Chem. Soc.* **1977**, 99, 602–603.
24 B. Hagenbruch, S. Hünig, *Liebigs Ann. Chem.* **1984**, 340–353.
25 G. Gapski, A. Kini, R. S. H. Liu, *Chem. Lett.* **1978**, 803–804.
26 A. Ishida, T. Mukaiyama, *Chem. Lett.* **1976**, 1127–1130.
27 G. Broszeit, F. Diepenbrock, O.

Graf, D. Hecht, J. Heinze, H. D. Martin, B. Mayer, K. Schaper, A. Smie, H. H. Strehblow, *Liebigs Ann./Recueil* **1997**, 2205–2213.
28 B. Jousselme, P. Blanchard, P. Frere, J. Roncali, *Tetrahedron Lett.* **2000**, *41*, 5057–5061.
29 W. E. Doering, Y. Q. Shi, D. C. Zhao, *J. Am. Chem. Soc.* **1992**, *114*, 10763–10766.
30 R. M. Cory, J. R. Walker, P. D. Zabel, *Synth. Commun.* **1994**, *24*, 799–807.
31 R. Gleiter, G. Fritzsche, O. Borzyk, T. Oeser, F. Rominger, H. Irngartinger, *J. Org. Chem.* **1998**, *63*, 2878–2886.
32 T. Ishii, T. Sawada, S. Mataka, M. Tashiro, T. Thiemann, *Chem. Ber.* **1996**, *129*, 289–296.
33 S. Warren, P. Wyatt, M. McPartlin, T. Woodroffe, *Tetrahedron Lett.* **1996**, *37*, 5609–5612.
34 S. Warren, P. Wyatt, *Tetrahedron: Asymmetry* **1996**, *7*, 989–992.
35 R. Daik, W. J. Feast, A. S. Batsanov, J. A. K. Howard, *New J. Chem.* **1998**, *22*, 1047–1049.
36 I. A. Khotina, V. A. Izumrudov, N. V. Tchebotareva, A. L. Rusanov, *Macromol. Chem. Phys.* **2001**, *202*, 2360–2366.
37 F. B. Mallory, K. E. Butler, A. Berube, E. D. Luzik, C. W. Mallory, E. J. Brondyke, R. Hiremath, P. Ngo, P. J. Carroll, *Tetrahedron* **2001**, *57*, 3715–3724.
38 S. Gupta, G. K. Kar, J. K. Ray, *Synth. Commun.* **2000**, *30*, 2393–2399.
39 M. Sander, E. V. Dehmlow, *Eur. J. Org. Chem.* **2001**, 399–404.
40 L. Castedo, J. M. Saa, R. Suau, G. Tojo, *J. Org. Chem.* **1981**, *46*, 4292–4294.
41 D. T. Witiak, P. L. Kamat, D. L. Allison, S. M. Liebowitz, R. Glaser, J. E. Holliday, M. L. Moeschberger, J. P. Schaller, *J. Med. Chem.* **1983**, *26*, 1679–1686.
42 G. Märkl, M. Hafner, P. Kreitmeier, C. Stadler, J. Daub, H. Nöth, M. Schmidt, G. Gescheidt, *Helv. Chim. Acta* **1997**, *80*, 2456–2476.
43 K. Agbaria, S. E. Biali, *J. Org. Chem.* **2001**, *66*, 5482–5489.
44 P. Blanchard, H. Brisset, B. Illien, A. Riou, J. Roncali, *J. Org. Chem.* **1997**, *62*, 2401–2408.
45 P. Blanchard, H. Brisset, A. Riou, R. Hierle, J. Roncali, *J. Org. Chem.* **1998**, *63*, 8310–8319.
46 E. H. Elandaloussi, P. Frere, P. Richomme, J. Orduna, J. Garin, J. Roncali, *J. Am. Chem. Soc.* **1997**, *119*, 10774–10784.
47 T. T. Nguyen, Y. Gouriou, M. Salle, P. Frere, M. Jubault, A. Gorgues, L. Toupet, A. Riou, *Bull. Soc. Chim. Fr.* **1996**, *133*, 301–308.
48 J. Larsen, K. Bechgaard, *Acta Chem. Scand.* **1996**, *50*, 71–76.
49 J. Larsen, K. Bechgaard, *Acta Chem. Scand.* **1996**, *50*, 77–82.
50 J. Cheng, J. E. Gano, A. R. Morgan, *Tetrahedron Lett.* **1996**, *37*, 2721–2724.
51 G. R. Newkome, J. M. Roper, *J. Org. Chem.* **1979**, *44*, 502–505.
52 H. Xie, D. A. Lee, D. M. Wallace, M. O. Senge, K. M. Smith, *J. Org. Chem.* **1996**, *61*, 8508–8517.
53 R. G. Khoury, L. Jaquinod, K. M. Smith, *Chem. Commun.* **1997**, 1057–1058.
54 P. M. Windscheif, F. Vögtle, *Synthesis* **1994**, 87–92.
55 R. Paolesse, R. K. Pandey, T. P. Forsyth, L. Jaquinod, K. R. Gerzevske, D. J. Nurco, M. O. Senge, S. Licoccia, T. Boschi, K. M. Smith, *J. Am. Chem. Soc.* **1996**, *118*, 3869–3882.
56 M. O. Senge, W. W. Kalisch, K. Ruhlandt-Senge, *Chem. Commun.* **1996**, 2149–2150.
57 W. W. Kalisch, M. O. Senge, K. Ruhlandt-Senge, *Photochem. Photobiol.* **1998**, *67*, 312–323.
58 L. Jaquinod, D. J. Nurco, C. J. Medforth, R. K. Pandey, T. P. Forsyth, M. M. Olmstead, K. M. Smith, *Angew. Chem. Int. Ed. Engl.* **1996**, *35*, 1013–1016.
59 L. Jaquinod, M. O. Senge, R. K. Pandey, T. P. Forsyth, K. M. Smith, *Angew. Chem. Int. Ed. Engl.* **1996**, *35*, 1840–1842.
60 T. A. Niemi, P. L. Coe, S. J. Till, *J. Chem. Soc., Perkin Trans. 1* **2000**, 1519–1528.

61 L. A. Paquette, G. J. Wella, G. Wickham, *J. Org. Chem.* **1984**, *49*, 3618–3621.

62 W. R. Roth, H. H. C. Horn, *Chem. Ber.* **1994**, 1781–1795.

63 A. Fürstner, G. Seidel, B. Gabor, C. Kopiske, C. Kruger, R. Mynott, *Tetrahedron* **1995**, *51*, 8875–8888.

64 A. Fürstner, A. Hupperts, *J. Am. Chem. Soc.* **1995**, *117*, 4468–4475.

65 Y. P. Wang, X. H. Lui, B. S. Lin, W. D. Tang, T. S. Lin, J. H. Liaw, Y. Wang, Y. H. Liu, *J. Organomet. Chem.* **1999**, *575*, 310–319.

66 B. Bildstein, P. Denifl, K. Wurst, M. Andre, M. Baumgarten, J. Friedrich, E. Ellmerer-Muller, *Organometallics* **1995**, *14*, 4334–4342.

67 P. Harter, K. Latzel, M. Spiegler, E. Herdtweck, *Polyhedron* **1998**, *17*, 1141–1148.

68 S. Pitter, G. Huttner, O. Walter, L. Zsolnai, *J. Organomet. Chem.* **1993**, *454*, 183–198.

69 S. O. Agustsson, C. H. Hu, U. Englert, T. Marx, L. Wesemann, C. Ganter, *Organometallics* **2002**, *21*, 2993–3000.

70 P. Toullec, L. Ricard, F. Mathey, *Organometallics* **2002**, *21*, 2635–2638.

71 A. W. Cooke, K. B. Wagner, *Macromolecules* **1991**, *24*, 1404–1407.

72 F. Cacialli, R. Daik, W. J. Feast, R. H. Friend, C. Lartigau, *Opt. Mater.* **1999**, *12*, 315–319.

73 F. Goldoni, R. A. J. Janssen, E. W. Meijer, *J. Polym. Sci. Polym. Chem.* **1999**, *37*, 4629–4639.

74 T. Itoh, H. Saitoh, S. Iwatsuki, *J. Polym. Sci. Polym. Chem.* **1995**, *33*, 1589–1596.

75 S. M. Reddy, M. Duraisamy, H. M. Walborsky, *J. Org. Chem.* **1986**, *51*, 2361–2366.

76 K. Kakiuchi, H. Okada, N. Kanehisa, Y. Kai, H. Kurosawa, *J. Org. Chem.* **1996**, *61*, 2972–2979.

77 F. D. Ayres, S. I. Khan, O. L. Chapman, *Tetrahedron Lett.* **1994**, *35*, 8561–8564.

78 L. A. Paquette, T. H. Yan, G. Wells, *J. Org. Chem.* **1984**, *49*, 3610–3617.

79 S. T. Jan, E. G. Rogan, E. L. Cavalieri, *Chem. Res. Toxicol.* **1998**, *11*, 408–411.

80 J. E. McMurry, L. R. Krepski, *J. Org. Chem.* **1976**, *41*, 3929–3930.

81 P. L. Coe, C. E. Scriven, *J. Chem. Soc., Perkin Trans. 1* **1986**, 475–477.

82 S. Gauthier, J. Y. Sanceau, J. Mailhot, B. Caron, J. Cloutier, *Tetrahedron* **2000**, *56*, 703–709.

83 J. Shani, A. Grazit, T. Livshitz, S. Biran, *J. Med. Chem.* **1985**, *28*, 1504–1511.

84 A. Detsi, M. Koufaki, T. Calogeropoulou, *J. Org. Chem.* **2002**, *67*, 4608–4611.

85 S. Gauthier, J. Mailhot, F. Labrie, *J. Org. Chem.* **1996**, *61*, 3890–3893.

86 M. J. Meegan, R. B. Hughes, D. G. Lloyd, D. C. Williams, D. M. Zisterer, *J. Med. Chem.* **2001**, *44*, 1072–1084.

87 V. N. Rubin, P. C. Ruenitz, F. D. Boudinot, J. L. Boyd, *Bioorg. Med. Chem.* **2001**, *9*, 1579–1587.

88 F. G. Zhang, P. W. Fan, X. M. Liu, L. X. Shen, R. B. van Breemen, J. L. Bolton, *Chem. Res. Toxicol.* **2000**, *13*, 53–62.

89 K. W. Ace, M. A. Armitage, R. K. Bellingham, P. D. Blacker, D. S. Ennis, N. Hussain, D. C. Lathbury, D. O. Morgan, N. O'Connor, G. H. Oakes, S. C. Passey, L. C. Powling, *Org. Process Res. Dev.* **2001**, *5*, 479–490.

90 S. Top, A. Vessieres, C. Cabestaing, I. Laios, G. Leclercq, C. Provot, G. Jaouen, *J. Organomet. Chem.* **2001**, *637*, 500–506.

91 S. Top, B. Dauer, J. Vaissermann, G. Jaouen, *J. Organomet. Chem.* **1997**, *541*, 355–361.

92 S. Top, E. B. Kaloun, A. Vessieres, I. Laios, G. Leclercq, G. Jaouen, *J. Organomet. Chem.* **2002**, *643*, 350–356.

93 G. Jaouen, S. Top, A. Vessieres, P. Pigeon, G. Leclercq, I. Laios, *Chem. Commun.* **2001**, 383–384.

94 A. L. Baumstark, C. J. McCloskey, K. E. Witt, *J. Org. Chem.* **1978**, *43*, 3609–3611.

95 J. E. McMurry, K. L. Kees, *J. Org. Chem.* **1977**, *42*, 2655–2656.

96 J. E. McMurry, T. Lectka, *J. Am. Chem. Soc.* **1993**, *115*, 10167–10173.
97 J. E. McMurry, T. Lectka, *J. Am. Chem. Soc.* **1990**, *112*, 869–870.
98 J. E. McMurry, T. Lectka, C. N. Hodge, *J. Am. Chem. Soc.* **1989**, *111*, 8867–8872.
99 J. E. McMurry, T. Lectka, C. N. Hodge, *J. Am. Chem. Soc.* **1984**, *106*, 6450–6451.
100 L. N. Lucas, J. van Esch, R. M. Kellogg, B. L. Feringa, *Tetrahedron Lett.* **1999**, *40*, 1775–1778.
101 J. A. Marshall, K. E. Flynn, *J. Am. Chem. Soc.* **1984**, *106*, 723–730.
102 E. Vogel, B. Nuemann, W. Klug, H. Schmickler, J. Lex, *Angew. Chem. Int. Ed. Engl.* **1985**, *24*, 1046–1048.
103 K. Yamamoto, S. Kuroda, M. Shibutami, Y. Yoneyama, J. Ojima, S. Fujita, E. Ejiri, K. Yanagihara, *J. Chem. Soc., Perkin Trans. 1* **1988**, 395–400.
104 J. Ojima, K. Yamamoto, K. Kato, K. Wada, Y. Yoneyama, E. Ejiri, *Bull. Chem. Soc. Jpn.* **1986**, *59*, 2209–2215.
105 H. Hopf, A. Kruger, *Chem. Eur. J.* **2001**, *7*, 4378–4385.
106 J. E. McMurry, G. J. Haley, J. R. Matz, J. C. Clardy, J. Mitchell, *J. Am. Chem. Soc.* **1986**, *108*, 515–516.
107 M. Eckhardt, R. Brückner, *Angew. Chem. Int. Ed. Engl.* **1996**, *35*, 1093–1096.
108 E. Rank, R. Brückner, *Eur. J. Org. Chem.* **1998**, 1045–1053.
109 M. Rucker, R. Brückner, *Tetrahedron Lett.* **1997**, *38*, 7353–7356.
110 F. Ferri, R. Brückner, *Liebigs Ann./Recueil* **1997**, 961–965.
111 M. Eckhardt, R. Brückner, *Liebigs Ann./Recueil* **1997**, 947–959.
112 H. F. Grutzmacher, G. Nolte, *Chem. Ber.* **1994**, *127*, 1157–1162.
113 H. Meier, M. Fatten, *Tetrahedron Lett.* **2000**, *41*, 1535–1538.
114 T. Kawase, N. Ueda, K. Tanaka, Y. Seirai, M. Oda, *Tetrahedron Lett.* **2001**, *42*, 5509–5511.
115 T. Yamato, K. Fujita, H. Tsuzuki, *J. Chem. Soc., Perkin Trans. 1* **2001**, 2089–2097.
116 T. Yamato, K. Fujita, T. Abe, H. Tsuzuki, *New J. Chem.* **2001**, *25*, 728–736.
117 T. Yamato, K. Fujita, K. Okuyama, H. Tsuzuki, *New J. Chem.* **2000**, *24*, 221–228.
118 T. Yamato, K. Fujita, K. Futatsuki, H. Tsuzuki, *Can. J. Chem.* **2000**, *78*, 1089–1099.
119 G. Dyker, J. Korning, W. Stirner, *Eur. J. Org. Chem.* **1998**, 149–154.
120 W. Y. Lee, C. H. Park, H. J. Kim, S. Kim, *J. Org. Chem.* **1994**, *59*, 878–884.
121 W. Y. Lee, C. H. Park, Y. D. Kim, *J. Org. Chem.* **1992**, *57*, 4074–4079.
122 H. Hopf, C. Mlynek, *J. Org. Chem.* **1990**, *55*, 1361–1363.
123 A. Arduini, S. Fanni, A. Pochini, A. R. Sicuri, R. Ungaro, *Tetrahedron* **1995**, *51*, 7951–7958.
124 P. Lhotak, S. Shinkai, *Tetrahedron Lett.* **1996**, *37*, 645–648.
125 A. E. Gies, M. Pfeffer, *J. Org. Chem.* **1999**, *64*, 3650–3654.
126 F. Dubois, M. Gingras, *Tetrahedron Lett.* **1998**, *39*, 5039–5040.
127 K. Tanaka, H. Suzuki, H. Osuga, *Tetrahedron Lett.* **1997**, *38*, 457–460.
128 K. Tanaka, H. Suzuki, H. Osuga, *J. Org. Chem.* **1997**, *62*, 4465–4470.
129 K. Yamamoto, T. Harada, Y. Okamoto, H. Chikamatsu, M. Nakazaki, Y. Kai, T. Nakao, M. Tanaka, S. Harada, N. Kassai, *J. Am. Chem. Soc.* **1988**, *110*, 3578–3584.
130 A. Fürstner, G. Seidel, C. Kopiske, C. Krüger, R. Mynott, *Liebigs Ann.* **1996**, 655–662.
131 M. Mirza-Aghayan, H. R. Darabi, L. Ali-Saraie, M. Ghassemzadeh, M. Bolourtchian, M. Jalali-Heravi, B. Neumuller, *Z. Anorg. Allg. Chem.* **2002**, *628*, 681–684.
132 A. Merz, A. Karl, T. Futterer, N. Stacherdinger, O. Schneider, J. Lex, E. Luboch, J. F. Biernat, *Liebigs Ann.* **1994**, 1199–1209.
133 M. Sukwattanasinitt, R. Rojanathanes, T. Tuntulani, Y. Sritana-Anant, V. Ruangpornvisuti, *Tetrahedron Lett.* **2001**, *42*, 5291–5293.
134 G. Märkl, R. Ehrl, H. Sauer, P. Kreitmeier, T. Burgemeister, *Helv. Chim. Acta* **1999**, *82*, 59–84.

135 G. Märkl, H. Sauer, P. Kreitmeier, T. Burgemeister, F. Kastner, *Tetrahedron* **1999**, *55*, 13407–13416.
136 G. Märkl, T. Knott, P. Kreitmeier, T. Burgemeister, T. Kastner, *Helv. Chim. Acta* **2000**, *83*, 592–602.
137 J. Nakayama, H. Machida, H. Saito, R. Hoshino, *Tetrahedron Lett.* **1985**, *26*, 1981–1982.
138 J. Nakayama, Y. Ikuida, Y. Murai, R. Hoshino, *J. Chem. Soc., Chem. Commun.* **1987**, 1072–1073.
139 J. X. Hu, X. Jiang, T. He, J. Zhou, Y. F. Hu, H. W. Hu, *J. Chem. Soc., Perkin Trans. 1* **2001**, 1820–1825.
140 E. Vogel, B. Binsack, Y. Hellwig, C. Erben, A. Heger, J. Lex, Y. D. Wu, *Angew. Chem. Int. Ed. Engl.* **1997**, *36*, 2612–2615.
141 J. L. Sessler, E. A. Brucker, S. J. Weghorn, M. Kisters, M. Schaefer, J. Lex, E. Vogel, *Angew. Chem. Int. Ed. Engl.* **1994**, *33*, 2308–2312.
142 I. Shimizu, H. Umezawa, T. Kano, T. Izumi, A. Kazahara, *Bull. Chem. Soc. Jpn.* **1983**, *56*, 2023–2028.
143 I. Shimizu, Y. Kamei, T. Tezuka, T. Izumi, A. Kazahara, *Bull. Chem. Soc. Jpn.* **1983**, *56*, 192–198.
144 M. A. Buretea, T. D. Tilley, *Organometallics* **1997**, *16*, 1507–1510.
145 J. E. McMurry, P. Kocovsky, *Tetrahedron Lett.* **1985**, *26*, 2171–2172.
146 J. E. McMurry, G. K. Bosch, *J. Org. Chem.* **1987**, *52*, 4885–4893.
147 C. B. Jackson, G. Pattenden, *Tetrahedron Lett.* **1985**, *26*, 3393–3396.
148 J. E. McMurry, G. K. Bosch, *Tetrahedron Lett.* **1985**, *26*, 2167–2170.
149 J. E. McMurry, J. R. Matz, *Tetrahedron Lett.* **1982**, *23*, 2723–2724.
150 J. E. McMurry, J. R. Matz, K. L. Kees, P. A. Bock, *Tetrahedron Lett.* **1982**, *23*, 1777–1780.
151 J. E. McMurry, J. R. Matz, K. L. Kees, *Tetrahedron* **1987**, *43*, 5489–5496.
152 N. Kato, K. Nakanishi, H. Takeshita, *Bull. Chem. Soc. Jpn.* **1986**, *59*, 1109–1123.
153 Y. Li, W. Li, Y. Li, *Synth. Commun.* **1994**, *24*, 721–726.
154 I. Jenny, H. J. Borschberg, *Helv. Chim. Acta* **1995**, *78*, 715–731.
155 T. Eguchi, K. Arakawa, T. Terachi, K. Kakinuma, *J. Org. Chem.* **1997**, *62*, 1924–1933.
156 T. Eguchi, K. Ibaragi, K. Kakinuma, *J. Org. Chem.* **1998**, *63*, 2689–2698.
157 W. G. Dauben, T. Z. Wang, R. W. Stephens, *Tetrahedron Lett.* **1990**, *31*, 2393–2396.
158 W. Li, J. Mao, Y. Li, Y. Li, *Org. Prep. Proc. Int.* **1994**, *26*, 445–457.
159 W. D. Z. Li, Y. Li, Y. Li, *Tetrahedron Lett.* **1999**, *40*, 965–968.
160 Z. Liu, W. Z. Li, Y. Li, *Tetrahedron: Asymmetry* **2001**, *12*, 95–100.
161 Z. Liu, W. Z. Li, L. Peng, Y. Li, Y. Li, *J. Chem. Soc., Perkin Trans. 1* **2000**, 4250–4257.
162 Z. Liu, T. Zhang, Y. Li, *Tetrahedron Lett.* **2001**, *42*, 275–277.
163 T. Zhang, Z. Liu, Y. Li, *Synthesis* **2001**, 393–398.
164 M. P. Mermet-Mouttet, K. Gabriel, D. Heissler, *Tetrahedron Lett.* **1999**, *40*, 843–846.
165 W. G. Dauben, K. L. Lorenz, D. W. Dean, G. Shapiro, I. Farkas, *Tetrahedron Lett.* **1998**, *39*, 7079–7082.
166 D. L. J. Clive, K. S. K. Murthy, A. G. H. Wee, J. S. Prasad, G. V. J. da Silva, M. Majewski, P. C. Anderson, C. F. Evans, R. D. Haugen, L. D. Heerze, J. R. Barrie, *J. Am. Chem. Soc.* **1990**, *112*, 3018–3028.
167 A. Fürstner, O. R. Thiel, N. Kindler, B. Bartkowska, *J. Org. Chem.* **2000**, *65*, 7990–7995.
168 U. Berlage, J. Schmidt, U. Peters, P. Welzel, *Tetrahedron Lett.* **1987**, *28*, 3091–3094.
169 F. E. Ziegler, H. Lim, *J. Org. Chem.* **1982**, *47*, 5229–5230.
170 W. G. Dauben, I. Farkas, D. P. Bridon, C. P. Chuang, K. E. Henegar, *J. Am. Chem. Soc.* **1991**, *113*, 5883–5884.
171 V. J. Wu, D. J. Burnell, *Tetrahedron Lett.* **1988**, *29*, 4369–4372.
172 H. F. Grutzmacher, A. Mehdizadeh, A. Mulverstedt, *Chem. Ber.* **1994**, *127*, 1163–1166.
173 H. R. Darabi, T. Kawase, M. Oda, *Tetrahedron Lett.* **1995**, *36*, 9525–9526.

174 E. Vogel, P. Koch, X.-L. Hou, J. Lex, M. Lausmann, M. Kisters, M. A. Aukoloo, P. Richard, R. Guilard, *Angew. Chem. Int. Ed. Engl.* **1993**, *32*, 1600–1604.

175 G. Märkl, H. Sauer, P. Kreitmeier, T. Burgemeister, F. Kastner, G. Adolin, H. Nöth, G. Polborn, *Angew. Chem. Int. Ed. Engl.* **1994**, *353*, 1151–1153.

176 G. Märkl, J. Stiegler, P. Kreitmeier, T. Burgemeister, F. Kastner, S. Dove, *Helv. Chim. Acta* **1997**, *80*, 14–42.

177 G. Märkl, R. Ehrl, P. Kreitmeier, T. Burgemeister, *Helv. Chim. Acta* **1998**, *81*, 93–108.

178 Z. Y. Hu, J. L. Atwood, M. P. Cava, *J. Org. Chem.* **1994**, *59*, 8071–8075.

179 T. Kawase, H. R. Darabi, R. Uchimiya, M. Oda, *Chem. Lett.* **1995**, 499–500.

180 M. Kozaki, J. P. Parakka, M. P. Cava, *J. Org. Chem.* **1996**, *61*, 3657–3661.

181 Z. Hu, C. Scordilis-Kelley, M. P. Cava, *Tetrahedron Lett.* **1993**, *34*, 1879–1882.

182 G. De Munno, F. Lucchesini, R. Neidlein, *Tetrahedron* **1993**, *49*, 6863–6872.

183 W. M. Dai, W. L. Mak, *Tetrahedron Lett.* **2000**, *41*, 10277–10280.

184 D. C. Miller, M. R. Johnson, J. A. Ibers, *J. Org. Chem.* **1994**, *59*, 2877–2879.

185 G. Märkl, D. Bruns, H. Dietl, P. Kreitmeier, *Helv. Chim. Acta* **2001**, *84*, 2220–2242.

186 T. Nussbaumer, C. Krieger, R. Neidlein, *Eur. J. Org. Chem.* **2000**, 2449–2457.

187 T. Nussbaumer, R. Neidlein, *Helv. Chim. Acta* **2000**, *83*, 1161–1167.

188 H. Kurata, H. Nakaminami, K. Matsumoto, T. Kawase, M. Oda, *Chem. Commun.* **2001**, 529–530.

189 S. P. Sabelle, J. Hydrio, E. Leclerc, C. Mioskowski, P. Y. Renard, *Tetrahedron Lett.* **2002**, *43*, 3645–3648.

190 J. E. McMurry, D. D. Miller, *J. Am. Chem. Soc.* **1983**, *105*, 1660–1661.

191 J. E. McMurry, D. D. Miller, *Tetrahedron Lett.* **1983**, 1885–1888.

192 M. Iyoda, T. Kushida, S. Kitami, M. Oda, *J. Chem. Soc., Chem. Commun.* **1987**, 1607–1608.

193 W. Li, Y. Li, Y. Li, *Chem. Lett.* **1994**, 741–744.

194 W. Li, Y. Li, Y. Li, *Synthesis* **1994**, 267–269.

195 W. Li, Y. Li, Y. Li, *Synthesis* **1994**, 678–680.

196 A. Fürstner, D. N. Jumbam, *Tetrahedron* **1992**, *48*, 5991–6010.

197 A. Fürstner, H. Weintritt, A. Hupperts, *J. Org. Chem.* **1995**, *60*, 6637–6641.

198 A. Fürstner, D. N. Jumbam, H. Weidmann, *Tetrahedron Lett.* **1991**, *32*, 6695–6696.

199 A. Fürstner, A. Hupperts, A. Ptock, E. Janssen, *J. Org. Chem.* **1994**, *59*, 5215–5229.

200 A. Fürstner, A. Ptock, H. Weintritt, K. C. Goddard, *Angew. Chem. Int. Ed. Engl.* **1995**, *34*, 678–681.

201 A. Fürstner, D. N. Jumbam, *J. Chem. Soc., Chem. Commun.* **1993**, 211–212.

202 A. Fürstner, A. Ernst, *Tetrahedron* **1995**, *31*, 773–786.

203 A. Fürstner, A. Ernst, H. Krause, A. Ptock, *Tetrahedron Lett.* **1996**, *52*, 7329–7344.

204 A. Fürstner, D. N. Jumbam, G. Seidel, *Chem. Ber.* **1994**, 1125–1130.

205 S. Talukdar, S. K. Nayak, A. Banerji, *Synth. Commun.* **1998**, *28*, 2325–2335.

206 S. K. Nayak, A. Banerji, *J. Org. Chem.* **1991**, *56*, 1940–1942.

207 J. E. McMurry, M. P. Fleming, *J. Org. Chem.* **1976**, *41*, 896–897.

208 B. E. Kahn, R. D. Rieke, *Chem. Rev.* **1988**, *88*, 733–745.

209 B. E. Kahn, R. D. Rieke, *Organometallics* **1988**, *7*, 463–469.

210 J. E. McMurry, M. P. Fleming, K. L. Kees, L. R. Krepski, *J. Org. Chem.* **1978**, *43*, 3255–3266.

211 S. Pogodin, S. Cohen, I. Agranat, *Eur. J. Org. Chem.* **1999**, 1979–1984.

212 A. Fürstner, G. Seidel, *Synthesis* **1995**, 63–68.

213 J. J. Eisch, X. Shi, J. Lasota, Z. *Naturforsch. B* **1995**, *50*, 342–350.

214 J. J. Eisch, X. Shi, J. R. Alila, S. Thiele, *Chem. Ber./Recueil* **1997**, *130*, 1175–1187.
215 J. J. Eisch, J. N. Gitua, P. O. Otieno, X. Shi, *J. Organomet. Chem.* **2001**, *624*, 229–238.
216 D. Lenoir, *Synthesis* **1977**, 553–554.
217 S. Rele, S. Chattopadhyay, S. K. Nayak, *Tetrahedron Lett.* **2001**, *42*, 9093–9095.
218 S. Talukdar, S. K. Nayak, A. Banerji, *J. Org. Chem.* **1998**, *63*, 4925–4929.
219 S. Rele, S. Talukdar, A. Banerji, S. Chattopadhyay, *J. Org. Chem.* **2001**, *66*, 2990–2994.
220 J. E. McMurry, N. O. Siemers, *Tetrahedron Lett.* **1993**, *34*, 7891–7894.
221 J. E. McMurry, J. G. Rico, Y. N. Shi, *Tetrahedron Lett.* **1989**, *30*, 1173–1176.
222 J. E. McMurry, R. G. Dushin, *J. Am. Chem. Soc.* **1989**, *111*, 8928–8929.
223 J. E. McMurry, N. O. Siemers, *Tetrahedron Lett.* **1994**, *35*, 4505–4508.
224 K. C. Nicolaou, J. J. Yang, Z. Liu, H. Ueno, P. G. Nantermet, R. K. Guy, C. F. Claiborne, J. Renaud, E. A. Couladouros, K. Paulvannan, E. J. Sorensen, *Nature* **1994**, 630–634.
225 T. Mukaiyama, I. Shiina, H. Iwadare, M. Saitoh, T. Nishimura, N. Ohkawa, H. Sakoh, K. Nishimura, Y. Tani, M. Hasegawa, K. Yamada, K. Saitoh, *Chem. Eur. J.* **1999**, *5*, 121–161.
226 K. C. Nicolaou, C. F. Claiborne, K. Paulvannan, M. H. D. Postema, R. K. Guy, *Chem. Eur. J.* **1997**, *3*, 399–409.
227 K. C. Nicolaou, J. J. Liu, Z. Yang, H. Ueno, E. J. Sorensen, C. F. Claiborne, R. K. Guy, C. K. Hwang, M. Nakada, P. G. Nantermet, *J. Am. Chem. Soc.* **1995**, *117*, 634–644.
228 K. C. Nicolaou, Z. Yang, J. J. Liu, P. G. Nantermet, C. F. Claiborne, J. Renaud, R. K. Guy, K. Shibayama, *J. Am. Chem. Soc.* **1995**, *117*, 645–652.
229 E. J. Corey, R. L. Danheiser, S. Chandrasekaran, *J. Org. Chem.* **1976**, *41*, 260–265.
230 T. Li, W. Cui, J. Liu, J. Zhao, Z. Wang, *Chem. Commun.* **2000**, 139–140.

231 T. Tsuritani, S. Ito, H. Shinokubo, K. Oshima, *J. Org. Chem.* **2000**, *65*, 5066–5068.
232 N. Balu, S. K. Nayak, A. Banerji, *J. Am. Chem. Soc.* **1996**, *118*, 5932–5937.
233 A. Clerici, L. Clerici, O. Porta, *Tetrahedron Lett.* **1996**, *37*, 3035–3038.
234 T. Mukaiyama, N. Yoshimura, K. Igarashi, A. Kagayama, *Tetrahedron* **2001**, *57*, 2499–2506.
235 A. Kagayama, K. Igarashi, T. Mukaiyama, *Can. J. Chem.* **2000**, *78*, 657–665.
236 S. Matsubara, Y. Hashimoto, T. Okano, K. Utimoto, *Synlett* **1999**, *9*, 1411–1412.
237 Y. Hashimoto, U. Mizuno, H. Matsuoka, T. Miyahara, M. Takakura, M. Yoshimoto, K. Oshima, K. Utimoto, S. Matsubara, *J. Am. Chem. Soc.* **2001**, *123*, 1503–1504.
238 D. Enders, E. C. Ullrich, *Tetrahedron: Asymmetry* **2000**, *11*, 3861–3865.
239 O. Maury, C. Villiers, M. Ephritikhine, *New J. Chem.* **1997**, *21*, 137–139.
240 T. A. Lipski, M. A. Hilfiker, S. G. Nelson, *J. Org. Chem.* **1997**, *62*, 4566–4567.
241 M. Bandini, P. G. Cozzi, S. Morganti, A. Umani-Ronchi, *Tetrahedron Lett.* **1999**, *40*, 1997–2000.
242 A. Bensari, J. L. Renaud, O. Riant, *Org. Lett.* **2001**, *3*, 3863–3865.
243 H. Handa, J. Inanaga, *Tetrahedron Lett.* **1987**, *28*, 5717–5718.
244 T. Hirao, B. A. Hatano, M. Y. Muguruma, A. Ogawa, *Tetrahedron Lett.* **1998**, *39*, 5247–5248.
245 Y. Yamamoto, R. Hattori, K. Itoh, *Chem. Commun.* **1999**, 825–826.
246 Y. Yamamoto, R. Hattori, T. Miwa, Y. Nakagai, T. Kubota, C. Yamamoto, Y. Okamoto, K. Itoh, *J. Org. Chem.* **2001**, *66*, 3865–3870.
247 M. C. Barden, J. Schwartz, *J. Am. Chem. Soc.* **1996**, *118*, 5484–5485.
248 A. Gansäuer, *Synlett* **1997**, 363–364.
249 G. Gansäuer, *Chem. Commun.* **1997**, 457–458.
250 A. Gansäuer, M. Moschioni, D. Bauer, *Eur. J. Org. Chem.* **1998**, 1923–1927.

251 A. Gansäuer, D. Bauer, *Eur. J. Org. Chem.* **1998**, 2673–2676.
252 M. S. Dunlap, K. M. Nicholas, *J. Organomet. Chem.* **2001**, *630*, 125–131.
253 T. L. Chen, T. H. Chan, A. Shaver, *J. Organomet. Chem.* **1984**, *268*, C1–C6.
254 S. Gambarotta, C. Floriani, A. Chiesi Villa, C. Guastini, *Organometallics* **1986**, *5*, 2425–2433.
255 H. Ledon, I. Tkatchenko, D. Young, *Tetrahedron Lett.* **1979**, *20*, 173–176.
256 M. A. Barteau, *Chem. Rev.* **1996**, *96*, 1413–1430.
257 H. Idriss, M. A. Barteau, *Langmuir* **1994**, 3693–3700.
258 H. Idriss, K. G. Pierce, M. A. Barteau, *J. Am. Chem. Soc.* **1994**, *116*, 3063–3074.
259 K. G. Pierce, M. A. Barteau, *J. Org. Chem.* **1995**, *60*, 2405–2410.
260 H. Bönneman, B. Korall, *Angew. Chem. Int. Ed. Engl.* **1992**, *31*, 1490–1492.
261 M. T. Reetz, S. A. Quaiser, C. Merk, *Chem. Ber.* **1996**, *129*, 741–743.
262 R. Dams, M. Malinowski, I. Westdorp, H. Y. Geise, *J. Org. Chem.* **1982**, *47*, 248–259.
263 I. E. Aleandri, B. Bogdanovic, A. Gaidies, D. J. Jones, S. Liao, A. Michalowicz, J. Roziere, A. Schott, *J. Organomet. Chem.* **1993**, *459*, 87–93.
264 I. E. Aleandri, B. Bogdanovic, P. Bons, C. Dürr, A. Gaidies, T. Hartwig, S. C. Huckett, M. Lagarden, U. Wilczok, *Chem. Mater.* **1995**, *7*, 1153–1170.
265 H. Bertagnolli, T. S. Ertel, *Angew. Chem. Int. Ed. Engl.* **1994**, *33*, 45–66.
266 I. E. Aleandri, S. Becke, B. Bogdanovic, D. J. Jones, J. Roziere, *J. Organomet. Chem.* **1994**, *472*, 97–112.
267 B. Bogdanovic, A. Bolte, *J. Organomet. Chem.* **1995**, *502*, 109–121.
268 M. Stahl, U. Pidun, G. Frenking, *Angew. Chem. Int. Ed. Engl.* **1997**, *36*, 2234–2237.
269 K. J. Covert, P. T. Wolczanski, S. A. Hill, P. J. Krusic, *Inorg. Chem.* **1992**, *31*, 66–78.
270 R. S. P. Coutts, P. C. Wailes, R. L. Martin, *J. Organomet. Chem.* **1973**, *50*, 145–151.
271 O. V. Ozerov, S. Parkin, C. P. Brock, F. T. Ladipo, *Organometallics* **2000**, *19*, 4187–4190.
272 D. P. Steinhuebel, S. J. Lippard, *J. Am. Chem. Soc.* **1999**, *121*, 11762–11772.
273 J. V. Kingston, O. V. Ozerov, S. Parkin, C. P. Brock, F. T. Ladipo, *J. Am. Chem. Soc.* **2002**, *124*, 12217–12224.
274 O. Maury, M. Ephritikhine, M. Nierlich, M. Lance, E. Samuel, *Inorg. Chim. Acta* **1998**, *279*, 210–216.
275 C. Villiers, R. Adam, M. Lance, M. Nierlich, M. Ephritikhine, *J. Chem. Soc. Chem. Commun.* **1991**, 1144–1145.
276 O. Maury, C. Villiers, M. Ephritikhine, *Tetrahedron Lett.* **1997**, *38*, 6591–6594.
277 O. Maury, C. Villiers, M. Ephritikhine, *Angew. Chem. Int. Ed. Engl.* **1996**, *35*, 1129–1130.
278 M. Ephritikhine, O. Maury, C. Villiers, M. Lance, M. Nierlich, *J. Chem. Soc., Dalton Trans.* **1998**, 3021–3027.
279 C. Villiers, M. Ephritikhine, *Angew. Chem. Int. Ed. Engl.* **1997**, *36*, 2380–2382.
280 C. Villiers, M. Ephritikhine, *Chem. Eur. J.* **2001**, *7*, 3043–3051.
281 C. Villiers, A. Vandais, M. Ephritikhine, *J. Organomet. Chem.* **2001**, *617*, 744–747.
282 F. N. Tebbe, R. L. Harlow, *J. Am. Chem. Soc.* **1980**, *102*, 6149–6151.
283 R. H. Grubbs, R. H. Pine, in *Comprehensive Organic Synthesis*, Vol. 5 (Eds.: B. M. Trost, I. Fleming, L. A. Paquette), Pergamon, New York, **1991**, p. 1115.
284 N. A. Petasis, E. I. Bzowej, *J. Am. Chem. Soc.* **1990**, *112*, 6392–6394.
285 K. Takai, Y. Kataoka, T. Okazoe, K. Utimoto, *Tetrahedron Lett.* **1988**, *29*, 1065–1068.
286 Y. Horikawa, M. Watanabe, T. Fujiwara, T. Takeda, *J. Am. Chem. Soc.* **1997**, *119*, 1127–1128.
287 M. D. Fryzuk, P. P. Duval, S. S. S. H. Mao, S. J. Rettig, M. J. Zaworotko, L. R. McGillivray, *J. Am. Chem. Soc.* **1999**, *121*, 1707–1716.

288 A. Caselli, E. Solari, R. Scopelliti, C. Floriani, *J. Am. Chem. Soc.* **1999**, *121*, 8296–8305.

289 Y. Fujiwara, R. Ishikawa, F. Akiyama, S. Teranishi, *J. Org. Chem.* **1978**, *43*, 2477–2480.

290 J. C. Bryan, J. M. Mayer, *J. Am. Chem. Soc.* **1990**, *112*, 2298–2308.

291 M. H. Chisholm, K. Folting, J. A. Klang, *Organometallics* **1990**, *9*, 602–606.

292 M. H. Chisholm, K. Folting, K. C. Glasgow, E. Lucas, W. E. Streib, *Organometallics* **2000**, *19*, 884–892.

293 E. Vedejs, *Org. React.* **1975**, *22*, 401–422.

294 W. B. Motherwell, *J. Chem. Soc., Chem. Commun.* **1973**, 935.

295 W. B. Motherwell, C. J. Nutley, *Contemp. Org. Synth.* **1994**, *1*, 219.

296 A. K. Banerjee, M. C. Sulbaran de Carrasco, C. S. V. Frydrych-Houge, W. B. Motherwell, *J. Chem. Soc., Chem. Commun.* **1986**, 1803–1805.

297 P. Denifl, A. Hradsky, B. Bildstein, K. Wurst, *J. Organomet. Chem.* **1996**, *523*, 79–91.

7
Asymmetric Carbonyl Olefination

Kiyoshi Tanaka, Takumi Furuta, and Kaoru Fuji

7.1
Introduction and Historical Aspects

Ylides, which are among the most structurally interesting of reactive species, were first recognized as synthetically versatile reagents with the birth of the Wittig reaction in 1953 [1]. Since then, the chemistry of ylides and related carbanions with respect to carbonyl electrophiles has grown rapidly and they have now become powerful and versatile synthetic tools in organic chemistry [2]. Thus, the stereo-, regio-, and chemoselectivities can be controlled to a great extent, and Wittig-type reactions have become one of the most valuable organic transformations for the creation of a carbon–carbon bond, introducing a new sp^2 carbon at the carbonyl group [3]. In addition to the phosphonium ylides of Wittig reactions, chemistry based on other types of ylides [2], such as sulfonium, heteroatom, ammonium, nitrile, and pyridinium species, has also been developed in recent years. Ylides can be regarded as special carbanions, which are stabilized by a neighboring positively charged heteroatom, and which undergo nucleophilic reactions to form a new C–C bond.

In analogy with phosphorus ylides, stabilized carbanions that are adjacent to an oxidized phosphorus atom, as in a phosphate or phosphine oxide, also show characteristic nucleophilicity towards carbonyl groups [4]. The specific Wittig-type reactivity of such species has been investigated alongside the development and refinement of the chemistry of the original Wittig reactions. Nowadays, the Wittig reaction and related processes are recognized as some of the most essential transformation methods permitting carbon–carbon bond formation, and are often used as a crucial step in the total synthesis of complex molecules such as natural products [5].

It is generally accepted that reactions of carbonyl compounds occupy a central position in organic synthesis and hence in asymmetric synthesis. Thus, the development of new methods for stereoselective or stereocontrolled synthesis involving carbonyl groups has been a major subject in organic chemistry, and in particular much attention has been focused on enantioselective transformations. Efficient and practical asymmetric versions of a variety of ordinary synthetic reactions have

Modern Carbonyl Olefination. Edited by Takeshi Takeda
Copyright © 2004 WILEY-VCH Verlag GmbH & Co. KGaA, Weinheim
ISBN: 3-527-30634-X

been widely sought by many research groups, with a view to establishing novel asymmetric methodologies.

In spite of the importance of the Wittig and related reactions with regard to C–C bond formation, no new sp^3 stereogenic carbon centers are created in these types of transformations, in contrast to ordinary asymmetric reactions involving the formation of a center of chirality. For this reason, progress in the application of Wittig and related reactions to asymmetric synthesis has long been significantly slow, and it is only in the last two decades that practical methods with high efficiency for asymmetric carbonyl olefination have been developed [6]. Applications of such processes to the enantioselective construction of complex or useful molecules have only just begun to emerge.

Various types of phosphorus reagents have been employed for asymmetric transformations of carbonyl compounds into alkenes [2, 4, 7–11]. Depending upon the structure of these reagents, different reaction names are given [3]. Phosphonium ylides and phosphine oxide are particularly popular, and are referred to as Wittig and Horner reagents, respectively. On the other hand, phosphonates and other phosphonic acid derivatives are termed Horner–Wadsworth–Emmons (HWE) reagents. In recent years, arsonium derivatives have often been used for similar reactions, and they are also covered herein.

Methods for the formation of carbon–carbon double bonds in an asymmetric manner through non-Wittig-type reactions [12, 13] have also been reported in recent years, including asymmetric induction by reactions with chiral sulfoxides [13], sulfones (Julia olefination) [14], sulfoximides [12a, 15], or selenides [16]; Pd-catalyzed allylic nucleophilic substitutions [17], as well as asymmetric deprotonation [18]. In the context of the topic of asymmetric carbonyl olefination, some of these asymmetric transformations are beyond the scope of this chapter, although a few of the transformations closely related to the Wittig-type reactions will be discussed in a later part of this chapter.

As discussed below, the strategies for asymmetric induction through olefination by Wittig-type reactions can be broadly classified into four groups. According to the type of these approaches to optically active olefinic compounds, asymmetric olefinations based on Wittig and related reactions will be reviewed in this chapter.

The historical aspect must first be mentioned here, prior to the main issue of asymmetric olefination by Wittig-type reactions. The first report concerning asymmetric Wittig-type reaction of a 4-substituted cyclohexanone appeared in 1962 [19], in which an optically active phosphonate **2** bearing *l*-menthol as a chiral auxiliary on the carboalkoxy moiety was used (Scheme 7.1). Although the authors reported that dissymmetric olefinic products **3** were obtained in high yields in optically active form, no precise degree of asymmetric induction (diastereomeric excess) was given, and an error in the optical purities of the products was subsequently pointed out by researchers of another group [6].

A few years after this first report, a chiral non-racemic phosphonium ylide **4** having a stereogenic phosphorus center was prepared and used for asymmetric olefination (Scheme 7.2). Here, it was reported that monocarbonyl substrates **1** were converted into the olefinic products in enantiomerically enriched form, al-

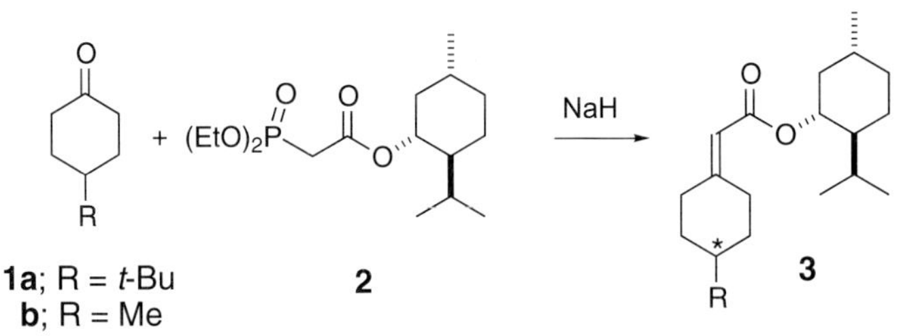

Scheme 7.1. The first report of asymmetric carbonyl olefination.

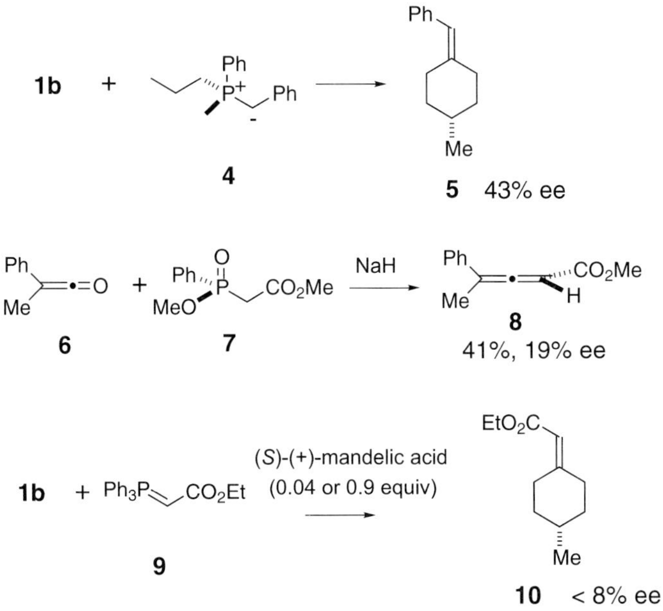

Scheme 7.2. Additional early examples of asymmetric carbonyl olefination.

though again the *ee* values of the olefins were not determined, except in the case of the 4-methylcyclohexyl derivative **5** (43% *ee*) [20].

The generation of dissymmetric non-racemic 1,2-dienes, i.e. allenic compounds, from the corresponding ketene **6**, was investigated in 1975 using optically active phosphinate-type esters such as **7** incorporating a stereogenic phosphorus center [21]. Although neither the chemical nor the optical yields of allenic products such as **8** were wholly satisfactory (41–80% yield and up to 23% *ee*), this is the only example of the use of a phosphinate ester in asymmetric olefination.

The first attempt to use a chiral catalyst in an asymmetric Wittig-type reaction was reported in 1970 [22]. A combination of the stabilized ylide **9** and a chiral organic acid as an external catalyst was evaluated in the olefination of 4-substituted cyclohexanone derivatives **1**. In this study, (S)-(+)-mandelic acid was found to be most effective in the generation of non-racemic cycloalkylidene derivatives such as 10, although the observed degree of asymmetric induction was quite low (up to 4% ee with 0.04 molar equivalents of catalyst) and the enhancement of the ee value was small even when 0.9 molar equivalents of chiral acid was used.

7.2
Strategies for Asymmetric Carbonyl Olefination

Asymmetric carbonyl olefination may be accomplished by means of any of the four general approaches outlined below, all of which are highly dependent on the structure of the carbonyl compound [7, 8, 10] (Scheme 7.3). Besides the four major approaches described here, other routes from achiral carbonyl compounds to non-racemic alkenes are available, and these topics will be mentioned separately towards the end of this chapter.

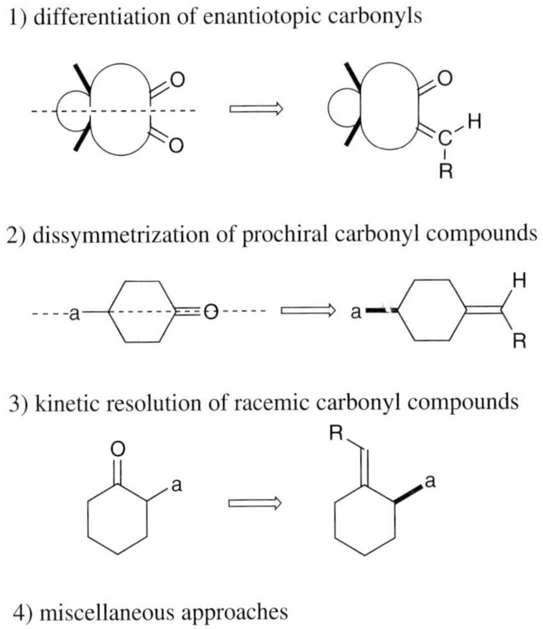

1) differentiation of enantiotopic carbonyls

2) dissymmetrization of prochiral carbonyl compounds

3) kinetic resolution of racemic carbonyl compounds

4) miscellaneous approaches

Scheme 7.3. Four general approaches for asymmetric carbonyl olefination.

The first approach is based on differentiation of enantiotopic carbonyl groups in symmetrical molecules such as *meso* compounds, and is therefore referred to as the desymmetrization of symmetric organic molecules. Discrimination of the π-faces of symmetrically substituted carbonyl groups constitutes the second type of approach leading to dissymmetric compounds having axial chirality, and is referred to as dissymmetrization. The third type of asymmetric carbonyl olefination is based on the kinetic resolution of racemic carbonyl compounds, and includes dynamic resolution and parallel kinetic resolution. The final category of approaches consists of miscellaneous alternative reactions, such as the stepwise construction of optically active olefinic compounds from achiral ketonic compounds, involving asymmetric dehydration or other elimination reactions. Recently, asymmetric induction caused by non-covalent bonding interactions with external chiral ligands has attracted much attention, and all such examples of asymmetric carbonyl olefination belong to one of the four approaches as classified above. The discussion in this chapter is ordered according to the class of strategy used for the transformation to optically active olefin compounds.

7.3
Optically Active Phosphorus or Arsenic Reagents Used in Asymmetric Carbonyl Olefination

A variety of achiral phosphorus reagents has appeared as a result of extensive research work on the Wittig and related reactions by many groups, and the phosphorus Wittig-type reagents [7, 8, 10] employed in these transformations can be classified into eight groups depending upon the kind of substituents attached to the phosphorus atom and/or its oxidation state, as shown in Scheme 7.4.

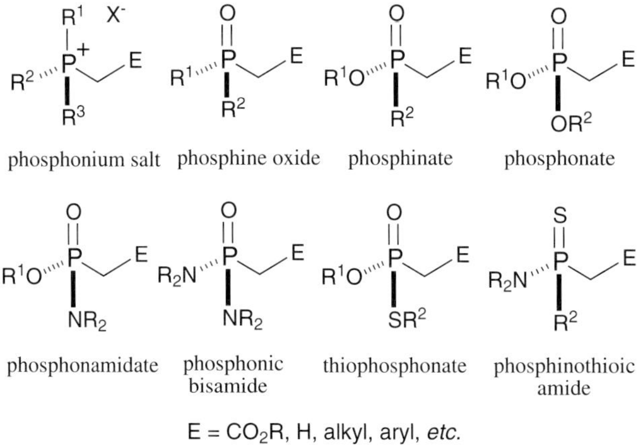

$E = CO_2R, H, alkyl, aryl, etc.$

Scheme 7.4. Wittig-type reagents used in carbonyl olefination.

Recently, chiral non-racemic organophosphorus compounds have received much attention. The major factor stimulating the extensive investigation of these compounds originates from their great practical value as ligands in catalysts for asymmetric synthesis and as efficient synthetic reagents as well as tools for the elucidation of biochemical mechanisms [9, 11, 23]. With the progress in comprehensive investigations of asymmetric carbonyl transformations, a considerable number of

7.3 Optically Active Phosphorus or Arsenic Reagents Used in Asymmetric Carbonyl Olefination

optically active chiral phosphorus reagents have emerged. The representative reagents are listed in Table 7.1, with an indication of the source of their chirality. The chiral phosphorus reagents are also divided into three main groups depending upon the location of the stereogenic chiral centers (Scheme 7.5). In type **I**, the tetrahedral phosphine atom is a stereogenic center; in type **II** it is the substituents connected to the phosphorus atom that are optically active, while in type **III** reagents the chirality exists at sites other than the phosphorus atom or its substituents, and the most widely used HWE reagents bearing an auxiliary group at the carboalkoxy moiety belong to this latter class. Although in the design of chiral phosphorus reagents, type **I** would seem to be most promising due to the proximity of the chiral center to the reaction site, the troublesome preparative work required to obtain this class of phosphorus compounds, as well as the difficulty in determining the absolute stereochemistry of the synthesized chiral phosphorus reagent, are reasons why this class of reagents has not been so widely utilized in asymmetric carbonyl olefinations. The olefination process consists of three successive steps, namely addition, ring closure to the four-membered ring, and elimination, and therefore the direct production of optically active olefinic products is possible with the chiral reagents of types **I** and **II**, through elimination of the phosphorus portion, whereas diastereomers are formed with the type **III** reagents.

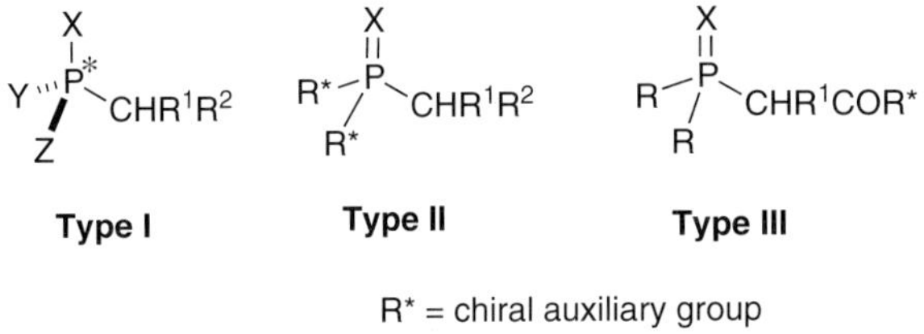

$$\underset{\text{Type I}}{\overset{X}{\underset{Z}{\overset{|*}{\underset{Y}{\overset{P}{\diagdown}}}}}\text{CHR}^1\text{R}^2} \qquad \underset{\text{Type II}}{\overset{X}{\underset{R^*}{\overset{||}{\underset{R^*}{\overset{P}{\diagdown}}}}}\text{CHR}^1\text{R}^2} \qquad \underset{\text{Type III}}{\overset{X}{\underset{R}{\overset{||}{\underset{R}{\overset{P}{\diagdown}}}}}\text{CHR}^1\text{COR}^*}$$

R* = chiral auxiliary group

Scheme 7.5. Three types of chiral phosphorus reagents employed in asymmetric carbonyl olefination.

The success and applicability of these reagents is largely dependent on their potential for asymmetric induction, easy prediction of the product stereochemistry, as well as their availability and accessibility. Since the anion of a HWE reagent can be regarded as a simple carbanion stabilized by two electron-withdrawing groups, and has the advantages of higher reactivity as well as easy separation of the reacted reagent from the reaction mixture, the chiral reagents that have been most widely used to date have been HWE-type reagents [24, 25]. Chiral HWE reagents containing optically active 8-phenylmenthol or binaphthol (BINOL) as auxiliary groups are especially popular. The effective function of these two moieties in HWE reagents as chiral auxiliaries for asymmetric induction in carbonyl transformations has been documented. Thus, in spite of the lower accessibility of both enantiomers, optically active 8-phenylmenthols can efficiently control the enantio(diastereo)-selectivity of products as well. For example, an ester bearing a (−)-8-phenylmenthyl auxiliary confers a high diastereofacial bias upon the derivatized anion, which

Tab. 7.1. Classification of chiral phosphorus or arsenic reagents used in asymmetric olefination.

Type of Reagents	Structure	Source of Chirality	Refs.
I, phosphonium ylide	Ph-P(Ph)=CHCH₂CH₃ **4**	(+)-(S)-benzyl(methyl)-phenyl-(propyl)phosphonium bromide	20
III, phosphonium ylide	Ph₃P=C(Me)CO₂R* **11a**; R* = (−)-menthyl **b**; R* = (−)-sec.octyl	(−)-menthol or (−)-sec.octanol	70
I, phosphonium ylide	Np-P(Ph)(Me)=C(Me)CO₂Et **12** Np = 1-naphthyl	racemic	70
I, phosphinate	Ph-P(=O)(OMe)-CH₂CO₂Me **7**	methylphenylphosphinylacetic acid	21
III, phosphonium ylide	Bu-CH=CH-CH=CH-PPh₃ **13** Fe(CO)₃	racemic	85
I, phosphonium ylide	(2-OMe-C₆H₄)(Me)(cyclohexyl)P⁺-CH₂-C(=O)- (2-methyl-1,3-dioxocyclopentyl) **14**	(+)-(R)-cyclohexyl-O-anisyl-methylphosphine (CAMP)	42

7.3 Optically Active Phosphorus or Arsenic Reagents Used in Asymmetric Carbonyl Olefination

Reagent type	Structure	Chiral auxiliary	Refs
I, phosphinothioic amide	**15**	(S)-N,S-dimethyl-S-phenyl-sulfoximine	78
I, phosphonamidate	**16a**; R = Ph **b**; R = SPh	(S)-3-hydroxybutyrate	54, 55
II, phosphonic bisamide	**17a**; R = Me **b**; R = Ph **c**; R = vinyl	(R,R)-1,2-diaminocyclohexane	6, 49, 50, 52a, 66
II, phosphonic bisamide	**17d**	(S,S)-1,2-diaminocyclohexane	6, 49, 50, 66
I, II, phosphonamidate	**18a**	camphorquinone	55
I, II, phosphonamidate	**18b**	camphorquinone	55

Tab. 7.1 (continued)

Type of Reagents	Structure	Source of Chirality	Refs.
I, II, phosphonamidate	**18c**	camphorquinone	55
I, II, thiophosphonate	**19a**	(−)-10-mercaptoisoborneol	56
I, II, thiophosphonate	**19b**	(−)-10-mercaptoisoborneol	56
III, phosphonate	**20**	(R,R)-1,2-diphenylaziridine	88
II, phosphine oxide	**21**	racemic	111

7.3 Optically Active Phosphorus or Arsenic Reagents Used in Asymmetric Carbonyl Olefination

Type	Structure	Chiral auxiliary	%
III, phosphonate	22a: R = Me; b: R = Et; c: R = i-Pr; d: R = Ph	(−)-benzopyrano[4,3-c]-isoxazolidine	57
II, phosphonate	23	D-(+)-mannitol	92
III, phosphonate	2	(−)-menthol	19
III, phosphonate	24	(+)-8-phenylnormenthol	67
III, phosphonate	25	(−)-trans-2-phenyl-1-cyclohexanol	60

Tab. 7.1 (continued)

Type of Reagents	Structure	Source of Chirality	Refs.
III, phosphonate	**26** (MeO)₂P(=O)CH₂C(=O)O-(8-phenylmenthyl), Ph	(+)-8-phenylneomenthol	60
III, phosphonate	**27a** (MeO)₂P(=O)CH₂C(=O)O-(8-phenylmenthyl), Ph	(+)-8-phenylmenthol	58, 59, 60
III, phosphonate	**27b**; R = Me **c**; R = Et **d**; R = iPr **e**; R = CF₃CH₂ **f**; R = o-Tolyl (RO)₂P(=O)CH₂C(=O)O-(8-phenylmenthyl), Ph	(−)-8-phenylmenthol	32, 39, 40, 45, 60, 61, 79, 80, 81, 88, 93, 96, 97, 98
III, phosphonate	**28** (CF₃CH₂O)₂P(=O)CH(Me)C(=O)O-(8-phenylmenthyl), Ph	(−)-8-phenylmenthol	79b, 88, 93
III, phosphine oxide	**29** Ph₂P(=O)CH₂C(=O)O-(8-phenylmenthyl), Ph	(−)-8-phenylmenthol	61

7.3 Optically Active Phosphorus or Arsenic Reagents Used in Asymmetric Carbonyl Olefination

III, arsenium	**30**	(−)-8-phenylmenthol	63, 84
II, phosphonate	**31a**: R = H **b**: R = Me **c**: R = SiMe₃ **d**: R = Ph	(S)-2,2'-BINOL	31, 33, 37, 46, 47, 68, 69, 74, 76
II, phosphonate	**31e**	(S)-2,2'-BINOL	8
II, phosphonate	**31f**	(S)-2,2'-BINOL	52b
II, phosphonate	**31g** ·MeOH	(S)-2,2'-BINOL	31

Tab. 7.1 (continued)

Type of Reagents	Structure	Source of Chirality	Refs.
III, phosphonate	**32**	(R)-2,2'-BINOL	8
II, phosphonate	**33**	(S)-8,8'-BINOL	99
III, phosphonate	**34**	(S)-8,8'-BINOL	99
II, arsenium	**35**	(S)-2,2'-BINOL	64

consistently displays highly diastereoselective alkylation to give enantiomerically enriched alkylated products after the subsequent removal of the chiral auxiliary [26]. The presence of a phenyl group in the auxiliary may play an essential role in differentiating the π-faces (*si*- or *re*-face) of the anionic species of the reagent in the transition state, presumably operating through π-π interaction with the π-systems of the enolate so as to effectively shield one of the faces from electrophilic approach of the carbonyl groups in an extended chair conformer, making it possible not only to induce a high level of diastereo- or enantioselectivity, but also to predict the stereochemistry of olefinic products (see, for example, Scheme 7.10).

On the other hand, BINOL and related derivatives [27], such as 2,2'-bis(diphenylphosphino)-1,1'-binaphthyl (BINAP), based on a C_2-symmetrical biaryl system with axial chirality, have been successfully used for a variety of asymmetric transformations, including catalytic processes. The conformation of one chiral, cyclic HWE reagent, **31g**, in the solid state has been elucidated by X-ray crystallographic analysis [4, 28], which revealed the efficient chiral environment created by the two naphthalene rings [8]. In addition to the aforementioned C_2-symmetric chiral phosphonates of BINOL derivatives and the HWE-type reagent bearing an 8-phenylmenthol auxiliary, the corresponding arsine analogues have also recently been developed.

It was observed that a small structural change in the phosphorus reagent can lead to the opposite sense of preferential absolute stereochemistry in the products, and this will be discussed later in more detail.

7.4
Discrimination of Enantiotopic or Diastereotopic Carbonyl Groups

7.4.1
Intermolecular Desymmetrization of Symmetrical Dicarbonyl Compounds

Though an approach of desymmetrization of symmetrical molecules to form optically active compounds had previously been exclusively studied in connection with the functions of biocatalysts [29], transformations by purely chemical methods rather than with the aid of enzymes or yeasts have in recent years received much attention as simple and efficient routes to synthetically useful non-racemic compounds [30]. Theoretically, differentiation between two enantiotopic or diastereotopic carbonyl groups, such as in *meso* compounds, can provide a single optically active enone in quantitative yield, if the stoichiometry is well controlled. Furthermore, this approach has the advantage that even if the minor monoalkene isomer is formed, which undergoes bis-alkylation to yield a diene much faster than the major monoalkene through kinetic resolution, the desired major monoalkene can be obtained with increased isomeric purity as a result [30]. Consequently, discrimination of enantiotopic carbonyl groups has been established as a promising and practical methodology for the preparation of functionalized molecules, specifically enones, without the need for a stoichiometric amount of a chiral reagent.

The concept of desymmetrization through intermolecular HWE reactions using chiral phosphonate reagents was independently demonstrated by two research groups in 1993 [31, 32]. Several bicyclic α-dicarbonyl compounds **36–41** were subjected to asymmetric olefination with anions of the HWE reagents (S)-**31a** or (S)-**33** to afford the corresponding E- and Z-olefinic products in good chemical yields and with excellent enantiomeric excesses [31, 33] (Scheme 7.6 and Table 7.2). The

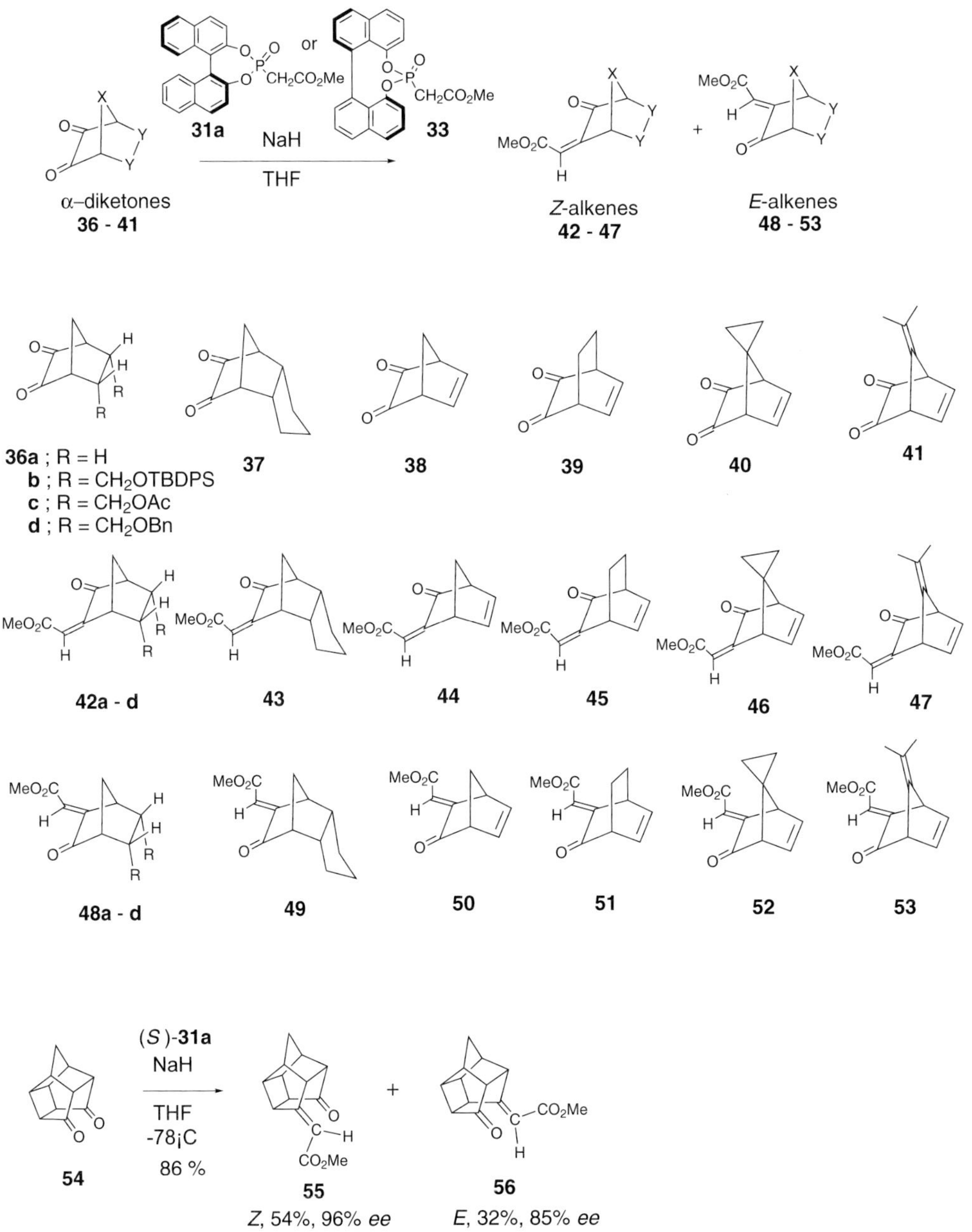

36a ; R = H
b ; R = CH₂OTBDPS
c ; R = CH₂OAc
d ; R = CH₂OBn

Scheme 7.6. Discrimination of enantiotopic carbonyls of α or γ-diketones.

7.4 Discrimination of Enantiotopic or Diastereotopic Carbonyl Groups | 301

Tab. 7.2. Asymmetric HWE reactions of **36–41** with **31a** or **33**.

Reagent	α-Diketones	Yield (%)	E/Z ratio	Products (% ee)	
31	36b	97	2/95	48b (30)	42b (98)
33	36b	59	1/97	48b (–)	42b (85)
31	36c	81	23/58	48c (23)	42c (90)
31	36d	86	22/64	48d (7)	42d (73)
31	37	72	15/57	49 (−1)	43 (93)
31	36a	50	37/13	48a (84)	42a (88)
31	38	83	58/25	50 (79)	44 (97)
31	40	97	62/35	52 (45)	46 (97)
31	41	83	53/30	53 (28)	47 (>99)
31	39	98	60/38	51 (75)	45 (89)

enantiotopic polycyclic γ-diketone **54** was also effectively differentiated by the chiral reagent with a base. In these reactions, a different carbonyl group is selected for the production of the respective Z- and E-isomers, so that the sense of chirality of the Z-product is opposite to that of the E-product. A higher degree of asymmetric induction is invariably observed for the Z-olefinic product. Furthermore, it was suggested in this investigation that reduction in optical purity of the olefinic products is caused by E/Z isomerization under the reaction conditions and/or during the work-up process.

It is likely that the initial addition step in the HWE reaction is rate-determining [34] when the Z-isomer is the major product, and energetically favorable approach of an anion of the chiral reagent must be invoked to account for the observed stereochemistry [25, 35]. In the planar anionic species of reagent (S)-**31a** chelated by a metallic cation, the *re*-face is sterically hindered due to the hydrogen atom at the 3-position of the naphthyl group, and nucleophilic attack on the carbonyl from the *exo* direction is favored (Scheme 7.7), as has been proven experimentally by sepa-

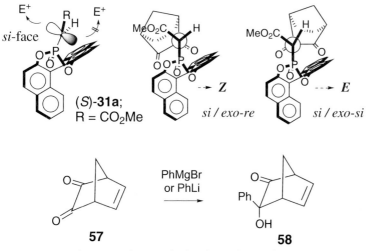

Scheme 7.7. Mechanistic explanation for the observed stereochemistry.

rate reactions, such as the transformation of **57** to **58**, and supported by theoretical calculations [36]. The *exo* face of the two carbonyl groups corresponds to the *re*- and *si*-face, respectively. Thus, only two possible combinations of the anion and the carbonyl substrate can be drawn (*si/exo-re* and *si/exo-si*) as reasonable transition states. The former combination does not suffer from the severe electronic repulsive interactions that exist in the latter, resulting in predominant production of the *Z*-isomer. The *E*-isomer obtained from the latter combination should have the opposite sense of absolute stereochemistry to the *Z*-isomer.

In a similar manner, asymmetric carbonyl olefination of *meso*-dicarbonyl compounds was extended to metallic arene or diene complexes [37], such as η^4-diene Fe or η^6-arene Cr complexes, to form planar complexes with high enantiomeric bias (Scheme 7.8). Since both complexation and decomplexation of these optically active compounds occur readily, these olefinic complexes are effective as stereocontrollers due to the presence of bulky metal tricarbonyl groups, and serve as useful reactants for obtaining optically active compounds of central chirality by appropriate chemical transformation.

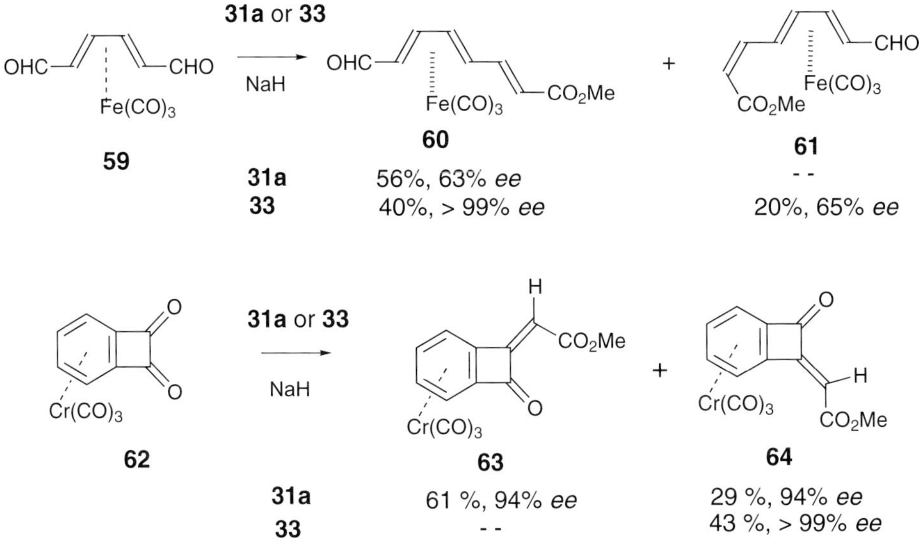

Scheme 7.8. Asymmetric carbonyl olefinations to give planar chiral alkenes.

Like the chiral reagents (*S*)-**31** and (*S*)-**33** having optically active 2,2'- or 8,8'-BINOL as auxiliaries at the phosphorus moiety, the chiral reagents (*S*)-**32** and (*R*)-**34** with the same auxiliaries at the carboalkoxyl portion also showed a similar selectivity towards symmetrical *meso*-dicarbonyl substrates, yielding the corresponding alkenes with high diastereoselectivity [8] (Scheme 7.9).

The enantiotopic aldehyde groups in the dialdehyde **67** were also efficiently discriminated by the chiral phosphonate **27b** to give monoolefination products with

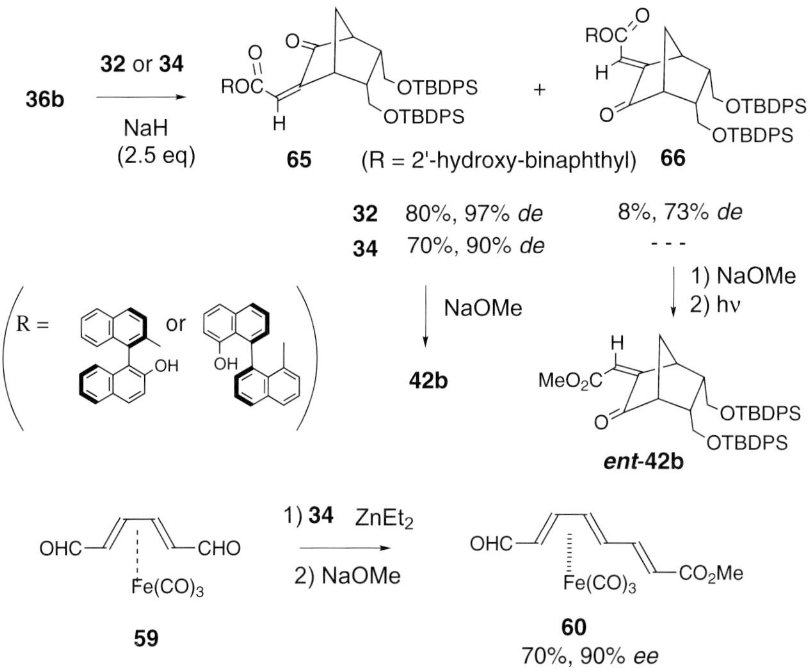

Scheme 7.9. Diastereoselective asymmetric carbonyl olefinations.

good diastereoselectivities (64–94% de) and high E-selectivity [32, 38] (Scheme 7.10). Various dialdehydes **69–71** were subjected to the asymmetric olefination and some of the products corresponded to partial structures of naturally occurring macrolides and could be used as building blocks for their total synthesis. It seemed likely that the reaction proceeded under thermodynamic control to give the (E)-alkenes, in spite of the reaction conditions of KHMDS, 18-crown-6, and THF, which tend to maximize kinetic control. Later, rationalization of product selectivity in this asymmetric transformation through the use of a new method for creating a transition state force field, based on quantum chemical normal-mode analysis, was reported [39, 40]. Here, two transition states for the addition step (TS1) and for the subsequent ring closure to an oxaphosphetane (TS2) were considered. According to this molecular mechanics study, the observed high E-selectivity was proposed to be caused by a greatly increased influence of TS2 due to a sterically demanding substrate, for example the dialdehyde **69**, blocking the path through the intermediate leading to the Z-product.

7.4.2
Intramolecular Discrimination Reactions

The most serious problems associated with the Wittig-type reactions stem from sensitivity to steric hindrance and enolization of the carbonyl substrates under the

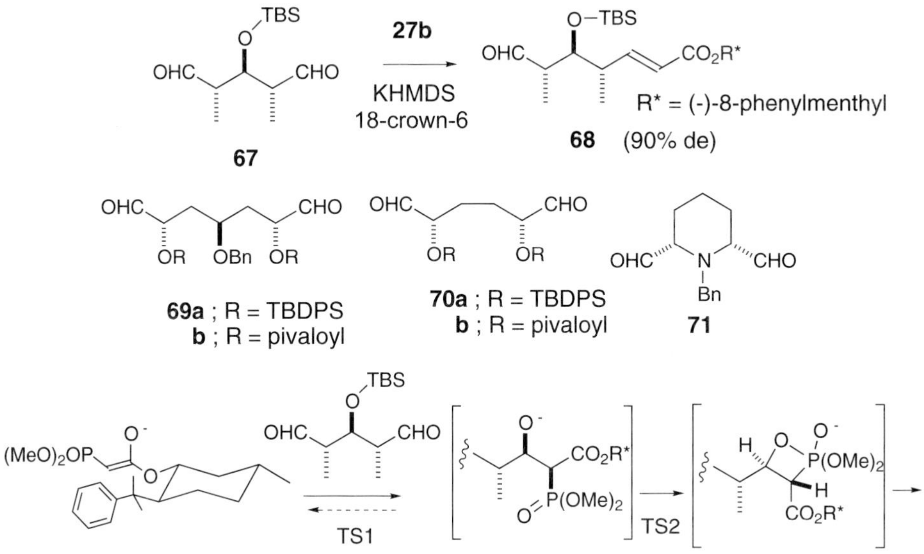

Scheme 7.10. Differentiation of enantiotopic dialdehydes.

basic conditions employed. On the other hand, an intramolecular process resulting in the construction of new ring systems can solve the problem of steric hindrance to some extent, because such a process is more entropically favored than an intermolecular one, especially when an energetically preferred 5- or 6-exo-trig process occurs. Intramolecular Wittig-type reactions [41] are effective not only for the construction of cyclopentene or cyclohexene derivatives, but also for larger ring systems.

The first example of the use of prochiral polycarbonyl compounds as substrates based on this idea appeared in 1980 [42, 43]. In this study, after examination of several chiral phosphorus ylides derived from triketone **72**, the most effective stereogenic phosphine was found to be R-(+)-CAMP (cyclohexyl o-anisylmethyl phosphine), which gave the cyclized diketone **73** (bis-nor-Wieland–Miescher ketone) with 77% ee (Scheme 7.11). The cyclized optically active ketone **73** was a useful chiral building block for the construction of some interesting polyquinan (polycondensed cyclopentanoid) natural products, such as coriolin [42]. Here, the stereoselectivity in discrimination of the two diastereotopic carbonyl groups was rationalized on the basis of a reversible initial nucleophilic addition step, but later another research group suggested that the initial step is likely to be irreversible even in the case of stabilized ylides [44].

A few additional examples of the use of intramolecular asymmetric HWE reactions to construct fused polycyclic ring systems containing a quaternary carbon have also been reported. The hydrindenone derivative **75** was successfully synthesized from the 1,3-cyclopentadione derivative **74** with high diastereoselectivity (98% de) [45]. The product **75**, containing a tetrasubstituted olefinic linkage, was

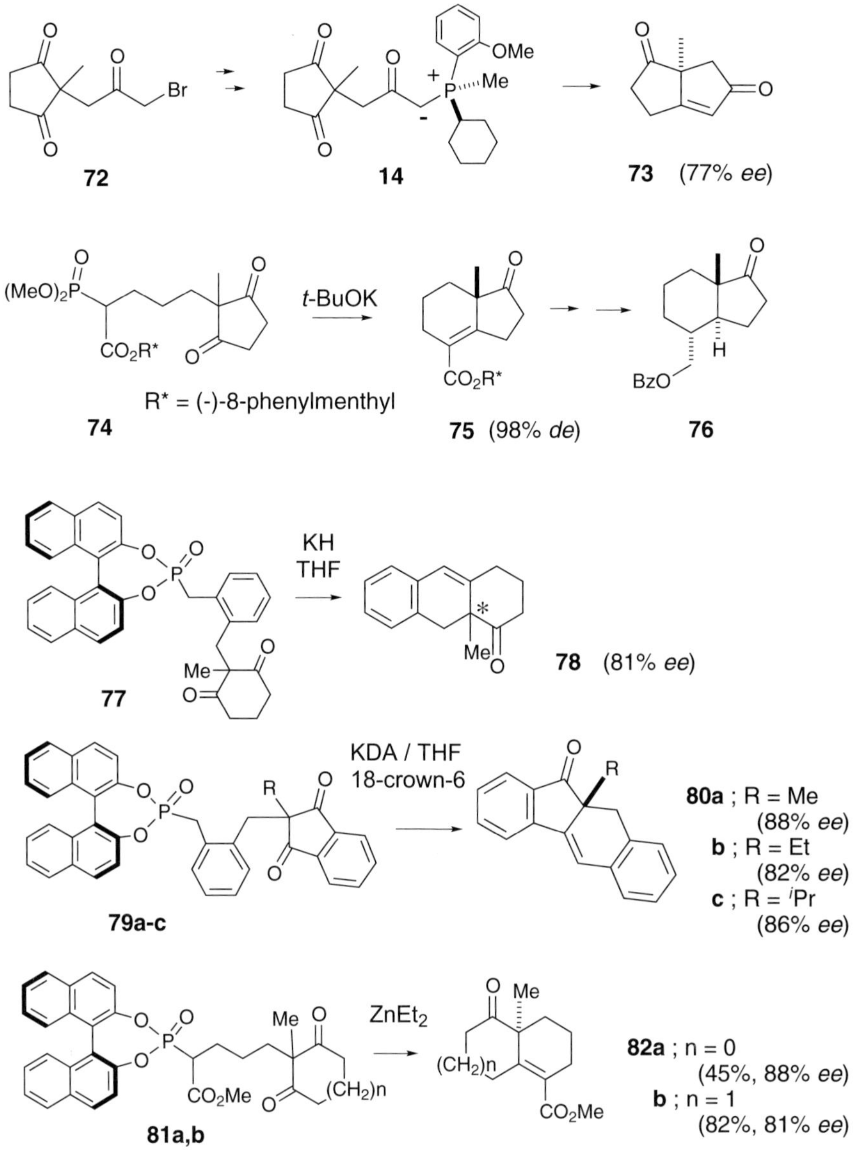

7.4 Discrimination of Enantiotopic or Diastereotopic Carbonyl Groups

Scheme 7.11. Intramolecular discrimination of carbonyl groups forming new five- or six-membered rings.

converted to compound **76**, which is useful as a new building block for 10,25-dihydroxyvitamin D$_3$. In a similar manner, novel dihydronaphthalene derivatives **78** and **80** [46], perhydroindanones **82a**, and perhydronaphthalenones **82b** [47] having a quaternary stereogenic carbon and/or a tetrasubstituted double bond have been enantioselectively prepared through differentiation of the diastereotopic carbonyl groups. A possible mechanistic explanation was given in these reports.

7.5
Discrimination of Enantiotopic or Diastereotopic Carbonyl π-Faces

7.5.1
Reactions with Prochiral Carbonyl Compounds

Some cycloalkylidene compounds possessing axial chirality but no asymmetric center are known to be optically active, and the first such compound to be identified was 4-methylcyclohexylideneacetic acid, obtained in 1909 through resolution with the alkaloid brucine [48]. Using symmetrically substituted monoketones as substrates, asymmetric carbonyl olefination causes dissymmetrization to give olefinic products with axial chirality. Since only two isomeric products are formed through four diastereomeric intermediates in this process, this transformation has often been used as a benchmark in evaluating newly developed chiral Wittig-type reagents. As already mentioned in the introduction, reactions of this kind became the first examples of asymmetric carbonyl olefinations.

After a couple of investigations of this approach using symmetrical 4-substituted cyclohexanone derivatives as substrates, a brilliant piece of work was reported in 1984 [6], in which a new chiral phosphonic bis(amide) of type **17** was exploited and used for the asymmetric carbonyl olefination of symmetrical ketones **1a** and **85a** (Scheme 7.12). The chiral molecules used here, trans-1,2-diaminocyclohexane derivatives, are well-known as effective chiral reagents and as ligands for catalysts used in asymmetric synthesis and molecular recognition [49, 50, 66]. Due to the C_2 symmetry of the diamine, the stereoelectronic requirements of the two nitrogen atoms attached to the phosphorus atom in a cyclic rigid core, and the overall topology of the resulting enantiomerically pure alkyl phosphonamides, the corresponding stabilized α-carbanions exhibit diastereofacial bias in their reactions with carbonyl electrophiles. In the case of the phosphonic bis(amide)-type reagent **17**, the intermediate anionic species do not undergo the subsequent elimination simultaneously, but can be isolated as the carbinols **83**, treatment of which with AcOH in a separate step results in elimination to give the olefinic products **84**. The product stereochemistry is thought to be governed by the initial kinetically controlled addition step of the anion. Later, the reagent **17** was successfully applied to the preparation of axially chiral dissymmetric alkenes, such as **86**. Interestingly, the olefinic products were examined as possible chiroptical triggers for a liquid-crystal based optical switch [51].

Reagents of this type have also proved to be effective for asymmetric Michael additions [52].

Another chiral phosphoramidate **16**, which possesses stereogenic centers at both phosphorus and carbon atoms, was examined in asymmetric HWE olefination reactions [53]. By fine-tuning of the N-substituents in the structure of the auxiliary,

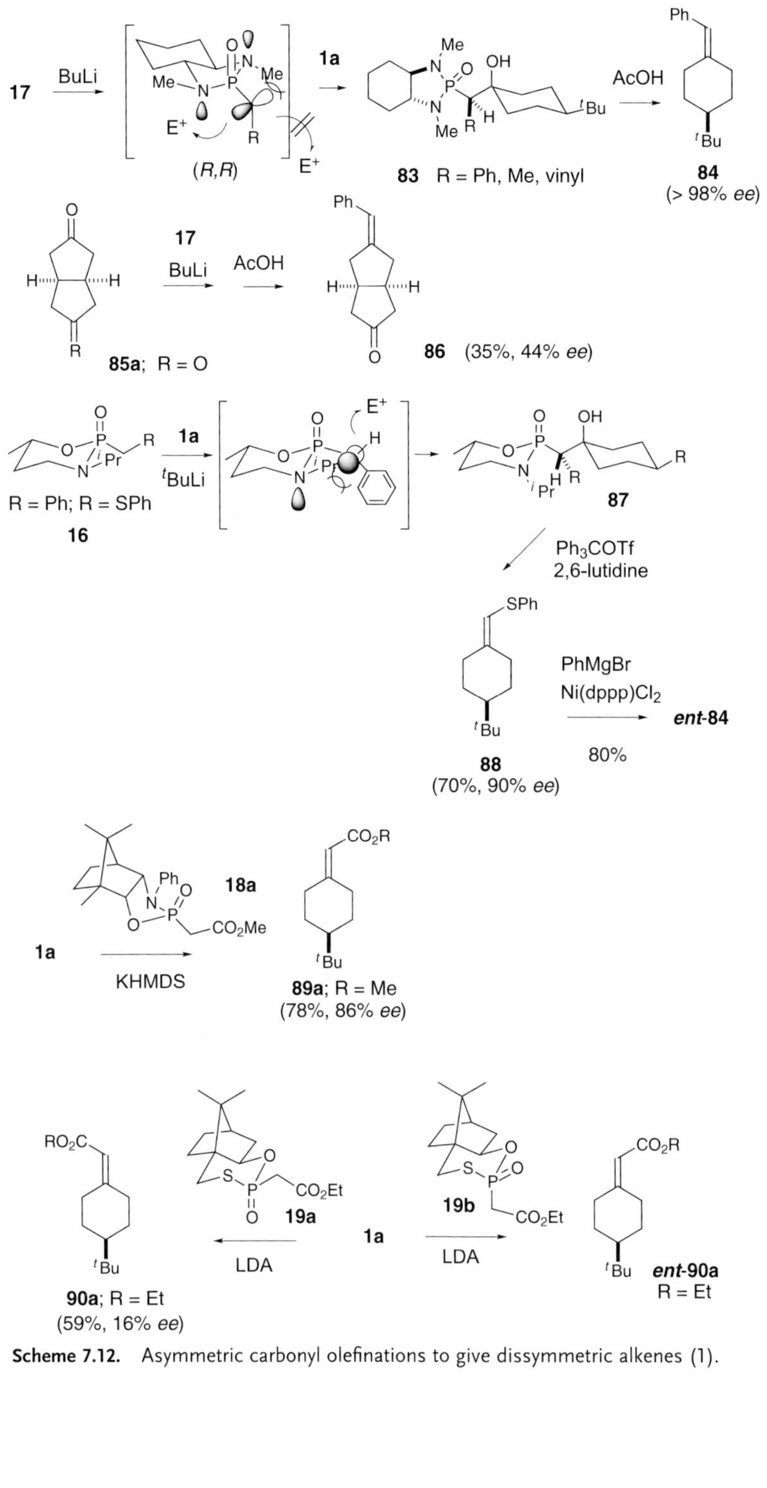

Scheme 7.12. Asymmetric carbonyl olefinations to give dissymmetric alkenes (1).

the cis-N-isopropyl phosphoramidate was found to give the best results. As with the phosphonic bis(amide) **17**, an additional elimination step to obtain the olefin **88** from the rather stable intermediate **87** was required [54]. This conversion was best achieved by the action of trityl triflate, and probably proceeded through the O-trityl phosphonium ion. Overall, the process of asymmetric olefination using a combination of reagents of this type furnished dissymmetric olefins with high enantioselectivity and in good chemical yield (Scheme 7.12). Furthermore, a closely related asymmetric conversion employing an alternative reagent of this class provided a convenient method for the preparation of a variety of substituted optically active dissymmetric alkenes through stereoselective Ni-catalyzed coupling with Grignard reagents [53].

In the case of phosphoramidate reagents bearing an anion-stabilized carboalkoxy group such as **18**, simultaneous elimination from the intermediate occurs, leading directly to the alkene without any further treatment. The well-designed compound **18a**, which contains a camphor ring and a carbomethoxymethylene group, gave a high asymmetric induction of up to 86% *ee* [55].

Two chiral phosphonic acid derivatives **19a,b**, containing a stereogenic phosphorus atom connected to a mercaptoisoborneol moiety, were prepared as a mixture, and were then chromatographically separated. Their ability in asymmetric carbonyl olefination was examined in the reaction with 4-*tert*-butylcyclohexanone **1a** [56]. The two lithium carbanions reacted with the carbonyl group of the substrate to give opposite enantiomers **90a**, although no remarkable degree of asymmetric induction was observed (up to 16% *ee*).

Chiral benzopyrano[4,3-*c*]isoxazolidine was incorporated into HWE phosphonate reagent **22a** and employed in asymmetric carbonyl olefinations of 4-substituted cyclohexanones to give the condensed products in good chemical yields and with high diastereoselectivity (Scheme 7.13); the products were efficiently converted to the corresponding allylic alcohol **90b**, unsaturated ketone **90c**, and unsaturated aldehyde **90d** by simple treatment with $LiBH_4$, a Grignard reagent, or DIBAL-H, respectively [57]. Again, both initial irreversible attack of the anion on the carbonyl from the equatorial direction and subsequent rapid elimination were considered in rationalizing the observed stereochemistry.

The chiral phosphonates **31a,e**, possessing optically active BINOL as an auxiliary, also demonstrated their ability as asymmetric inducers in the dissymmetrization of carbonyl compounds. In order to achieve both high enantioselectivity and good chemical yield, addition of zinc chloride was quite effective in these transformations [8]. It is known that bicyclo[3.3.0]octane derivatives usually adopt either W-, S-, or V-shaped conformations, and the observed stereochemistry of the alkene **92a** was best explained by considering an initial approach of the nucleophile to the W-shaped bicyclo[3.3.0]octanone in the direction in which steric interaction between the reagent and the substrate is minimized.

The reagents most widely applied to the dissymmetrization of ketones are chiral HWE phosphonate reagents possessing an 8-phenylmenthyl auxiliary on their carboalkoxy portion. Using reagents of this type, high diastereoselectivity has been achieved [58–60] (Scheme 7.14), and this may be due to the presence of the phenyl

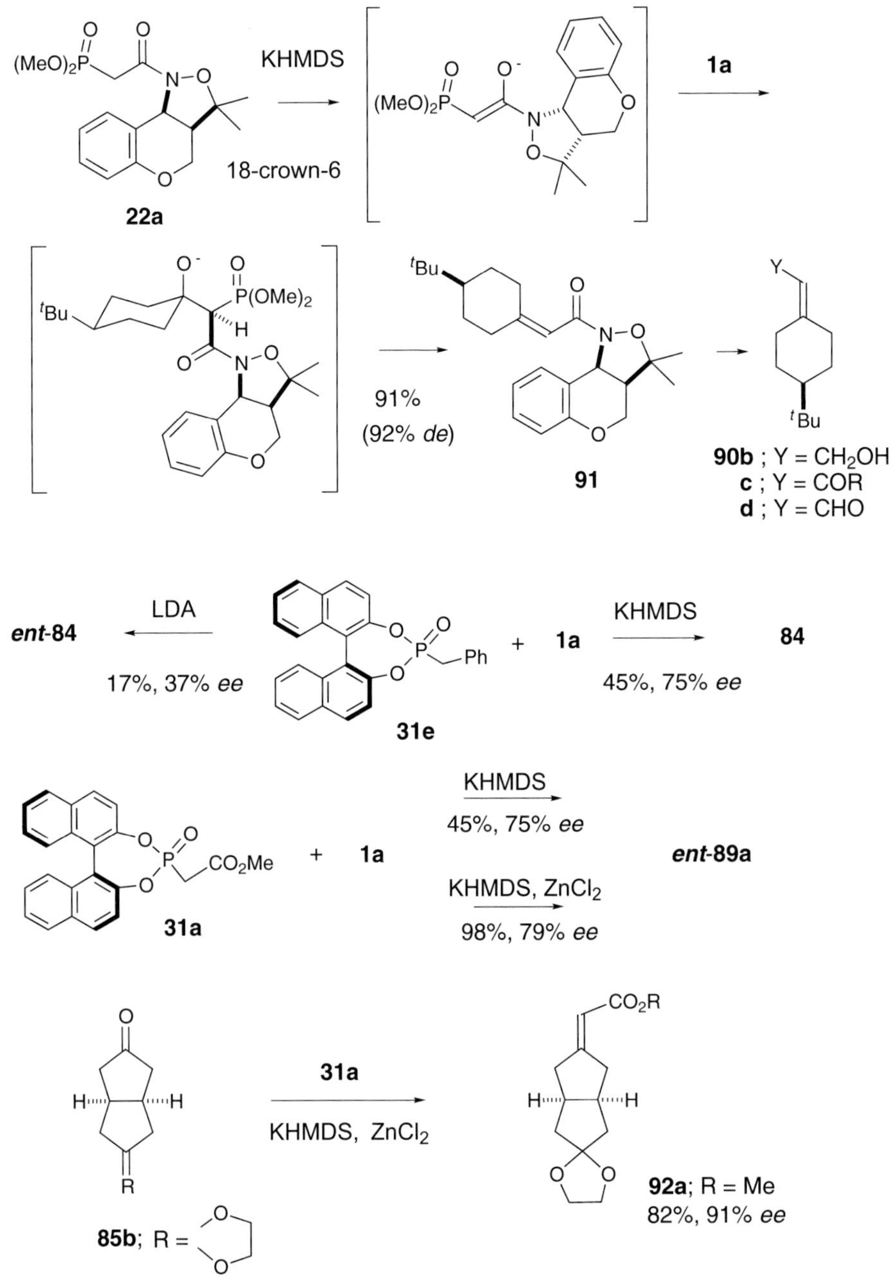

7.5 Discrimination of Enantiotopic or Diastereotopic Carbonyl π-Faces | 309

Scheme 7.13. Asymmetric carbonyl olefinations to give dissymmetric alkenes (2).

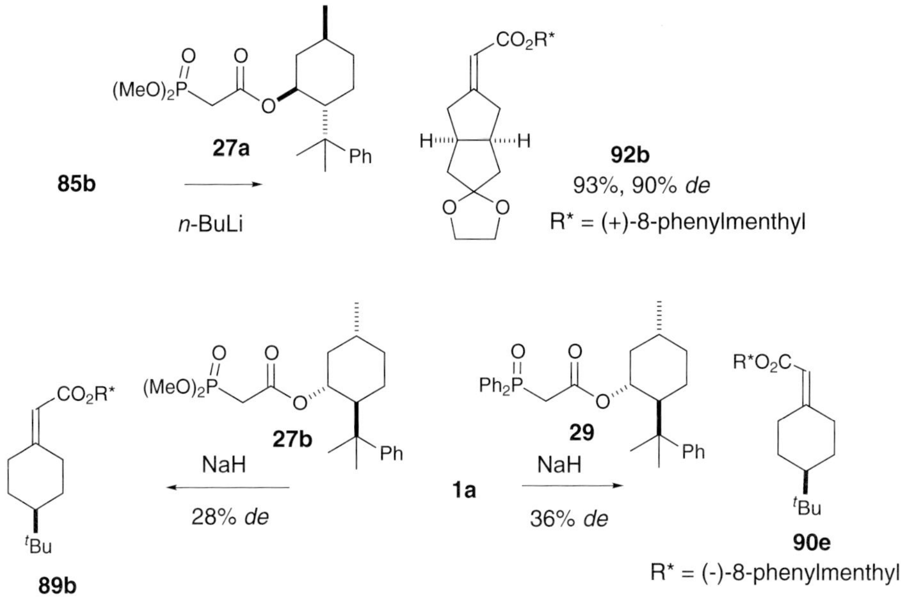

Scheme 7.14. Asymmetric carbonyl olefinations to give dissymmetric alkenes (3).

group in the chiral 8-phenylmenthyl auxiliary, which plays an important role in the addition of the phosphonate carbanion by effectively shielding one of the diastereotopic faces from approach of the electrophile.

The Horner-type reagent **29** has also been used for the same asymmetric carbonyl olefination [61], and the results were compared with those obtained using the corresponding HWE chiral reagent **27b**. The level of asymmetric induction with the Horner-type reagent **29** was not so remarkable, but the absolute stereochemistry of the olefinic product was opposite to that obtained from the reaction with the HWE reagent **27b**, despite the use of the same chiral auxiliary with the same absolute stereochemistry at the carboalkoxy moiety. These experiments clearly indicate that the substituents at phosphorus can significantly affect not only the level but also the mode of enantio- or diastereoselectivity. Consideration of the differences in the enolate geometry or in the rate-determining step of the aforementioned reaction (TS1 or TS2, *vide supra*) might lead to a reasonable explanation for the observed results.

Generally, arsonium ylides [62] are more reactive but less accessible than phosphonium ylides. Recently, the chiral arsonium reagent **30** has appeared, and has been applied in asymmetric Wittig-type carbonyl olefinations. This first chiral arsonium reagent also bears 8-phenylmenthyl as a chiral auxiliary on its carboalkoxy portion [63], and gave moderate chemical yields and diastereoselectivities in the conversion of 4-substituted cyclohexanone derivatives to axially chiral non-racemic alkylidene cyclohexanes under the same reaction conditions as used for the related reactions with phosphorus reagents (Scheme 7.15). On the other hand, the corre-

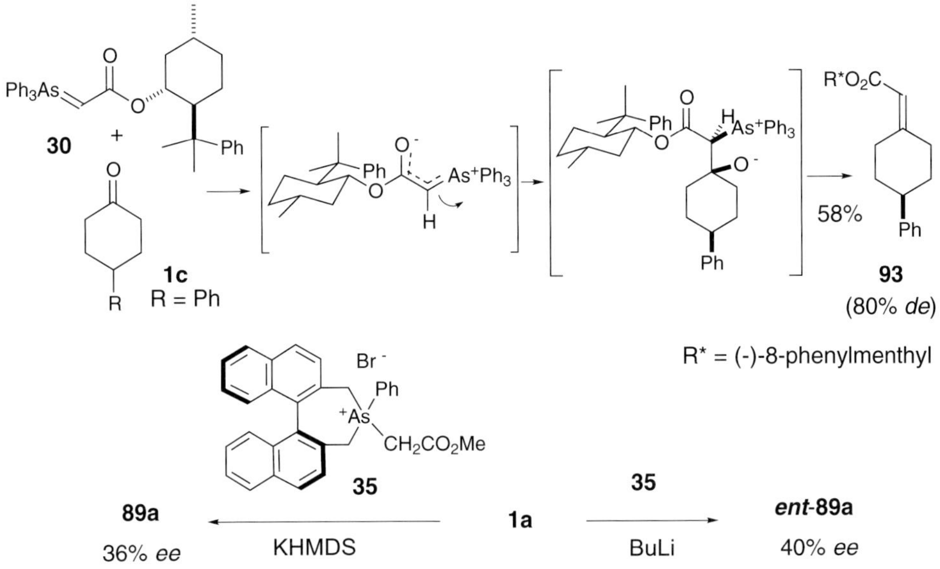

Scheme 7.15. Asymmetric carbonyl olefinations to give dissymmetric alkenes (4).

sponding HWE-type arsonium reagent exhibited lower diastereoselectivity. The observed stereochemistry of the products can again be rationalized by considering the initial nucleophilic attack to occur on the equatorial face of the cyclohexyl carbonyl group under kinetically controlled conditions from one of the favorable π-faces of the reagent created by the specific circumstance associated with the 8-phenylmenthyl auxiliary.

Recently, a novel C_2-symmetric chiral arsine **35** has been prepared from (S)-BINOL and employed in the enantioselective olefination of 4-substituted cyclohexanones to give the alkenes with moderate enantioselectivities of up to 40% *ee*. Moreover, a reversal in the stereochemistry of the products was observed simply by changing the counter cation of the base from lithium to potassium [64].

7.5.2
Reactions with Chiral Non-Racemic Carbonyl Compounds

Since double asymmetric synthesis often represents a promising transformation method for obtaining organic compounds in high optical purity [65], this method has been applied to the highly stereoselective conversion of a chiral unsymmetrical ketone to an alkene with a single geometry. In this context, the phosphoramides **17** were demonstrated to afford *E*- or *Z*-olefinic products **95** and **96** from (3*R*)-cyclohexanone (**4**) with excellent stereoselectivity, depending upon which enantiomer of the reagent was employed [66]. In the reaction with (2*R*,5*R*)-dihydrocarvone **97**, contrasting results were observed in that complete stereoselectivity in

7 Asymmetric Carbonyl Olefination

favor of the E-isomer **98** was achieved as a result of a matched combination and little selectivity was shown in the mismatched one.

The anions of the HWE reagents **27a,b** were reacted with the chiral monoketone **99** to afford the corresponding Z- and E-olefins **100** and **101** with high diastereomeric excesses, depending upon which enantiomer of the chiral phosphonate was employed. The olefinic products thus obtained served as key intermediates in the synthesis of prostacyclin derivatives [59, 60]. A closely related chiral reagent, **24**, bearing 8-phenylnormenthol [67], both enantiomeric forms of which are readily accessible, provided an improved diastereoselectivity in favor of the E-isomer **101** (Scheme 7.16).

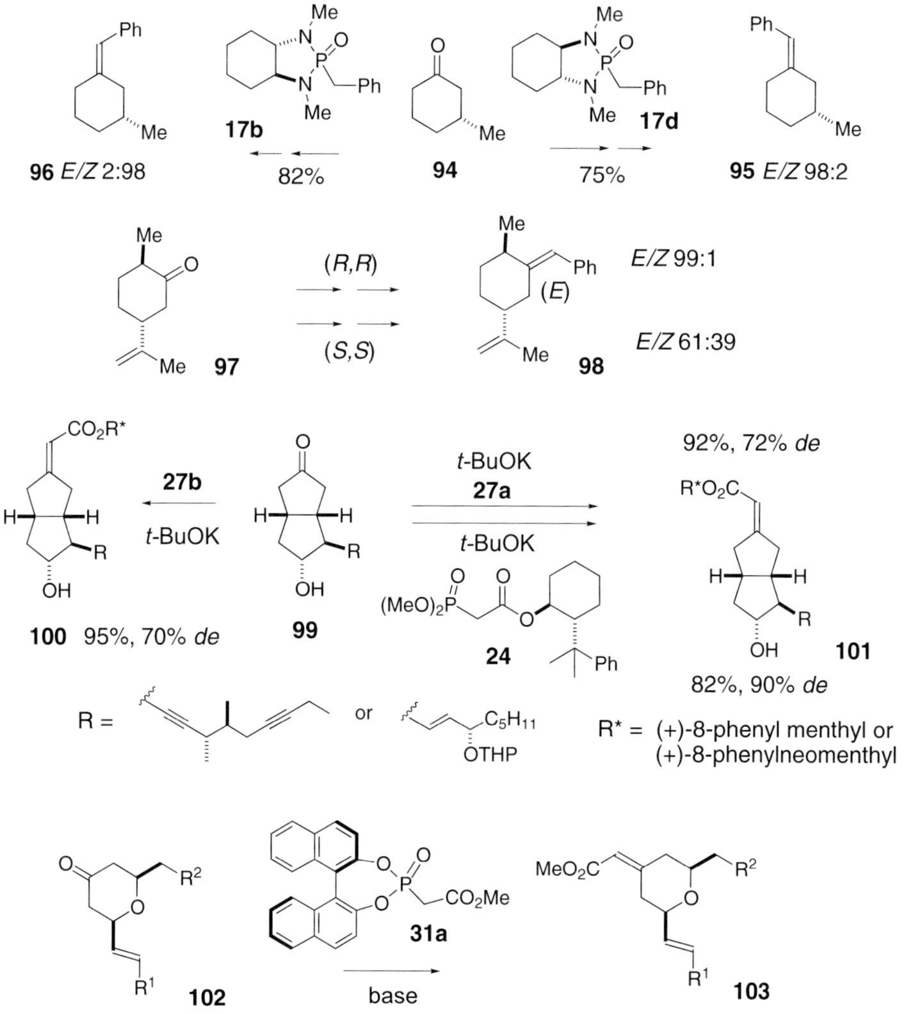

Scheme 7.16. Asymmetric olefinations with non-racemic chiral carbonyl compounds.

Recently, by selection of the appropriate enantiomer of the chiral HWE reagent **31a**, the concept of double asymmetric induction in an asymmetric carbonyl olefination step has been applied in controlling the geometry of the alkenic intermediate **103** in the total synthesis of structurally complex macrolides [68, 69] (*vide infra*).

7.5.3
Reactions with Prochiral Ketenes to give Dissymmetric Allenes

The idea that the facile condensation reaction between the non-enolizable and sterically less hindered carbonyl group of a ketene with a Wittig-type reagent provides a possible means of assembling compounds with axial chirality has long been recognized. Several attempted reactions based on this idea were reported during the 1960s. However, these early investigations had only limited success in that only low *ee* values were obtained [20, 70]. The most serious problem associated with the preparation and manipulation of ketenes is their inherently high reactivity and lability, which often leads to polymerizations or undesired side reactions and hence to a significant reduction in the chemical yield of the desired products. Therefore, general methods for the preparation of optically active allenic compounds have relied on transformation from propargyl derivatives having built-in stereogenic chiral centers [16d, 71].

In order to solve these troublesome problems of ketenes and to develop a new preparative method based on carbon–carbon bond formation, exploratory experiments were carried out. The most practical and successful protocol proved to be a one-pot procedure involving HWE olefination of in situ generated ketenes with a phosphonate reagent. By using this convenient one-pot procedure, which was first developed for the preparation of racemic conjugate allenecarboxylate (alka-2,3-dienoate) derivatives [72], troublesome handling of labile ketenic compounds can be avoided. Thus, treatment of the anion of the HWE reagent with in situ generated ketenes from enolates of 2,6-di-*tert*-butyl-4-methylphenyl (BHT) esters **104** gave the desired allenecarboxylates **105** in high yields. Thereafter, by using α,β-unsaturated BHT esters as substrates, tandem Michael–HWE reactions were performed [73]. Thus, the Michael-initiated ketenes, generated in situ from organolithium reagents and α,β-unsaturated BHT esters such as **106**, were effectively reacted with the anion of an achiral HWE reagent to give δ-branched allenecarboxylates **107** and derivatives thereof in good yields, although the diastereoselectivity was poor (Scheme 7.17).

These reactions were successfully applied to the enantioselective preparation of axially chiral allenecarboxylates [74]. In these one-pot transformations, replacement of the metallic counter cation by a somewhat less electropositive one, such as Zn^{2+} or Sn^{2+}, was found to be effective in controlling the nucleophilicity of the reagents and hence led to improved chemical yields. Additionally, the substituent at the 3-position of the naphthalene ring of the HWE reagent **31** significantly affects the degree of asymmetric induction, and a methyl substituent was found to give the best results. Treatment of BHT esters **104** as ketene precursors with a base in the presence of zinc ions generated the ketene, which was subsequently reacted

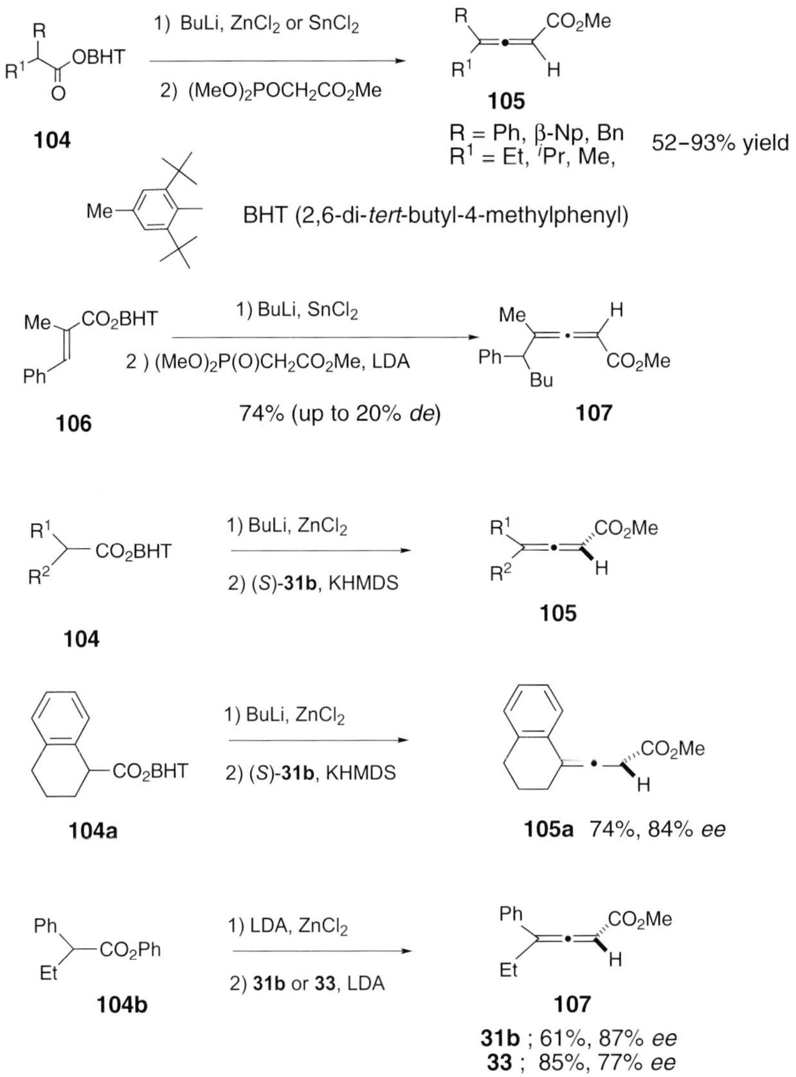

Scheme 7.17. Preparation of allenic compounds from ketenes.

with the anion of chiral phosphonate **31b** to form a dissymmetric allenecarboxylate in good chemical yield and with satisfactory optical purity (Table 7.3). A simple phenyl ester such as **104b** can also serve as a substrate for effective asymmetric transformation in place of the BHT esters [75] of lower availability [76]. Conjugated allenecarboxylates are interesting building blocks showing a versatile reactivity, and the selective transfer of axial chirality in these molecules to central chirality by ap-

Tab. 7.3. Preparation of optically active allenecarboxylates with the anion of (S)-**31b**.

BHT esters		allenecarboxylate	
R^1	R^2	yield (%)	% ee (config.)
Ph	Me	94	62 (Ra)
^{i}Pr	Ph	71	81 (Sa)
cyclo-Hex	Ph	81	79 (Sa)
⌬		74	84 (Ra)
2-Naphthyl	Et	91	72 (Ra)
Ph$_2$CH	Me	85	61 (Ra)

propriate transformation provides an efficient preparative method for a variety of optically active compounds.

The enantioselectivity observed by using the chiral HWE reagent (S)-**31b** is best understood by consideration of the favorable transition state rather than the stability and reactivity of the phosphoxetane intermediate. Thus, the addition of Zn^{2+} ions leads to a rigidly chelated phosphate anion bound by Zn^{2+}, as depicted in Figure 7.1, where the axially dissymmetric binaphthyl group dictates the orientation of the approach to the electrophile from the less hindered *si*-face of the reagent (route a). Assuming $R^1 > R^2$ in terms of bulkiness, it is likely that the nucleophile approaches the LUMO of the ketene carbonyl from the face including the less bulky R^2, so as to avoid the more severe steric repulsive interaction with R^1. Consequently, the substituent at the 3-position of the naphthalene ring exerts its influence as the bulkiness of the R^2 group increases.

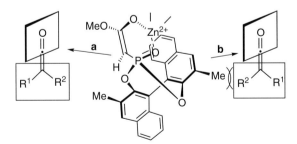

Fig. 7.1. Possible mechanistic explanation for the observed enantioselectivity in the asymmetric HWE reaction with (S)-**31b**.

7.6
Kinetic Resolution

7.6.1
Resolution of Racemic Carbonyl Compounds

When there exists a considerable difference in activation energy between diastereomeric transition states derived from each enantiomer of a racemic compound and a chiral reactant, kinetic resolution [77] becomes possible based on the difference in relative reaction rates, to give enantiomerically enriched product together with recovery of starting material in non-racemic form. In some aspects, this concept resembles the discrimination of enantiotopic carbonyl groups discussed above, and indeed both transformations are realized with biocatalysts, such as enzymes or yeasts. Although kinetic resolution has an inherent and practical problem in that theoretically only up to 50% of chemical yield can be obtained, this chemical approach to optically active compounds has been widely used as one of the most convenient resolution methods for racemates. Thus, for complete conversion to a single alkene in an asymmetric carbonyl olefination, at least a twofold excess of the racemic carbonyl substrate is necessary for the reaction with the chiral Wittig-type reagent. An early study [21] along these lines, employing the chiral phosphinate reagent **7**, demonstrated that the reaction of racemic 2-methylcyclohexanone **108** resulted in both the *E*- and *Z*-olefinic compounds **109** and **110** with low enantiomeric excess (Scheme 7.18), and it was found that the absolute configurations at the stereocenters were opposite to each other, suggesting preferential reaction with the opposite enantiomer for the production of the two stereoisomers in this kinetic resolution process.

The asymmetric olefination of the racemic ketone **111** with the anion of the phosphinothioic amide **15** can be regarded as an example of kinetic resolution;

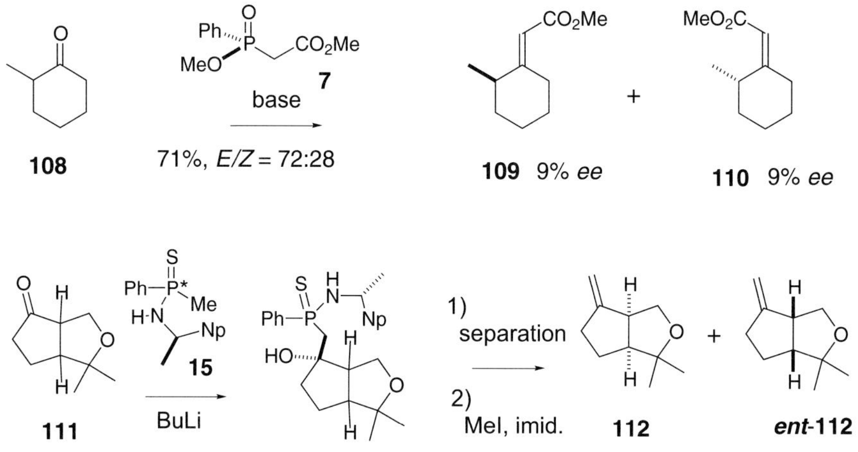

Scheme 7.18. Early examples of asymmetric olefination through kinetic resolution.

however, the transformation involving the enantioselective preparation of both enantiomers of the irridoid monoterpene hop ethers **112** from the corresponding diastereomers relied on successful separation of the diastereomeric intermediates, formed in a 3:2 ratio, rather than on a difference in their relative rates of formation [78].

Some years later, the first investigations [19, 49, 66] of the preparative value of this method were carried out by using chiral reagents **17**. In the reaction of two equivalents of racemic *cis*-2,4-dimethylcyclohexanone **113** with the anion of chiral phosphonate **17b**, kinetic resolution took place to give the *E*-alkene **114** with excellent enantioselectivity in 60% yield based on the reagent employed, after treatment of the adduct with AcOH (Scheme 7.19).

Later, the method of kinetic resolution was extended to aldehyde substrates [79, 80]. Here, acrolein dimer **115** was used as a racemic substrate, and 2.1–3 equivalents thereof was efficiently resolved to give the corresponding *E*- or *Z*-alkenes with high levels of diastereoselectivity by reactions with a series of chiral phosphates **27** or **28**. In this study, the change in structure of the connecting groups to the phosphorus atom significantly affected the E/Z selectivity of the olefinic products. Since modification at the phosphonate moiety allowed control of the product stereochemistry, it was possible to synthesize either (R, E)- or (S, Z)-alkenes in good yields and with high diastereomeric excesses. It is noteworthy that optically active *E*- and *Z*-alkenes were produced from different enantiomers of the racemic substrates, as in the preferential discrimination of opposite carbonyl groups in the differentiation of enantiotopic carbonyl groups mentioned above. The trisubstituted *Z*-alkenes **118** and **120** were prepared in this way using phosphate **28** [79b].

In a similar way, racemic 3-substituted cyclohexanones **121** were reacted with the anion of the chiral phosphoamidate reagent **18a** to give two isomeric optically active alkenes, that is (S, E)- and (R, Z)-alkenes possessing the opposite absolute configuration at the homoallylic carbon centers, with high enantiomeric excess [55].

In a closely related study on kinetic resolution, it was shown that not only the structure of the chiral phosphorus reagent, but also the structure of the racemic carbonyl substrates and the reaction conditions significantly affect the product stereoselectivity. All four possible isomeric alkenes **125–128** could be preferentially obtained from the racemic *N*-diphenylphosphoryl-protected aldehyde **124** as the major products by appropriate selection of the base, the solvent, and the substituent R on the phosphonates **27** and **28** (geometric selectivities from 66:34 to 98:2; diastereomer ratios between 93:7 and >99:1) (Scheme 7.20) [81]. Mechanistically, a switch between Felkin–Anh–Eisenstein and chelation control [82] at the transition state of addition to the aldehyde occurs, depending upon the reaction conditions used, and the combination with the influence of substrate stereocenters is responsible for the bias in the product distribution. Insight into the mechanism of the reaction between the chiral reagents and the aldehyde substrates was gained from a molecular mechanics modeling study, which supported the hypothesis of kinetic stereoselectivity in the addition step, as previously assumed by many research groups. Preferential production of an alkene from one of the eight (2^3) the-

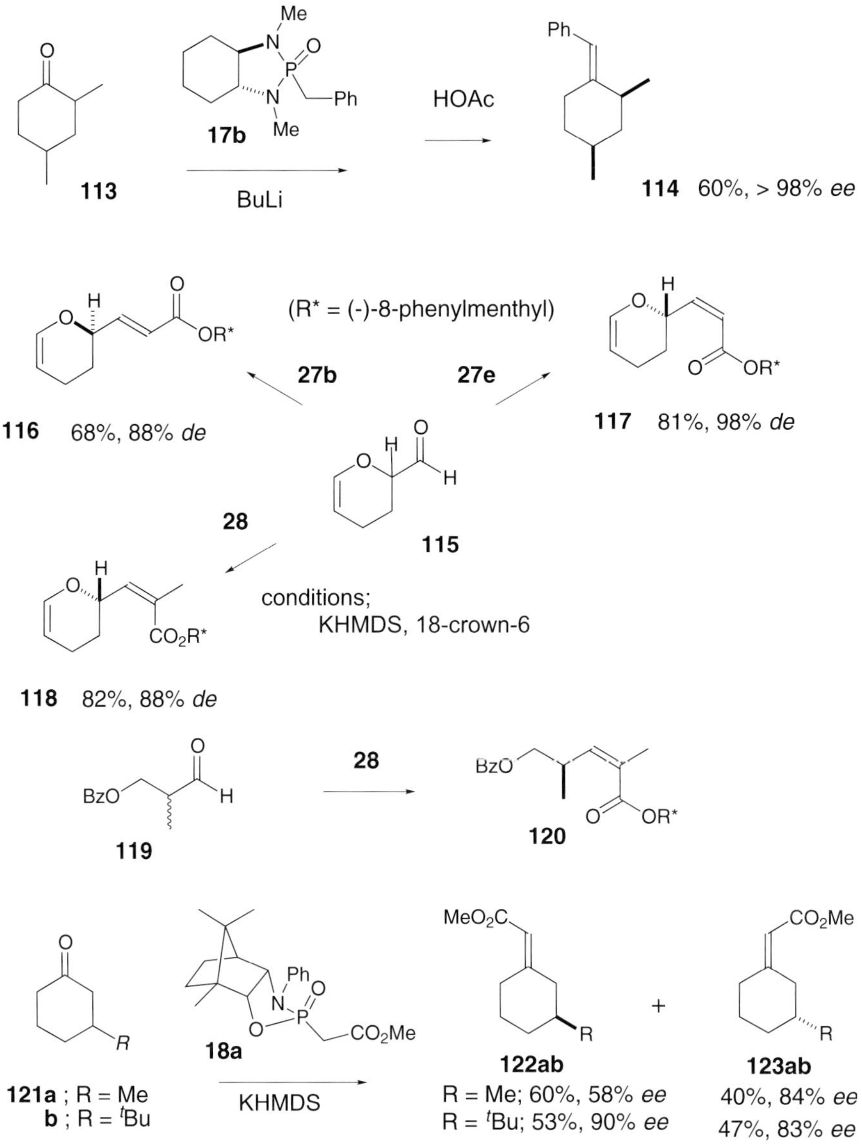

318 | *7 Asymmetric Carbonyl Olefination*

Scheme 7.19. Asymmetric carbonyl olefinations through kinetic resolution.

oretically possible intermediates, as well as the formation of the opposite absolute configuration at the allylic stereocenter of the products from a reaction involving kinetic resolution, can be rationalized by considering the following factors. Thus, the selectivity is determined by the influence of three major factors: the chiral auxiliary, the nature of the R group connected to the phosphorus atom, and the α-stereocenter in the substrate. The chiral auxiliary dictates the face of the phospho-

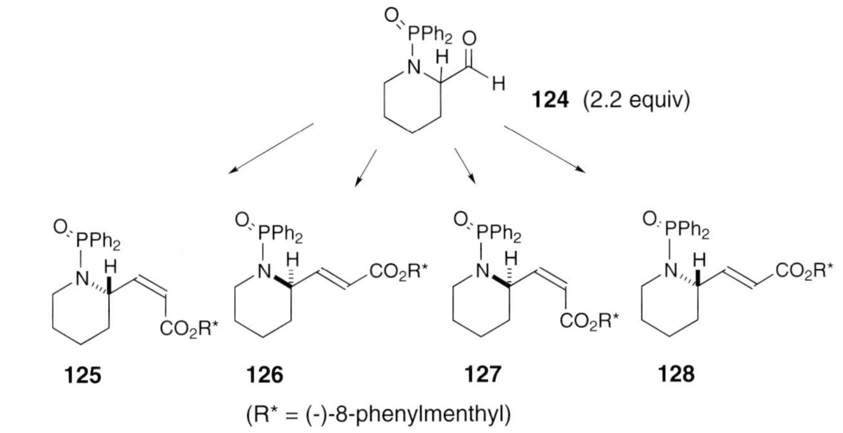

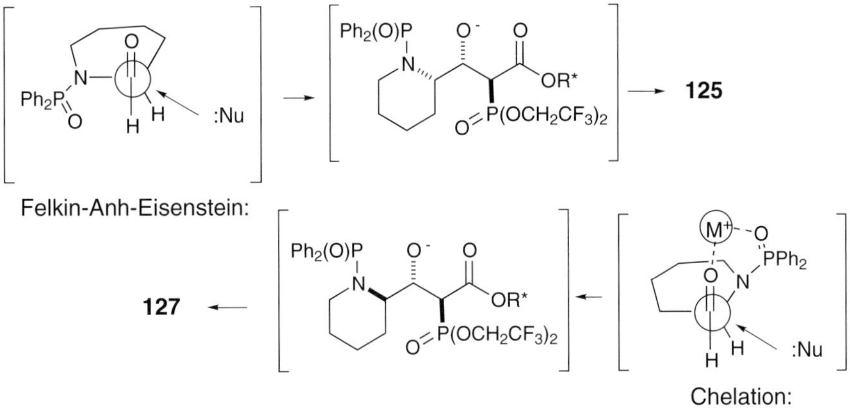

Scheme 7.20. Preferential formation of one of the four possible diastereomers.

Alkenes	125	126	127	128
Reagents	27e	27f	27e	27b
Base	KHMDS / 18-crown-6	KHMDS / 18-crown-6	NaHMDS	KHMDS
Solvent	THF or EtCN	THF	CH_3CN	THF

nate enolate that is attacked, hence the absolute configuration at C-2. The R group on the phosphonate and the substrate α-stereocenter affect the relative configurations at C-2 and C-3, and C-3 and C-4, which determine the E/Z stereochemistry of the alkene and the absolute stereochemistry induced by the chiral reagent, respectively (Figure 7.2). Additionally, a modeling study indicated considerable effect of the substrate stereocenter not only on the addition step but also on the elimination in some cases.

Asymmetric olefination based on kinetic resolution was then directed to the use

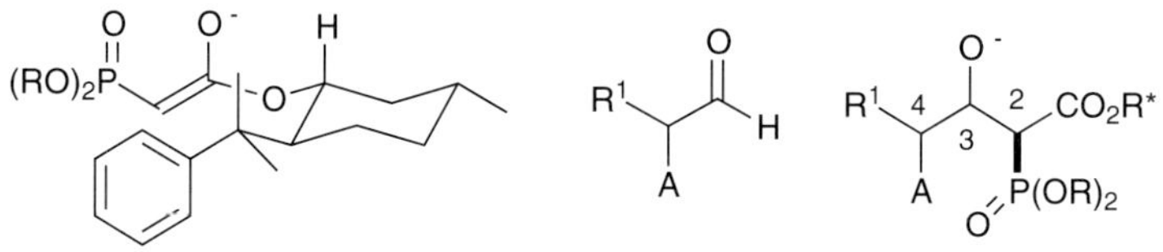
(RO)2P
H
R1
A
R1 4 3 2 CO2R*
P(OR)2

Fig. 7.2. Mechanistic consideration of the reaction with reagents **27**.

of α-amino aldehyde **129** as a substrate (Scheme 7.21), whereby the preferential production of opposite enantiomers **130** and **131** was again observed by a small modification of the structure of the chiral reagent using the same enantiomer as a chiral auxiliary, yielding (*R*, *E*)- and (*S*, *E*)-alkenes with **27b** and **29**, respectively [61]. This method constitutes a complementary route to optically active non-proteinogenic amino acid derivatives of great interest [83].

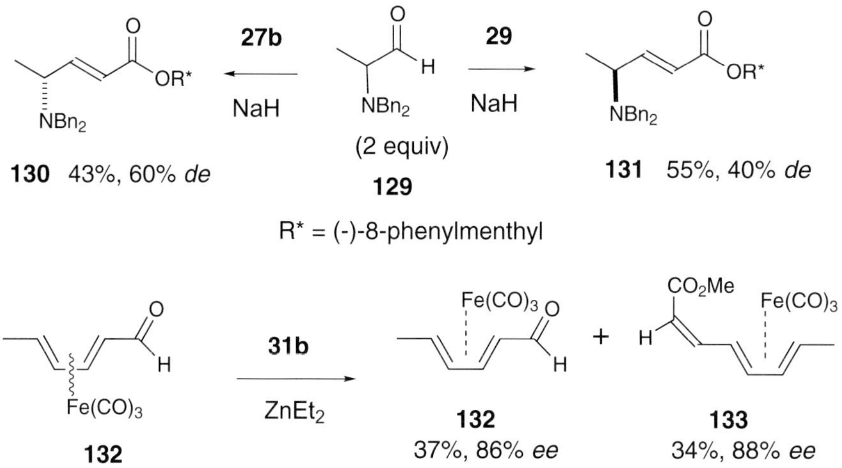
27b
NaH
29
NaH
OR*
NBn2
NBn2
(2 equiv)
130 43%, 60% de
129
131 55%, 40% de
R* = (-)-8-phenylmenthyl
Fe(CO)3
31b
ZnEt2
CO2Me
132
132
37%, 86% ee
133
34%, 88% ee

Scheme 7.21. Additional examples of kinetic resolution of racemic aldehydes.

An example of kinetic resolution of a racemic aldehyde of the η^4-Fe complex **132** was reported to give the corresponding *Z*-olefin **133** together with the starting aldehyde with a satisfactory level of optical purity [37]. Provided that the conjugated aldehyde exists in the *s-trans* conformation, the initial rate-determining nucleophilic attack from the *si*-face of the reagent anion is energetically favored, and this addition occurs on the less hindered *anti* side of the iron carbonyl groups in the substrate. Consequently, the anion may approach the *re*-face of the aldehyde carbonyl rather than the *si*-face, leading to the preferential formation of the observed product **133** along with recovery of the non-racemic **132**.

The kinetic resolution of racemate **134** possessing axial chirality was examined with the chiral arsonium ylide **30**, which afforded the non-racemic atropisomer **135** with a satisfactory *de* of up to 76% [84]. In this reaction, dynamic kinetic resolution was operative to some extent (*vide infra*). The same kind of kinetic resolution through discrimination of axial chirality of the racemic substrate **136** was also examined with the axially chiral phosphorus reagent **31b**; this gave the optically active atropisomers **136–138** with satisfactory enantiomeric purity, although the yields were low (Scheme 7.22).

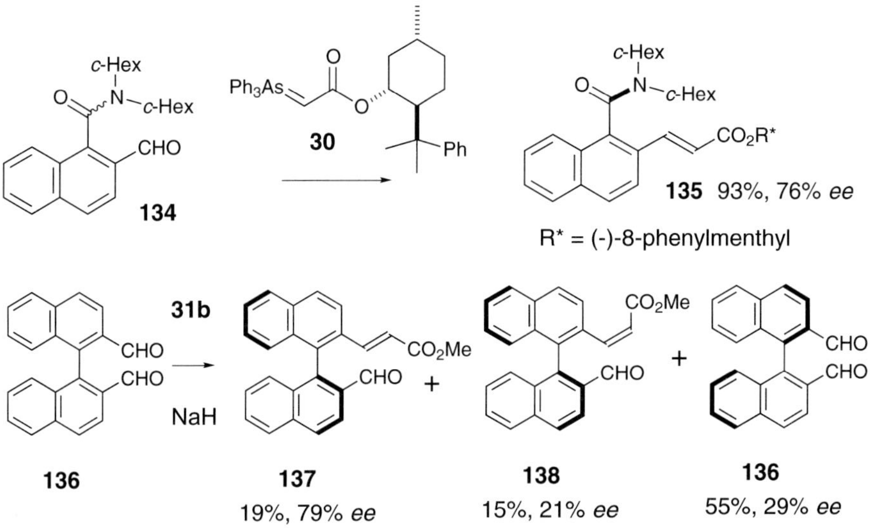

Scheme 7.22. Kinetic resolution of racemic aldehydes to give axially chiral compounds.

7.6.2
Resolution of Racemic Phosphorus Reagents

Kinetic resolution of a racemic Wittig-type reagent through reaction with a chiral carbonyl substrate constitutes a reverse strategy to that described above. A preliminary report described the reaction of the racemic stabilized phosphonium ylide **12** with an optically active alkanoyl chloride **139** to furnish a chiral allenic compound **140**, albeit with only a modest level of stereoselectivity [70]. More recently, the optically active epoxy aldehyde **141** was treated with racemic phosphonium ylide **13** having a η^4-dienyl Fe(CO)$_3$ group to afford the Z-alkene **142** with 60% de in moderate chemical yield [85] (Scheme 7.23). These two examples are clearly suggestive of kinetic resolution of the ylide, with a matched double asymmetric induction leading to preferred formation of the major isomer, and a mismatched combination suffering from serious non-bonded interactions leading to the minor products.

7.6.3
Parallel Kinetic Resolution

Although kinetic resolution is an established method for the preparation of chiral compounds, it requires a large difference in the rate constants of the enantiomers of the substrate in order to obtain the product and recovered starting material with

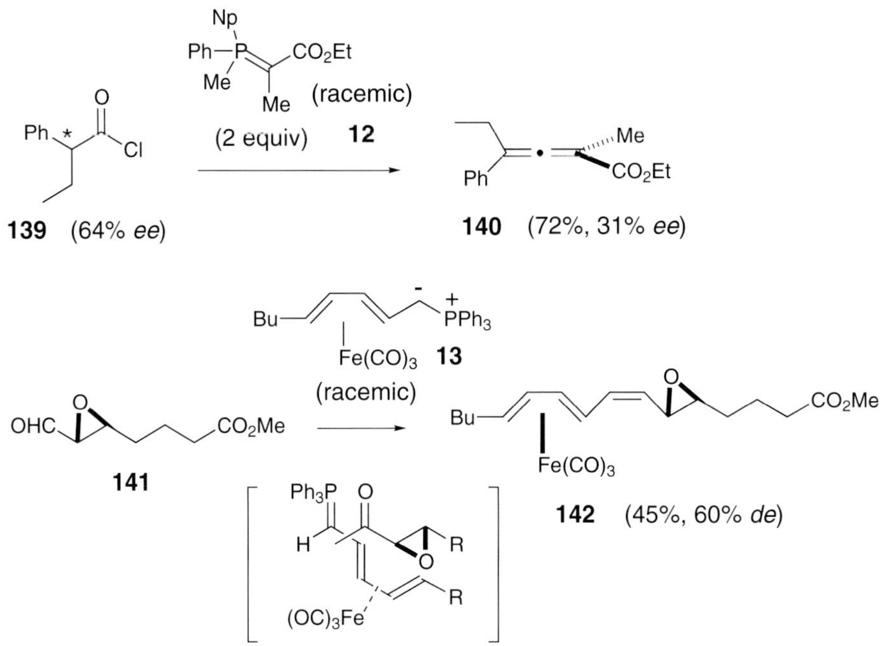

Scheme 7.23. Kinetic resolution of racemic reagents.

high optical purity and in good chemical yield. Therefore, an exceptionally high selectivity factor s ($s = k$(fast-reacting enantiomer)/k(slow-reacting enantiomer)) is required to meet these conditions (for example, $s = 500$, 98% ee in 50% yield for both the slow- and fast-reacting enantiomers). Clearly, there will be a continuous increase in the relative concentration (and therefore the relative rate of reaction) of the less reactive substrate as the faster-reacting substrate is consumed. To overcome this concentration effect, a new strategy for the optical resolution of racemic substrates, termed "Parallel Kinetic Resolution (PKR)", has recently been proposed [86, 87] and developed as an improved version of simple kinetic resolution.

Under PKR conditions, two enantiomeric substrates are simultaneously converted into two structurally and configurationally different chiral products by reaction with chiral reagents or catalysts. It has been shown that to achieve the same selectivity, the selectivity factor s can be significantly lower for PKR than for a traditional kinetic resolution. As yet, there has been only one report of an asymmetric HWE reaction under PKR conditions [88], in which one equivalent of racemic aldehyde 143 was converted into alkene products 144 and 145 by reaction with half an equivalent each of two chiral phosphonates 28 and 20 bearing different chiral auxiliaries (Scheme 7.24). These alkene products, 144 and 145, were readily separable as a result of the difference in polarity between the two auxiliaries. It was clearly shown that the diastereoselectivities of the alkene products were dramatically improved compared to those obtained in the respective individual kinetic resolutions, especially in the case of alkene 145.

The alternative approach to PKR using asymmetric HWE reactions, in which

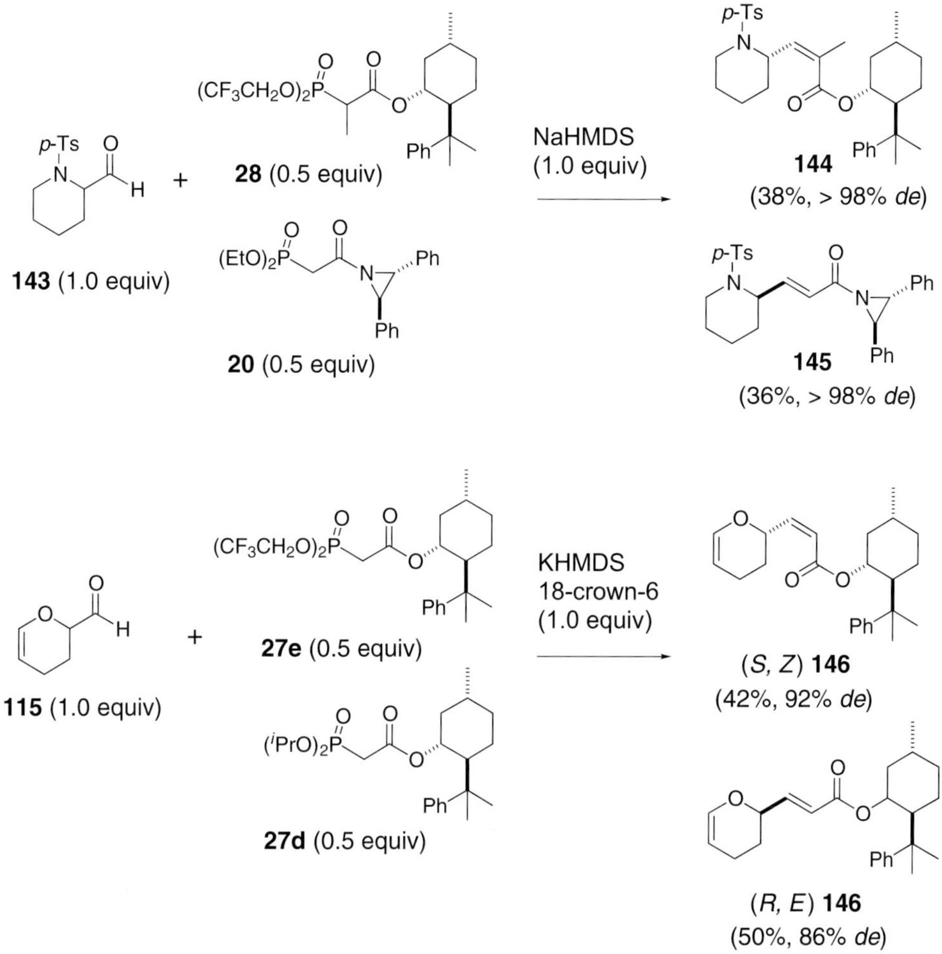

Scheme 7.24. Asymmetric carbonyl olefinations through parallel kinetic resolution.

one E- and one Z-selective reagent bearing the same auxiliary are used in combination, has been examined in the reactions of acrolein dimer **115** and phosphonates **27e** and **27d**. The products, (R, E)- and (S, Z)-**146**, were obtained with synthetically useful diastereoselectivities and in excellent chemical yields. Moreover, the alkene products could be readily separated by chromatographic means.

7.7
Dynamic Resolution

Kinetic resolution has long been recognized as an effective tool for the preparation of enantiomerically enriched compounds, but the inherent drawback of this methodology is that the maximum yield of one enantiomer is 50%. In addition, the enantiomeric purity of the recovered substrate and product is profoundly affected by the extent of conversion. This limitation can be overcome if the stereogenic center of a racemic substrate can racemize rapidly during the course of the reaction, which can, in principle, lead to quantitative conversion of the substrate into a single stereoisomer of the product [89, 90].

Several efforts have been made to prepare chiral alkene products by the use of an asymmetric Wittig-type reaction in a dynamic process [91]. The dynamic kinetic resolution of racemic 2-benzylcyclohexanone (**147**) with an equimolar amount of chiral cyclic phosphonate **23**, prepared from mannitol, afforded the chiral (*E*, *S*)-alkene product **148** with excellent optical purity in good chemical yield [92] (Scheme 7.25). In the presence of excess LDA, racemization of the substrate led to recovery of the starting ketone with low optical purity. Molecular mechanics calculations aimed at evaluating conformational and constitutional equilibria of the oxyanion obtained by addition of the phosphonate carbanion to the carbonyl group of **147** were performed to clarify the stereoselectivity. The three energetically most favored isomers from the eight possible diastereomers considered theoretically were selected. It was shown that two of these three isomers, that have much better molecular geometries than the other, undergo ring closure to the oxaphosphetane and then to the alkene **148**. This result indicates that the stereoselectivity is not only controlled by the direction of addition of the phosphonate carbanion to the

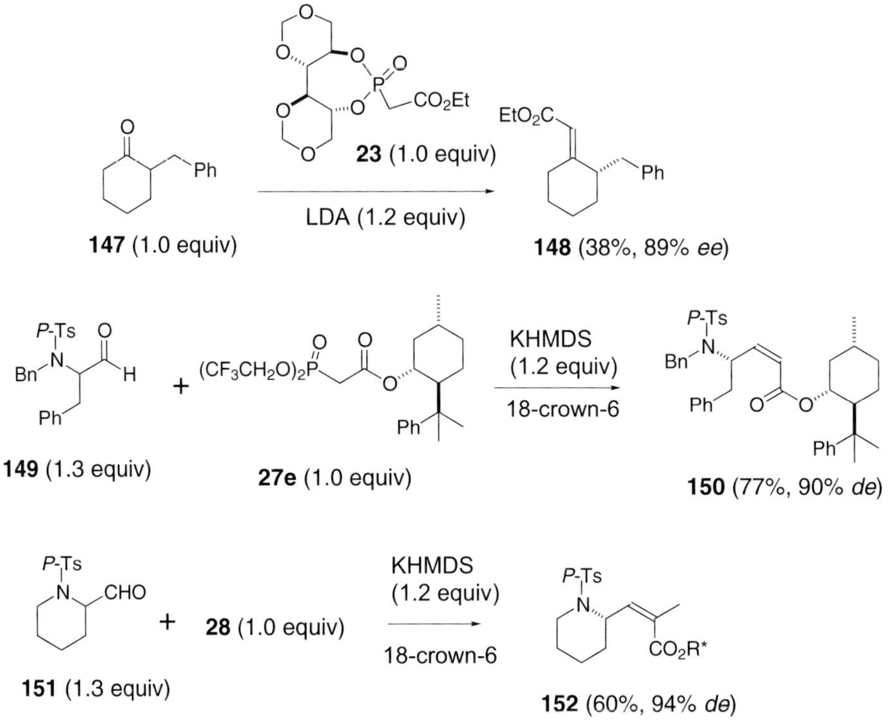

Scheme 7.25. Examples of dynamic kinetic resolution.

carbonyl group, but also by the relative rates for the ring closure and elimination to give the products.

Another example of this category is the first dynamic kinetic resolution of a racemic α-amino aldehyde [93]. It has been shown that the N-tosyl-protected aldehyde **149** reacts with a near equimolar amount of chiral phosphonate **27e** to afford vinylogous amino acid esters (R, E)-**150** with excellent diastereoselectivity and chemical yield. Similarly, the N-tosyl-protected piperidine **151** was converted into the trisubstituted alkene **152** with good diastereoselectivity and in high chemical yield by reaction with chiral phosphonate **28**. In several cases, it has been shown that the selectivity obtained under dynamic conditions exceeds that obtained by a traditional kinetic resolution. The HWE products obtained appear to be attractive precursors for non-proteinogenic amino acids [83] as well as various alkaloids.

The aforementioned transformation of racemic **134** to **135** belongs to the category of dynamic kinetic resolution [84] (Scheme 7.22). It was shown that the diastereoselectivity of the alkene product is lower than that of the product obtained from the corresponding simple kinetic resolution (*vide supra*).

7.8
Further Application of Asymmetric Wittig-Type Reactions in Enantioselective Synthesis

7.8.1
Use of Asymmetric Wittig-Type Reactions in the Total Synthesis of Natural Products

The usefulness of the Wittig and related reactions in facilitating crucial carbon–carbon bond-formation steps in multi-step chemical syntheses has been demonstrated by many examples of the construction of useful complex molecules such as natural products. Some successful constructions of optically active functionalized building blocks with chiral HWE reagents for the total synthesis of natural products have already been discussed in this chapter [12a, 45, 59–61, 78], and additional examples, which have recently been reported, are mentioned here.

The first example is the stereoselective introduction of the α side chain into 3-oxacarbacyclin and 3-oxaisocarbacyclin molecules by reaction with chiral HWE reagents. Reaction of the THP-protected ketone **153** with three equivalents of chiral phosphonate **24** in the presence of LiCl gave an E/Z mixture of the α,β-unsaturated ester ($E:Z = 95:5$) [67] (Scheme 7.26). The desired E-isomer **154** was isolated after deprotection of the THP groups and converted into the 3-oxacarbacyclin **155** as well as 3-oxaisocarbacyclin [94].

In the total synthesis of bryostatins, chiral HWE reagents have been used for the stereocontrolled transformation of the C13 ketone to the C13–C30-unsaturated enoate [68]. The reaction of macrocycle **102a** with chiral phosphonate **31a** provided the α,β-unsaturated ester of Z-stereochemistry, **103a**, with a diastereoselectivity of $Z/E = 85:15$ (75% isolated yield of the Z-isomer, **103a**). The obtained Z-isomer **103a** was successfully converted into bryostatin by several subsequent chemical

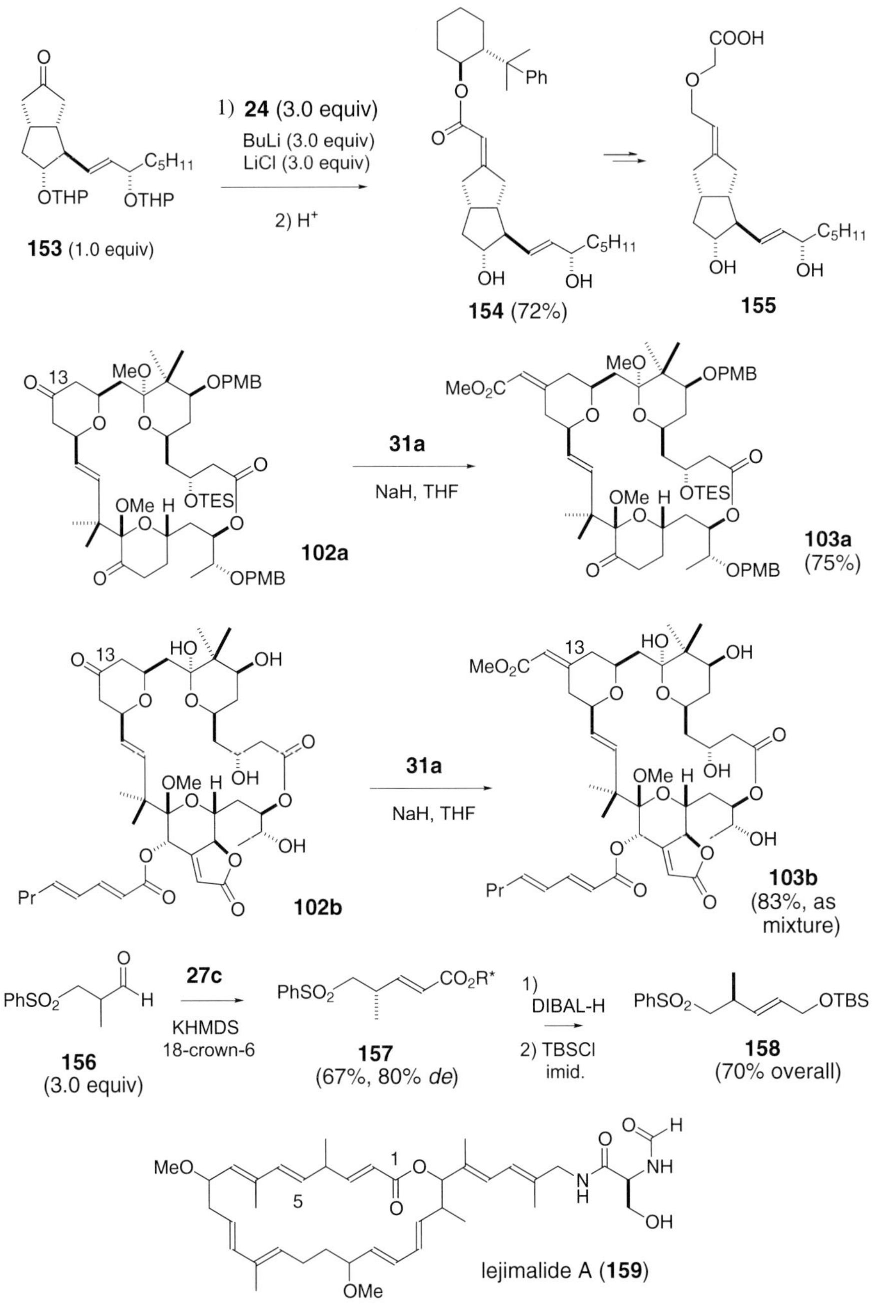

Scheme 7.26. Some examples of the application of asymmetric carbonyl olefination to natural product synthesis.

transformations. Similarly, another research group reported that the C13 ketone of compound **102b** was transformed into a mixture of α,β-unsaturated esters containing the Z-isomer **103b** in 83% yield ($Z/E = 89:11$) by using the same phosphonate **31a** as a key step in the total synthesis of bryostatin [69].

Kinetic resolution of racemic aldehyde **156** was employed in the construction of the subunit of iejimalide A (**159**), a marine macrolide exhibiting high cytotoxicity. Thus, the chiral *E*-alkene **157** was obtained with 80% de using **27c**, and then converted into the C(1)–C(5) fragment of iejimalide A [80, 95].

7.8.2
Sequential HWE and Pd-Catalyzed Allylic Substitutions

The preparation of the same chiral target molecule from both enantiomers of a racemate via the isomeric synthetic intermediates (enantioconvergent reaction sequence) has been studied as an interesting and useful synthetic method. In the case of asymmetric carbonyl olefination, the success of such an enantioconvergent strategy is highly dependent on the combination of an asymmetric HWE reaction and subsequent Pd-catalyzed allylic substitution. Reaction of the racemic aldehyde **160**, bearing a diphenylphosphonyl group as a protecting group for the alcohol, which acts as a leaving group in the subsequent Pd-catalyzed allylic substitution reaction, with the anion of the chiral phosphonate **27f** afforded a near equimolar mixture of alkenes (*R*, *E*)-**161** and (*S*, *Z*)-**162**, with good to excellent diastereoselectivity (Scheme 7.27). In the subsequent step, the mixture of alkenes was subjected to Pd-catalyzed allylic substitution with carbon nucleophiles. Thus, reaction of the η^3-allylpalladium complex derived from alkene mixture (*R*, *E*)-**161** and (*S*, *Z*)-**162** with sodium dimethylmalonate gave a single, optically active product (*E*)-**163** in good overall yield and with satisfactory diastereoselectivity [96]. This stereoconvergent strategy relied on opposite stereoselectivities of the allylic substitutions of (*R*, *E*)-**161** and (*S*, *Z*)-**162**, the substrates reacting with inversion and retention of configuration, respectively. This strategy has also been applied for the preparation of the chiral building blocks in the synthesis of iejimalides (**159**).

The construction of chiral tetrahydrofuran (THF) and tetrahydropyran (THP) derivatives as building blocks for the total synthesis of natural products containing these moieties, such as mucocin, has also been successfully achieved by application of the asymmetric HWE reaction with subsequent cyclization (Pd-catalyzed intramolecular allylic substitution) strategy. Thus, desymmetrization of the *meso*-dialdehyde **70b**, derived from *cis*-2-cyclohexane-1,4-diol, with the chiral phosphonates **27d** and **27e** gave the (*E*)-alkene **164** and (*Z*)-alkene **165**, respectively, with excellent diastereoselectivities. Reduction of the remaining aldehyde groups in these alkenes and subsequent acyl group migration gave the allylic carboxylates (*E*)-**166** and (*Z*)-**167**, respectively. Intramolecular Pd(0)-catalyzed allylic substitution of (*E*)-**166** in the presence of neocuproine as a ligand gave the optically active THF derivatives **168** with excellent stereoselectivities and overall retention of configuration at the allylic stereocenter. On the other hand, the ring closure of (*Z*)-**167**

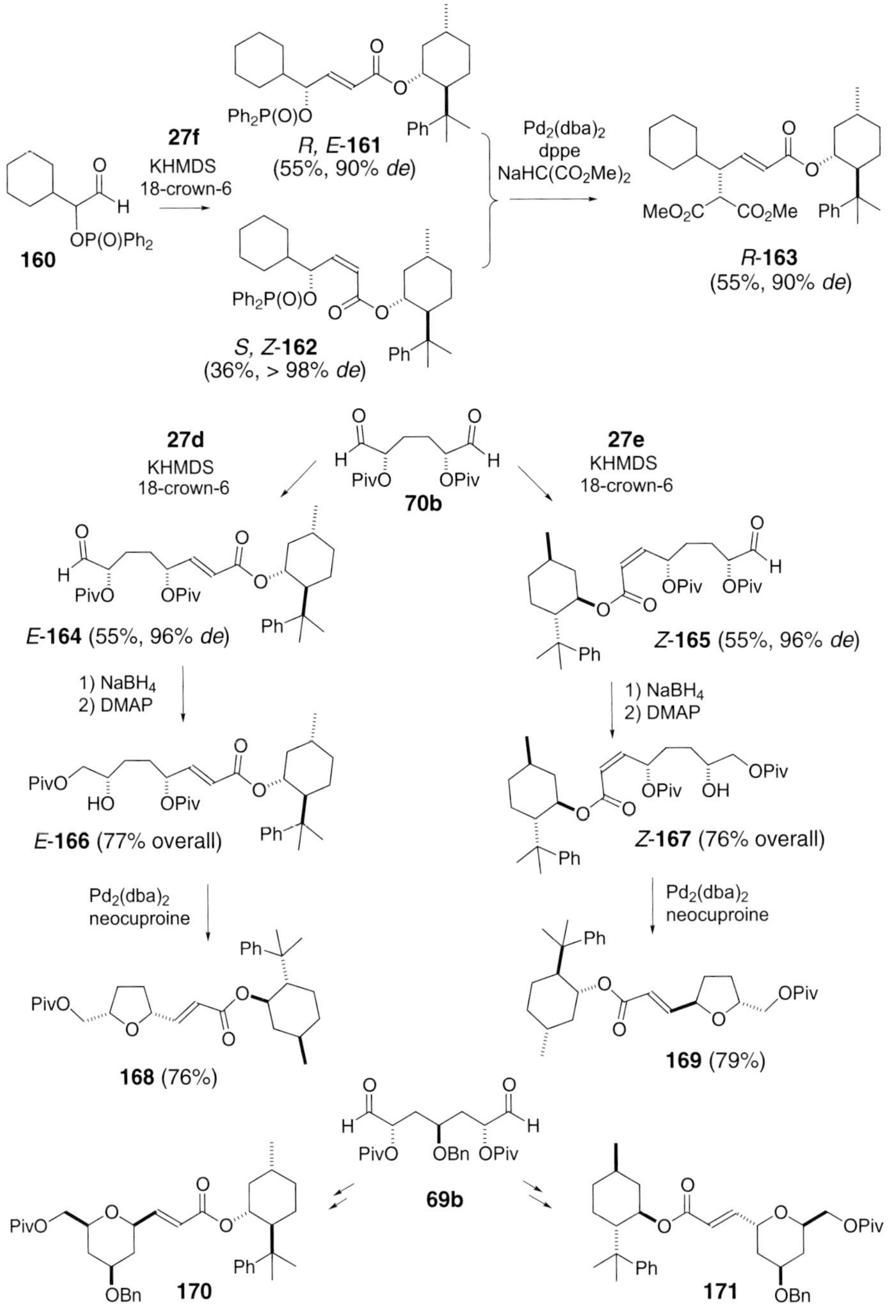

Scheme 7.27. Examples of asymmetric sequential HWE and Pd-catalyzed allylic substitutions.

7.9 Asymmetric Carbonyl Olefinations Without Usage of Optically Active Phosphorus Reagents | 329

proceeded under conditions of moderate heating to give the optically active THF derivatives **169** with overall inversion of configuration [97].

In an analogous manner to the route to the THF derivatives described above, the chiral THP derivatives **170** and **171** were successfully prepared from the *meso*-dialdehyde **69b** by asymmetric HWE reaction followed by Pd-catalyzed intramolecular cyclization [98].

7.8.3
Tandem Michael–HWE Reaction

Recently, asymmetric carbonyl olefination has been introduced into tandem-type reaction sequences for the construction of one or two asymmetric centers in a reaction involving the formation of two carbon–carbon bonds in a one-pot procedure [99]. Thus, the tandem Michael–HWE reaction was carried out starting from a substrate of the arylidene derivative **172**, which was derived from a chiral HWE reagent hitherto used for asymmetric olefination. The enolate derived from the initial Michael addition was reacted with benzaldehyde to give α,β-unsaturated ester **173** with high enantiomeric excess. When the dialdehyde **59** was reacted as an electrophile in the second step, double asymmetric induction took place to afford both the (Z)- and (E)-alkenes **174** with high diastereoselectivity. Furthermore, dynamic kinetic resolution was observed in the tandem reaction with acrolein dimer **115** to yield the α,β-unsaturated ester **175** having two stereogenic carbon centers in the branched substituents in more than 50% yield (85%) (Scheme 7.28). A change of the metal counterion to Sn^{2+} or Zn^{2+} was quite effective to achieve both high diastereoselectivity and good chemical yield in these interesting asymmetric transformations.

7.9
Asymmetric Carbonyl Olefinations Without Usage of Optically Active Phosphorus Reagents

Recently, asymmetric induction mediated by external chiral ligands that are not covalently bonded to the reagent has attracted much attention, and it is believed that the information obtained from these studies will prove useful in developing a novel system for efficient catalytic asymmetric transformation. In order to explore the possibilities, a variety of reaction systems capable of effective asymmetric induction at a specific site in the course of a reaction have been devised, and several investigations have also been directed towards Wittig-type olefination. An early study using an optically active organic acid with stabilized ylides [22] was unfruitful, as discussed in the introductory section.

The HWE reactions of 4-*tert*-butylcyclohexanone (**1a**) with reagent **176** in the presence of the alkoxide of chiral amino alcohol **178** as a base resulted in the formation of dissymmetric alkene **179** in good yield with up to 52% *ee* [100] (Scheme 7.29). In this study, it was suggested that the addition step is reversible and that the

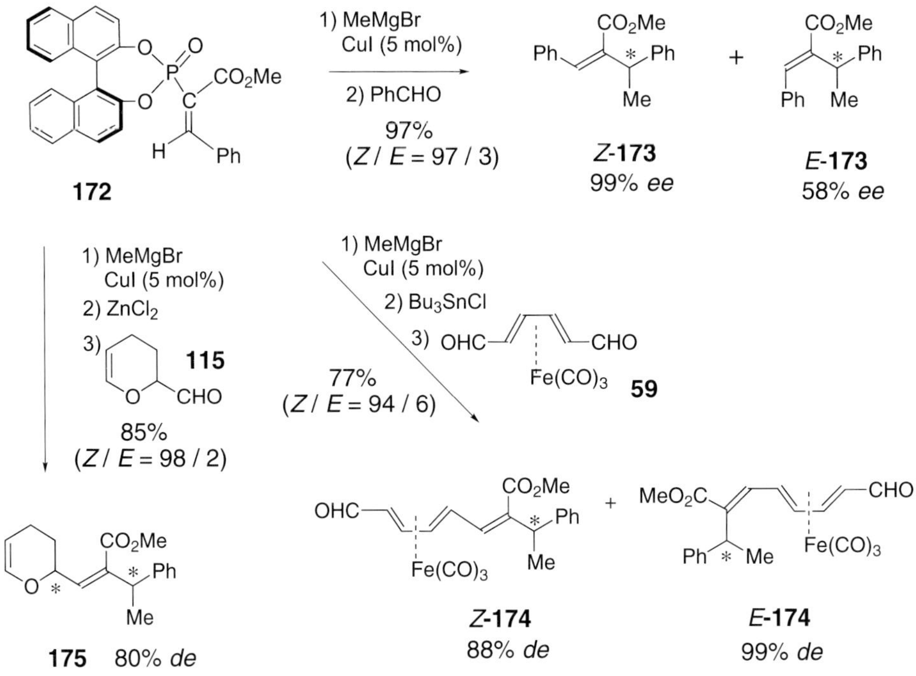

Scheme 7.28. Asymmetric tandem Michael–HWE reactions.

origin of asymmetric induction is probably a difference in elimination rate from the diastereomeric intermediates. In a closely related study, the lithiated achiral HWE reagent **177** was reacted with 4-*tert*-butylcyclohexanone (**1a**) in an asymmetric manner in the presence of chiral ligand **180** to give the isolable adduct **181**, from which the alkene **84** was obtained with a high level of optical purity in a discrete elimination step [101].

When the achiral phosphonate **182** was reacted with the ketone **1a** in the presence of Sn(II) triflate and *N*-ethylpiperidine, the chiral diamine **183** was shown to act as a good asymmetric inducer in generating the tetrasubstituted dissymmetric alkene **184** with a high level of enantiomeric excess [35j, 102]. In order to achieve high levels of asymmetric induction, stoichiometric amounts of ligands in relation to the external chiral source were required in all of the aforementioned asymmetric carbonyl olefinations.

The use of a substoichiometric amount (20 mol%) of an external chiral source was first demonstrated in the asymmetric olefination of a 4-substituted cyclohexanone [103] by using a chiral phase-transfer catalyst **186** derived from cinchonine; here, a combination of the chiral phase-transfer catalyst and rubidium hydroxide as a base was essential in generating the dissymmetric alkene **187** with 57% *ee* after a re-esterification step (Scheme 7.30). Though the problem of low turnover still has to be solved, this result provided an informative concept for the

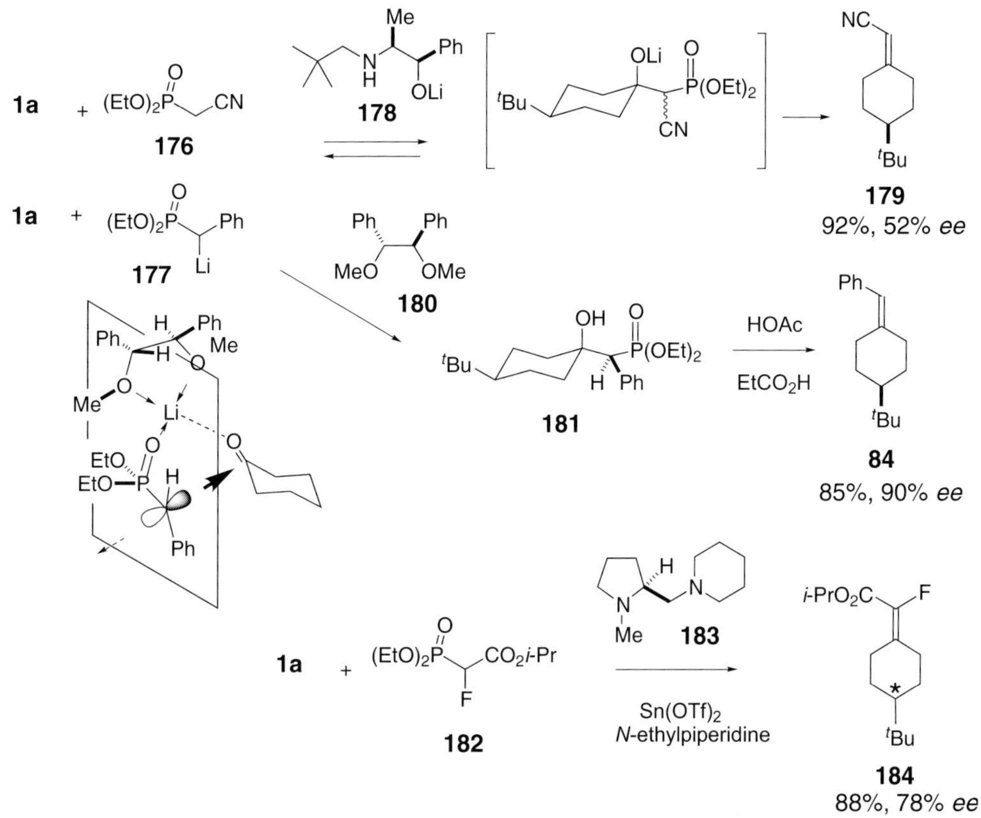

Scheme 7.29. Examples of asymmetric olefinations mediated by external chiral ligands.

development of new catalytic asymmetric olefination methods based on achiral Wittig-type reagents.

Besides the use of chiral bases or catalysts in solution, a rather interesting and unique approach that belongs to the present category involves the utilization of inclusion complexes of the stabilized ylides [104]. In the solid state, an achiral stabilized ylide such as **190** is reacted with a symmetrically substituted prochiral cyclohexanone such as **189** in the presence of a chiral host molecule. The best result was obtained using the chiral host molecule **191**, which gave the dissymmetric alkene **192** with up to 57% ee.

7.10
Asymmetric Carbonyl Olefination by Non-Wittig-Type Routes

Asymmetric carbonyl olefination methods by routes other than the Wittig and related olefination reactions are available, and some precedents belonging to this

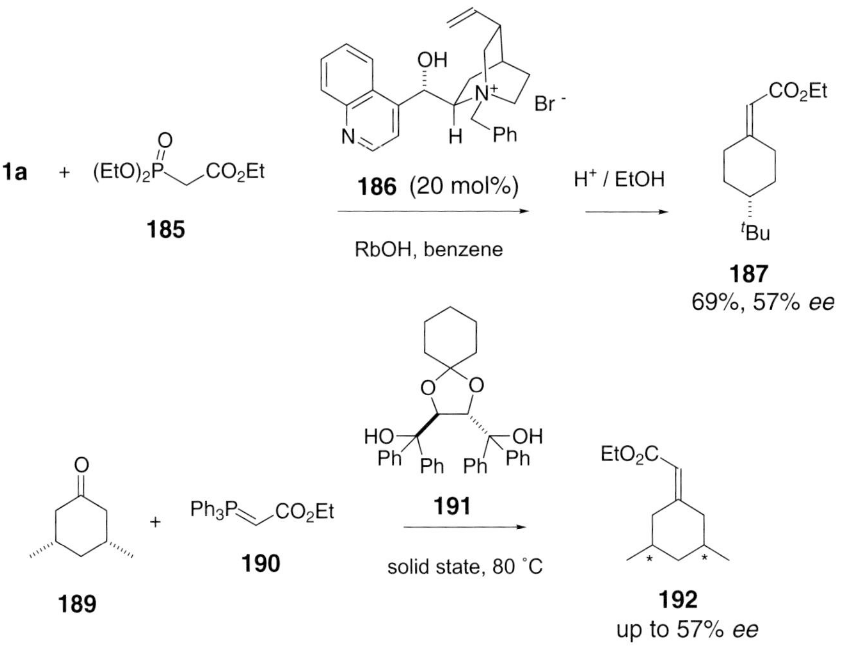

332 | *7 Asymmetric Carbonyl Olefination*

Scheme 7.30. Additional examples of asymmetric olefinations mediated by external chiral PTC or host molecules.

category will be emphasized in this section. There are some reports dealing with reactions in which carbon–carbon bond-forming reactions are not directly involved in simultaneous asymmetric induction, but where the overall process from the carbonyl compounds can be regarded as an asymmetric olefination. Consequently, a few of these approaches to dissymmetric alkenes will be briefly described here.

An example of asymmetric construction of alkylidenecyclohexane derivatives using an optically active chiral Horner-type reagent **193** has been reported [105]. At first glance, this transformation might appear to belong to Wittig-type asymmetric olefinations, but is essentially independent of the hitherto discussed strategies in spite of the use of a phosphorus reagent. Thus, in the overall transformation process to alkenes, an additional sp^3 unit was incorporated into **196** and **198** prior to Horner-type elimination to give **197** and **199**, respectively (Scheme 7.31).

In a total synthesis of the ginseng sesquiterpene (−)-β-panasinsene, ketone methylenation with optical resolution was reported [12a]. Thus, a kinetic resolution was operative in the reaction of the lithium carbanion of chiral sulfoximide **201** with the racemic ketone **200**, giving a separable mixture of two compounds, (+)-**202** and (+)-**203**. The latter diastereomer was converted to the natural product.

By using the same lithium salt (S)-**201**, asymmetric elimination to give **205** with high diastereotopic differentiation without loss of chirality was reported [15] (Scheme 7.32). This asymmetric carbonyl olefination allowed the selective synthesis of both the (Z)- and (E)-alkenylsulfoximides **208** and **210**, which are useful

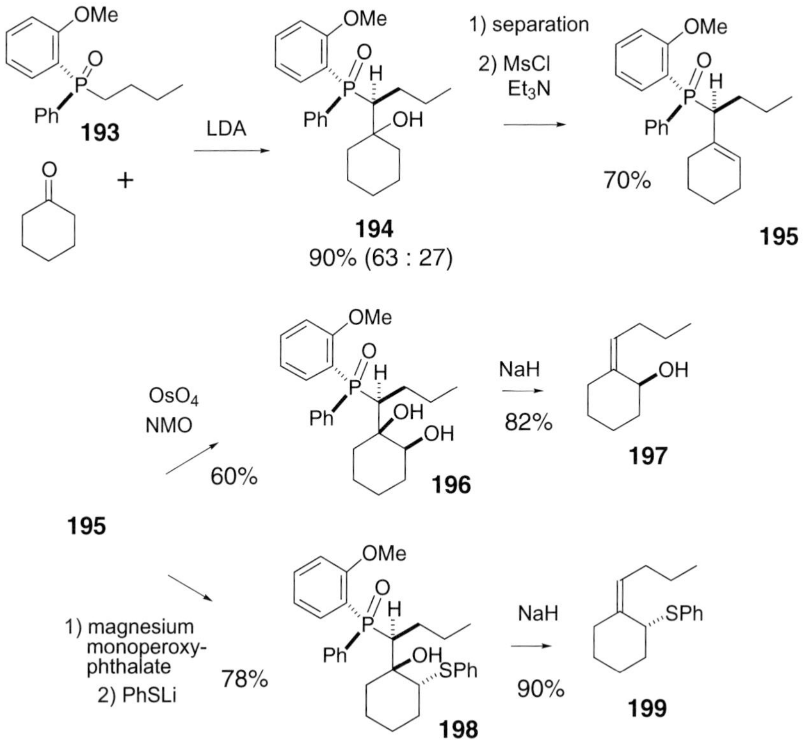

Scheme 7.31. Miscellaneous examples of asymmetric carbonyl olefinations (1).

precursors for carbaprostacyclin derivatives. A possible explanation for the observed asymmetric elimination was provided, together with a proposed model intermediate from which elimination would be more rapid. The same chiral carbanion (s)-**201** was also used in asymmetric elimination to give 1-alkenylsulfoxide **212** with axial and central chirality through the adduct **211**.

The Peterson reaction of an α-silanyl carbanion with a carbonyl compound has been widely utilized as a powerful alternative and complementary synthetic tool for the preparation of substituted alkenes [106], but little effort has been devoted to the asymmetric version of this useful reaction. The Peterson reaction of the bicyclic ketone **85** with the silyl enolate of **213** bearing an 8-phenylmenthol moiety as a chiral auxiliary proceeded with the same sense and degree of asymmetric induction as the corresponding HWE reactions giving the alkene **92c** as described above [60]. The stereochemistry of the reaction can be rationalized in terms of a high substrate- and auxiliary-induced facial selectivity in the addition of an E-configured silyl enolate to the carbonyl group [59]. In the closely related transformations with the lithium enolates of 8-phenylmenthylacetate **214**, as well as its analogous enolates **217a, b**, the corresponding β-hydroxy esters **215a, 218ab** were obtained, which were dehydrated with moderate to high diastereoselectivity upon treatment with Martin's sulfurane **216** to give the alkene **92c** and the corresponding alkenes

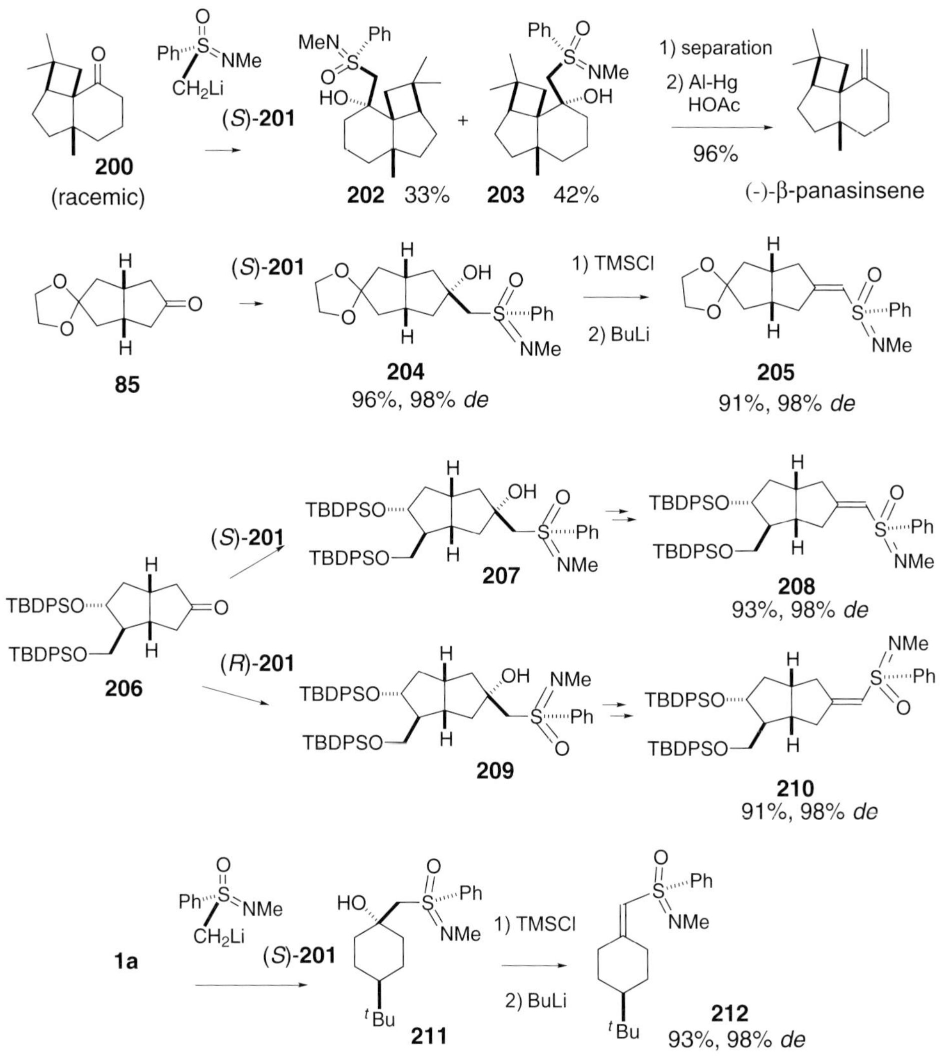

Scheme 7.32. Miscellaneous examples of asymmetric carbonyl olefinations (2).

(Scheme 7.33). The dehydration proceeds according to an E1-mechanism, and the carbenium ions exist as contact ion pairs with the base [OC(Ph)(CF$_3$)$_2$]$^-$ as the counter ion; the H-atom, the empty p-orbital, and the p-orbital of the carbonyl are most likely aligned in a mutually periplanar fashion, as in **215b**. Attack of the base on the pro-*R* H atom leads to the *Z*-alkene from the transition state that is hindered neither by the auxiliary nor by the bicyclic ring system. This successful extension of the two-step asymmetric olefination of prochiral cycloalkanone derivatives proved to offer an additional route to axially chiral alkenes.

Asymmetric Peterson reaction of symmetrically substituted cyclohexanone derivatives with the lithium enolate of silanylacetate **219** in the presence of chiral

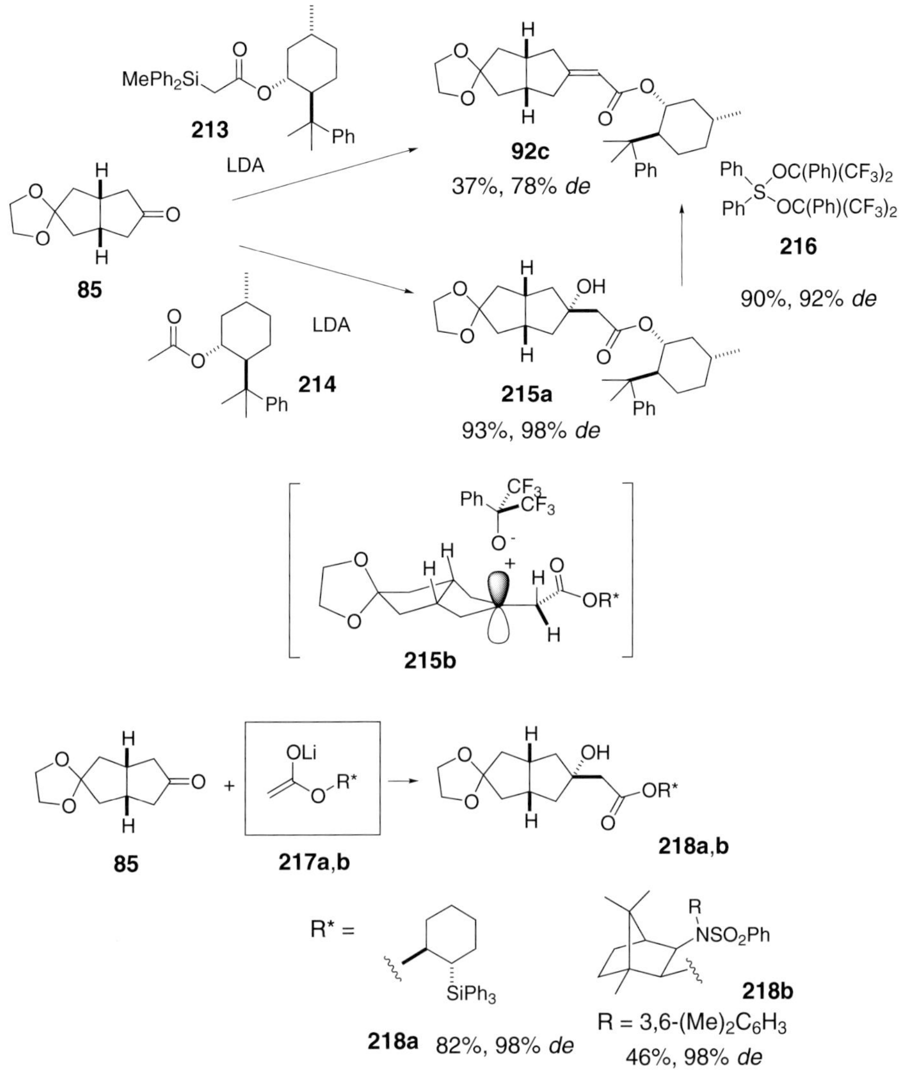

Scheme 7.33. Miscellaneous examples of asymmetric carbonyl olefinations (3).

ligand **220** has recently been demonstrated to give the corresponding alkylidene derivatives **221** with enantioselectivities of up to 85% [107]. An external chiral tridentate amino diether was used for this asymmetric transformation, where three equivalents of the lithium enolate–chiral ligand complex was necessary for satisfactory asymmetric induction. Stereospecific transfer of axial chirality to central chirality in **222** was achieved by treatment of non-racemic alkylidene derivatives **221** with LDA (Scheme 7.34).

Although elimination of chiral sulfoxides [13] or selenoxides [16], as well as asymmetric deprotonation [18] with chiral bases, offer alternative routes to axially

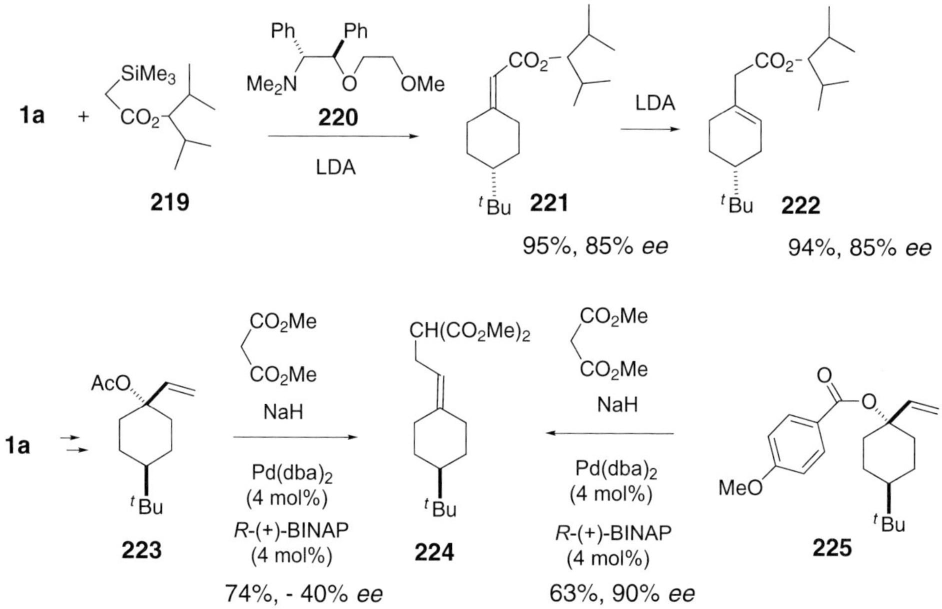

Scheme 7.34. Miscellaneous examples of asymmetric carbonyl olefinations (4).

chiral non-racemic alkenes, the starting compounds in these transformations are not necessarily carbonyl compounds or their equivalents. Therefore, these reactions will not be emphasized in this chapter.

On the other hand, the approach based on Pd-catalyzed asymmetric synthesis of axially chiral alkenes must be briefly discussed here, since the starting esters of allylic alcohol can be essentially derived from carbonyl compounds through nucleophilic addition of the vinyl anion or its equivalent followed by esterification. An early study of this asymmetric transformation showed that trans-4-tert-butyl-1-vinylcyclohexyl acetate (223) gave optically active (up to 40% ee) dimethyl 2-(4-tert-butylcyclohexylidene)methylmalonate (224) through Pd-catalyzed allylic substitution with sodium dimethylmalonate in the presence of 4 mol% of a chiral phosphine ligand such as (R)-(+)-BINAP [17a]. The enantioselectivity was greatly improved by both electronic and steric tuning of the structure of the benzoate substrate. Use of allylic benzoates such as 225, bearing electron-donating p-substituents, in this case a methoxy group, led to efficient asymmetric induction of up to 90% ee under the same reaction conditions [17c].

7.11
Concluding Remarks and Future Perspectives

Since the pioneering contributions of a few early investigations, work on asymmetric olefination has been extensively developed over the past 20 years, and in particular many exciting and useful applications of this transformation have been conceived only in the last decade. As can be seen from the foregoing discussion, it is fair to say that this field of chemistry has only just been born.

Despite the inherent low atom economy of the Wittig and related reactions from the viewpoint of so-called green chemistry [108], asymmetric transformation with this class of phosphorus reagents seems to be quite an attractive method since the reactions proceed with concomitant elimination of the phosphorus groups, resulting either in the direct production of optically active olefinic compounds or, more usually, a separable mixture with a high diastereomeric excess in a stereochemically predictable manner.

Most optically active olefinic products possess axial or planar chirality, which can be easily converted into central chirality by further appropriate chemical transformation without any serious loss of optical purity. The products obtained by the discrimination of enantiotopic carbonyl groups or kinetic resolution already have central chirality as well as reactive functional groups such as olefinic or unsaturated carbonyl systems. Consequently, asymmetric olefination provides an efficient methodology for the construction of useful chiral synthons; applications along these lines in the asymmetric construction of useful and complex chiral molecules have just started and will be extensively investigated in the future.

Though several different approaches to optically active olefinic compounds through asymmetric Wittig and related transformations have hitherto been demonstrated, their success still depends on appropriate choice of the carbonyl compounds used as substrates. This restriction often limits the applicability of asymmetric olefination in the construction of complex chiral molecules. Moreover, the substituents and functionalities on both the substrates and the reagents have a significant influence on the degree and success of the asymmetric induction, the course of the stereochemistry, as well as the reactivity of the ylides or carbanions. The nature of the metallic cation and the reaction conditions, including the nature of the solvent, might also affect the stereochemistry and yields of the olefinic products. In this context, more detailed and extensive mechanistic investigations are required for a full understanding of the aforementioned phenomena, which, in turn, should lead to the further development of asymmetric olefination methodologies based on new designs and concepts. More detailed mechanistic studies should also help to predict the stereochemistry of the olefinic products. Apart from the few examples of established chiral phosphorus reagents employed for asymmetric olefinations, improved availability of chiral sources is necessary for economic and practical reasons; efforts to overcome these limitations are sure to make asymmetric olefination an attractive and useful methodology for the creation of optically active compounds.

External catalysts play a role not only in rate enhancement but also in creating favorable bias in the stereochemistry at the transition state. Consequently, the design of catalyst systems occupies an essential and central position in the future study of asymmetric transformations, including olefination. The use of external chiral sources with a combination of achiral or racemic reagents has been developed in recent years, and this approach holds promise for the future development of efficient asymmetric catalytic procedures. As yet, however, the reported reaction systems still suffer from slow turnovers. Since the asymmetric methods studied to date have utilized expensive and difficult to obtain chiral sources, a reduction in the amount of such chiral sources to catalytic levels would be highly desirable for practical application to asymmetric transformations. The development of asymmetric reactions requiring only catalytic amounts of the chiral source, and that generate the product with predictable stereochemistry should make asymmetric olefination more popular and familiar. There are some publications in which catalytic cycles are proposed for catalytic Wittig-type reactions [109, 110]. In these studies, novel arsenium or telluride reagents have been designed and used to give olefinic compounds in high yields. It seems likely that the establishment of novel redox systems along with the design of chiral catalysts and reagents will be central to achieving catalytic asymmetric olefination through Wittig-type reactions.

In conclusion, it is clear that the development and successful investigation of chiral catalyst systems for this class of asymmetric transformation is still in its infancy, and there is no doubt that the establishment of effective catalytic systems of economic value remains a challenging goal for future research. Major investigations on asymmetric olefination are thus sure to unfold.

References

1 a) G. WITTIG, G. GEISSLER, *Liebigs Ann. Chem.* **1953**, *580*, 44; b) G. WITTIG, *Science* **1980**, *210*, 600.
2 A.-H. LI, L.-X. DAI, V. K. AGGARWAL, *Chem. Rev.* **1997**, *97*, 2341.
3 E. MARYANOFF, A. B. REITZ, *Chem. Rev.* **1989**, *89*, 863.
4 O. MOLT, T. SCHRADER, *Synthesis* **2002**, 2633.
5 a) K. C. NICOLAOU, M. W. HÄRTER, J. L. GUNZNER, A. NADIN, *Liebigs Ann./Recueil*, **1997**, 1283; b) S. J. AMIGONI, L. J. TOUPET, Y. J. LE FLOC'H, *J. Org. Chem.* **1997**, *62*, 6374; c) H. J.-M. GIJSEN, C.-H. WONG, *Tetrahedron Lett.* **1995**, *36*, 7057; d) A. KRIEF, T. OLLEVIER, W. DUMONT, *J. Org. Chem.* **1997**, *62*, 1886; e) M. KALESSE, M. QUITSHALLE, C. P. KHANDAVALLI, A. SAEED, *Org. Lett.* **2001**, *3*, 3107.
6 S. HANESSIAN, D. DELORME, S. BEAUDOIN, Y. LEBLANC, *J. Am. Chem. Soc.* **1984**, *106*, 5754.
7 T. REIN, T. M. PEDERSEN, *Synthesis* **2002**, 579.
8 K. TANAKA, K. FUJI, *J. Synth. Org. Chem. Jpn.* **1998**, *56*, 521.
9 O. I. KOLODIAZHNYI, *Tetrahedron: Asymmetry* **1998**, *9*, 1279.
10 T. REIN, O. REISER, *Acta Chem. Scand.* **1996**, *50*, 369.
11 D. F. WIEMER, *Tetrahedron* **1997**, *53*, 16609.
12 a) C. R. JOHNSON, N. A. MEANWELL, *J. Am. Chem. Soc.* **1981**, *103*, 7667; b) L. DUHAMEL, A. RAVARD, J.-C. PLAQUEVENT, D. DAVOUST, *Tetrahedron Lett.* **1987**, *28*, 5517; c) I. ERDELMEIER, H.-J. GAIS, *J. Am. Chem. Soc.* **1989**, *111*, 1125; d) N. J. S. HARMAT, S. WARREN, *Tetrahedron Lett.* **1990**, *31*, 2743; e) Z. CHEN, R. L. HALTERMAN, *J. Am. Chem. Soc.* **1992**, *114*, 2276; f) M. S. VAN NIEUWENHZE, K. B. SHARPLESS, *J. Am. Chem. Soc.* **1993**, *115*, 7864; g) J. MULZER, T. SPECK, J. BUSCHMANN, P. LUGER, *Tetrahedron Lett.* **1995**, *36*, 7643; h) T. F. J. LAMPE, H. M. R. HOFFMANN, *J. Chem. Soc., Chem. Commun.* **1996**, 2637.
13 G. SOLLADIÉ, R. ZIMMERMAN, R. BARTSCH, *Synthesis* **1985**, 662.
14 a) L. ERMOLENKO, N. A. SASAKI, P. POTIER, *J. Chem. Soc., Perkin Trans. 1*

2000, 2465; b) J. M. Harris, G. A. O'Doherty, *Tetrahedron* **2001**, *57*, 5161.

15 I. Erdelmeier, H.-J. Gais, H. J. Lindner, *Angew. Chem. Int. Ed. Engl.* **1986**, *25*, 935.

16 a) N. Komatsu, S. Matsunaga, T. Sugita, S. Uemura, *J. Am. Chem. Soc.* **1993**, *115*, 5847; b) N. Komatsu, T. Murakami, Y. Nishibayashi, T. Sugita, S. Uemura, *J. Org. Chem.* **1993**, *58*, 3697; c) Y. Nishibayashi, J. D. Singh, S. Uemura, *Tetrahedron Lett.* **1994**, *35*, 3115; d) Y. Nishibayashi, J. D. Singh, S.-I. Fukazawa, S. Uemura, *J. Org. Chem.* **1995**, *60*, 4114.

17 a) J. C. Fiaud, J. Y. Legros, *Tetrahedron Lett.* **1988**, *29*, 2959; b) J. C. Fiaud, J. Y. Legros, *J. Organomet. Chem.* **1989**, *370*, 383; c) J. Y. Legros, J. C. Fiaud, *Tetrahedron* **1996**, *50*, 465.

18 M. Amadji, J. Vadecard, D. Cahard, L. Duhamel, P. Duhamel, J.-C. Plaquevent, *J. Org. Chem.* **1998**, *63*, 5541.

19 I. Tömösközi, G. Janzso, *Chem. Ind.* **1962**, 2085.

20 a) H. J. Bestmann, J. Lienert, *Angew. Chem. Int. Ed. Engl.* **1969**, *8*, 763; b) H. J. Bestmann, E. Heid, W. Ryschka, J. Lienert, *Liebigs Ann. Chem.* **1974**, 1684.

21 a) S. Musierowicz, A. Wróblewski, H. Krawczyk, *Tetrahedron Lett.* **1975**, *16*, 437; b) S. Musierowicz, A. Wróblewski, *Tetrahedron* **1980**, *36*, 1375.

22 H. J. Bestmann, J. Lienert, *Chem. Zeit.* **1970**, *94*, 487.

23 a) U. Verfürth, I. Ugi, *Chem. Ber.* **1991**, *124*, 1627; b) P. Kielbasinski, R. Zurawinski, K. M. Pietrusiewicz, M. Zablocka, M. Mikolajczyk, *Tetrahedron Lett.* **1994**, *35*, 7081; c) N. Serreqi, R. J. Kazlauskas, *J. Org. Chem.* **1994**, *59*, 7609; d) P. G. Devitt, T. P. Kee, *Tetrahedron* **1995**, *51*, 10987; e) J. M. Brown, J. C. P. Laing, *J. Organomet. Chem.* **1997**, *529*, 435; f) S. E. Denmark, R. L. Dorow, *Chirality* **2002**, *14*, 241.

24 P. Brandt, P.-O. Norrby, I. Martin, T. Rein, *J. Org. Chem.* **1998**, *63*, 1280.

25 K. Ando, *J. Org. Chem.* **1999**, *64*, 6815.

26 D. B. Berkowitz, M. K. Smith, *J. Org. Chem.* **1995**, *60*, 1233.

27 a) R. Noyori, H. Takaya, *Acc. Chem. Res.* **1990**, *23*, 345; b) R. Noyori, *Chem. Soc. Rev.* **1989**, *18*, 187.

28 R. Larsen, G. Aksnes, *Phosphorus and Sulfur* **1983**, *15*, 219 and 229.

29 a) C.-H. Wong, G. M. Whiteside, *Enzymes in Synthetic Organic Chemistry*, Elsevier Science Inc., New York, 1994; b) R. Porter, S. Clark (Eds.), *Enzymes in Organic Synthesis*, Pitman, London, 1985.

30 S. L. Schreiber, T. S. Schreiber, D. B. Smith, *J. Am. Chem. Soc.* **1987**, *109*, 1525.

31 K. Tanaka, Y. Ohta, K. Fuji, *Tetrahedron Lett.* **1993**, *34*, 4071.

32 N. Kann, T. Rein, *J. Org. Chem.* **1993**, *58*, 3802.

33 K. Tanaka, Y. Ohta, K. Fuji, *Tetrahedron Lett.* **1997**, *38*, 8943.

34 S. K. Thompson, C. H. Heathcock, *J. Org. Chem.* **1990**, *55*, 3386.

35 a) W. C. Still, C. Gennari, *Tetrahedron Lett.* **1983**, *24*, 4405; b) A. D. Buss, S. Warren, *Tetrahedron Lett.* **1983**, *24*, 3931; c) M. L. Morin-Fox, M. A. Lipton, *Tetrahedron Lett.* **1993**, *34*, 7899; d) G. Hutton, T. Jolliff, H. Mitchell, S. Warren, *Tetrahedron Lett.* **1995**, *36*, 7905; e) K. Kokin, S. Tsuboi, J. Motoyoshiyama, S. Hayashi, *Synthesis* **1996**, 637; f) R. Kiu, M. Schlosser, *Synlett* **1996**, 1195; g) F. Rubsam, A. M. Evars, C. Michel, A. Giannis, *Tetrahedron* **1997**, *53*, 1707; h) K. Ando, *J. Org. Chem.* **1997**, *62*, 1934; i) S. Kojima, R. Takagi, K. Akiba, *J. Am. Chem. Soc.* **1997**, *119*, 5970; j) S. Sano, K. Yokoyama, M. Fukushima, T. Yagi, Y. Nagao, *J. Chem. Soc., Chem. Commun.* **1997**, 559; k) K. Kokin, J. Motoyoshiya, S. Hayashi, H. Aoyama, *Synth. Commun.* **1997**, *27*, 2387; l) K. Ando, *J. Org. Chem.* **1999**, *64*, 6815; m) K. Ando, *J. Synth. Org. Chem. Jpn.* **2000**, *58*, 869.

36 T. Ohwada, unpublished results.

37 K. Tanaka, T. Watanabe, K. Shimamoto, P. Sahakitpichan,

K. Fuji, *Tetrahedron Lett.* **1999**, *40*, 6599.

38 J. S. Tullis, L. Vares, N. Kann, P.-O. Norrby, T. Rein, *J. Org. Chem.* **1998**, *63*, 8284.

39 P.-O. Norrby, P. Brandt, T. Rein, *J. Org. Chem.* **1999**, *64*, 5845.

40 T. Rein, L. Vares, I. Kawasaki, T. M. Pedersen, P.-O. Norrby, P. Brandt, D. Tanner, *Phosphorus, Sulfur and Silicon* **1999**, *144–146*, 169.

41 a) J.-P. Gourves, H. Couthon, G. Sturtz, *Eur. J. Org. Chem.* **1999**, 3489; b) M. R. Elliott, A.-L. Dhimane, L. Hamon, M. Malacria, *Eur. J. Org. Chem.* **2000**, 155; c) D. L. Comins, C. G. Ollinger, *Tetrahedron Lett.* **2001**, *42*, 4115; d) M. Pipelier, M. S. Ermolenko, A. Zampella, A. Olesker, G. Lukacs, *Synlett* **1996**, 24.

42 a) B. M. Trost, D. P. Curran, *J. Am. Chem. Soc.* **1980**, *102*, 5699; b) B. M. Trost, D. P. Curran, *J. Am. Chem. Soc.* **1981**, *103*, 7380.

43 B. M. Trost, D. P. Curran, *Tetrahedron Lett.* **1981**, *22*, 4929.

44 E. Vedejs, M. J. Peterson, *Top. Stereochem.* **1994**, *21*, 1.

45 T. Mandai, Y. Kaihara, J. Tsuji, *J. Org. Chem.* **1994**, *59*, 5847.

46 A. V. Bedekar, T. Watanabe, K. Tanaka, K. Fuji, *Tetrahedron: Asymmetry* **2002**, *13*, 721.

47 J. Yamazaki, A. V. Bedekar, T. Watanabe, K. Tanaka, J. Watanabe, K. Fuji, *Tetrahedron: Asymmetry* **2002**, *13*, 729.

48 a) W. H. Perkin, W. J. Pope, *J. Chem. Soc.* **1908**, *93*, 1075; b) W. H. Perkin, W. J. Pope, *J. Chem. Soc.* **1909**, *95*, 1789; c) W. H. Perkin, W. J. Pope, *J. Chem. Soc.* **1911**, *99*, 1511.

49 a) Y. L. Bennani, S. Hanessian, *Chem. Rev.* **1997**, *97*, 3161; b) S. Hanessian, Y. L. Bennani, *Synthesis* **1994**, 1272; c) S. Hanessian, Y. L. Bennani, D. Delorme, *Tetrahedron Lett.* **1990**, *31*, 6461 and 6465.

50 a) V. J. Blazis, K. J. Koeller, C. D. Spilling, *J. Org. Chem.* **1995**, *60*, 931; b) K. J. Koeller, C. D. Spilling, *Tetrahedron Lett.* **1991**, *32*, 6297.

51 a) R. P. Lemieux, G. B. Schuster, *J. Org. Chem.* **1993**, *58*, 100; b) Y. Zhang, G. B. Schuster, *J. Org. Chem.* **1994**, *59*, 1855; c) M. Suarez, G. B. Schuster, *J. Am. Chem. Soc.* **1995**, *117*, 6732.

52 a) S. Hanessian, A. Gomtsyan, A. Payne, Y. Hervé, S. Beaudoin, *J. Org. Chem.* **1993**, *58*, 5032; b) K. Tanaka, Y. Ohta, K. Fuji, *J. Org. Chem.* **1995**, *60*, 8036; c) T. Arai, H. Sasaki, K. Yamaguchi, M. Shibasaki, *J. Am. Chem. Soc.* **1998**, *120*, 441.

53 S. E. Denmark, C.-T. Chen, *J. Am. Chem. Soc.* **1992**, *114*, 10674.

54 a) S. E. Denmark, J. Amburgey, *J. Am. Chem. Soc.* **1993**, *115*, 10386; b) S. E. Denmark, C.-T. Chen, *J. Org. Chem.* **1994**, *59*, 2922.

55 S. E. Denmark, I. Rivera, *J. Org. Chem.* **1994**, *59*, 6887.

56 T. Takahashi, M. Matsui, N. Maeno, T. Koizumi, *Heterocycles* **1990**, *30*, 353.

57 A. Abiko, S. Masamune, *Tetrahedron Lett.* **1996**, *37*, 1077 and 1081.

58 H.-J. Gais, G. Schmiedl, W. A. Ball, J. Bund, G. Hellmann, I. Erdelmeier, *Tetrahedron Lett.* **1988**, *29*, 1773.

59 H.-J. Gais, G. Schmiedl, R. K. L. Ossenkamp, *Liebigs Ann./Recueil* **1997**, 2419.

60 H. Rehwinkel, J. Skupsch, H. Vorbrüggen, *Tetrahedron Lett.* **1988**, *29*, 1775.

61 T. Furuta, M. Iwamura, *J. Chem. Soc., Chem. Commun.* **1994**, 2167.

62 a) Z.-Z. Huang, X. Huang, Y.-Z. Huang, *Tetrahedron Lett.* **1995**, *36*, 425; b) Z.-Z. Huang, X. Huang, Y.-Z. Huang, *J. Chem. Soc., Perkin Trans. 1* **1995**, 95; c) C. Liévre, S. Humez, C. Fréchou, G. Damailly, *Tetrahedron Lett.* **1997**, *38*, 6003; d) C. M. Moorhoff, *Tetrahedron* **1997**, *53*, 2241; e) C. M. Moorhoff, *Synlett* **1997**, 126.

63 W.-M. Dai, J. Wu, X. Huang, *Tetrahedron: Asymmetry* **1997**, *8*, 1979.

64 W.-M. Dai, A. Wu, H. Wu, *Tetrahedron: Asymmetry* **2002**, *13*, 2187.

65 S. Masamune, W. Choy, J. S. Petersen, L. R. Sita, *Angew. Chem. Int. Ed. Engl.* **1985**, *24*, 1.

66 S. Hanessian, S. Beaudoin, *Tetrahedron Lett.* **1992**, *33*, 7655.

67 I. Vaulont, H.-J. Gais, N. Reuter, E. Schmitz, R. K. L. Ossenkamp, *Eur. J. Org. Chem.* **1998**, 805.

68 a) K. D. A. Evans, P. H. Carter, E. M. Carreira, J. A. Prunet, A. B. Charette, M. Lautens, *Angew. Chem. Int. Ed.* **1998**, *37*, 2354; b) K. D. A. Evans, P. H. Carter, E. M. Carreira, A. B. Charette, J. A. Prunet, M. Lautens, *J. Am. Chem. Soc.* **1999**, *121*, 7540.

69 K. Ohmori, Y. Ogawa, T. Obitsu, Y. Ishikawa, S. Nishiyama, S. Yamamura, *Angew. Chem. Int. Ed.* **2000**, *39*, 2290.

70 a) I. Tömösközi, H. Bestmann, *Tetrahedron Lett.* **1964**, *5*, 1293; b) I. Tömösközi, H. Bestmann, *Tetrahedron* **1968**, *24*, 3299.

71 a) J. Tsuji, T. Mandai, *Angew. Chem. Int. Ed. Engl.* **1995**, *34*, 461; b) J. A. Marshall, K. G. Pinney, *J. Org. Chem.* **1993**, *58*, 7180; c) Y. Naruse, H. Watanabe, Y. Ishiyama, T. Yoshida, *J. Org. Chem.* **1997**, *62*, 3862.

72 K. Tanaka, K. Otsubo, K. Fuji, *Synlett* **1995**, 933.

73 K. Tanaka, K. Otsubo, K. Fuji, *Tetrahedron Lett.* **1995**, *36*, 9513.

74 K. Tanaka, K. Otsubo, K. Fuji, *Tetrahedron Lett.* **1996**, *37*, 3735.

75 R. Häner, T. Laube, D. Seebach, *J. Am. Chem. Soc.* **1985**, *107*, 5396 and 5403.

76 J. Yamazaki, T. Watanabe, K. Tanaka, *Tetrahedron: Asymmetry* **2001**, *12*, 669.

77 H. B. Kagan, *Tetrahedron* **2001**, *57*, 2449; H. B. Kagan, J. C. Fiaud, *Top. Stereochem.* **1988**, *18*, 249.

78 a) C. R. Johnson, R. C. Elliott, N. A. Meanwell, *Tetrahedron Lett.* **1982**, *23*, 5005; b) C. R. Johnson, R. C. Elliott, *J. Am. Chem. Soc.* **1982**, *104*, 7041.

79 a) T. Rein, N. Kann, R. Kreuder, B. Gangloff, O. Reiser, *Angew. Chem. Int. Ed. Engl.* **1994**, *33*, 556; b) T. Rein, J. Anvelt, A. Soone, R. Kreuder, C. Wulff, O. Reiser, *Tetrahedron Lett.* **1995**, *36*, 2303.

80 M. T. Mendlik, M. Cottard, T. Rein, P. Helquist, *Tetrahedron Lett.* **1997**, *38*, 6375.

81 R. Kreuder, T. Rein, O. Reiser, *Tetrahedron Lett.* **1997**, *38*, 9035.

82 a) A. Mengel, O. Reiser, *Chem. Rev.* **1999**, *99*, 1191; b) B. W. Gung, *Tetrahedron* **1996**, *52*, 5263; c) E. P. Lodge, C. H. Heathcock, *J. Am. Chem. Soc.* **1987**, *109*, 3353.

83 a) C. Grison, S. Genéve, P. Coutrot, *Tetrahedron Lett.* **2001**, *42*, 3831; b) J. Gante, *Angew. Chem. Int. Ed. Engl.* **1994**, *33*, 1699.

84 W.-M. Dai, C. W. Lau, *Tetrahedron Lett.* **2001**, *42*, 2541.

85 P. Pinsard, J.-P. Lellouche, J.-P. Beaucourt, R. Grée, *Tetrahedron Lett.* **1990**, *31*, 1137.

86 a) E. Vedejs, X. Chen, *J. Am. Chem. Soc.* **1997**, *119*, 2584; b) J. Eames, *Angew. Chem. Int. Ed.* **2000**, *39*, 885.

87 F. Bertozzi, P. Crotti, F. Macchia, M. Pineschi, B. L. Feringa, *Angew. Chem. Int. Ed.* **2001**, *40*, 930.

88 T. M. Pedersen, J. F. Jensen, R. E. Humble, T. Rein, D. Tanner, K. Bodmann, O. Reiser, *Org. Lett.* **2000**, *2*, 535.

89 a) R. Noyori, M. Tokunaga, M. Kitamura, *Bull. Chem. Soc. Jpn.* **1995**, *68*, 36; b) R. S. Ward, *Tetrahedron: Asymmetry* **1995**, *6*, 1475.

90 S. Caddick, K. Jenkins, *Chem. Soc. Rev.* **1996**, *25*, 447.

91 M. Yamaguchi, M. Hirama, *Chemtracts Org. Chem.* **1994**, *7*, 401.

92 K. Narasaka, E. Hidai, Y. Hayashi, J.-L. Gras, *J. Chem. Soc., Chem. Commun.* **1993**, 102.

93 T. Rein, R. Kreuder, P. von Zezschwitz, C. Wulff, O. Reiser, *Angew. Chem. Int. Ed. Engl.* **1995**, *34*, 1023.

94 a) P. A. Aristoff, *J. Org. Chem.* **1981**, *46*, 1954; b) W. Skuballa, E. Schillinger, C.-St. Stürzebecher, H. Vorbrüggen, *J. Med. Chem.* **1986**, *29*, 313; c) A. Takahashi, M. Shibasaki, *J. Org. Chem.* **1988**, *53*, 1227.

95 K. Bodman, S. Has-Becker, O. Reiser, *Phosphorus, Sulfur and Silicon* **1999**, *144–146*, 173.

96 T. M. Pedersen, E. L. Hansen, J. Kane, T. Rein, P. Helquist, P.-O. Norrby, D. Tanner, *J. Am. Chem. Soc.* **2001**, *123*, 9738.

97 L. Vares, T. Rein, *Org. Lett.* **2000**, *2*, 2611.
98 L. Vares, T. Rein, *J. Org. Chem.* **2002**, *67*, 7226.
99 D. Monguchi, T. Furuta, K. Tanaka, manuscript in preparation.
100 T. Kumamoto, T. Koga, *Chem. Pharm. Bull.* **1997**, *45*, 753.
101 a) M. Mizuno, K. Fujii, K. Tomioka, *Angew. Chem. Int. Ed.* **1998**, *37*, 515; b) K. Tomioka, M. Hasegawa, *J. Synth. Org. Chem. Jpn.* **2000**, *58*, 848.
102 S. Sano, *Yakugaku Zasshi* **2000**, *120*, 432.
103 S. Arai, S. Hamaguchi, T. Shioiri, *Tetrahedron Lett.* **1998**, *39*, 2997.
104 F. Toda, H. Akai, *J. Org. Chem.* **1990**, *55*, 3446.
105 a) N. J. S. Harmat, S. Warren, *Tetrahedron Lett.* **1990**, *31*, 2743; b) J. Clayden, S. Warren, *Angew. Chem. Int. Ed. Engl.* **1996**, *35*, 241.
106 a) D. J. Peterson, *J. Org. Chem.* **1968**, *33*, 780; b) D. J. Ager, *Org. React.* **1990**, *38*, 1.
107 M. Iguchi, K. Tomioka, *Org. Lett.* **2002**, *4*, 4329.
108 P. T. Anastas, J. C. Warner, Green Chemistry, *Theory and Practice*, Oxford University Press, Oxford, **1998**.
109 L. Shi, W. Wang, Y. Wang, Y.-Z. Huang, *J. Org. Chem.* **1989**, *54*, 2027.
110 a) Z.-Z. Huang, S. Ye, Y. Tang, *Chem. Commun.* **2001**, 1384; b) Y.-Z. Huang, L.-L. Shi, S.-W. Li, X.-Q. Wen, *J. Chem. Soc., Perkin Trans. 1* **1989**, 2397; c) Y.-Z. Huang, S. Ye, W. Xia, Y.-H. Yu, Y. Tang, *J. Org. Chem.* **2002**, *67*, 3096; d) Z.-Z. Huang, Y. Tang, *J. Org. Chem.* **2002**, *67*, 5320.
111 R. P. Polniaszek, A. L. Foster, *J. Org. Chem.* **1991**, *56*, 3137.

Index

a

acid anhydrides 160
acid halides 160
2-acyl-1,3-dithianes 45
addition
– of α-silyl carbanion 23
– of α-sulfoxide carbanions 112
aldehydes, with functional heteroatom groups 228
alkenes
– bearing an adjacent leaving group 117
– sterically hindered 224
– strained 224
N-1-alkenyl amides 65
alkenylboranes 215
alkenyl carbamates 40
(1-alkenyl)iminophosphoranes 50
(Z)-alkenylphosphonates 49
N-1-alkenyl phthalimides 65
alkenyl sulfides 41, 69, 180
alkenylphosphonates 48
alkenylsilanes 59, 174, 212, 215
– germane 212
– borane 212
alkenylstannanes 215
1-alkenyl sulfinamides 46
1-alkenyl sulfones 46
1-alkenyl sulfoxides 45
alkenylsulfoximides 332
1-alkenyl trifluoromethyl sulfones 47
α-alkoxy aldehydes 217
alkyl halide-titanocene(II) system 176
alkylidenation 162, 172, 185, 208
– aldehydes 179, 211, 215
– carboxylic acid derivatives 179
– esters 211
– ketones 179
– on solid supports 181

alkylidenecyclopropanes 58
2-alkylidene-1,3-dithianes 44
3-alkylidene-β-lactams 55
α-alkylidene lactones 53
α-alkylidenetitanacyclobutanes 164
alkylidene-tungsten 192
alkylidene-β-sultams 47
S-alkyl α-(trimethylsilyl)dithioacetates 53
O-alkyl α-(trimethylsilyl)thioacetates 53
allenation 164, 176
allenecarboxylates 313
allenes 64, 164, 176
allenylidenecyclopropanes 58
allylic chromium reagent 57
allylic sulfones 115
allylsilanes 5, 35, 181
allylstannanes 5
allyltitanium 144
1-amidoethenylsilane 69
2-aminoalkenyl sulfones 46
α-amino aldehyde 325
Ando method 6
antiperiplanar orientation 122
Arbuzov reaction 5
η^6-arene Cr complexes 302
arsonium ylides 310
asymmetric carbonyl olefinations, diastereoselective 303
asymmetric dihydroxylation 76
asymmetric elimination 333
asymmetric HWE reactions, intramolecular 304
asymmetric olefinations, with non-racemic chiral carbonyl compounds 312
asymmetric Peterson reaction 334
asymmetric Wittig-type reaction 287
axial chirality 289
2-aza-1,3-dienes 66

Modern Carbonyl Olefination. Edited by Takeshi Takeda
Copyright © 2004 WILEY-VCH Verlag GmbH & Co. KGaA, Weinheim
ISBN: 3-527-30634-X

b

Barbier-type conditions 137, 140
Barton-McCombie radical deoxygenation 130
benzothiazole derivatives 137
benzylidenation 172
benzylidene acetals 229, 258
benzyltrimethylsilane 38
betaine 2
BINAP 299
BINOL 299
bis(cyclopropyl)titanocene 174
bis(iodozincio)methane 161, 205 f
bis(phenylthio)methane 180
1,3-bis(phenylthio)propene 178
α,α'-bis(trimethylsilyl)alkyl phenyl sulfides 59
bis(trimethylsilyl)methane 64
α,α-bis(trimethylsilyl)methanes 60
bis(trimethylsilyl)methyl isothiocyanate 66
bis(trimethylsilyl)methylamine 65
bis(trimethylsilylmethyl)titanocene 174
1,3-bis(trimethylsilyl)propyne 58
BT-sulfones 137
1,2,3-butatrienes 80
butterfly mechanism 28

c

calixarenes 242
carbene complexes, α-heteroatom-substituted 182
carbenes 274
carbenoid 272, 274
carbometallation 211
carbonyl methylenation-Claisen rearrangement strategy 170
chelation control 26, 317
chiral arsenic reagents 292
chiral host molecule 331
chiral HWE reagents 291
chiral phase-transfer catalyst 330
chiral phosphonate reagents 299
chiral phosphorus reagents 291
chloro(trimethylsilyl)methyllithium 51
chlorosilane-modified Clemmensen reduction 274
$(\eta\text{-}C_6H_5Me)_2Ti$ 266
$(\eta\text{-}C_6H_6)_2Ti$ 266
circulene 243
Claisen rearrangement 157
classic Wittig reaction 2
$(\eta\text{-}C_6Me_6)Ti(AlCl_4)_2$ 265
conjugated olefins 115
conjugated polyenes 226
C,O,O-tris(trimethylsilyl)ketene acetals 61
$[Cp_2TiCl]_2$ 265
$Cp_2Ti(CO)_2$ 266
$Cp_2Ti(PMe_3)_2$ 266
$Cp_2TiCl_2\text{-}i\text{PrMgCl}$ 265
$CpTiCl_2(THF)_{1.5}$ 266
$CpTiCl_3\text{-}LiAlH_4$ 265
cross-tandem coupling 252
crown ether-calixarenes 243
crownophanes 243
cyclic enol ethers 182
cyclic ethers 158
cyclic polyenes 241
cyclic siloxane 74
cycloalkanones 229
cycloalkenes 240
cyclophanes 242, 250
cyclopropane 79
cyclopropenes 63
cyclopropylidenation 174

d

deprotonation, of alkylsilanes 37
desilylation 60
 – fluoride ion induced 62
desymmetrization 289
 – intermolecular 299
 – of *meso*-dialdehyde 327
dialkenyltitanocenes 176
dialkyltitanocenes 172, 177
1,2-diastereocontrol 109
diastereoselectivity 211
dibenzyltitanocene 172
1,1-dichloroalkenes 51
dichloromethylenation 185
η^4-diene Fe 302
1,3-dienes 75, 141
2-dienylphosphonates 49
1,3-dienylphosphonium salts 50
differentiation
 – of enantiotopic carbonyls 289
 – of enantiotopic dialdehydes 304
1,2-dihydroxyalkylsilane 76
dimethylphenylsilyllithium 73
dimethyltitanocene 166
N,N-dimethyl 2-trimethylsilylacetamide 54
1,2-diol 263
diphenyl thioacetals 178
discrimination, intramolecular 303
1,1-disilylalkenes 215
dissymmetric alkenes 306
dissymmetric allenes 313
1,2-disubstituted olefins 109
N,S-disubstituted alkenes 42
dithienylethylenes 230
double asymmetric induction 321

double asymmetric synthesis 311
double deprotonation, of sulfoximines 144
dynamic kinetic resolution 320, 324, 329
dynamic resolution 290

e
elimination
– of β-hydroxy pyridinic sulfones 139
– of trimethylsilanol 68
α-elimination 166, 172
β-elimination 172
1,4-elimination 80
enamidines 41
enamines 172, 180
enantioconvergent reaction 327
enantiotopic carbonyl groups 317
enol ethers 39, 156, 172, 180, 187, 254
enolization 111
enones 52
– non-conjugated 227
enyne 58
epimerization 155
α,β-epoxy silanes 72
external chiral ligands 329
E/Z isomerization 301

f
Felkin-Anh-Eisenstein 317
ferrocenophanes 245
furans 255
gem-dichromium species 214
gem-dihalides 184
gem-dihaloalkanes 208
gem-dimetal compound 200, 211
gem-dizinc compound 202
germanium-Peterson reaction 85
germenes 91
Grubbs' reagents 159

h
haloalkene 215
α-haloalkylzinc compound 209
halohydrins 73
halomethyl stannanes 141
halomethylmagnesium halides 141
helicene 243
heteroatom-substituted olefins 167
heterocyclic sulfones 136
1,3,5-hexatrienes, 1,6-dihalo 229
1(2H)-isoquinolones 67
homo-Brook Rearrangement 77
homo-Peterson Reaction 79
Horner-type reagent 310
Horner-Wadsworth-Emmons reaction 5

Horner-Wittig (HW) reaction 9
(HWE) reagents 287
HWE-type arsonium reagent 311
Hydroboration, of 1-silylallenes 74
β-hydroxyalkylsilanes 18 f, 70
hydroxy ketones 155, 183, 190
o-(hydroxymethyl)benzylsilane 80
β-hydroxy phosphine oxide 9
β-hydroxy sulfoximines 144
β-hydroxy-α-vinylsilanes 56

i
imines 88, 187
indole 38, 256
π-π interaction 299
intermolecular coupling 224
– aldehydes 224
– dialdehydes 235
– diketones 235
– ketones 224, 226
– unsaturated aldehydes 226
intermolecular cross-coupling 236
internal olefin 156
intramolecular carbonyl olefination 182, 192
– esters 182
– thiol esters 183
intramolecular coupling
– aldehydes 240
– dialdehydes 241
– ketones 240
– oxoamides 256
intramolecular cyclization 38
– of acyloxycarbonyl compounds 255
intramolecular McMurry coupling, of keto esters 255
isomerization, under basic conditions 34

k
ketene dithioacetals 44
ketene O,S-acetals 42
ketenes 313
keto ester coupling 254
α-keto-β-hydroxyalkylsilanes 61
ketones
– enolizable 109, 154
– functional heteroatom groups 228
– having a chiral center 155
– possessing a leaving group 154
– sterically hindered 154
– α-substituted 155
β-keto phosphonates 8
ketyl radicals 265
kinetic resolution 290, 316
– of a racemic Wittig-type reagent 321

l

lead-Peterson reaction 88
α-lithioalkenylsilanes 64
2-lithio-2-trimethylsilyl-1,3-dithiane 44
Luche reduction 13

m

macrocyclic ether lipids 247
McMurry coupling
– acylsilane 261
– benzothiophene-2-carbaldehydes 231
– 2-bromobenzaldehyde 228
– ferrocenylketones 235
– 3-fluorobenzaldehyde 228
– formylferrocene 235
– in polymer synthesis 235
– 3-pyridine carbaldehydes 231
McMurry pinacol reaction 263
McMurry reaction 224
– of acylsilanes 234
mechanism
– carbonyl methylenation with dimethyltitanocene 171
– fluoride ion induced Peterson-type reaction 65
– Horner-Wadsworth-Emmons reaction 6
– Horner-Wittig reaction 10
– McMurry reaction 266
– Na(Hg)-promoted reductions 123, 129
– Peterson reaction 19
– reduction of vicinal oxygenated sulfoxides 134
– reductive coupling 268
– second-generation Julia reactions 138
– SmI$_2$-promoted reverse reductions 132
– Wittig reaction 2
medium-sized rings 157
meso compounds 289
meso-dicarbonyl compounds 302
metallaoxetane 272, 274
metallopinacolate 223
metallopinacols 267
methoxy(trimethylsilyl)methyllithium 39
methylenation 180, 189, 202
– aldehydes 155, 206
– amides 157
– anhydrides 168
– carboxylic acid derivatives 156, 167
– chemoselective 157
– dimethyl acetals 36
– enolizable ketones 203
– ester 204, 208
– α-hydroxy ketone 202
– imides 157, 168
– ketones 154, 203, 205
– lactams 169
– lactones 169, 205
– optically active polyketones 208
– small-ring diones 169
– thioanhydrides 168
– thioesters 157
– α,β-unsaturated esters 156
methyl ketones 36
methyl trimethylsilylacetate 23
methyl xanthate 131
Michael additions, asymmetric 306
Michaelis-Arbuzov reaction 8
1,3-migration, of a silyl group 77
molybdenum-alkylidene 158, 191
molybdenum carbene complexes 189
molybdenum-methylidene 189

n

Na(Hg) 122
Na(Hg) elimination 125
niobium carbene complexes 188
non-basic reductant 126
non-stabilized ylides 3
nucleophilicity, of dizinc species 202
Nysted reagent 202, 205, 206

o

olefin metathesis 162, 193
– carbonyl olefination sequence 164
olefination, of acid-sensitive substrates 167
oligoindoles 257
one-step Julia reaction 137
organozinc reagents 201
ortho-lithiation 118
oxacephem 67
1,2-oxagermetanide 85
oxaphosphetane 2, 303
1,2-oxasiletanide 20, 22, 25, 78
1,2-oxastannetanides 86
oxatitanacyclobutane 153

p

para-coupling 269
parallel kinetic resolution 290, 322
Pd-catalyzed allylic substitution 327, 336
Peterson elimination 18
Peterson reaction
– concerted mechanism 25
– convergently stereoselective 31
– β-hydroxyalkylsilanes 19
– Z-selective 23, 49
– α-silyl aldimines 53
phenanthrenes 243
α-(phenylthio)alkyl ketones 42
phenyl trimethylsilylmethyl sulfone 46

phosphaalkenes 89
phospha-Peterson reactions 89
phosphonates 8
phosphonate-stabilized carbanion 5
phosphonium salts 5
phosphonium ylide 2
phosphoramidate 306
pinacol rearrangement 261
pinacols 224
poly(arylene vinylene)s 236
polyadamantane 236
polyepoxyannulenes 243
polyquinan 304
porphyrin analogues 251
porphyrins 245, 250
precursors, of the Julia reaction 106
preparation, of phosphine oxides 15
PT-sulfones 140
pyrometallurgy grade zinc 204
pyrroles 256
pyrrolo[2,1,5-cd]indolizines 245

q
quaternary stereogenic carbon 305
o-quinodimethane 82

r
racemization 324
reaction pathways, for the methylenation of esters with the Tebbe reagent 153
reduction
– β-keto phosphine oxide 13
– α-silyl ketones 71
– vinyl sulfones 136
– of vicinal oxygenated sulfoxides 133
reductive coupling
– of aromatic aldehydes 223
– ketones 223
reductive desulfonylation, of vinyl sulfones 127
reductive elimination 120
– β-acetoxy sulfones 127
– aliphatic sulfones 105
– β-hydroxy sulfoxides 133
– β-hydroxy sulfoximines 145
reductive lithiation
– bis(phenyl)sulfones 114
– α-silyl sulfides 58
Reformatsky-Peterson reactions 37
retroaldolization 113
retrograde fragmentation 109, 122
retro-Wittig decomposition 3
reverse Julia olefination 131
reverse reductions 130
ring-closing metathesis 158

ring-opening
– cyclic esters 73
– cyclic ethers 73
– oxiranes 72

s
samarium diiodide 114
samarium-Peterson reaction 35
Schlosser modification 3
Schrock-type carbene complexes 151, 185
selectivity factor 322
selenophenes 244
[2,3]sigmatropic rearrangement, of sulfoximines 144
silametallation 82
sila-Peterson reactions 90
silatropic migration 234
silenes 90
β-silyl alkoxides 70
α-silyl alkyllithium reagents 32
α-silylalkylphosphine oxides 83
α-silylalkyl selenides 59
γ-silyl allylboranes 75
α-silyl amidines 41
α-silyl benzyl anions 24
α-silyl carbanion 19
– bearing a heteroaromatic group 38
– bearing an ester function 21
β-silyl carboxylic acid 71
1-silylcyclopropyllithiums 58
2-silyl-1,3-diol 71
α-silyldibromomethyllithium 79
α-silyl esters 52, 71
1-silylethenyl ketones 70
α-silyl Grignard reagents 32
α-silyl ketones 52
α-silyl-α-ketimines 54
silyl(methoxy)benzotriazol-1-ylmethane 40
1,3-silyl migration 21
α-silyl nitriles 55
α-silyl phosphonates 48
α-silyl-α-sulfanylalkylphosphonate 49
silylsulfinamide 89
SmI$_2$-promoted reduction 126
stabilized ylides 3
α-stannylalkylsilanes 60
stepwise mechanism, of the Peterson elimination 21
stereochemistry, of the Peterson elimination 19
stereocontrol, of the Horner-Wittig reaction 14
stereoselective synthesis, of 1,3-dienes 56
stereoselectivity
– benzylation 173
– Wittig reaction 2

stereospecific preparation, of β-hydroxyalkyl-
    silane 23
stilbene
– crowded 226
– 1,2-disilylated 234
– uncrowded 226
Still-Gennari modification 6
styrene derivatives 172
N-styryl isothiocyanate 66
α-sulfanyl alkyllithium 41
sulfides 83
sulfineimines 89
sulfines 91
sulfones
– bearing vicinal hydroxyl groups 122
– bearing vicinal leaving groups 127
sulfoxide-metal exchange 134
sulfoximines 143
syn-dihydroxylation 76
syn-elimination 3, 76

t

tandem McMurry reaction 249
tandem Michael-HWE reactions 313, 329
tandem olefin metathesis-carbonyl olefination
    191
tandem Peterson-Michael reaction 83
tantalum carbene complex 188
Tebbe reagent 152
– in situ preparation of 153
– zinc and magnesium analogues of 161
template effect 261
terminal alkenes 33
terminal dienes 80
terminal olefins 107
tetrahydrofuran 327
tetrahydropyran 327
tetrasubstituted alkenes 7
tetrasubstituted olefins 114, 184
theoretical study, on the Peterson reaction 29
thiaheterohelicene 243
thioacetals 178
thioacetal-titanocene(II) system 178
thioesters 211
thiophenes 244
Ti(biphenyl)$_2$ 266
β-TiCl$_3$ 206
TiCl$_3$(DME)$_{1.5}$-Zn(Cu) 260, 263
TiCl$_3$-C$_8$K 255, 260
TiCl$_3$-K 260
TiCl$_3$-Li 258, 261
TiCl$_3$-LiAlH$_4$ 255, 260
TiCl$_3$-LiAlI$_4$-Et$_3$N 255
TiCl$_3$-Li-naphthalene 262

TiCl$_3$-Mg system 223
TiCl$_3$-Na 260
TiCl$_3$-Zn(Cu) 260
TiCl$_4$-nBu$_4$NI 263
TiCl$_4$(THF)$_2$-Zn 263
TiCl$_4$-Li(Hg) 264
TiCl$_4$-Mg(Hg) 263
TiCl$_4$-Zn 260
TiCl$_4$-Zn system 225
TiCl$_4$-Zn-pyridine 260
Ti-Me$_3$SiCl 266
tin-lithium transmetallation 60
tin-Peterson reaction 86
titanacyclobutanes 155, 159
– preparation of 161
titanacyclobutenes, dialkyl-substituted 165
titanium pinacolate 269
titanium(0) colloid 266
titanium-graphite 248
titanocene-alkylidene 162, 163, 178
– formation of 172
titanocene-methylidene 159
– aluminum-free 160
– formation of 166
– higher homologues of 161alkenyltitanium
    complex 162
titanocene-trimethylsilylmethylidene 174
trimethylenc thioacetals 178
(trimethylgermyl)acetate 85
1-trimethylsilylallenes 75
α-(trimethylsilyl)allyl anion 56
N-trimethylsilylamide 88
α-trimethylsilyl ketones 35
trimethylsilylmethanesulfinamides 46
(trimethylsilyl)methylenephosphorane 50
(trimethylsilyl)methylenation 174
(trimethylsilylmethyl)iminophosphoranes 50
trimethylsilylmethyllithium 33
trimethylsilylmethylmagnesium chloride 33
trimethylsilylpotassium 73
α-trimethylsilyl sulfonamides 47
(triphenylstannyl)methyllithium 86
tris(trialkylsilyl)silyllithium 90
tris(trimethylsilyl)germyllithium 91
tris(trimethylsilyl)titanacyclobutene 174
tris(trimethylsilylmethyl)titanocene 174
trisubstituted olefins 112
tungsten alkylidynes 193
tungsten carbene complexes 192

u

UCl$_4$-M(Hg) 270
α,β-unsaturated aldehydes 61
α,β-unsaturated amides 54

α,β-unsaturated carboxylic acids   61
α,β-unsaturated esters   22, 52, 61, 68
– α-chloro   51
α,β-unsaturated imines   53, 68
α,β-unsaturated nitriles   37, 55
α,β-unsaturated sulfonamides   47
α,β-unsaturated γ-lactone   61
β,γ-unsaturated carboxylic acids   63

v
vinyl ethers   186, 213
vinyl sulfides   213
vinyl sulfones   105, 115
vinylcarbene complexes   178
vinylidene complexes   164, 176

vinylogous Peterson olefination   80
vinylsilanes   174

w
Warren's method   12
[2,3]-Wittig rearrangement   83

z
zinc-copper couple   202
zinc-lead couple   202
zirconium   1,1-bimetallic reagents   187
zirconium carbene complexes   185
Zn-CH_2X_2-$TiCl_n$ 203
Zn-$RCHBr_2$-$TiCl_4$-TMEDA   211
Z-selective olefination   27